FIRE FIGHTING APPARATUS AND PROCEDURES

Third Edition

GLENCOE PUBLISHING COMPANY FIRE SCIENCE SERIES

Bryan: **Fire Suppression and Detection Systems**
Bush/McLaughlin: **Introduction to Fire Science,** Second Edition
Carter: **Arson Investigation**
Clet: **Fire-Related Codes, Laws, and Ordinances**
Erven: **Fire Fighting Apparatus and Procedures,** Third Edition
Erven: **First Aid and Emergency Rescue**
Erven: **Handbook of Emergency Care and Rescue,** Revised Edition
Erven: **Techniques of Fire Hydraulics**
Gratz: **Fire Department Management: Scope and Method**
Meidl: **Explosive and Toxic Hazardous Materials**
Meidl: **Flammable Hazardous Materials,** Second Edition
Meidl: **Hazardous Materials Handbook**
Robertson: **Introduction to Fire Prevention,** Second Edition

About the author . . .

Lawrence W. Erven, now retired, was a pioneer in fire science education, a battalion chief of the Los Angeles City Fire Department with over 30 years of fire fighting experience. In 1955 he was instrumental in developing the first complete fire science curriculum in the country. From 1955 until 1969 he taught fire fighting, emergency rescue, and fire apparatus courses at Los Angeles Harbor College. He is the author of *First Aid and Emergency Rescue, Techniques of Fire Hydraulics,* and *Handbook of Emergency Care and Rescue.*

FIRE FIGHTING APPARATUS AND PROCEDURES

THIRD EDITION

LAWRENCE W. ERVEN
Battalion Chief (Retired)
Los Angeles City Fire Department

(A Glencoe Book)

MACMILLAN PUBLISHING CO., INC.
New York
COLLIER MACMILLAN PUBLISHERS
London

This book is dedicated to the drivers
and operators of fire fighting apparatus.

These individuals must be among the most skilled,
competent, and knowledgeable in the
fire fighting profession.

The first and second editions of this book were published under the title *Fire Company Apparatus and Procedures.* Copyright © 1969, 1974 by Lawrence W. Erven.

Printed in the United States of America

Glencoe Publishing Co., Inc.
17337 Ventura Boulevard
Encino, California 91316
Collier Macmillan Canada, Ltd.

Library of Congress Catalog Card Number: 77-095330

PRINTING 10 YEAR 9

ISBN 0-02-472470-X

Contents

Preface

Modern fire fighting is a complex operation. Old-time fire fighters depended mainly on brawn and guts. These are still essential attributes of fire department personnel, but highly developed intelligence, technical knowledge, and manipulative skills are becoming increasingly important. The strategy required to combat a major blaze may be compared to the battle strategy of modern warfare: Knowledge must be gleaned and accumulated beforehand, personnel must be thoroughly educated and trained, appropriate strategy must be devised during the initial assessment of the fire, and then the proper fire fighting tactics must be put into action. The extent of knowledge about one item of fire fighting equipment or apparatus can determine whether a fire is controlled with a minimum of damage or whether a small blaze accelerates into a major conflagration. This third edition of FIRE FIGHTING APPARATUS AND PROCEDURES attempts to provide the knowledge and information that apparatus operators must possess to drive, operate, and maintain their fire fighting equipment competently, regardless of whether they use the most modern equipment or apparatus that is many years old. It also offers suggestions and guidelines for developing sound fire fighting procedures and expanded coverage of modern fire fighting strategy and tactics.

Aside from trained and knowledgeable personnel, fire fighting apparatus and equipment are the principal assets of any fire department. Technological advances have resulted in major vehicle improvements; operation of fire fighting vehicles requires far greater technical knowledge than ever before. In addition, trucks have grown so much in size and weight that they can be very unwieldy. To minimize traffic hazards, safety rules must be understood and defensive driving emphasized. It is also important that every fire fighter, whether a new recruit or the highest ranking officer, have a thorough working knowledge of fire fighting equipment and know how to use fire fighting appliances and vehicles efficiently and effectively. It is mandatory, for example, that all fire fighters know the multitude of factors that affect the production and application of effective fire fighting streams of water. The information necessary for proper handling of fire apparatus and equipment is provided in this text.

If a department has standard operating procedures, the officer in command will not have to make routine decisions during the initial hectic stages of a fire. Some standard operating procedures are applicable to every fire department, regardless of size and location; others, however, must be designed to meet the requirements of a specific organization. A company might, for example, develop a plan for controlling a fire in a local chemical plant. Preplanned operating procedures are an essential element of successful fire fighting. This text helps fire fighters and departments plan standard procedures.

New industrial processes, increasingly hazardous materials, and the construction of larger and taller buildings have multiplied the fire fighter's problems. The fire service has attempted to combat these hazards with advanced methods of fire fighting and training, and technological improvements in equipment and apparatus. No longer are complicated fire hazards confined to the larger urban and manufacturing centers, however; major industries are increasingly moving to rural areas. In addition, vast quantities of hazardous commodities are daily transported by truck, train, ship, and airplane all across the country. Because of these changes, even the most remote volunteer organization now may be suddenly confronted with a complex and potentially disastrous holocaust. It is thus important that fire department personnel everywhere be knowledgeable and well trained to lessen the loss of life and property and prevent calamity. This book should prove an invaluable guide.

Fire fighting training must first and foremost be practical. Information is useless if it is too complex or too exacting to be applied quickly on the fire ground. This book focuses not on theory but on practical solutions that can be easily and quickly applied to fire ground problems. Key terms are defined in a comprehensive glossary at the end of the text.

Because every facet of fire fighting technology is expanding and changing so rapidly today, even the newest equipment may quickly become obsolete. Fire service manuals and textbooks thus must constantly be revised. During the preparation of the manuscript for the third edition of this book, the entire text and all illustrations were reviewed; manufacturers were contacted to ascertain whether the portions of the book concerned with their products were still valid and reflected their current and past models; new and improved equipment, appliances, and other devices were researched and studied. This book, then, is not the result of one person's work but is a compilation of the knowledge and experience of a multitude of fire departments, manufacturers, and individuals. The author's every request for specific technical information and illustrations met with nothing but friendly and cooperative attention. Without this mutual participation and collaboration, any attempt to compile a book such as this would have been impossible.

This new edition was reviewed in its entirety by Deputy Robert Moore, Office of the State Fire Marshal, Fire Prevention Division, Klamath Falls, Oregon; and Center Associate Frank Oberg, Fire Information, Research, and Education Center, University of Minnesota. Many of the photographs and illustrations were prepared by Eunice E. Erven. The author thanks all of these people for their help and advice.

(Courtesy of the San Francisco Fire Department)

1 Evolution of Fire Apparatus

The earliest known water pump, invented by Ctesibius of Alexandria in about 200 B.C. (Figure 1-1A), consisted of two brass cylinders with carefully fitted pistons. Water was admitted through valves at the base of the cylinder and forced through outlet valves into a chamber. As water rose in the chamber, it compressed the air inside, forcing the water to be ejected in a steady stream through a pipe and nozzle. The pistons were operated by long handles; in many ways this early machine looked similar to hand engines that first appeared in England and America in the early 1700's (Figure 1-1B, C, D). This type of engine was also used in ancient Greece and Rome.

Following the fall of Rome, the fire engine of the ancient civilizations was forgotten. Religious leaders taught their followers that fires were the willful vengeance of the Lord, and they discouraged any attempt to make apparatus that would extinguish flames.

It was not until the sixteenth century, when the writings of the ancient Greeks, Vitruvius and Hero, were translated, that the principles of the ancient fire engines were brought to light. The discovery was not put to use and more than a century passed before people began, once again, to experiment with fire extinguishing apparatus.

Many attempts were made to build effective "water engines," but the people were suspicious of such appliances. The great turning point in fire fighting history occurred because of the 1666 fire in London, England. This conflagration convinced a great number of people that fire fighting methods and equipment needed a revolutionary change. The world's first fire insurance was written less than a year after the London fire by Dr. Nicholas Barton; his business mushroomed until it became the Phoenix Fire Insurance Company, which is functioning to this day.

The first specialized fire fighting equipment in America was the leather fire bucket, usually of three-gallon capacity. This was used in the bucket brigade. Two human lines were formed between the burning structure and the nearest source of water; one line was used to pass full buckets of water up to the blaze, while the other line returned the empties. Whenever there was an outbreak of fire the citizens would put their faith in a

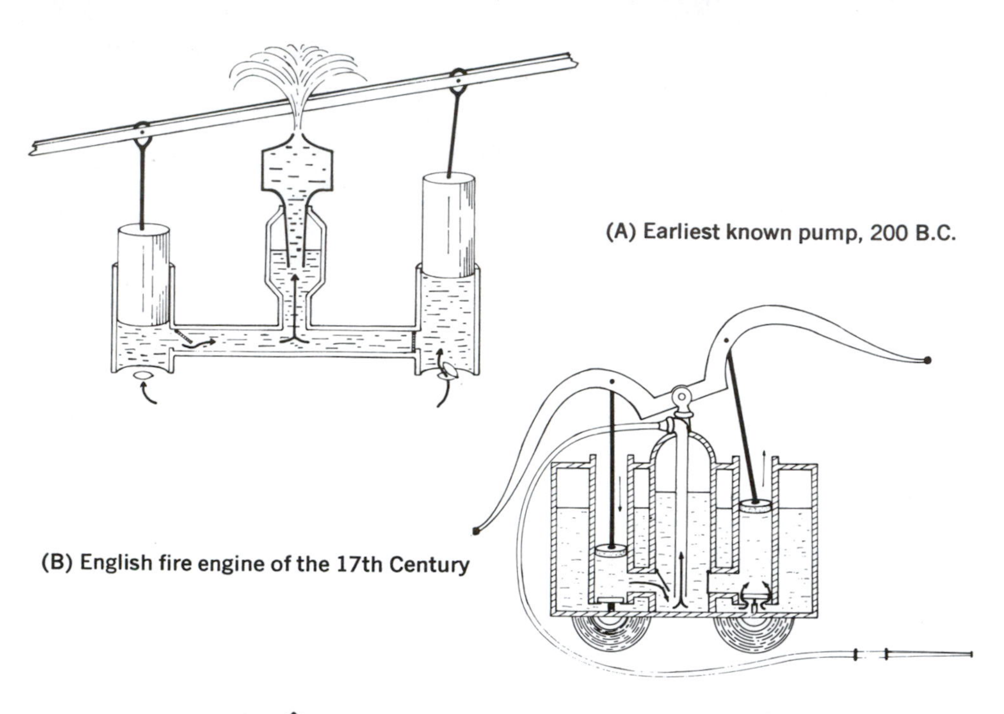

(A) Earliest known pump, 200 B.C.

(B) English fire engine of the 17th Century

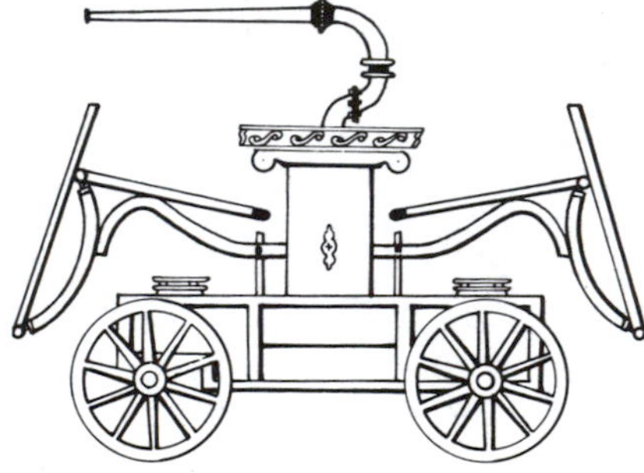

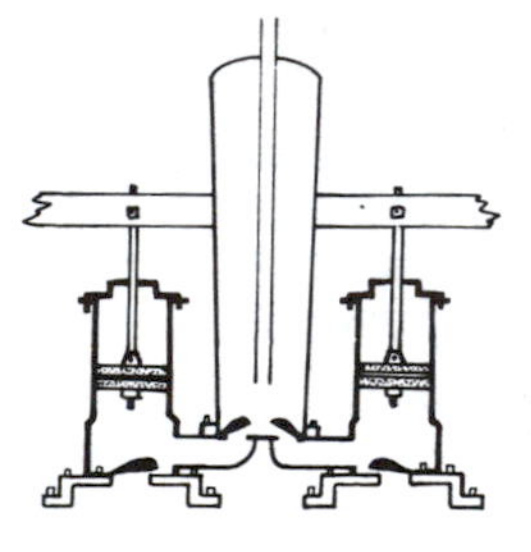

(C) American fire engine of the 18th Century

(D) American-built hand pumper, 1740

Figure 1-1. Early versions of fire apparatus

bucket brigade and pray that the fire would not become so intense that they would be unable to approach it with their water buckets.

Organized fire fighting began in New York about 1650, during the administration of Governor Peter Stuyvesant. Almost all of the early fires recorded in Boston, New Amsterdam, and Philadelphia were the result of faulty and soot-filled chimneys–with dry thatched roofs a contributing factor. Governor Stuyvesant proposed an ordinance–which was quickly adopted–that levied a fine for having a dirty chimney; this money was utilized for the maintenance of buckets, hooks, and ladders.

In 1678, Boston bought its first piece of mechanical fire apparatus, which was called a water engine. It was a rectangular box, or tub, into which the bucket brigade poured water while workers on the pump handles supplied pressure. On the engine was mounted a big nozzle which fulfilled the purpose of a wagon battery on a modern pumper; this nozzle allowed water to be applied more directly and under greater pressure than was possible by throwing it from buckets.

In 1731, two hand-drawn pumpers built by Richard Newsham were imported from England by New York; within a few years Americans were manufacturing their own. A hand-operated pumper with optional snow runners is shown in Figure 1-2.

In later years, rudimentary water mains, cisterns, and hydrants supplied water, doing away with bucket brigades; but pumpers still depended upon volunteers to pull apparatus to the fire and operate the pumps. In those days buildings were relatively low and engines were small and light. Hand-drawn hose wagons (Figure 1-3) were built in the railroad shops in Carson City, Nevada, for fire fighting service in the frontier town. With the advent of automobiles, the larger cart was equipped with a tongue so it could be towed with a car.

In those times, before hydrants and fire hoses, hand pumpers would arrive at a fire and have to wait until a bucket brigade could assemble and pour water into the tub before they could start to pump and get a stream of water on the burning structure. An enterprising fire fighter conceived the idea of keeping the engine tub filled with water as it stood in the fire station awaiting an alarm. The fire company would start pumping immediately, hoping the bucket brigade would get water to them before their tub was empty. This practice was later abandoned because of the disadvantages which include rust, rot and freezing in cold weather.

The volunteer fire fighters dropped whatever they were doing and headed for the fire house whenever an alarm was sounded. The first arrivals took hold of the drag ropes and got the machine rolling; as more volunteers caught up and got a grip on the ropes, they worked up to a dead run. To see a hand tub come down the street with fifty people on the ropes, bell ringing and everyone shouting was an unforgettable sight. If any of them stumbled and fell, they would have to roll out of the way or be run over by the engine. Journals of old companies tell of many persons who were injured or killed while rushing to a fire.

All the human energy was needed to work the handles or brakes, as they were then called, of the pumping engine. As the design was improved over the years, the small handles on early machines were lengthened in later models until they ran the entire length of the engine, providing room for fifteen or more operators on each side. They worked quickly to give the strongest stream possible, a back-breaking job that usually wore the crew down at a rapid rate. When one crew was completely exhausted, a fresh crew was put on the brakes allowing the first crew to rest and have refreshments. Liquor was considered indispensable at every fire. At first many of the fire personnel carried their own flasks to every blaze; if the fire turned out to

Figure 1-2. Large hand-operated pumper with optional snow runners (Courtesy of American LaFrance)

Figure 1-3. Hand-drawn hose wagons

be a long one, they would elect someone to go to the nearest tavern or store and replenish the supply. However, by 1800, nearly every company employed a steward whose job it was to promptly answer all alarms with a keg slung over one shoulder.

Volunteer organizations were well-organized, politically active, and admired by the local youngsters. Every fire was an occasion for wild partisan activity by the backers of the individual companies. Just getting to the fire and pulling the pumpers took great stamina and courage; it was a breakneck foot race. The volunteers were often impeded by the tricks and skullduggery of volunteers of competing companies and their followers. A fire alarm and the attendant responses of the volunteer units was more often the occasion for pandemonium than for a display of efficient fire extinguishing. A meeting of volunteer companies at a crossroads or narrow street would often wind up in a lively brawl; usually more energy was expended preventing the response of a rival company than in putting out the fire. As competition became greater, companies would send runners on ahead to camouflage the nearest hydrant to the fire—or otherwise prevent other volunteers from beating them to it.

As communities grew and buildings became taller and more closely congregated, larger and more powerful engines were needed to cope with fires. The spirit of rivalry existing in the days of "vollies," naturally resulted in inefficiency. Insurance companies and business people started campaigns to develop more effective fire engines and to replace the large crews of volunteers with paid fire crews.

In 1840, the New York insurance companies commissioned Paul Hodge to build a steam pumper. A year later it was completed. The apparatus had a horizontal boiler, like a locomotive, and it went to the fire under its own power without the need of horses or an uproarious gang of volunteers to pull it. The engine seemed to be a success when it put a stream of water from a 1½-in. nozzle over the cupola of the New York city hall. And yet it failed, for the vast army of volunteer fire fighters refused to cooperate with it in any way at fires. They saw to it that the steamer did not get the hose or water supply it needed, and by a dozen or more sly, covert tricks they made this mechanical intruder look poor in its performance. Because fire fighters controlled the vote, politicians were reluctant to offend them in any way and the steamer was not used.

One of the earliest improvements in hand-drawn fire apparatus was the replacement of hand pumpers with chemical tanks (Figure 1-4). Dumping a bottle of sulphuric acid into a sodium bicarbonate solution and then agitating the mixture with the front crank developed sufficient pressure to discharge a fire stream through the hose on the reel.

A steam pumper, the "Jeanie Jewell," won a three-day pumping competition in Chicago in 1878. The builder, Asa LaFrance, a

Figure 1-4. Early version of chemical tank

pioneer in fire apparatus, is shown with his engine in Figure 1-5.

In 1852, Moses Latta invented the first successful steam fire engine and it went into service in Cincinnati, Ohio, on January 1, 1853. The completed pumper weighed 22,000 lb and required four horses, ridden artillery style, plus the propelling power of the steam engine to move it. This fire engine pumped water through a 3-in. hose and a 1½-in. nozzle to attain a reach of 225 ft and it could supply sufficient water for four hose streams. At its first major fire the Latta engine gave excellent service until the volunteer fire crew launched an attack on it, cut its hose, threw rocks, and threatened to put it out of commission. A large group of citizens came to the rescue of the new steamer. After a tremendous battle at the fire scene, the volunteers were repulsed by the enraged citizens and the Latta engine continued to work.

Many early steamers were pulled to the fires by gangs of workers, but this new machine was a different proposition from the lightly built hand tub. It was simply too heavy for any group to hustle along the streets unless there were 70 or 80 in attendance. To get that many volunteers to report speedily to the fire station proved very difficult. It seemed the best way to get this steamer to the fire was with a team of three big horses. For another half century, the streets clattered beneath the steel hooves of gallant fire horses, who seemed even more magnificent when pulling a smoke-belching nickel-plated steamer or lengthy ladder truck (Figure 1-6).

The horses were kept in stalls behind the apparatus, and they were specially trained for their work. When the alarm tapper hit a box to which the engine responded, the man on watch at the desk threw a switch so that an electric mechanism flung open the door

Figure 1-5. The "Jeanie Jewell," champion steam pumper (Courtesy of American LaFrance)

Figure 1-6. Horse-drawn steam pumper (Courtesy of the Los Angeles City Fire Department)

in front of each stall; another switch activated a mechanical whip in the rear of each stall that gave the horses a smart flick over the rump. Instantly the well-trained team came thundering out of the stalls and galloped past the apparatus to take their places under a complete rig of harness hanging from the ceiling. The driver pulled a cord, the harness dropped onto the backs of the horses, and the driver had no more to do than snap the collars and be ready to go. The driver sprang to the seat, took the reins, reached up and pulled another cord so the front doors of the fire station could swing open, and the apparatus rolled (Figure 1-7). This whole uproar took no more than thirty seconds.

Horses that became too old for the rigorous life of a fire horse were retired to a farm or sold to delivery men. Many stories are related of retired fire horses, relegated to pulling a bread or milk wagon, becoming unmanageable when they heard a fire bell. Apparatus responding to a fire would occasionally be followed closely by an old fire horse pulling a milk wagon at breakneck speed.

The problem of how to have the steam fire engine arrive at the fire scene ready to start pumping was solved by connecting the steamer to a stationary boiler in the station; thus hot water was circulated at all times through the engine's boiler. It was the stoker's duty to disconnect the steam lines before the apparatus rolled. To keep the boiler hot on the way to the fire, kindling wood was laid in the firebox; the stoker lit this kindling and had a fire before the steamer was around the first corner. If there was a working fire, the stoker fed the firebox with coal from boxes built into the platform-like step at the rear of the boiler.

Soon after the development of the steamer, the first chemical engine appeared. A two-horse vehicle carried two 50-gal tanks of bicarbonate of soda solution. It gave a

Figure 1-7. Early fire-house scene (Courtesy of the Los Angeles City Fire Department)

strong stream through a 1-in. hose when the operator turned a crank that tipped a sulphuric acid bottle; the resultant carbon dioxide gas developed sufficient pressure to expel the contents of the tank. These chemical engines were convenient for extinguishing small fires.

Introduction of the steam fire engine foretold the ultimate doom of the volunteer companies in large cities. Steam apparatus eliminated the need for hand pumping of water, and horses ended the problem of hauling the engines by hand. Also, the horses, being controlled by a driver, did not stop on the way to the fire to indulge in brawling with rival companies. Ladder trucks were developed as buildings became taller.

Figures 1-8 and 1-9 illustrate some of the fire apparatus of the day. The Waterous steam pumper in Figure 1-8 has a 600-gpm (gallons per minute) capacity. Equipment on the engine consists of: 20 ft of suction hose, copper strainer, large fuel tender of ample capacity, driver's seat with cushion, one 12-in. gong with foot-trip attachment, foot brake, two brass hand lanterns, brass torch for engineer's use, oil cans, poker, fire shovel, and a full assortment of wrenches and tools.

The first paid fire fighters worked a continuous tour of duty, with three hours a day off for meals and one day off a month. Up until 1907, when a two-platoon plan was first introduced in a fire department, the continuous duty system was the universal working schedule for fire crews. Agitation for elimination of the barracks system, which kept fire fighters on duty twenty-one hours a day (three hours being allowed for meals), spread throughout the country and ultimately resulted in the two-platoon system.

The speed with which a team of horses could get the apparatus rolling was almost unbelievable. It was not at all unusual for apparatus to be rolling within thirty seconds after the alarm. So rapid were the horses that when automobiles were first placed in the same quarters, the animals usually arrived at the fire first. It was only after the self-starter was invented that automotive apparatus began to reach the fire before the horses. Some examples of the horse-drawn apparatus of the period are shown in Figures 1-10 through 1-12.

Figure 1-8. Waterous steam pumper (Courtesy of Waterous Company)

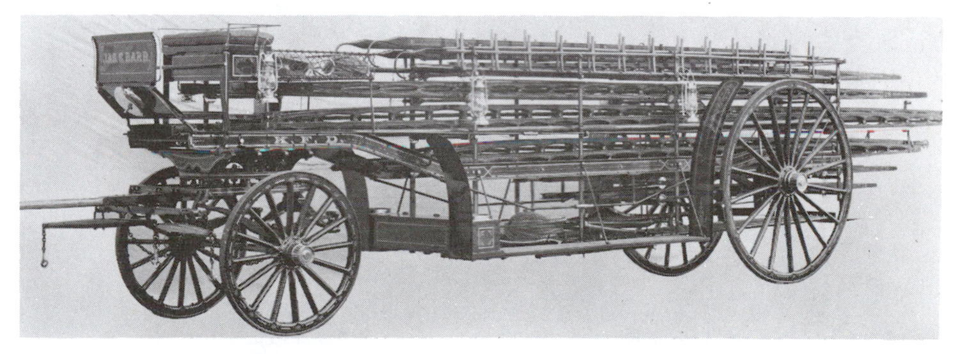

Figure 1-9. Pirsch ladder truck, 1895 (Courtesy of Peter Pirsch & Sons Company)

Figure 1-10. Horse-drawn hook and ladder truck. Note the tiller operator's position under the aerial ladder. (Courtesy of the Los Angeles City Fire Department)

Figure 1-11. The team of horses, men, and apparatus were well groomed and efficient. Note the complexity of gages, piping, pulsation dome, and other devices on the steam engine. (Courtesy of the Los Angeles City Fire Department)

Figure 1-12. Truck personnel were usually teamed up for height so that ground ladders could be raised more evenly. Notice that the horses were also teamed up according to size and color. (Courtesy of the Los Angeles City Fire Department)

By the dawn of the twentieth century, the doom of the fire horse was apparent. The cost of maintaining early self-propelled steamers was claimed to be $28 per month, compared with $62 for a horse-drawn fire engine. A completely self-contained front-wheel-drive gasoline engine conversion unit was attached to the front end of steam fire engines and ladder trucks to take the place of horses. During a transition period, steam, gasoline, and horses were all utilized for power (Figures 1-13 through 1-19). But the industrial revolution was soon completed in the fire service. The advent of the internal combustion engine doomed the horses and the colorful steamers to extinction.

Figure 1-13. Howe horse-drawn piston pumper driven by Rutenber 6-cylinder engine (Courtesy of the Howe Fire Apparatus Company)

Figure 1-14. The first gasoline engines replaced the horses but not the steam engines.

Figure 1-15. Christie motorized conversion of steamer, 1900 (Courtesy of American LaFrance)

Figure 1-16. The first pumper with a single engine providing both driving and pumping power was delivered by Waterous Engine Works Co. to Alameda, California, in 1907. The pump had a 600-gpm capacity. (Courtesy of Waterous Company)

Figure 1-17. 1917 Model "T" Ford pumper with piston pump

Figure 1-18. "Gasoline is now king of the streets!" (Courtesy of the Los Angeles City Fire Department)

Figure 1-19. 1920 Mack Bulldog city service ladder truck. Note the two soda acid chemical tanks and hose reel. (Courtesy of Mack Trucks, Inc.)

(Courtesy of American Fire Apparatus Company)

2 Fire Fighting Apparatus: Construction and Equipment

Fire fighting apparatus are the various types of vehicles whose primary purposes are fire extinguishment, rescue and care of people with endangered lives, and the salvage operations necessary to reduce property damage. The apparatus in use today have undergone a dramatic metamorphosis in the last few decades, and important improvements are still being made. They bring new flexibility and efficiency to the fire service, and greatly reduce the numbers of fire fighters that were historically required to control large blazes.

Variety in Modern Apparatus

Fire fighting apparatus in use today have come a long way–in both variety and improved design–from the days of the early hand-drawn vehicles discussed in Chapter 1. Although most of the modern apparatus require specially trained operators, all fire company personnel should be familiar with their functions.

Pumpers

Pumpers are the most important apparatus for the control of fire (Figures 2-1–2-5), and the engine company personnel operating these pumpers are the backbone of fire ground operations. Without their skillful teamwork in effectively applying water and other extinguishing agents in sufficient quantities to the correct location, fire extinguishment would seldom be possible. No fire department could function without pumpers. The pump operator of an engine company, who drives the vehicle, is also called an engineer, and the pumper may be referred to as an engine; this is a nostalgic carryover from the steam engine era.

The primary function of a fire department pumper is to obtain and deliver water to the fire through hoses at the proper pressure so the nozzle operators will have effective fire streams. The majority of pumpers are called triple combinations because they have three main features: fire pumps with a capacity of at least 750 gpm (gallons per minute); hose compartments; and water tanks. A few cities that require extraordinarily huge volumes of water at high pressures operate some apparatus with a sole purpose of pumping; fire hose is carried on a separate tender or hose wagon. Some departments have pumpers with a mounted elevating platform, turret or monitor nozzle, or water tower so a quick attack with a large fire stream can be launched in the minimum amount of time with the least amount of manpower. In addition, engine companies carry an extension ladder and a roof ladder, together with a variety of other tools and equipment, so they can be effectively operated when not closely accompanied by a truck company or other fire department units.

Quadruple Pumpers

A quad is a long-wheelbase triple combination pumper that also carries a full complement of ground ladders. It receives its name from the fact that it has four main features: fire pump, hose, water tank, and ground ladders. This type of apparatus is no longer very common.

Figure 2-1. Modern triple combination pumper. See Figure 2-2 for equipment identification. (Courtesy of Ward LaFrance Truck Corporation)

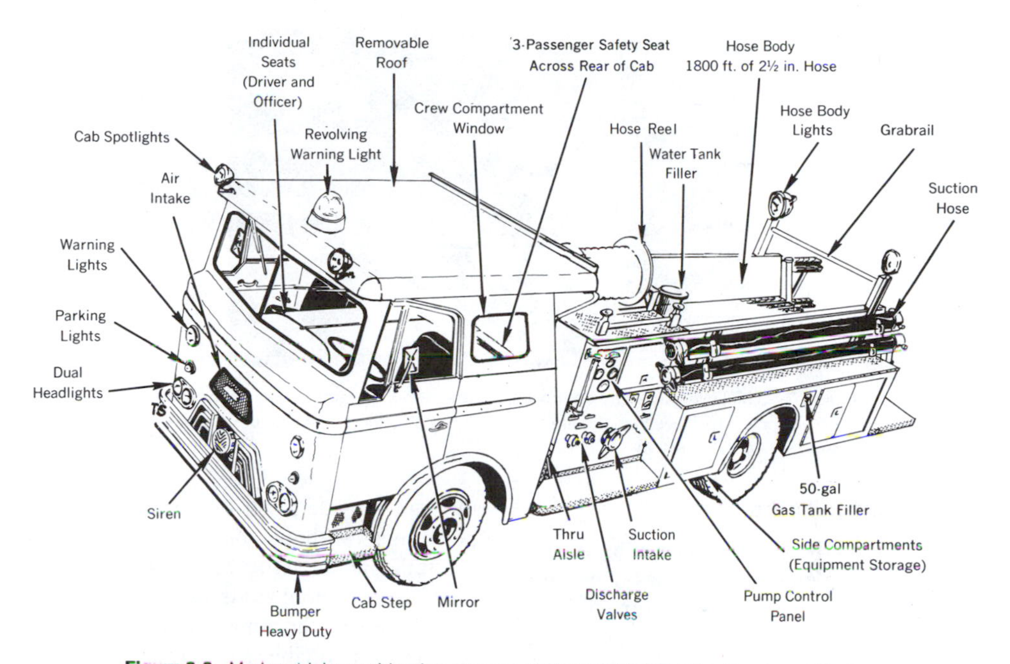

Figure 2-2. Modern triple combination pumper equipment identification (Courtesy of Ward LaFrance Truck Corporation)

Figure 2-3. Top console for the pump operator permits full visibility and protection. This pumper is equipped with a 1000-gpm pump and roll midship pump with a 750-gal tank, making it ideal for combating wildland fires. (Courtesy of W. S. Darley & Company)

Figure 2-4. Modern triple combination pumper has sufficient seating within the cab for the entire crew. Traverse hose compartments allow a rapid attack on a fire with small hose lines. (Courtesy of Crown Fire Coach Corporation)

Figure 2-5. Multiple gages and controls allow the pump operator to obtain maximum pressures and capacities from the fire pump without abusing the apparatus. (Courtesy of Crown Fire Coach Corporation)

Quintuple Pumpers

A quint (Figure 2-6) consists of a triple combination pumper that has a mounted aerial ladder or elevating platform and a full complement of ground ladders. Quints are a combination of an engine company and a truck company. The main drawback of this type of apparatus at a fire is deciding where the most effective use can be made of its features and equipment: pumping from a hydrant or spotted in front of the fire where the aerial ladder or elevating platform can be utilized.

City Service Truck

A city service truck carries a full complement of ground ladders, tools, and other truck company equipment. This type of apparatus is now rarely seen.

Aerial Ladders

Aerial ladder trucks (Figures 2-7 and 2-8) are an important part of the fire fighting teams of modern departments, especially in those cities that have many tall buildings where proper access cannot be provided with ground ladders. The most common lengths of aerial ladders are 65, 75, 85, and 100 ft. Once an aerial ladder has been positioned and stabilized, one person has full control of all aerial ladder operations; this can provide a self-supporting base for personnel without depending on the strength and integrity of the structure on fire. Because aerial ladders can be operated by one person, the remainder of the crew is released for forcible entry, rescue, ventilation, salvage, and other traditional truck company duties.

An aerial ladder can quickly place fire fighters on a roof for effective ventilation, provide access for the rapid rescue of trapped or endangered persons (including fire fighters), allow a convenient means of swiftly removing glass from the windows on the upper floors of a building to release heat and smoke, and furnish a direct path into the fire area so hose lines can be stretched up the most convenient route to the fire. In addition, a ladder pipe attached to the tip of the ladder will allow a master stream of water to be applied directly on the fire. Truck companies are also equipped with a large variety of tools and other equipment that enable the fire fighters to make a rapid and aggressive attack on the fire.

Elevating Platforms

Elevating platforms (Figures 2-9 and 2-10) are also commonly referred to as aerial platforms, cherry pickers, and Snorkels; the term *Snorkel* is actually a trade name for one manufacturer's products. An elevating

Figure 2-6. A quint is a triple combination pumper with a mounted aerial ladder or elevating platform and a full complement of ground ladders. (Courtesy of Peter Pirsch & Sons Company)

Figure 2-7. Two-axle aerial ladder truck (Courtesy of Mack Trucks, Inc.)

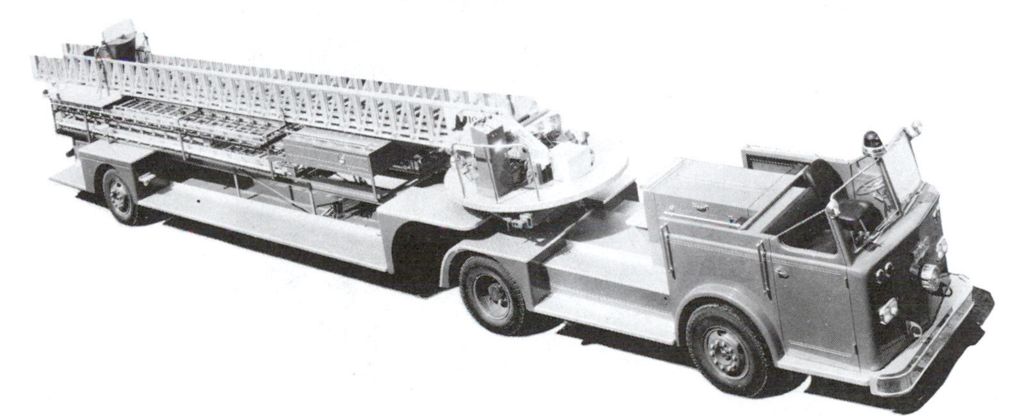

Figure 2-8. 100-ft aerial on a tractor-trailer chassis (Courtesy of Peter Pirsch & Sons Company)

platform consists primarily of two or more booms or sections with a passenger-carrying assembly on the tip.

The principal purpose of an elevating platform is to quickly provide a convenient elevated base so a master fire stream can be directed on a fire from the most effective location. They are also helpful for rescue operations from the upper floors of a structure, for aerial reconnaissance of the fire ground, and for gaining rapid access to the higher levels of a building. Elevating platforms are usually equipped with a complement of ground ladders and other truck company equipment or with a fire pump and hose. The heights to which they can extend range from 50 to 150 ft.

The first elevating platforms were of the articulating type with two or three booms that could be raised, lowered, and rotated. Of later manufacture was a second type that telescoped like an aerial ladder and could be

Figure 2-9. A ground ladder to the fire escape, a 100-ft aerial ladder, and an 85-ft elevating platform are used to provide fire fighting access to a blaze on the fourth floor of a building. (Courtesy of the Los Angeles City Fire Department)

Figure 2-10. Telescoping-boom elevating platform provides three fire fighters with a convenient access to the fire. (Courtesy of the New York City Fire Department)

extended and retracted in a straight line. A third type combines both the articulating and the telescoping principles. Some apparatus incorporate a ladder on top of the telescopic sections that may also serve as an aerial ladder.

The elevating platforms with the longer reaches have a tendency to be massive vehicles; the effective height is measured from the bottom of the basket to ground level. Departments contemplating the purchase of one should survey the geography of the city to ascertain if there are any bridges, tunnels, steep grades, sharp turns, or other restrictions or obstructions that would forbid the passage of the selected apparatus. Some of these vehicles are 12 ft high, up to 50 ft long, and may weigh over 25 tons, so it is imperative to give careful study to the specifications of any proposed purchase to avoid complications after delivery. Fire departments have been known to purchase a large piece of fire fighting apparatus and then discover to their dismay that it would not fit in their fire station. Size limitations could also be true for aerial ladders.

Departments purchasing their first unit for the purpose of reaching high fires need to know and consider all of the merits and limitations of aerial ladders versus elevating platforms. The ideal apparatus for one community could be useless in another city.

Water Towers

Water towers are not a new fire fighting concept, but the older versions were heavy, unwieldy, and provided satisfactory service at very few fires. The new hydraulically operated water towers (Figure 2-11) are highly flexible, mobile, and are equipped with a large-capacity nozzle that can be controlled by one person on the ground; this releases the other members of the company for other fire ground duties. The nozzle can be operated by remote control in a full range of discharge patterns from a straight stream for maximum reach to a 90-degree fog pattern for maximum heat absorption; the fire stream can be rotated a full 260 degrees horizontally and the fire stream can be applied anywhere from ground level to the full reach of the booms.

Water towers bring a new flexibility to fire ground tactics and, to some extent, they have replaced the manual positioning and operation of fire streams in some of the more dangerous locations during fire fighting. Water can be applied to the correct location in effective quantities for the maximum fire fighting efficiency because the tower allows a direct attack on the fire with a heavy stream without endangering fire fighting personnel. The monitor nozzle can be positioned in a window to supply handlines on the upper floors of a building without the time loss of

Figure 2-11. Water tower mounted on a triple combination pumper allows a rapid and massive attack on a fire with maximum efficiency. (Courtesy of General Safety Equipment Corporation)

stretching hose lines up interior stairways or an aerial ladder.

Tankers

The primary purpose of mobile water supply tankers is to transport water to fires in areas that are not adequately served by hydrants. Normally, a tanker is supplemental to a pumper. In most rural areas a maximum capacity of 1000 gal is recommended, because a greater load may seriously limit the mobility of the vehicle. It is suggested that where over 1500 gal of water are to be transported to fires, serious consideration be given to the load-bearing capacity of roads, bridges, and the effect of grades in the area to be served. In many cases, it has been found that two smaller tankers will have greater utility and will cost little, if any, more than one extra large vehicle.

Rescue Apparatus

Rescue apparatus fall into three categories: ambulances and paramedic units, light rescue units, and heavy rescue units. Fire departments are inheriting a greater responsibility than in the past for this country's emergency medical care of the sick and injured, and the trend in this direction is growing at an accelerated pace.

Fire department ambulances and paramedic units are operated by fire personnel who are trained paramedics and emergency medical technicians. The light and heavy rescue vehicles carry a large complement of special tools and equipment for rescue operations at fires and other emergencies.

Aircraft Crash Trucks

Aircraft crash trucks (Figures 2-12 and 2-13) are built to perform specific duties at a certain airport; their characteristics and the functions that they may perform will depend upon the size of aircraft to be protected, the number of aircraft, life hazards, additional fire fighting apparatus in the area, and many other variables. This type of apparatus is equipped to conduct self-sustained operations until additional fire fighting equipment can arrive. Aircraft crash trucks and crash fire fighting are an entire subject by themselves. They are covered thoroughly in pamphlet No. 414 of the National Fire Protection Association, "Aircraft Rescue Fire Fighting Vehicles."

Figure 2-12. Crash truck combating fire (Courtesy of Ward LaFrance Truck Corporation)

Figure 2-13. Modern U.S. Air Force crash and rescue vehicle is driven by two 430-HP diesel engines. Gross vehicle weight with 80 gal fuel, 6000 gal water, 515 gal AFFF agent, crew and equipment totals over 62 tons. Full loaded, it will accelerate from 0 to 50 mph in 70 sec. (Courtesy of Oshkosh Truck Corporation)

Figure 2-14. Fireboats extinguishing waterfront fire (Courtesy of Long Beach, California, Fire Department)

Fire Boats

Cities that have extensive waterfront structures or marine installations usually have fire boats (Figure 2-14) to help provide protection in these areas. The size of the boat and its pumping capacity will vary, depending upon the type of hazards.

To fully utilize the large pumping capacity of fire boats when combating fires in shore installations, many fire departments dispatch a hose-carrying tender on all waterfront alarms. This type of apparatus carries a large amount of hose and is equipped with one or more large turret nozzles.

Purchasing Fire Apparatus

Purchasing the correct apparatus that is capable of doing the best job for the least amount of money is a tremendous responsibility for fire department officers. As these vehicles may involve an investment of $50,000 to $80,000 or more for pumpers and $100,000 to $200,000 or more for ladder trucks and elevating platforms, caution must be exercised when writing specifications, evaluating bids, and awarding the contracts.

Competitive Bid Purchase

Generally, purchases are made by specifying the features desired in a fire apparatus and asking for bids. This method is known as purchasing by competitive bids based on adequate specifications. It is designed to eliminate favoritism or personal influence, insure delivery of equipment that will perform satisfactorily and provide the purchaser with maximum utility and economy.

However, such benefits are only realized through the use of proper standards; the apparatus will be no better than the specifications. Few municipalities are able to employ personnel with the qualifications and ability to draw up specifications that will adequately cover all phases of construction and performance of fire apparatus. Frequently, in attempting to draw up adequate specifications, the result becomes so excessive or restrictive in some requirements that it increases the cost unnecessarily or prohibits bidding entirely. Often important requirements are omitted and irrelevant and unduly costly provisions are included.

To provide general standards, a committee was formed by joint action of the International Association of Fire Chiefs and the National Fire Protection Association. Their recommendations are available from the National Fire Protection Association in the publication entitled *Standard for Automotive Fire Apparatus* (NFPA No. 1901). Specifications are complete except for such items as the capacity required in pumpers, the length of ladders on ladder trucks, and other items that vary according to local conditions or operation or preference.

Replacement Purchase

It is wasteful economy for a city not to provide apparatus and equipment of the best and most dependable type. The largest item of expense for a fire department is the cost of personnel. The cost of apparatus, which has a service life of at least 20 years, is proportionally small in the overall budget.

The number of miles traveled and hours of pumping operation do not normally provide a basis for determining the need for replacement. Many other factors limit the effective and economical life of an apparatus and make replacement desirable: advancements in design of fire fighting equipment; inability of vehicle to compete with modern traffic because of inadequate braking and slow acceleration, which may result in a driver's tendency not to slow up at intersections; inadequate protection for driver and fire fighters; structurally weakened chassis because of overloading; increased maintenance costs; parts replacement difficulties with old apparatus; and lack of reliability under the stress of emergency service. Some of these drawbacks increase the dangers to the public and to fire fighters because of the increased chance of accidents. Apparatus relieved from first-line service may be retained as reserve equipment; this should also be considered when assessing replacement costs.

Writing the Specifications

Determining exactly what type, size, and model to purchase is the first step in writing specifications for a fire department apparatus. As the department will probably be either blessed or stuck with this equipment for 20 or more years, a great amount of thought must be devoted to acquiring the best vehicle for the job. A large-capacity pumper should not be purchased just for the prestige. A longer ladder than necessary should not be ordered. A tanker or pumper with an excessively large water tank may be a liability. Consideration should be given to the fire hazards, terrain, roads and highways, weather and climatic conditions, building heights and areas, water supply, fire station locality, mutual aid arrangements, and every other character of the response area that this apparatus will be expected to protect. The growth possibilities of the city should also be considered. Higher structures and larger buildings will undoubtedly be constructed. Rural areas may be annexed. After all of these variables are analyzed, then a definite idea may be formed of what size and type of apparatus can do the task the best.

A department that has to primarily protect rural areas must be concerned about not ordering a vehicle that is too heavy for the unimproved roads. Highly maneuverable apparatus are needed in cities with narrow and winding streets. A city with large industrial factories may require a pumper with a 1500-gpm capacity and a 500-gal tank, while a rural department would be better equipped with a 1000-gpm pumper with a 1000-gal tank.

Caution must be exercised when attempting to tailor the specifications for apparatus too closely to the area in which it will be operated. One large fire department that was responsible for protecting a huge section composed of highly industrialized and residential areas, flat sections of agricultural land, and very mountainous forest and brush terrain was ordering a large number of pumpers. They were concerned with acquiring the largest number of pumpers for the least amount of money, so they wrote two sets of specifications, one for mountain apparatus and one for flat-land rigs. Because the pumpers assigned regularly to the mountainous section needed far more power than the flat-land rigs—and there were other differences—they received three of the lower-powered apparatus for roughly the same price as two mountain rigs. This was a substantial money saving, but the mistake was never made again. Because the flat-land apparatus was seriously underpowered, the dispatchers were forced to be extremely selective about what greater alarm companies they sent to large fires in the mountainous districts.

It is best to order by performance specifications, as detailed in NFPA 1901. They allow the manufacturer greater latitude in selecting the best and most modern components and equipment for the vehicle. A deviation from this principle may be made to designate a diesel engine over gasoline powered, an automatic transmission instead of a manual gear box, a certain model of chassis because of the availability of repair facilities, or other definite preferences.

Awarding the Contract

Most governmental agencies are closely controlled when writing specifications, advertising for bids, and awarding the contracts for any purchase of a substantial amount. Because these are primarily legal processes, local laws play a fundamental role in the apparatus and equipment acquisition process. If the estimated amount of the contract exceeds a certain specified sum of money ($10,000 in some localities), sealed bids must be solicited by public notice in the particular manner and subject to the requirements of the law.

When any agency calls for bids for the purchase of apparatus or equipment, specifications should not be prepared so as to exclude all but one type or kind, but should include competitive supplies and equipment. Writing specifications with the intent of securing one certain model and make of apparatus is discouraged. Fire department officers occasionally are so convinced that one manufacturer builds vastly superior equipment that they will use the product's advertising specifications to write the bid specifications. This practice smothers competition. It is also generally illegal to designate one certain make and model or to use the wording, "equivalent to Jones Spray Nozzle, Model #165."

Specifications should be drawn as precisely and as definitely as possible so bidders will have a reasonable basis on which to submit a proposal. Court cases seem to indicate that courts will uphold imprecise specifications only if the item specified is of a nature that requires an imprecise bid and if the specifications do not prejudice any bidder's opportunity to submit proposals.

The courts will not uphold specifications that improperly stifle competition; specifications must be drawn in a manner that will permit more than one bidder to have a chance of success. If any portion of a specification is restrictive, there should be a clear reason for the restrictive provisions. When only one bid is received, or there are similar difficulties, many agencies will request a withdrawal and rewriting of the original specifications and then issue a new invitation to bid.

The underlying principle of the bidding process is that the governmental body awards the bid to the lowest responsible bidder meeting the terms and conditions of the bid invitation. To determine if a bidder is truly responsible and capable of fully performing the desired services or furnishing the wanted equipment or vehicle, it is a legitimate obligation of the agency to investigate the bidders to determine that they do have the skills, abilities, and record of past performances to ensure that the specified item will be delivered at the correct time. The absolutely low bid does not usually have to be accepted if it can be clearly shown that a higher priced apparatus is a better buy for the money. There are many legitimate questions that should be answered before a bid is awarded.

If a bidder takes exception to (or deviates from) any of the terms of the specifications, the legal question is whether the variance is material. If the variance or exception gives the bidder a substantial advantage or benefit not enjoyed by other bidders, then the exception is important. The bidder has made a new offer and has not really bid. In this case, the bid should be rejected.

After a contract for an apparatus has been awarded, fire department representatives should visit the factory to see that the manufacturer is meeting specifications. When the equipment or vehicle is delivered, the finished product should be inspected, tested, and otherwise measured against specifications before the delivered item is accepted.

Only after the correct type and size of apparatus has been decided on, proper specifications written, bids solicited from as many manufacturers as possible who have the experience and facilities necessary to build a quality product, bids analyzed and the contract awarded, supervision during construction given attention, and a thorough inspection and testing period observed upon delivery of the equipment or vehicle can a department have any degree of certainty that they are selecting the apparatus that will do the best job for the least amount of money.

Engine Capability

Engines may be either gasoline or diesel fuel type, depending on the preference of the purchaser. The engine should be capable of performing the specified pumping and road tests without exceeding its maximum no-load governed speed as shown on a certified brake horsepower curve for the type of engine used, without accessories. The size of the engine is usually determined by the driving performance and the carrying capacity of the vehicle rather than by the power required to operate the fire pump, with the possible exception of large-capacity pumps.

Road Tests

Road tests are designed to prove that the engine supplied with the apparatus is of adequate size and that the transmission and differential gear ratios are correct to assure that the vehicle will satisfactorily conquer the steep grades and other terrain problems at sufficient speed to enable fire fighters to respond to fires and other emergency incidents with the least delays or hazards. For these reasons, it is best if these tests are conducted in the areas in which the vehicles will be stationed so that they will be subjected to identical elevations, climatic conditions, terrain, and other aspects of emergency response that the apparatus must be able to overcome.

Test Conditions

All of these tests should be performed with the apparatus loaded with a complete complement of personnel, hose, equipment, and a full water tank. If the normal personnel and equipment loading cannot be provided, compensatory loading to the anticipated maximum vehicle weight will allow a practical test to be performed.

Cities in which there are many steep grades in excess of 10% should include this information in the purchase specifications; the occasional exposure to excessive grades is of less concern than an everyday problem. A combination of steep grades and narrow winding roads will warrant specific changes in the standard road tests to verify that the new apparatus can satisfactorily accelerate and maneuver in the area to which it will be assigned.

The road tests shall be performed on a dry paved road with two runs in opposite directions over the same route. From a standing start and shifting up through the gears, pumpers should attain a true speed of 35 miles per hour (mph) within 25 seconds (sec); vehicles carrying over 800 gal of water or those equipped with an aerial ladder or an elevating platform shall attain a true speed of 35 mph in 30 sec. In addition, from a steady speed of 15 mph in direct drive, the vehicle should accelerate to 35 mph within 30 sec without shifting gears; this is an important point to check because it verifies the ability of the vehicle to quickly regain speed after slowing down at a street intersection, which is a desirable safety feature.

The vehicle should attain a top speed of not less than 50 mph, unless a higher speed is specified; an apparatus that is required to respond for long distances on open roads, throughways, freeways, or other high-speed highways should be designed to be capable of keeping up with the normal traffic flow.

Vehicle Carrying Capacity

The carrying capacity of a vehicle is apparently one of the least understood features of design and one of the most important. All vehicles are designed for a rated gross vehicle weight (rated GVW), which is the sum of weights for the chassis, body, cab, equipment, fuel, crew, and all other load. This should not be exceeded by the manufacturer or the purchaser after the apparatus has been placed in service. Many factors make up the rated GVW, including the design of the springs or suspension system, the rated axle capacity, the rated tire loading, and the distribution of weight between the front and rear wheels.

One of the most critical factors is the size of the water tank; because water weighs 8.35 lb per gallon, a 500-gal tank adds 2 tons to the vehicle. A value of 10 lb per gallon is used when the additional weight of the tank is being considered, making about 2½ tons for a 500-gal tank of water.

The distribution of weight between the front and rear wheels is also a factor for major consideration because improper design will seriously affect the handling characteristics. Manufacturers vary their weight distribution to some extent, but generally they adhere to an approximate ratio of 25% on the front wheels and 75% on the rear wheels of a conventional chassis, and 33⅓% on the front wheels and 66⅔% on the rear wheels of a cab-ahead chassis. Weight distribution is particularly important in water tankers.

Too little weight on the front wheels may cause a front-end skid; over bumpy roads, it may cause the front of the vehicle to veer from side to side. At the very least, it will be difficult to keep the vehicle under control. Too much weight on the front wheels will reduce the traction of the rear wheels and may result in a rear-end skid or in difficulty in traveling over unpaved roads or in mud.

Fire departments frequently upset these critical weight distribution factors by overloading the apparatus. Most commonly this is a result of putting additional hose and equipment on the rear of the rig. Overloading the vehicle will materially reduce the life of the vehicle and increase maintenance costs. It will also affect acceleration, deceleration, stability characteristics, and handling sufficiently to make steering and operation in general particularly difficult.

Water Tanks

Water tanks should have a minimum capacity of 300 gal; on apparatus equipped with either an aerial ladder or an aerial platform, the capacity should be at least 150 gal. For normal municipal service, the maximum capacity for pumpers should be 500 gal.

The design of a water tank can be a very critical factor in the handling characteristics of fire apparatus. If water is free to travel either longitudinally or laterally in a tank, as would be the case if a tank were half full, a tremendous amount of inertia can build up; this will tend to force the vehicle in the direction the water has been traveling. When the water reaches the end of the tank, this sudden application of force can throw the vehicle out of control and has been known to cause fire apparatus to turn over or to skid when going around a curve or coming to a sudden stop. The only method of preventing such an accident is to impede the motion of the water so that inertia will not build up; this is done through the installation of swash partitions (baffle plates) designed to contain water in small compartments within the tank, as shown in Figure 2-15.

An overflow outlet is necessary so that unwanted pressure or vacuum will not occur within the tank when it is filling or emptying. Water is likely to spill out of the overflow when the apparatus is turning a corner, stopping, or accelerating. Therefore, it is essential that this overflow outlet be arranged so that water is not spilled in front of the rear wheels where it could cause a skid or a loss of traction.

Figure 2-15. Water tank construction showing surge plates and baffles (Courtesy of the Howe Fire Apparatus Company)

Discharge piping should provide a minimum flow of 250 gpm so as to supply two 1½-in. lines or one 2½-in. line from the tank for an initial attack on a fire. When water tanks are of larger than 800-gal capacity, and adequate pump capacity is supplied, a minimum flow rate of 500 gpm from tank should be provided for; this will permit rapid unloading when the vehicle is being used to transport water to the scene of a fire. The most common design fault in rural fire apparatus has been inadequate piping from the water tank.

Quick-dump valves of large size on the rear of the tank will permit the rapid release of the water.

Hose Compartment

A hose compartment, or compartments, of at least 55 cu ft should be provided on apparatus designed to carry 2½-in. and 1½-in. hose. It should have a capacity, for double-jacketed rubber-lined fire hose, of not less than 1500 ft of 2½-in. hose on pumpers and hose trucks (40-cu ft and 1000-ft capacity on pumper-ladder trucks) and 400 ft of 1½-in. hose.

Pumper Size Capability

The modern pumper is the most important piece of fire fighting apparatus in use today. Only seven sizes or capacities are suitable for municipal use: 500 gpm, 750 gpm, 1000 gpm, 1250 gpm, 1500 gpm, 1750 gpm, and 2000 gpm. Minipumpers, midipumpers, and other quick-attack units are gaining in popularity. These pumpers have small or medium-sized pumps, and their lighter weight will sometimes permit a faster response and attack on a fire.

Any pump mounted permanently on the apparatus and rated less than 500 gpm is called a booster or auxiliary pump.

When tested at draft, main fire pumps (Figure 2-16) should deliver the percentage of rated discharge at pressures (in pounds per square inch, or psi) indicated below (though pumps may discharge considerably more water from a good hydrant supply):

100% at 150 psi net pump pressure.

70% at 200 psi net pump pressure.

50% at 250 psi net pump pressure.

Figure 2-16. Between the engine and the booster tank of a modern pumper lies the pump with a necessary maze of valves and pipes.

A large-capacity apparatus will do everything a small one will do, but there are many situations at serious fires in which a small-capacity pumper may prove inadequate or be seriously overloaded. When a 750-gpm pumper is connected to a good hydrant near a fire, it often results in inefficient operation because other pumpers may have to lay long hose lines to more distant hydrants. A large pumper at the same hydrant permits a more efficient operation when multiple lines or heavy fire streams are needed.

The larger pumps can usually supply the required flow with much greater power reserve; this results in considerably less strain on the apparatus. For example, a 1500-gpm pumper is only working at 50% of capacity when delivering 750 gpm, while a 750-gpm pumper is operating at full capacity.

When selecting a pump, consider what the apparatus will do at its rated 150 psi net pump pressure and how much water it can pump at higher pressures. Wagon batteries, ladder pipes and fog nozzles all call for relatively high pressures. Unless the pumper can be placed very close to the nozzle so that discharge pressures can be held to a minimum, it is also important to consider its output at 200 psi pump pressure. Because high discharge pressures are needed at most fires, the quantity of pump discharge at 200 psi pressure is often the true measure of a pumper's ability. Thus, a 1000-gpm rated pumper is only about a 700-gpm pumper when operating at 200 psi pressure, unless aided by an unusually strong hydrant.

Pump Drives

Fire pumps may be mounted on an apparatus chassis in several ways (Figure 2-17). Front mounting requires pumps which can be driven through a clutch on the front end of the engine crank shaft; such pumps are utilized on rural apparatus constructed on a commercial chassis, but are also easily mounted on conventional vehicles.

Booster pumps and small, high-pressure pumps are usually driven by a power takeoff from the side of the apparatus road transmission. This type of drive permits the apparatus to be driven at the same time it is pumping; front mounted pumps and pumps with pump transmission ahead of the road transmission can also fight running fires while the apparatus is moving.

Main fire pumps are usually driven by one of three different methods: pump transmission ahead of the road transmission, pump transmission behind the road transmission, or power takeoff from an automatic transmission.

A few pumps have individual engines whose only duty is to drive a pump.

Pump Shifting Devices

Pumpers are provided with various levers and valves to control their operation. The operation and location of pump shift levers vary with make, age, and design of the apparatus. Regardless of the method of operation and the location of the pump shift lever, its operation involves changing the engine power from driving to pumping. To accomplish this change, the engine power must be removed from the road transmission by releasing the clutch so that the gears will not rotate during the change. If the pump transmission is behind the road transmission, the road transmission must be in gear, usually direct drive (fourth or fifth gear, depending on transmission). When the pump transmission is ahead of the road transmission, or a power takeoff or front mounted pump is used, the road transmission should be placed in neutral. When the pump shift lever is located in the cab near the driver's seat, the clutch may be released with a foot pedal or by a vacuum clutch control. If the pump shift control is located by the pump operator's panel, a mechanical, air, or vacuum clutch control must be provided to release the clutch.

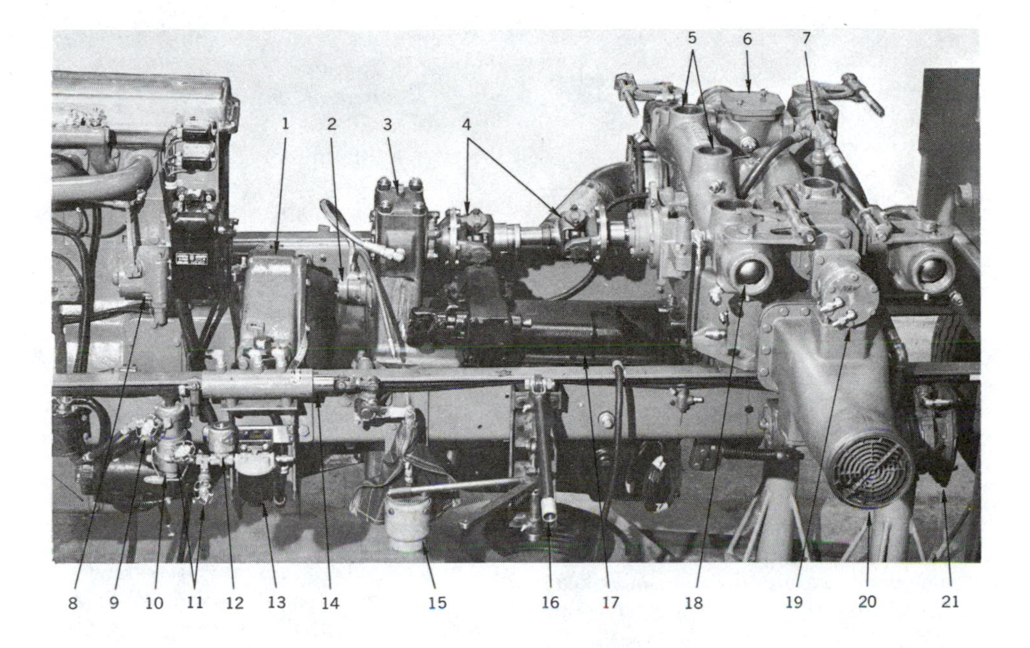

1 Clutch housing
2 Lubricating pump for pump transmission
3 Pump transmission
4 Universal joints in pump drive
5 Pump outlets
6 Pump outlet to wagon battery
7 Priming valve (lines go to priming pump and to eyes of front and rear impellers)
8 Engine governor
9 Fuel shut-off valve
10 Electric fuel pump
11 Fuel shut-off valves
12 Fuel solenoid valve
13 Fuel filter
14 Hydraulic clutch booster cylinder (utilizes hydraulic pressure from the power steering pump to aid in disengaging the clutch when pedal is depressed)
15 Air cylinder connected to pump transmission shift lever (disengages the clutch when shift lever is raised)
16 Handle for pump transmission shift lever
17 Road transmission
18 Discharge gate with ball valve
19 Relief valve
20 Strainer in pump suction
21 Electric priming pump

Figure 2-17. Chassis with the pump transmission ahead of the road transmission (Courtesy of the Los Angeles City Fire Department; photo by Art Gluskoter)

Instruments

Engineers on steam pumpers had few gages to watch because their fire apparatus was less complicated. About all the pump operator had to do was watch the boiler water-level sight glass and steam pressure gage, both usually located on the rear of the boiler. Two other gages on the sides of the steamer measured pump pressure and vacuum.

On modern equipment, however, the number of measuring and indicating gages has been multiplied several times. Such instruments are necessary for the effective and safe operation of engines and pumps, and it is imperative that apparatus operators understand their instruments.

Pressure Gages

Three types of pressure gages are used on fire apparatus:

Vacuum gages are used to measure any pressure below atmospheric (negative pressure). Calibration will be in inches of mercury; this is sometimes referred to as inches of vacuum. Late model equipment may use a vacuum gage on the dash to show the intake manifold pressure of the engine or the reserve tank pressure of vacuum-assisted brakes.

Straight pressure gages are used to indicate positive pressure above atmospheric and will be calibrated in pounds per square inch.

Compound gages indicate both positive and negative pressures. This type of gage is designed to measure in inches of mercury any pressure below atmospheric pressure that may be developed by the pump. The zero reference point on a pump gage actually represents atmospheric pressure. The same gage is also calibrated in pounds per square inch above the zero. Positive pressures are indicated by a clockwise rotation of the hand, negative pressures by a counterclockwise movement.

Preventing gage damage. Centrifugal main pumps use compound gages on the suction side to measure the incoming pressure, positive or negative. The pump discharge pressure may be indicated by either a straight pressure gage or a compound gage. A centrifugal pump presents a continuous waterway from the suction inlet, through the casing, to the discharge gate. If a vacuum is applied to the pump, such as when taking suction, the vacuum exists in the entire pump until the pump is primed. Under ordinary circumstances, a straight pressure gage would be damaged by such use. Possible gage damage may be prevented by either locating a valve in the tubing between the pump and the gage, or by using a vacuum protection device, which consists of a metal stop on the movement that prevents the hand from being rotated below zero.

Where there is a gage dampening valve that has an exterior handle, the valve should be closed whenever drafting. When pumping, this valve should be adjusted to allow the gage to promptly indicate a change of pressure and yet give a good steady reading without undue vibration. The valve should be kept open in cold weather to prevent water in the line from freezing.

Gage construction. Except for gages used for special tests and experiments, all pressure gages are of the Bourdon type, named after its inventor. While less sensitive than the more complicated mercury tube gages used to indicate pressure in more painstaking experimental work, the Bourdon type is sufficient for all normal operations.

Bourdon tube gages are of two types: single-spring and double-spring. In older fire apparatus, the double-spring type was used almost exclusively because of its more rugged construction and greater stability; however, because of advancements in gage design and the development of more durable parts and construction, modern gages generally are of the single-spring type (Figure 2-18).

Both types operate on the same principle. Pressure enters through the threaded gage fitting and passes into the Bourdon tube (or tubes), which is thin, curved, and hollow. Its closed end is attached to a movable quadrant, or sector, which meshes with a pinion gear carrying the indicating pointer. A pressure increase creates a tendency for the tube to straighten out a definite amount for each increment in pressure. The movement of the Bourdon tube tip is transferred to the sector, which turns the pinion and pointer. The tip movement is relatively small and must be considerably multiplied to make the hand travel around the dial. A hairspring attached to the pinion shaft holds the assembly together tightly and serves to dampen the movement of the pointer. Vacuum in the tube tends to increase the curvature and cause the hand to move in a counterclockwise direction.

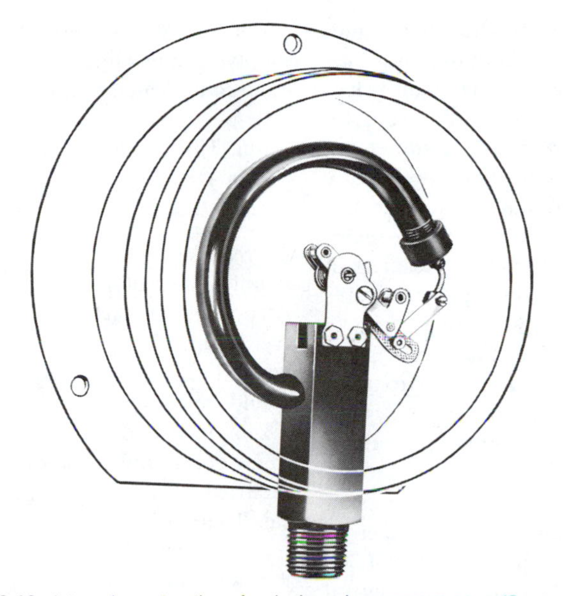

Figure 2-18. Internal construction of a single-spring pressure gage (Courtesy of Marsh Instrument Company)

Individual Discharge Gages

The inclusion of individual gages for each discharge gate on a pumper has become very popular. When used with modern controlling gates, this arrangement permits a pump operator greater ease in supplying various hose lengths and nozzle tips at one time. Instead of averaging hose line requirements and setting the pump pressure to meet the average, an individual gage and a controlling gate permit each line to be supplied at optimum pressure. This results in increased operating efficiency, as well as greater safety for the nozzle operators.

Flow Meters

Some manufacturers are now supplying pumpers with flow meters, instead of individual discharge gages. A flow meter will indicate the quantity of water flowing through a specific hose line. It is a more complex instrument than a normal pressure gage and is calibrated in gallons per minute instead of pounds per square inch. This is not a new instrument. Many years ago they were being installed on pumpers to measure the amount of water flowing through each discharge outlet. They were fantastic when they worked, but they were gradually replaced on the apparatus with individual pressure gages because they were undependable after a few years and failed because of internal corrosion, in spite of frequent flushing. Manufacturers now claim to have conquered this defect.

Interpreting flow meter readings. Flow meters do not make the knowledge of fire fighting hydraulics, and the necessary mathematical calculations, unnecessary; they do, however, greatly simplify the pump operator's duties on the fire ground. With pressure readings, the pump operator must take into account the friction losses in hose layout, fittings, nozzles, and other sources that cause a depletion of the original pump discharge pressure. Engine pressure is also further reduced or increased by a difference in elevation. Because of these myriad pressure losses, many of them incalculable, getting an exact nozzle pressure reading at a fire is more a result of luck and an educated guess than it is of deliberate calculation.

With a flow meter, on the other hand, unless there is a leaking coupling or a broken hose, there will be no water quantity loss between the pump discharge outlet and the nozzle; if 200 gpm flow into the pumper end of the hose line, 200 gpm will flow out of the nozzle coupling. Friction loss and difference of elevation will not affect the flow meter reading. Of course the engine pressure will have to be increased for these pressure losses to obtain the desired flow reading.

All that is necessary for the pump operator to know is the gpm discharge from the specific nozzle being used at the desired nozzle pressure. If a 300-gpm spray tip is being used on a fire on the tenth floor of a building, the pump discharge pressure should be increased until the flow meter on that discharge outlet indicates a flow of 300 gpm. If the pumper is supplying several different hose lines equipped with a variety of nozzles, the engine pressure must be sufficient to assure that the gpm discharge gage on the outlet that demands the greatest flow attains that quantity; then the other discharge gates should be slowly closed until each is delivering the correct quantity. It is not unusual at times for a smaller nozzle with a high pressure loss from friction and head to have a greater effect on the necessary pump discharge pressure than a hose line with a greater flow.

Flow meters really are a benefit to pump operators when they are supplying a heavy stream, such as for an aerial ladder or an elevating platform, through several hose lines which are all of the same size and length; merely divide the nozzle discharge by the number of hose lines. Thus, if four 2½-in. hose lines were supplying a ladder pipe equipped with a 1000-gpm spray nozzle, each individual flow gage would register 250 gpm. If there were considerable length variations, and if exactly equal flows through each line were desired, then adjustment of discharge outlet valves would be necessary to compensate for varying friction loss effects.

But then if a variety of hose sizes were supplying one large tip, it would become more complicated as it is very apparent that a 3½-in. line should carry more water than a 2½-in. This would require computing what percentage of the total flow would be carried by each line.

Correct use of flow meters. The flow through parallel lines of hose is controlled by the friction loss in each line. The flow through two 2½-inch hose lines of the same lengths would be similar because the friction losses would be similar. If one 2½-inch and one 3½-in. line were being used to supply a single nozzle, the flows would stabilize when the friction losses in both lines were identical. Therefore, to correctly utilize the flow meters to supply a single nozzle through multiple hose lines of varying sizes, the friction factors must be considered when calculating the percentage of the total gallonage that should flow through each hose line. In this case, the pump discharge pressure gage may be the most practical solution.

The correct use of these flow meters requires that the pump operator know specifically what nozzle is on each hose line and what quantity of water will be discharged from the tip at the correct nozzle pressure. Obviously, it could be disastrous if a pumper were supplying two lines, one with a 1-in. straight tip and the other equipped with a 350-gpm spray nozzle, and the engineer did not know which of the two lines was supplying which nozzle.

Oil Pressure Gage

The oil pressure gage indicates whether or not the engine lubricating system is operating. It is mounted on the dash panel of an apparatus and may be duplicated on the pump panel. There are two types: mechanical and electrical. The mechanical unit consists of a Bourdon tube with a pointer attached. Pressure is introduced through small tubing connected to the engine oil system.

The electrical gage consists of a sending unit mounted on the engine and connected by wire to a panel-mounted receiving unit. Oil under pressure is admitted to the sending unit where it exerts force on a diaphragm. Connected to the diaphragm is a sliding contactor that moves across a resistor according to the force exerted. This in turn regulates the flow of current through the wire to two electromagnets in the receiving unit.

With no oil pressure in the engine, all of the electric current passes through one of the magnets, holding the dial pointer, which is pivoted between both magnets, at zero. As the pressure builds up, the current is divided between the two magnets by the resistor, and the pointer moves according to the greatest attraction.

With the engine running, the pressure indication should be close to that stated in the manufacturer's instruction manual. It is possible that a lower pressure may be recorded if the engine speed is less than 1000 rpm.

Rapid fluctuations of the pointer should be investigated because they are usually caused by a low oil level. Any abnormally low pressure reading is a warning to stop the engine immediately and to investigate the failure.

Some manufacturers utilize warning lights to indicate low oil pressures and low oil levels. A burned-out lamp bulb or defective sending unit will render the warning lights inoperative.

Fuel Gage

The same principle of electrical sending and receiving units is used for gasoline gages, although the sending unit consists of a tank float connected by an arm to the resistor contactor. The position of the tank float on the fuel level is reflected in the calibrated magnetic attraction exerted on the pivoted pointer of the gage.

Ammeter

Because of the radio, warning lights, and other devices that use electricity, the ammeter is one of the most important instruments on fire apparatus. It is placed in the circuit connecting the battery to the generator or alternator and indicates the direction and amount of current flow. It indicates when the generating system is charging, how much electricity is consumed, when the battery is supplying the requirements of the system; it is also a rough indicator of the battery's condition.

A fairly high charging rate when the engine is first started and a reduction in the charging rate to zero after the engine has been in operation for a short while indicate a normal condition; the battery is discharged a small amount when starting the engine, and the generating system rapidly recharges the battery. A discharged or defective battery is indicated when the time required for the charging rate to drop back to normal is unusually long. A defective alternator, generator, or voltage regulator is indicated when the charging rate is low or discharging, and the battery is not fully charged.

When the engine is stopped and the ignition is turned off, the pointer should return to zero. If it continues to show a discharge, it is a warning that electrical equipment has been left on, that a short exists in the system, or that the cutout points may be stuck together. It is good practice to check the ammeter for an indication of discharge when the engine and switches are turned off; this may prevent the embarrassment of later discovering a discharged battery. At times the pointer will rest slightly off center with no current flowing in either direction. An off-center needle can be checked by disconnecting the batteries.

Heat Indicators

Temperature gages indicate the operating temperatures of engine coolant, lubricating oil, and pump transmissions. Continued operation of an overheated engine, without using auxiliary cooling, may permanently damage its working parts. A cold internal combustion engine does not evaporate condensates or substances that dilute fuel. Sludge deposits are therefore formed throughout the engine.

Tachometers and Speedometers

Tachometers and speedometers are essentially the same instrument; the differences between the two are the point of the connection on the vehicle and the dial calibration. A tachometer indicates the speed of the engine crankshaft in revolutions per minute. It is used to prevent lugging or overspeeding while driving. To find out if there is reserve power and if the pump is operating correctly, check the tachometer while pumping. Speedometers indicate the road speed of the vehicle in miles per hour.

There are several types of tachometers and speedometers. One type utilizes a long drive cable to drive a permanent magnet. In this type, a nonmagnetic (speed cup) disc is influenced by the magnet and attempts to follow it, while a coiled spring retards the motion. Any movement of the disc is accompanied by a corresponding movement of an indicating hand over a calibrated dial.

Another type uses a long cable to spin flyballs; as the engine speed increases, centrifugal force will throw the weights outward and move the needle.

Electric speedometers and tachometers are of several types and prevent problems caused by long drive cables. One type uses a small generator as a sending unit and an electric motor in the instrument to spin a magnet. The faster the engine speed, the more current the sending generator will produce, which causes the slave electric motor to rotate more rapidly.

In another type of electric speedometer, an alternating current voltmeter measures the ignition primary interruption speed at the coil or distributor.

Heavy Stream Appliances

Most fires are brought under control and extinguished by discharges from small hand lines or portable extinguishers. When these are insufficient to absorb the heat generated by a fire, heavy streams are used (Figures 2-19 and 2-20).

The terms *master* or *heavy* fire streams usually refer to a water discharge, either as a spray or solid stream, in excess of 400 gpm. For average use, the generally recognized standard nozzle pressure is 80 psi for solid streams and 100 psi for spray and fog streams. Heavy stream appliances include deluge nozzles, portable monitors, wagon batteries (Figure 2-21), combination portable monitor and wagon batteries, ladder pipes (Figure 2-22), and water towers or elevating platform turrets. Figure 2-23 shows a three-inlet portable monitor, which is capable of discharging 1200 gpm. The low profile places the point of fire stream discharge so low to the ground that the unit can be left unattended. The nozzle will automatically adjust to any volume of water between 350 and 1000 gpm, and will discharge that amount at an effective pressure. The gage is handy for determining if the pumper is delivering sufficient water at an effective pressure.

Automatic Nozzles

Nozzles, pumpers, pressure control devices, and other fire fighting equipment are becoming increasingly more automated and self-governing. Automatic, or thinking, nozzles make the most effective use of whatever water is available, because they spontaneously adjust to the volume of water being furnished and discharge that amount at an adequate pressure. The quantity being discharged from a nozzle may vary, but the nozzle pressure will remain constant. Possibly the simplest way to describe the operation of these nozzles is to compare them with a pressure relief valve. They have an orifice that opens and closes with only a small variation of pressure, according to the availability of water.

Figure 2-19. Heavy streams are required to absorb large quantities of heat. (Courtesy of the Los Angeles City Fire Department; photo by Paul L. Garns)

Figure 2-20. Water curtain set up by wagon battery protects fire fighters on handlines (see Figure 2-19) at oil refinery fire. (Courtesy of the Long Beach, California, Fire Department)

A pumper can deliver available water at any desired pressure up to its maximum capacity and pressure limits. It cannot, however, deliver more water than is attainable. If a pumper that has a limited quantity of water has a hose lay with a fixed orifice nozzle that demands more water than is available, neither the pump nor the nozzle can build up a pressure sufficient to produce an effective fire stream.

The greatest advantage of an automatic nozzle is the decrease in the surge or nozzle reaction that develops in other nozzles when one or more nozzles are suddenly shut down. Automatic nozzles react faster than a relief valve in a pumper because they are at

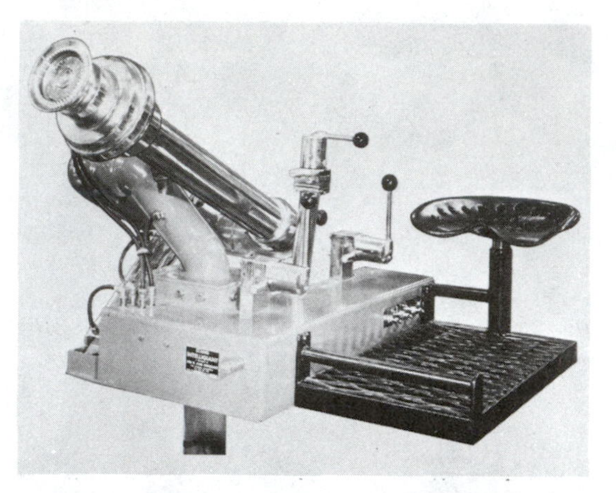

Figure 2-21. "Intelligiant" hydraulically operated wagon battery (Courtesy of John W. Stang Manufacturing Company)

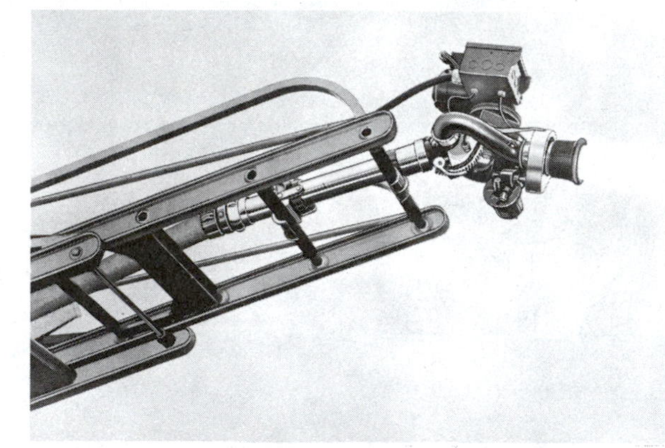

Figure 2-22. Remote control ladder pipe on aerial ladder (Courtesy of Akron Brass Company)

Figure 2-23. Three-inlet portable monitor (Courtesy of Akron Brass Company)

the ends of the lines; they automatically open up to relieve shock to the nozzle operators.

The automatic nozzle offers another advantage to those departments, such as New York City, which have adopted the Rapid Water System that allows 1¾-in. hose to carry as much water as 2½-in. hose because the slippery water additive reduces the friction loss. If the introduction of the rapid water additive ceases, an automatic nozzle spontaneously compensates for the change of flow.

Electrical Power Outlets

Many departments specify that apparatus be equipped with outlets that supply 110-volt (V) current. Some officers are later disappointed that these outlets are not as useful and versatile as they had expected. Outlets are valuable for supplying 110-V current to floodlights and electrical equipment close to the vehicle, but the voltage drop resulting from using long extension cords does impose a severe limitation on their usage. Furthermore, depending upon the large apparatus engine to drive the alternator at a speed necessary for producing the 110-V current imposes extra wear on an expensive machine. An alternator begins to produce a usable amount of current at rotor speeds of 800 rpm and produces the maximum current, depending upon the alternator rating, at speeds of 1400 to 1600 revolutions per minute (rpm). However, the alternator current produced has a frequency in cycles per second of one tenth the alternator speed; therefore, the frequency varies from 80 to 160 cycles per second (Hz) or more. These outlets will never render the small, light, portable, self-contained generator obsolete and unnecessary for all around fire ground electrical requirements.

An alternator will only produce a limited amount of current, much of which must be used for charging the battery and operating lights, radio, and other vehicle equipment; the only current available for operating floodlights and power tools is that over and above the vehicle requirements. It is important that the alternator on the engine has sufficient capacity to supply the current required in addition to the normal demands of the vehicle. For an output from the apparatus charging system sufficient to operate a 110-V, 500-watt (W) flood light, divide the 500 W by 12 V; it can then be discerned that the light will require an additional alternator output of 41.6 amperes (A) just for the lamp. There are two different methods of converting the 12-V alternating current to 110 V. When a power inverter is used, the frequency remains constant at 60 Hz at all engine speeds.

A transformer will boost the 12-V alternating current to 110-V power, but the cycling will depend upon the alternator speed; an alternator produces current which cycles at one-tenth of the rotor speed. If the rotor is turning at 2000 rpm, the current reverses its polarity 200 times each second; this would be 200-Hz alternating current. It may be theoretically possible to figure out the ratio between the alternator speed and the engine rpm and then set the engine throttle at the correct speed to produce 60-Hz current, but it would not be practical because speed fluctuations would cause a varying cycling.

Electrical motors designed for 110-V, 60-Hz current will not operate on other cycles; they will just hum and burn out. If saws, drills, and other tools must be powered by 110-V current that is not a steady 60 Hz, they necessarily will require either direct current motors or universal motors designed for use with either a.c. or d.c. current.

Warning: With the increased use of electrical power plants and outlets, caution should be exercised in the use of power transmission and consuming devices because a rubber-tired vehicle is not normally grounded, and any electrical malfunction could transmit an electrical shock to a person standing on the ground using electrical devices or in contact with the vehicle. When additional lights are added to the top of aerial ladders, it is recommended that 110-V systems be avoided since there is a certain amount of danger involved. The extension and retraction of the sliding sections could possibly result in an exposed wire. With a 12-V system, an exposed wire could not be considered a serious problem.

Eductors

Eductors, which are also called suction boosters, syphons, and syphon ejectors, satisfy requirements that are not met by any other device. An example of an eductor is illustrated in Figure 2-24. They are light, portable, and easily connected up to allow a pumper to secure water from a source which is too far away or too far below the apparatus to be available for drafting with a hard suction hose. As eductors push, not lift, the water up, the normal limits of drafting heights do not apply. In addition, these devices may be used to dewater a basement, boat, ship, or other flooded area without the danger of the fire pump becoming clogged or damaged from drafting water that is full of debris or contamination. When a building is flooded, the premises may often be relieved of water without resorting to the use of eductors or pumps. The desired result may be obtained by opening a trap or soil line, removing a toilet, or clearing a clogged drain so the trapped water can escape down a sewer line.

Eductors are manufactured by several different companies; all operate similarly, but they do have construction differences.

Inlet

Water under pressure enters the eductor through an inlet nozzle; the diameter of the nozzle orifice depends on the eductor capacity. Some of the larger eductors have an adjustment that moves the position of the jet tube to its most efficient location; this depends on the incoming pressure. The restriction provided by the inlet nozzle converts the incoming water pressure into a high-velocity jet stream.

Suction

When water flow is started through the eductor, air in the suction chamber is entrained by the jet stream and is forced from the discharge, which lowers the air pressure in the suction chamber. As the pressure in the suction chamber is lower than the atmospheric pressure on the surface of the water source, water is forced through the strainer into the suction chamber.

Discharge

Water from the suction mixes with the high-velocity jet stream and acquires part of its energy. In the jet tube the velocity of the water is converted to a pressure greater than the suction pressure but lower than the inlet motivating pressure.

Foot Valve

Some eductors have a foot valve in the suction; others do not. Foot valves are not a necessity, but they can make some operations more simple. A foot valve prevents the loss of water, or backflow, through the suction and can avoid losing a prime if the operation is shut down temporarily. When a pumper is supplying the pressure for operating an eductor, water will not be syphoned from the booster tank if the pump is shut down and the tank dump valve is left open. Also, the pump discharge water will not flow out through the eductor suction if the discharge line is kinked or shut down.

Hose Layout

Inlet pressure, the length and size of the discharge hose, and the height between the eductor and the point of discharge are factors that affect the efficiency of this device. In consideration of these variables in all operations, it is not practicable to establish a

Figure 2-24. Eductor, or siphon ejector, used for dewatering or obtaining water from a static source from which water cannot be readily drafted (Courtesy of Akron Brass Company)

method of calculating the most effective inlet pressures for different conditions. Instead, it has been substantiated that as a general rule, a figure of 150 psi inlet pressure can be recommended to cover all situations.

Eductor sizes refer to the inlet hose diameter. As the length and size of the discharge lines have a major effect on eductor operation, they should be at least twice the internal area of the supply line; this can be accomplished by either using a discharge line a larger size than the supply hose or by siamesing two lines of the inlet size. Thus, a 2½-in. eductor would accommodate either one 3½-in. or two 2½-in. hose lines to handle the discharge most effectively.

The lengths of the supply and discharge lines should be limited as much as possible to reduce friction loss. Efficiency drops drastically when the length of the discharge line exceeds 100 ft; this may be remedied by increasing the diameter or the number of discharge hoses. Test data with one model of eductor showed that the quantity of waste water removed from a basement was reduced one-third when the length of the discharge hoses was doubled and the height and distance remained the same.

Every precaution should be taken to avoid kinks in the discharge lines. The suction should be well below the surface to prevent drawing air and floating debris into the educator. Avoid placing the strainger on the bottom to prevent clogging from sand and rocks. If there is a foaming white discharge, it is an indication of inefficiency or nonfunctioning of the device. One or more of the following conditions may be the cause: air entering the jet, static head too great, educator clogged, discharge line too long or insufficient diameter, engine pressure insufficient, educator intake not completely submerged, or the discharge line kinked or blocked.

Some educators are threaded to permit removing the strainer and substituting a section of hard suction hose where working space is severely limited to preclude submerging the educator itself below the water surface. This arranement will function without any appreciable loss of efficiency.

3 Pump Theory and Construction

Pumps are designed to fit many different purposes. In order to understand the proper application of a pump to any given workload or situation, engine company personnel must be familiar with the different pump types and the theory of pump capacity.

Positive Displacement Pumps

The theory of positive displacement is based on the principle that water is a noncompressible fluid and will, therefore, occupy a volume directly proportional to the weight.

Positive displacement pumps are mechanical devices capable of increasing the volume of a chamber. Water is drawn into the chamber as the volume is increased and in turn is forced out as the volume is decreased. Theoretically, the quantity discharged is equal to the change in the volume of the chamber into which the water is drawn; hence, the term positive displacement. Actually, discharge is somewhat less, due to slip. Slip is leakage between internal parts of the pump. The amount of slippage is dependent on the condition of the pump and the operating pressure.

Because of the close fit of its internal working parts, air as well as water can be drawn into and expelled by a positive displacement pump. These pumps are, therefore, capable of producing sufficient vacuum to prime themselves.

Since positive displacement pumps require considerable power, selecting a gear ratio that allows the engine to run at the same speed or faster than the pump is very important.

A positive displacement pump delivers more water during one part of its stroke revolution than during other parts. The slip of water back between rotors and past pistons is greater at some parts of the revolution than at others. Consequently, positive displacement pumps produce a pulsating flow. On pumps large enough to cause pulsation to become a problem, an *air dome* composed of trapped air in a container is used. The air expands and contracts, which reduces the pulsation of the streams of water leaving the pump.

Piston Pumps

The piston pump (Figures 3-1 and 3-2) creates a pressure differential by enlarging the cylinder capacity as the piston is moved upward by force. It is evident that as cylinder space is enlarged, pressure is decreased;

Figure 3-1. This 1917 Model "T" Ford pumper has a three-cylinder piston pump.

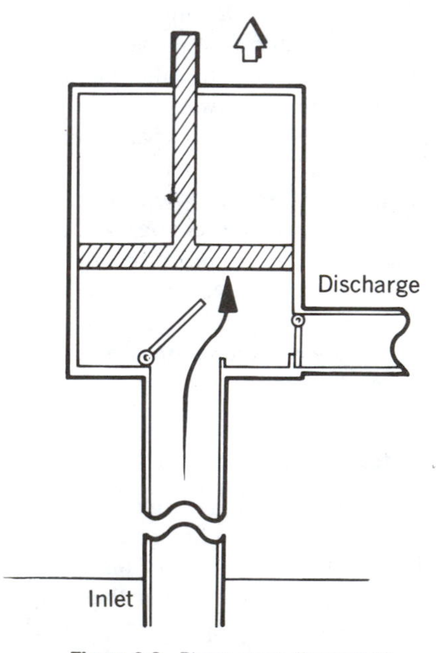

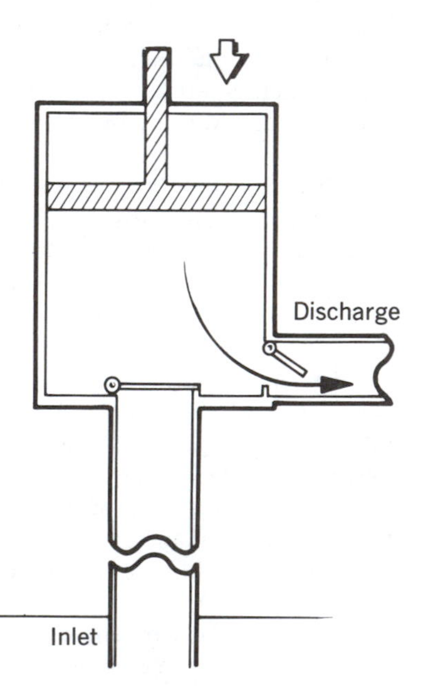

Figure 3-2. Piston pump (force type)

thus, air or water is forced into the cylinder. The check valve on the discharge gate prevents any backward flow.

When the pistons move downward, cylinder volume is decreased, pressure is applied to the contents, the check valve on the suction closes, and the flow is out through the discharge gate. This type of pump is mainly used for pumping small volumes of water at extremely high pressures.

Rotary-Type Pumps

Rotary-type positive displacement pumps consist of a casing divided into separate suction and discharge chambers by a rotor. The rotor is designed and located so that travel from the suction chamber to the discharge chamber displaces a definite quantity of water per revolution.

Operating Characteristics

In the operation of a rotary-type positive displacement pump, water forced into the discharge chamber is prevented from returning to the suction chamber by the barrier formed by the rotor, and fills the chamber. With the discharge chamber full, a quantity of water equal to the amount drawn into the chamber will be forced through the discharge outlet. In a positive displacement pump, as speed increases, the amount of water pumped increases proportionately. If a rotary pump is pumping 400 gpm when it is turning at 300 rpm, then at 600 rpm the pump will put out about 800 gpm.

Raising or lowering the pressure raises or lowers the load on the engine proportionately. A positive displacement pump delivering 500 gpm at 100 psi pressure takes one-half the engine power that it takes to pump 500 gpm at 200 psi pressure.

When available, hydrant pressure tries to rotate the pump, helping the engine to turn the pump. An engine driving a rotary pump at 500 gpm and 120 psi pressure from draft has to work twice as hard as it would if it were delivering 500 gpm at 120 psi pressure from a 60-psi hydrant, because the hydrant is doing half the work. In other words, rotary pumps take pressure advantage of hydrants. The rotary pump does not take volume advantage of a hydrant unless the pump and engine are run faster. The only way to get more water out of a pump is to increase its speed.

Displacement Factors

The principle measure of the condition of a positive displacement pump is the volumetric efficiency. Every time a pump turns over once, it moves a certain amount of water from the suction side to the discharge side of the pump. This space is called the theoretical displacement. The theoretical displacement is never reached during pumping because pressure on the discharge side pushes some water back into the suction. In other words, water slips back through pump clearances. The result is that the pump never delivers per revolution its theoretical displacement, but a smaller figure called the actual displacement. The difference between these two figures is slip. The ratio between actual and theoretical displacement is called volumetric efficiency:

$$\text{Volumetric efficiency} = \frac{\text{actual displacement} \times 100}{\text{theoretical displacement}}$$

To find the percent of slip, the formula is

$$100 - \frac{\text{actual displacement} \times 100}{\text{theoretical displacement}} = 100 - \text{volumetric efficiency}$$

Theoretical displacement is always marked on the pump in gallons per revolution (gpr). Actual displacement is found by test, and it is the actual gallons pumped per revolution. That is:

$$\text{Actual displacement} = \frac{\text{Gallons per minute}}{\text{Pump rpm}}$$

Example: A pump at 120 psi is putting out 625 gpm and is running at 652 rpm. The theoretical displacement from the pump name plate is 1.139 gpr. What is the volumetric efficiency and the percent slip? Using the above formulas,

Actual displacement

$$= \frac{\text{gpm}}{\text{rpm}} = \frac{625}{652}$$

$$= 0.959 \text{ gallons per revolution}$$

Volumetric efficiency

$$= \frac{\text{actual displacement} \times 100}{\text{theoretical displacement}}$$

$$= \frac{959 \times 100}{1.139} = 84.2\%$$

Percent slip

$$= 100 - \text{volumetric efficiency}$$

$$= 100 - 84.2 = 15.8\%$$

A new pump will run about 5 to 10% slip at 120 psi, 10 to 20% slip at 200 psi, and 15 to 25% at 250 psi. Rotary pumps are not good for high pressures; most manufacturers do not recommend pressures higher than 300 psi.

In positive displacement pumps, the quantity of discharge is directly proportional to the speed of the pump. Pump speed has no direct bearing on the effective pressure developed. Effective pressure is developed when the flow is restricted by the fittings and hoses through which it passes. Pressure gages of rotary gear booster pumps will normally fail to indicate any effective pressure when charging dry lines until the hose is filled and the restriction of flow established.

Rotary Gear Pumps

One type of rotary pump consists of two intermeshing gears or cams revolving in opposite directions within a close-fitting casing (Figure 3-3). Design variations occur mainly in the shape of rotors and in transmission of power to gears or cams. The rotors move away from each other on the suction side and toward each other on the discharge side, each moving from suction to discharge along walls of the casing. Air or water is trapped on the suction side of the pump in pockets formed by two adjacent teeth and the casing wall; as the teeth pass by the suction inlet, the water is carried along the wall of the casing toward the discharge outlet. Air or water trapped in the pocket is released, and pressure, built up by the meshing of the rotor teeth as they return toward the suction side, forces air or water out through the discharge manifold. Meshing of the rotor teeth also prevents the air or water under pressure in the discharge side of the casing from returning to the suction side. Although discharge from each rotor is intermittent, rotors are so arranged that as the discharge is being completed from the space between two teeth on one rotor, a corresponding space on the other rotor is just beginning to discharge. This has a tendency to produce a more nearly continuous discharge from the pump and reduces pulsation.

In another type of rotary pump (Figure 3-4), the rotor lobes mesh in the same manner as a pair of gears, except that they normally do not touch each other, and are driven by steel gears. During operation, the rotor lobes disengage at the suction side of pump, trapping liquid between two lobes of each rotor, the pump body, and heads. As the rotors turn, trapped water is moved from the suction side of the pump to the discharge side. The rotor lobes reengage at the discharge side of the pump, forcing the liquid between them to the pump discharge opening.

Because a rotary pump must positively displace or discharge a nearly fixed amount of liquid for every revolution of the rotors, extremely high pressures will develop if a discharge valve is closed while the pump is in operation. For this reason, a relief valve is necessary to prevent excessive pressure from stalling engines or damaging pumps. When the discharge pressure rises to a point higher

Figure 3-3. Rotary gear pump. The eccentric bearing wear in the end plate indicates that higher pressure on the discharge side of the pump was forcing the gears and shafts towards the suction side.

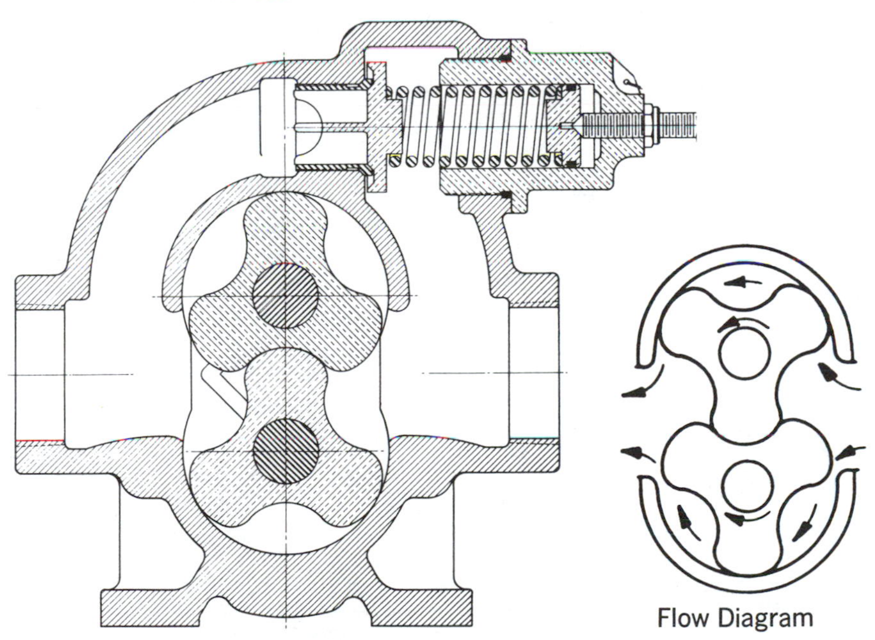

Figure 3-4. Waterous rotary fire pump (Courtesy of Waterous Company)

than that for which the relief valve is set, the valve opens and bypasses water from the discharge side of the pump back to the suction side. As soon as pressure recedes below the point at which the relief valve is set, the valve will close. The pump should never be operated at pressures higher than the rated pressure given on the serial plate. Prolonged bypassing of water by the relief valve will result in an excessively high pump temperature.

Rotary pumps vary in the number of lobes or teeth on the rotors, but the principle of operation is exactly the same.

Rotary Vane Pumps

A rotary vane pump (Figure 3-5) consists of a single rotor within, and eccentric to, a casing. A series of slots in the rotor permits sliding vanes to move in and out, maintaining contact with the surface of the casing by means of centrifugal action in rotation of the rotor. As each vane passes the suction inlet, the space between the vane, rotor, wall of the casing, and the next vane is constantly increasing, drawing air or water into the space. As the motion of the rotor brings this vane to a point opposite the discharge opening, the space is constantly decreasing, and air or water is forced out through the discharge opening. This process is repeated four times during each revolution of the rotor.

Centrifugal Pumps

The word centrifugal means to proceed away from the center, to develop outward; centrifugal force is that energy which tends to impel an object outward from a center of rotation. This is the principle of centrifugal pump design and construction.

Operating and Construction Characteristics

In theory, the operation of centrifugal pumps is based on the principle that a rapidly revolving disk will tend to throw

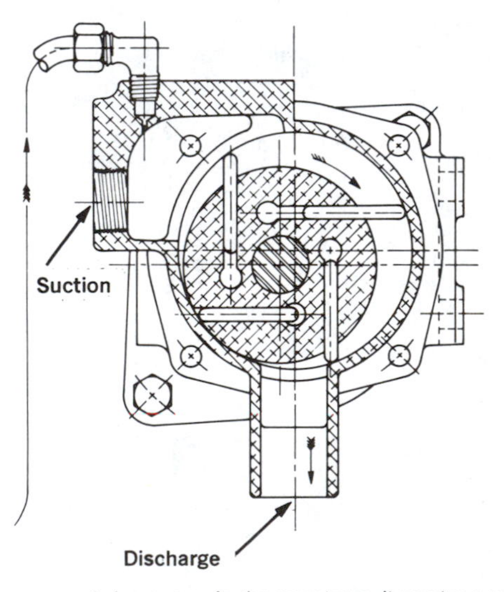

Figure 3-5. Rotary vane priming pump. As the rotor turns, it creates a vacuum on the side of the pump, pulling the air from the main pump at the same time oil is drawn in through the top tubing fitting to lubricate the pump and bearings. (Courtesy of Hale Fire Pump Company)

water introduced at the center toward the outer edge of the disk. A similar action may be demonstrated by swinging a pail of water in a circular motion over the head. It will be seen that the so-called centrifugal action creates a tendency to hold the water to the enclosed bottom of the pail, which describes the outer arc of the circle, and that therefore no water is spilled from the open end of the pail, which describes the inner arc of circle. If a small hole is cut in the bottom of the pail, however, it will be found that the stream of water that emerges through this hole gains in intensity and reach as the pail is swung faster or, in other words, as the velocity of the pail increases.

In a centrifugal pump, a revolving disk known as an impeller (Figure 3-6), rotates very rapidly within a casing. As the impeller revolves in the direction indicated by the arrow, water introduced through a suction tube enters the impeller at the inlet eye, is thrown to the outer edge of the impeller by centrifugal action, and is hurled through openings into the open space in the casing. Since the circumference of the impeller disk is greater at the outer edge of the blades than it is at the inner edge, the outer rim of the impeller travels at a greater linear speed than the hub. The velocity of water passing through the inlet eye and hurled into the open casing is increased as it passes through the impeller. Likewise, as the rate of impeller rotation increases, the velocity with which water is thrown from the impeller increases.

The internal shape of casings may be any one of three different types (Figure 3-7):

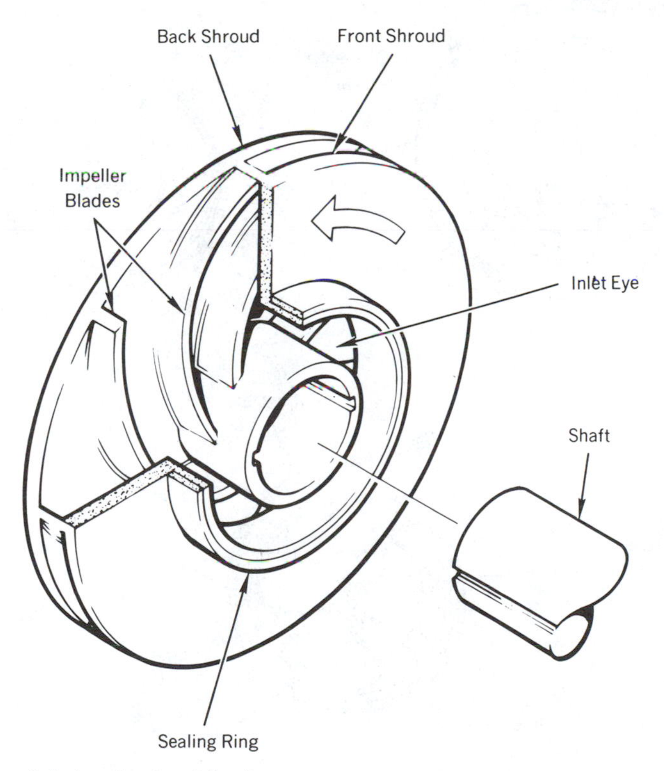

Figure 3-6. Impeller of centrifugal pump

1. Diffusion type in which the flow of water is directed toward the discharge outlet by means of a series of stationary diffusion vanes fastened to, or a part of, the inner wall of pump casing.

2. Volute type which has a spiral shaped casing.

3. Combination diffusion and volute type which has a spiral shaped casing with stationary diffusion vanes.

The Volute Principle

The majority of centrifugal pumps are constructed on the volute principle. It will be noted that the cross-sectional area between the outer edge of the impeller and the wall of the casing is constantly increasing as it approaches the discharge outlet, producing a volute, or spiral. This spiral is needed because water thrown from the impeller around the entire circumference (and the total quantity of water passing through the casing) is increasingly greater toward the discharge outlet. The action of the volute is to to enable the pump to handle this increasing quantity of water, at the same time permitting the velocity of the water to remain constant or to decrease gradually and maintain the required continuity of flow. Actually, movement of the impeller simply creates a velocity in the water which is

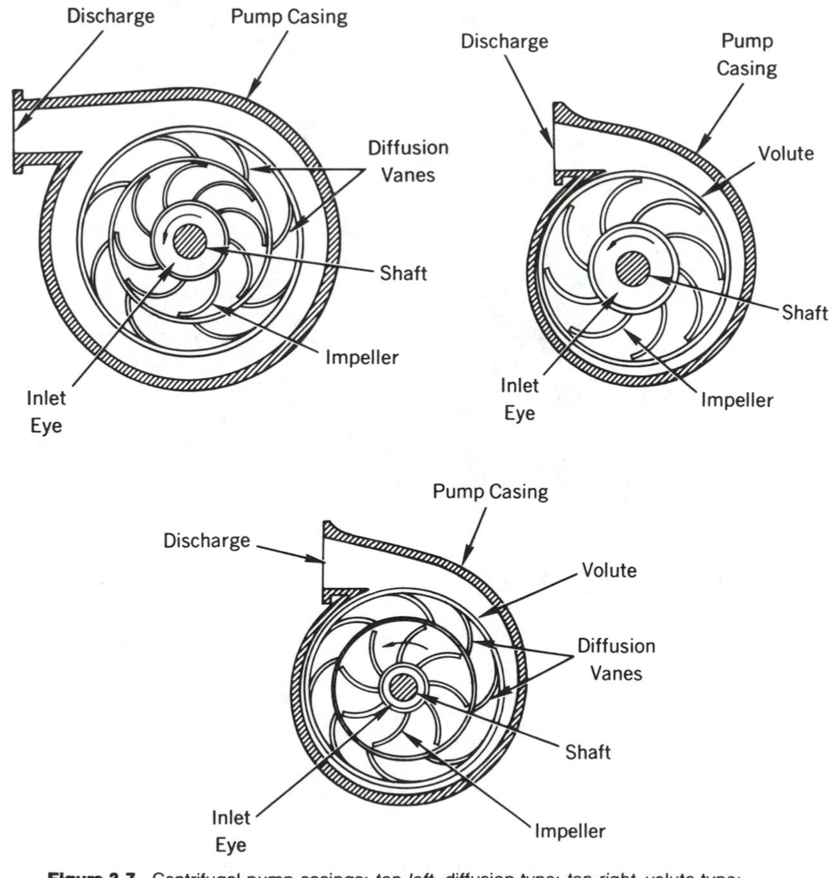

Figure 3-7. Centrifugal pump casings: *top left,* diffusion type; *top right,* volute type; *bottom,* combination diffusion and volute type

converted into pressure as it approaches the confining space of the discharge pipe. Water under pressure on the discharge side of the casing is prevented from flowing back into the pump by the close fit between the casing and the impeller at the entrance to the suction inlet, by the rapid movement of the impeller, and by the unequal pressure in the pump casing.

In operation, atmospheric or other pressure continuously forces water into the suction eye, the impeller forces water outward, and the volute conducts the water out. Figure 3-8 shows the path a single drop of water takes through a centrifugal pump. The pump discharge pressure and capacity can be regulated by adjusting the pump speed, the transfer valve setting, the discharge valve opening, and the nozzle size.

Displacement Factors

Two theories involving centrifugal pumps, which may be difficult to understand, are:

1. *When pumping at a constant pressure, the quantity of water is directly proportional to speed of the pump.* This principle can be shown by formula. As pressure is constant, it does not enter the problem. The formula is:

$$\text{New quantity} = \frac{\text{new pump speed}}{\text{old pump speed}} \times \text{old quantity}$$

Example: If a pump is discharging 500 gpm at a pressure of 100 psi while turning at a speed of 1000 rpm, what amount of water will it discharge if pressure remains the same and speed is increased to 1500 rpm? The formula to find this is:

$$\begin{aligned}\text{New quantity} &= \frac{1500}{1000} \times 500\\ &= 1.5 \times 500\\ &= 750 \text{ gpm}\end{aligned}$$

Therefore, if a pump is discharging 500 gpm at a pressure of 100 psi while turning at a speed of 1000 rpm, and speed is increased to 1500 rpm with pressure remaining at 100 psi, discharge will increase to 750 gpm.

2. *When pumping at a constant quantity, pressure is directly proportional to the square of the speed.* This principle can also be shown by formula. Since quantity is constant, it does not enter the problem. The formula is:

$$\text{New pressure} = \left(\frac{\text{new pump speed}}{\text{old pump speed}}\right)^2 \times \text{old pressure}$$

Example: If a pump is discharging 500 gpm at a pressure of 100 psi while turning at a speed of 1000 rpm, at what pressure will it discharge water if quantity remains at 500 gpm and speed is increased to 1500 rpm? To find this, the formula is:

$$\begin{aligned}\text{New pressure} &= \left(\frac{1500}{1000}\right)^2 \times 100\\ &= (1.5)^2 \times 100\\ &= 2.25 \times 100\\ &= 225 \text{ psi}\end{aligned}$$

Therefore, if a pump is discharging 500 gpm at a pressure of 100 psi while turning at a speed of 1000 rpm, and the speed is increased to 1500 rpm with discharge remaining at 500 gpm, pressure will be increased to 225 psi.

The quantity of discharge from a centrifugal pump is related to the speed of the pump and develops pressure in such a way that if any one of three factors–quantity, speed, or pressure–remains constant, an increase or a decrease in one or the other of the remaining two factors will produce a change in the third factor. Within certain limits governed by the design of the pump, with pressure constant, an increase in speed will produce an increase in quantity, or vice versa; with quantity constant, an increase in speed will produce an increase in pressure, or vice versa; or with speed constant, an

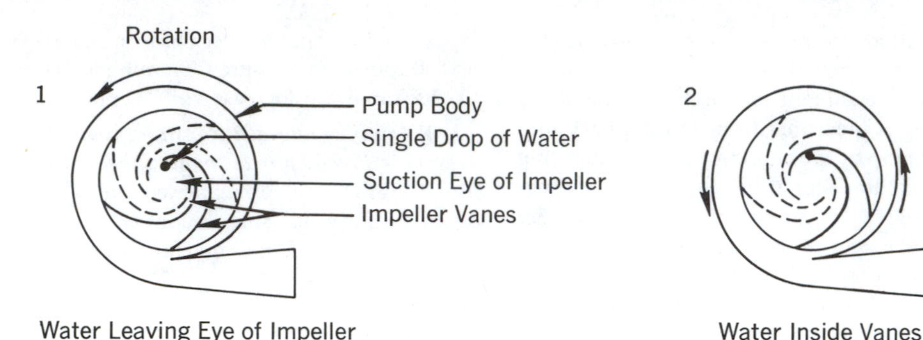

Water Leaving Eye of Impeller

Water Inside Vanes

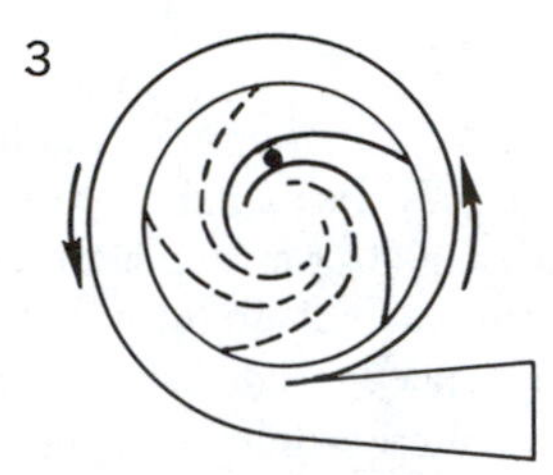

Beginning Outward Travel

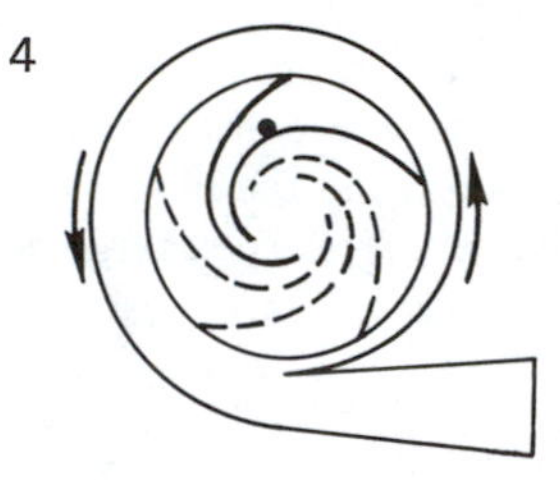

Picking Up Outward Speed

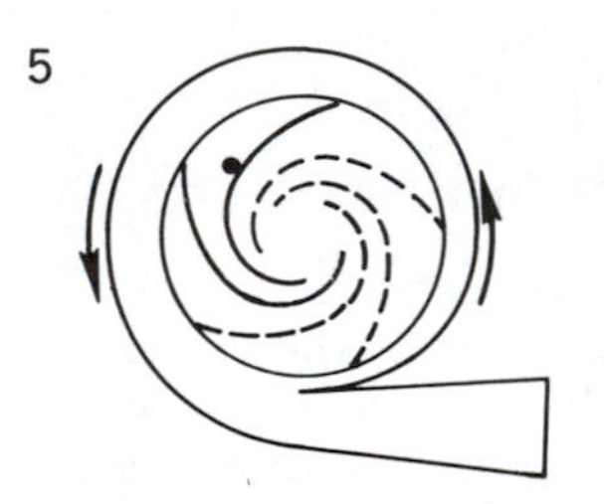

Beginning to Rotate with Impeller

6

Reaching Outside Edge of Impeller

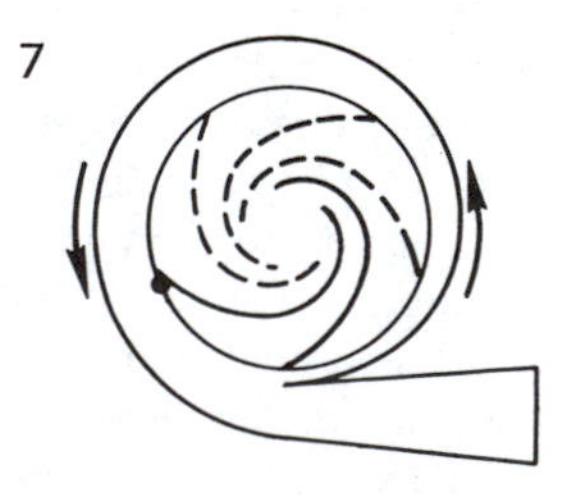

Water Being Hurled Out of Impeller

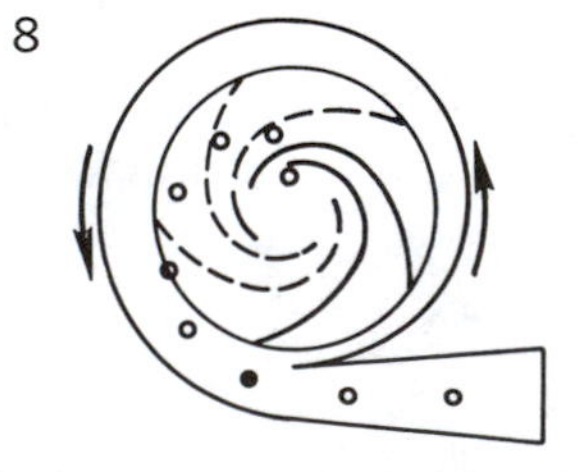

Path Taken Through Pump Body and Out Discharge Passageway

Figure 3-8. Operating sequence of centrifugal pump (Courtesy of Waterous Company)

increase in pressure will produce a decrease in quantity, or vice versa.

Slippage

Unlike positive displacement pumps, slippage is not a factor with centrifugal pumps. A shutdown on operating lines will cause a momentary increase in pressure and an automatic slippage of 100% within the pump, which prevents pump or engine damage. Friction from this slippage will heat water trapped in the pump casing. A centrifugal pump has a smooth flow, which is much easier on hoses than the action of a rotary pump, and therefore needs no air dome.

Design and Construction Features

The design features of pumps by different manufacturers are illustrated in Figures 3-9 through 3-12.

Figure 3-9. Single-stage fire pump (Courtesy of Hale Fire Pump Company)

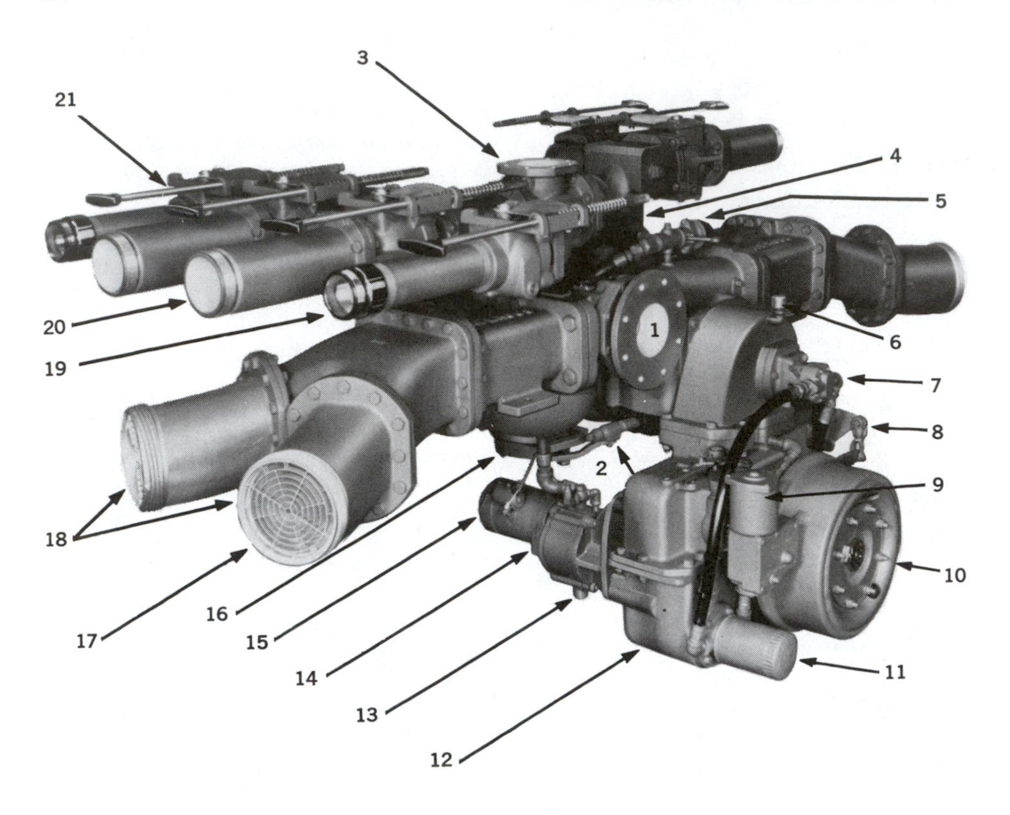

1 Booster tank to pump suction connection. A check valve is provided to prevent damage to the tank if the tank-to-pump (dump) valve is left open when connected to a hydrant.
2 Electric transfer valve actuator assembly.
3 Discharge outlet to wagon battery, water tower, rear discharge outlets, etc.
4 Fitting to direct the flow from the pump discharge to the discharge valves.
5 Priming valve.
6 Pump transmission breather.
7 Pump transmission oil pump.
8 Parking brake linkage.
9 Electric pump transmission shift assembly.
10 Parking brake drum.
11 Pump transmission oil filter.
12 Chain drive pump transmission.
13 Priming pump discharge outlet.
14 Rotary gear priming pump.
15 Priming pump electric motor.
16 Main pump body.
17 Suction inlet strainer.
18 Siamese pump suction inlets.
19 2½-in. discharge outlet.
20 3½-in. discharge outlet.
21 Discharge valve linkage.

Figure 3-10. Nomenclature of a Model CMUYB 2000-gpm two-stage series-parallel Waterous pump. The siamese-type inlet fittings allow two suction hoses to be attached so that the full capacity of the pump can be attained while drafting. The 2½- and 3½-in. discharge valves allow the larger diameter hose to be conveniently attached. (Courtesy of Waterous Company)

Impeller shaft assembly. Centrifugal pumps may have from one to four impellers, each within its own casing. Multiple impellers are keyed to the same shaft and rotate at identical speeds. Generally, impellers are bronze, the shaft stainless steel.

Pump drive. Power is transmitted from the drive shaft, through a pump transmission and intermediate gear, to the impeller shaft. The transmission allows engaging and disengaging of the pump drive and establishes a proper engine-to-pump gear ratio.

Pump casing or body. The pump contains water passageways and the volutes, separated by an interstage seal. Water enters the eye of the impeller, passes through the impeller into the volute, and is conducted out to the discharge. Water under pressure in the volute is prevented from returning to the suction side of the pump by the close fit of the impeller hub to a stationary clearance (wear) ring at the eye of the impeller. Failure of the pump to deliver rated capacity and pressure may be an indication that the pump wear rings are badly worn. Excessive wear will allow severe internal leakage from the discharge side of the pump back to suction, which will reduce net pump capacity.

Multistage Centrifugal Pumps

Connecting of two or more single-stage centrifugal pumps in a series, with the discharge

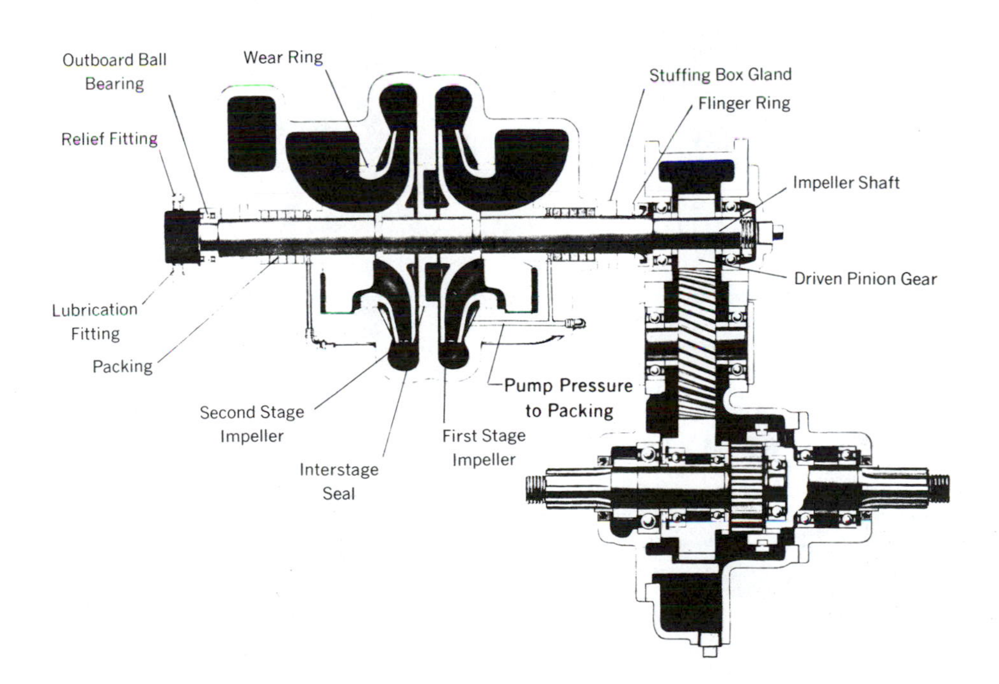

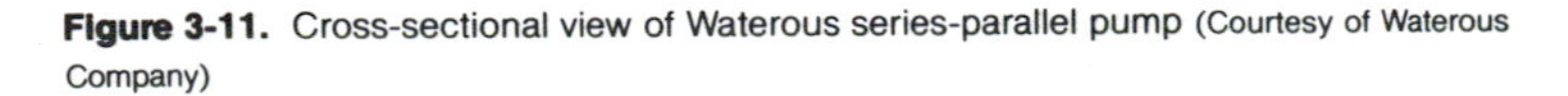
Figure 3-11. Cross-sectional view of Waterous series-parallel pump (Courtesy of Waterous Company)

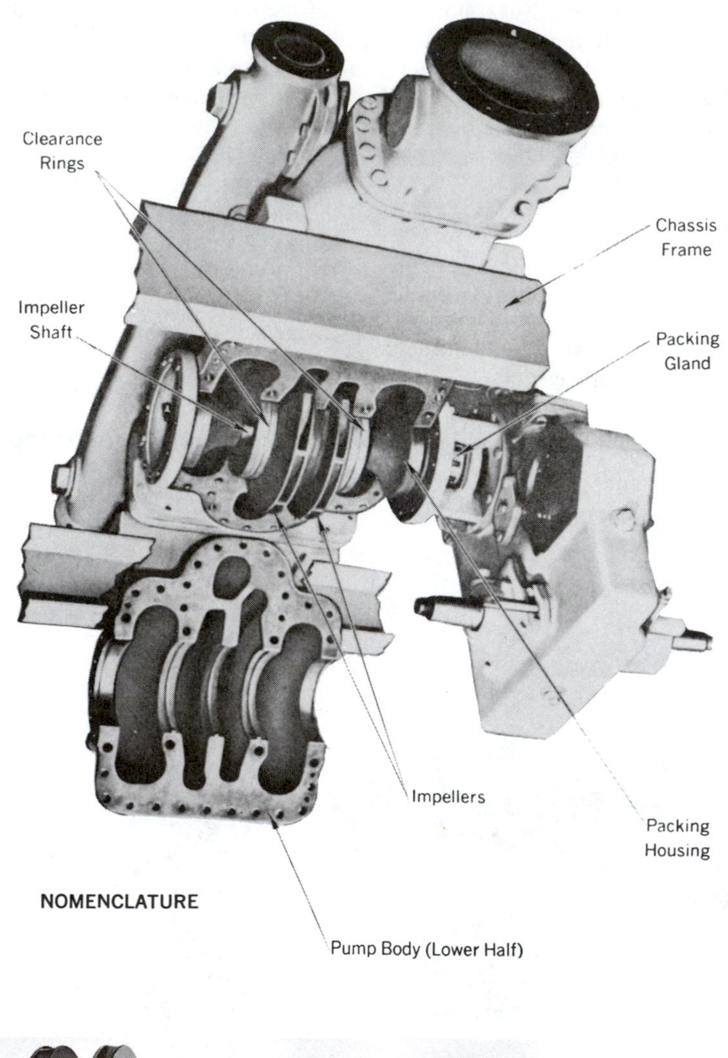

Impeller shaft assembly from Hale series-parallel pump

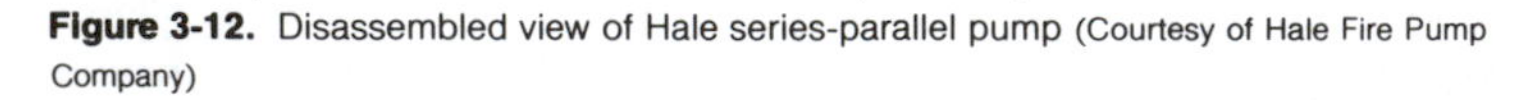

Figure 3-12. Disassembled view of Hale series-parallel pump (Courtesy of Hale Fire Pump Company)

of the first impeller, or stage, connected directly to the inlet of the second pump, creates what is called a multistage centrifugal pump. (Review Figures 3-10–3-12.)

Since the quantity of water passing through the second stage is limited by the amount delivered to it by the first pump, the quantity passing through both stages will be the same. In other words, the quantity taken in through the inlet in the first stage will be the same as the quantity discharged from the second pump. However, each stage will add an equal amount to the pressure created if they are of identical design and both are operating at the same speed.

The inherent advantage of this type of pump over the single-stage pump lies in its ability to deliver greater quantities of water at higher pressures without requiring excessive speed, particularly in pumps of large capacity. In practice, multistage pumps are generally built as single units with the impellers all mounted on a single shaft and within a single divided casing, each impeller unit division representing an additional stage of pumping. A stage refers to the number of impellers that are operating in series. Some pumps utilize four impellers, two connected in parallel and two in series, but they are not true four-stage pumps.

Series-Parallel Operation

When greater variations in capacity and pressure are required than can be conveniently obtained from a single impeller, series-parallel pumps are used. These pumps have two or more impellers, each enclosed in its own volute, which is usually part of a common body. The combination of each impeller and volute is called a stage.

The operation of a series-parallel pump is demonstrated in Figure 3-13. A transfer (changeover) valve at the outlet of the first stage directs the water to discharge or to the second stage, depending upon how the pump is being used. When the transfer valve is in *volume* (parallel) position, each impeller pumps half the total volume at total discharge pressure. The water enters both volutes at the same time from a common suction and leaves through a common discharge.

If the transfer valve is turned to *pressure* (series) position, the first stage pumps its full volume and pressure directly to the second stage, instead of to discharge. The second stage then pumps this same volume of water to discharge, but at twice the first-stage pressure. With the transfer valve in this position, the first-stage pressure also closes its clapper valves, which prevents water from entering the second stage directly from suction. At a given speed, a series-parallel pump in series position will develop twice the pressure with half the volume delivered by parallel operation.

Consider a 500-gpm series-parallel centrifugal pump. Each stage is designed to pump one-half the total capacity at 150 psi pressure; in this case it amounts to 250 gpm at 150 psi. When both stages are pumping in parallel, the total discharge will be 250 plus 250, or 500 gpm.

When the transfer valve is switched to series operation, each stage still pumps 250 gpm at 150 psi, but one stage pumps into the other; total discharge is 250 gpm. Each stage is still exerting 150 psi pressure; the second stage adds 150 psi to the first stage, so the total pressure is 300 psi. (This theoretical pressure is not actually obtained because of pump inefficiency at higher pressures.) This changeover from parallel to series can be done without any net change in the engine speed or the throttle opening.

The pressure and volume of each stage are the same. The engine speed and throttle opening are the same. The net horsepower output and input are the same. The pump, however, changes from 500 gpm at 150 psi to 250 gpm at 300 psi; that is, from capacity to pressure with no change in engine speed or pump efficiency.

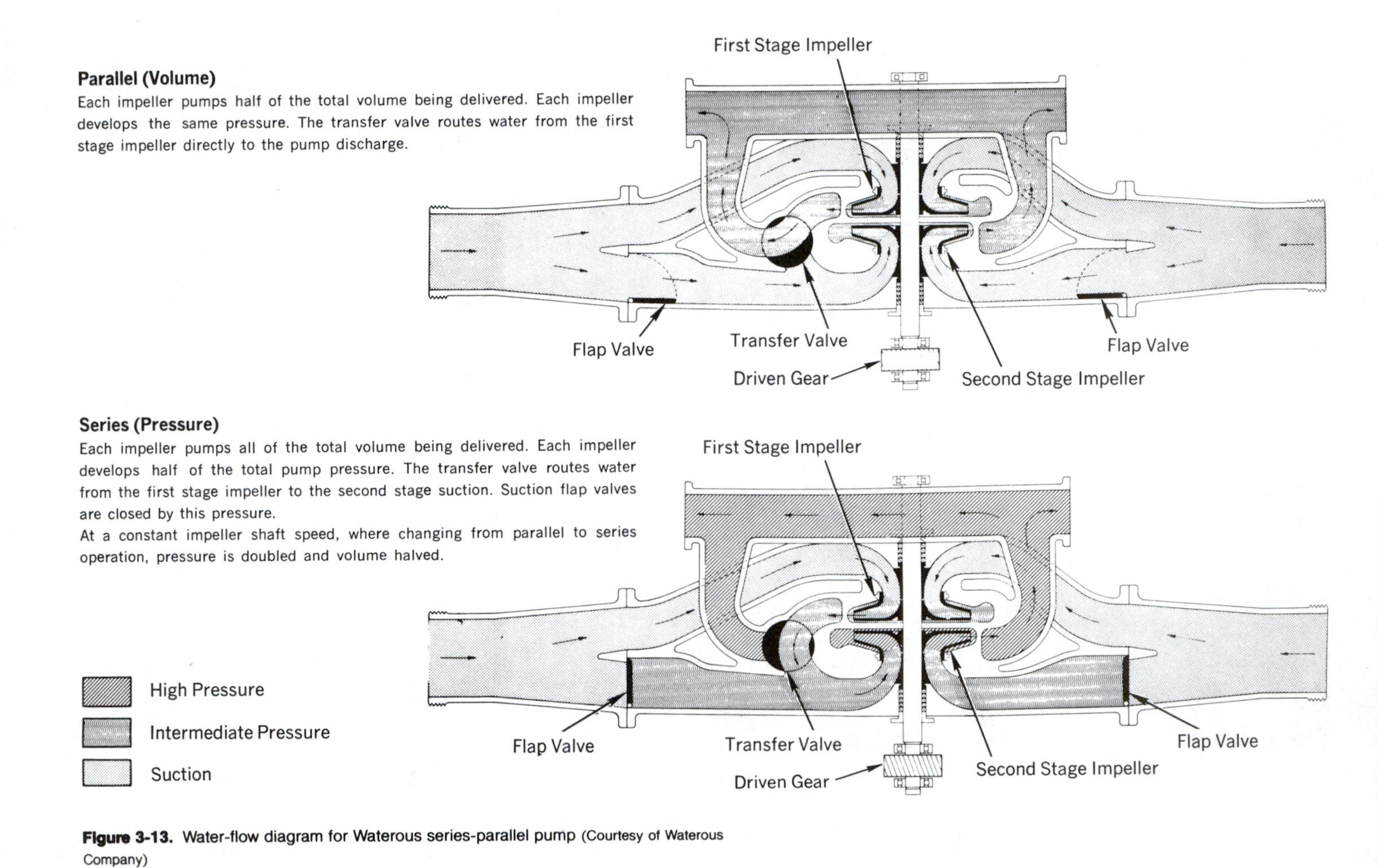

Figure 3-13. Water-flow diagram for Waterous series-parallel pump (Courtesy of Waterous Company)

Transfer Valve

No hard-and-fast rule can be used to decide when to pump in series and when to pump in parallel. In general, for long, single lines, use pressure, and for short lines with big tips, use volume. Another guide is to use the arrangement, parallel or series, which gives the desired result at the *lowest* engine speed. This will vary from apparatus to apparatus and also will vary from condition to condition (i.e., pumping from a hydrant or pumping at draft.)

When testing pumpers at pressures of 150 psi, 200 psi, and 250 psi, it has been found that the volume position is most efficient for 150 psi and that the pressure position is best for 250 psi. When the pump must discharge 70% of capacity at 200 psi net pump pressure, some pumps require pressure position and others do better in volume. Therefore, a good general rule is to place the transfer valve in the pressure position for quantities up to 70% of the pump's capacity; for a 1000 gpm pump, this would be 700 gpm.

The transfer valve should be carried in the position most likely to be used at the next fire; usually this will be in pressure.

Care must be taken to switch at a low pressure. This is particularly critical when changing from parallel to series. In theory, switching from parallel to series with a two-stage pump would cause the pressure to double. This could be dangerous if pumping at high pressure. Some changeover valves require that pressure be reduced to switchover. Others do not. Safety requires good judgment.

Variations of Series-Parallel Pumps

The majority of series-parallel pumps use two impellers attached to one shaft, have two volutes within a single pump body, and route water between stages through a transfer valve.

A variation in cities which require high pressure is a four-stage series-parallel pump that has three transfer valve positions: series, parallel and series-parallel. Normally, this type is operated in the series-parallel position. Series position is only used when pumping water to about the 50th floor of a building. The parallel position is reserved until the pump is delivering more than one-half of its rated capacity.

Another type is the duplex-multistage pump (Figure 3-14). The two impellers are mounted on separate shafts, allowing individual operation. Because of independent pump and gear-train design, three distinct operating characteristics or service conditions are obtainable. First, large capacities at normal pressures are delivered by a capacity pump operating with its own individual set of gears. Second, normal high pressures are delivered by a high-pressure pump operating with its own individual set of gears. Third, extremely high pressure is delivered with both pumps in series. There is no transfer valve on this type of pump; two self-closing poppet type valves fulfill the purpose.

Another variation is to attach a single-stage pump to a conventional series-parallel pump. Usually, this third stage rotates whenever the main pump is in gear. Extremely high pressures are available by routing water from the discharge of the main pump through the third stage.

Theoretical Pump Capacity

A pump, like any engine, is limited in the amount of work it can do. In pumps, this workload is based on the pump's capacity in pound-gallons. Theoretical pound-gallons of a pump can be determined by multiplying its rated capacity in gallons by its rated pressure. As an example, a pump rated at 1000 gallons per minute at 100 pounds pressure would have a theoretical pound-gallons capacity of 1000 × 100, or 100,000 pound-gallons. This is the theoretical maximum amount the pump can produce. The actual amount that may be produced can exceed this when the hydrant is helping the pump.

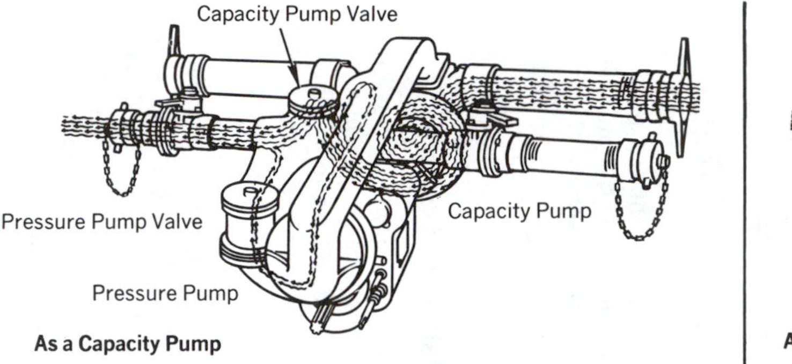

As a Capacity Pump

In the diagrams the main water stream is shown by the multiple lines; the bypass flow, which ensures priming and prevents freezing, is shown by two lines.

With only the capacity pump operating, the main water stream goes directly into the pump and out through its automatic valve into the discharge manifold. A small portion of the water is forced through the pressure pump, which is standing idle, and into the discharge manifold — keeping the pressure pump primed for immediate service, and preventing danger of freezing.

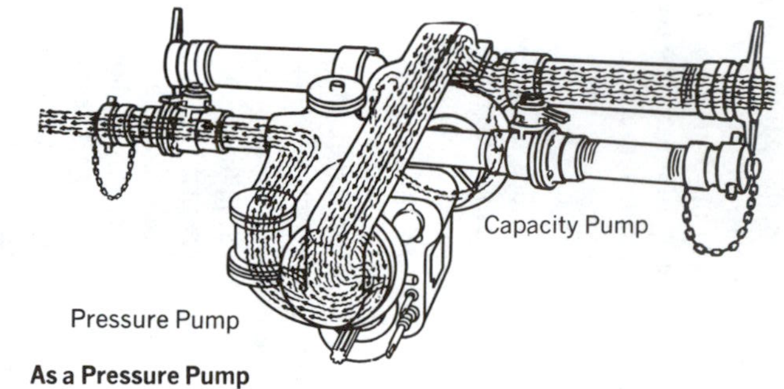

As a Pressure Pump

A feature of this pump is the suction cup which can be removed so that the suction hose may be connected while pumping from the booster pump.

With only the pressure pump operating, the main water stream goes across the top manifold into the pump and out through its automatic valve into the discharge manifold. A small portion of the water is drawn through the capacity pump and into the suction of the pressure pump to keep the capacity pump primed and to prevent freezing — the capacity pump automatic valve is closed due to pressure produced by the pressure pump.

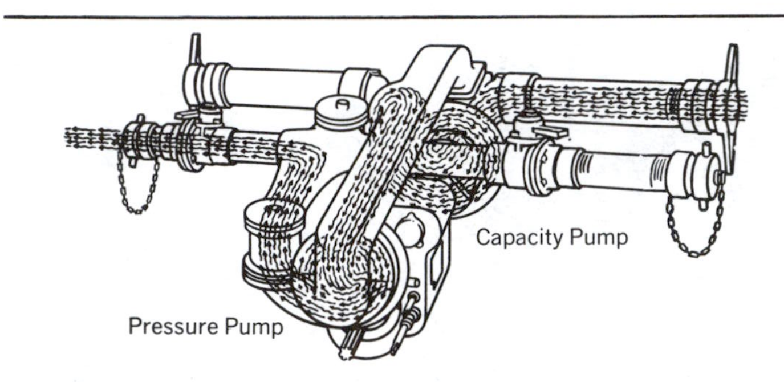

In Series Operation

With both pumps running, the capacity pump discharges directly into the top manifold leading to the pressure pump suction — the capacity pump automatic valve being closed due to the higher pressure created by the pressure pump. The pressure pump then more than doubles the incoming pressure produced by the capacity pump and discharges into the discharge manifold.

Figure 3-14. Water diagram for American duplex-multistage pump (Courtesy of American Fire Apparatus Company)

This theory is based on pump performance at net pump pressure.

Any time that a pump exceeds its rated net pump pressure, a reduction in quantity of available water will occur. This might be illustrated by the following example:

Pump rated at 1000 gpm at 100 psi.

Pump is then rated at 100,000 pound-gallons.

It is desired to pump at 200 psi.

How much water can be obtained from the pump? To find this, the formula is:

$$\text{Amount of water} = \frac{\text{total pound-gallons available}}{\text{desired pumping pressure}} = \frac{100{,}000}{200} = 500 \text{ gpm}$$

Pump Ratings

Pumpers are manufactured to perform as follows:

100% rated capacity at 150 pounds net pump pressure;

70% rated capacity at 200 pounds net pump pressure;

50% rated capacity at 250 pounds net pump pressure.

They are manufactured in various capacities: 500 gpm, 750 gpm, 1000 gpm, 1250 gpm, 1500 gpm, 1750 gpm, 2000 gpm, and occasionally larger. Discharges for the most popular sizes vary as the net pump pressures change—as shown in Table 3-1.

The larger pumpers constitute a small minority found chiefly in larger cities having special problems, such as extensive industrial areas where heavy streams or large volumes must be supplied through long lines of hose. In these cities, large-capacity pumps may be purchased, not so much for their rated output as for their ability to provide a desired volume at 200 psi.

Prior to 1956, many lower capacity (Class "B") pumpers were built. These provided 100% of their rated capacities at 120 psi net pump pressure, 50% at 200 psi, and 33⅓% at 250 psi. This class of pumper was rendered obsolete because taller buildings made necessary large quantities of water at higher pressures.

Because pumper ratings are based on 250-gpm increments, this shows the number of 250-gpm hose streams that various sizes of pumpers will deliver at various pressures when operating from draft.

It is generally true that modern pumpers have less difficulty meeting specified pumping requirements than they do in meeting road tests.

Cavitation in Fire Pumps

In any fire pump, pronounced cavitation (Figure 3-15) can be destructive. Pump operators should, therefore, study the principles of cavitation and learn to recognize when a pump is cavitating. Cavitation can

TABLE 3-1

RATED OUTPUT AT 150 psi	RATED OUTPUT AT 200 psi	RATED OUTPUT AT 250 psi
500	350	250
750	525	375
1000	700	500
1250	875	625
1500	1050	750
1750	1225	875
2000	1400	1000

occur in pumps of all types, sizes, and designs.

If the pressure at any point inside a pump drops below the vapor pressure corresponding to the temperature of the liquid, the liquid will vaporize and form cavities of vapor. Vapor bubbles are carried along with the stream until a region of higher pressure is reached, at which point they collapse or implode with a tremendous shock on adjacent walls. This is called cavitation.

Causes of Cavitation

Cavitation in fire pumps is caused by reducing the absolute pressure on water below the boiling point. It occurs at the point of lowest pressure—that is, highest vacuum—which is usually at the impeller eye in back of the leading edge of the impeller blade. The temperature at which water boils varies with the absolute pressure, as shown by Table 3-2.

There is always atmospheric pressure on the surface of the earth, land and water, caused by the weight of the atmosphere above us. Atmospheric pressure is higher in areas below sea level, and lower at high altitudes; it also changes with the weather. Normal atmospheric pressure at sea level is 14.7 psi, or approximately the same as 29.92 in. of mercury or 34 ft of water.

When a pump is cavitating, water in the fire pump suction flashes into steam, or

Figure 3-15. When a centrifugal pump is allowed to cavitate for an extended period of time, the impeller or impellers may become damaged. The front disk and vanes of this impeller have completely separated from the hub and rear disk due to billions of implosions. (Courtesy of Hale Fire Pump Company)

TABLE 3-2

TEMPERATURE AT WHICH WATER BOILS, °F	when under	ABSOLUTE PRESSURE, lb PER sq. in.	CORRESPONDING VACUUM, in. MERCURY
101.83		1	27.88
126.15		2	25.85
153.01		4	21.78
170.06		6	17.73
182.86		8	13.64
193.22		10	9.57
201.96		12	5.50
209.55		14	1.87
212.00	Atmospheric	14.7	0.00

water vapor which is the same thing. Bubbles of water vapor may be as large or larger than a cubic inch, but normally there are thousands of small bubbles, resulting in a mixture of water and water vapor. As the mixture flows through the impeller there is a rapid, almost instantaneous change from a vacuum to substantial pressure. Vapor instantly condenses under the high pressure, and water rushes in at high velocity to fill the void. Suddenly the bubble is full, and the inertia of the water produces an intense shock. This shock strikes walls surrounding the water with a forceful impact and can cause severe damage to the metal, especially if continued for too long a time.

Cavitation will damage piston and rotary pumps more rapidly than centrifugals because the increase in absolute pressure on water passing from a vacuum to 120 or more psi is almost instantaneous. In a centrifugal pump, pressures rise more gradually.

Conditions Creating Cavitation

Cavitation can occur whenever a pump is used improperly for the existing conditions. This situation will happen when an attempt is made to deliver more water out of a pump than can readily enter. Problems may develop at any tank, hydrant, relay, or drafting operation.

Pressure is reduced (and conversely, vacuum is increased) on water within the suction of the pump by excessive lift, too small or too long a suction hose, a blocked or inadequate suction strainer, pumping abnormally warm water, pumping at low atmospheric pressure (high altitude) and frequently by a combination of several of these conditions, all of which tend to produce cavitation.

Operating to Avoid Cavitation

A reliable way to determine the cavitation point of a fire apparatus pump under any suction condition is to open the throttle gradually (always good practice), while watching the engine tachometer and pressure gage. *An increase in engine rpm without a corresponding increase in pressure indicates cavitation* (assuming the clutch is not slipping). When this condition occurs, nothing is to be gained by running at a higher rpm. Cavitation will become more severe and no more water will be put on the fire. The pump is said to be running away from the water.

Most compound gages are difficult to read and are often inaccurate on the vacuum scale. Experienced operators can judge when the pump is approaching cavitation by watching the compound gage. A better way is to note the point at which speed increases without an increase in pressure.

Hard suction hose (noncollapsible under vacuum) must be used when pumping from draft. Soft suction has an advantage when pumping from hydrants. Fire pumpers occasionally have the capacity to pump more water than a hydrant outlet will supply. Under this condition, the pump can cavitate. Soft suction will collapse under vacuum and will thus indicate that the hydrant can supply no more water, and that the pump is cavitating or is approaching cavitation. *Cavitation when pumping from a hydrant is not uncommon.*

Cavitation is usually more severe at high pump pressures than at low pressures. It is possible to damage a fire pump seriously by trying to pump more water from the booster tank than will flow into the pump through booster piping. This would be especially severe if the pump were operating at 250 psi or higher pressure.

Remember that cavitation will gradually damage a fire pump. *To avoid cavitation, engine and pump speed should not be increased beyond that speed at which pump pressure ceases to rise.*

4 Pump Accessories

This chapter covers the various accessories used in conjunction with the pumping equipment discussed in the preceding chapter. The design and operating principles of accessory equipment from different manufacturers are presented, and detailed operating and maintenance instructions are given.

Priming Devices

Despite common fire service terminology concerning pumping at draft, it is not quite true to say that a pump is lifting water from the source. Actually, the pump is creating a partial vacuum within its suction chambers–that is, a pressure lower than atmospheric. Atmospheric pressure on the surface of the water being drafted then forces water up through suction hose and into the pump.

Positive displacement pumps can expel air as well as water because of the close fit between internal parts. Centrifugal pumps must be filled with water before they can create a vacuum because they have an open waterway from the suction to the discharge side of the pump.

When the pump is filled with water, it will create quite a high vacuum, but when the pump has been drained and contains only air, it is a very poor vacuum pump. In order to obtain water, it is necessary to create a vacuum inside the pump by some other means. A priming device that expels air by one of three methods is used:

1. A positive displacement pump
2. The engine intake manifold
3. The engine exhaust gases.

Priming devices withdraw air from the centrifugal pump and are intended to reduce pressure within its casing to substantially below atmospheric pressure. Priming devices should develop a vacuum of at least 22 in. of mercury.

Whether pump impellers should be rotating or stationary while the pump is being primed depends upon several factors.

If the pumper has three positions in the pump transmission–Road, Pump and Prime–it is evident that the priming pump is driven by the pump transmission. It would be impossible to mesh both pump and prime gears at the same time; thus the impellers will be stationary.

Some priming pumps are driven through a clutch by the impeller shaft. The main pump must be rotating before the priming pump can be operated.

Except for mechanical reasons such as stated above, whether the impellers are rotating will largely depend upon where the priming device is connected into the main pump. For consistently reliable priming, the impellers should be rotating when the pump is primed from the suction side near the impeller eye, and the impellers should be stationary when the priming device is connected to the top of the pump discharge manifold.

Figure 4-1 illustrates the reason for this. For simplicity, no discharge valves, relief valves, priming pump, or other accessories are shown. A single-stage pump is sketched, the principle being the same with two or more stages. The source of the vacuum may be a pump, exhaust ejector, or engine intake manifold.

When priming from the top with the pump stationary, the primer is operated until a steady stream of water flows from the primer discharge. It is then shut off and the main pump is put in gear. When priming from the suction of the pump, with pump impellers rotating, the primer is not shut off until a steady stream of water is flowing from the priming device. Caution must be exercised when terminating the use of a priming device because the prime may be lost if a shutdown is made immediately upon observing the first emission of spray; this water usually contains a large amount of air.

Any unavoidable air lock or air trap in the pump or suction hose is more easily removed by priming from pump suction. Without interfering with pump rotation or speed, air is scavenged by momentarily engaging the priming device.

Air leaks in a pump make priming uncertain. If the pump is primed in spite of air leaks, operation will be rough and efficiency will be lowered. Maximum capacity of the pump will be substantially reduced, especially on lifts higher than normal and at high altitudes. While it is hard to define the size of an air leak, any leakage that drops vacuum faster than 10 in. in 10 min will cause reduced performance, rough operation, hose pulsation, and a ragged stream.

When pumping from a draft or booster tank through a ¾-in. or 1-in. booster hose, flow through the pump is so small compared to pump capacity that it is often difficult to scavenge all air from the pump; this results in unexpectedly low pressure. It is good practice when this occurs to open momentarily a 2½-in. discharge gate. The larger flow for a few seconds while the priming device is still operating completes removal of air.

Exhaust Priming System

This exhaust priming system is operated with the main pump engaged so that the impellers are turning during the priming operation.

Figure 4-2 illustrates the operating sequence of the system. The exhaust conejector is connected to the exhaust pipe; the butterfly valve in this unit is operated by a knob on the control panel. When this butterfly valve is closed, exhaust gases, as indicated by arrows, are directed through the conejector priming unit. The greatly reduced outlet area of the conejector causes a high velocity at the ejector outlet which creates a vacuum in the line connected to the top of pump suction intake. The vacuum opens check valve A and closes check valve B.

Since the exhaust conejector operates only while air is being exhausted from the main pump, a cut-off (priming valve) is provided between the priming unit and the main pump. This priming valve is operated by the same knob that operates the butterfly valve in the conejector. Discharge from the conejector is made directly on the ground;

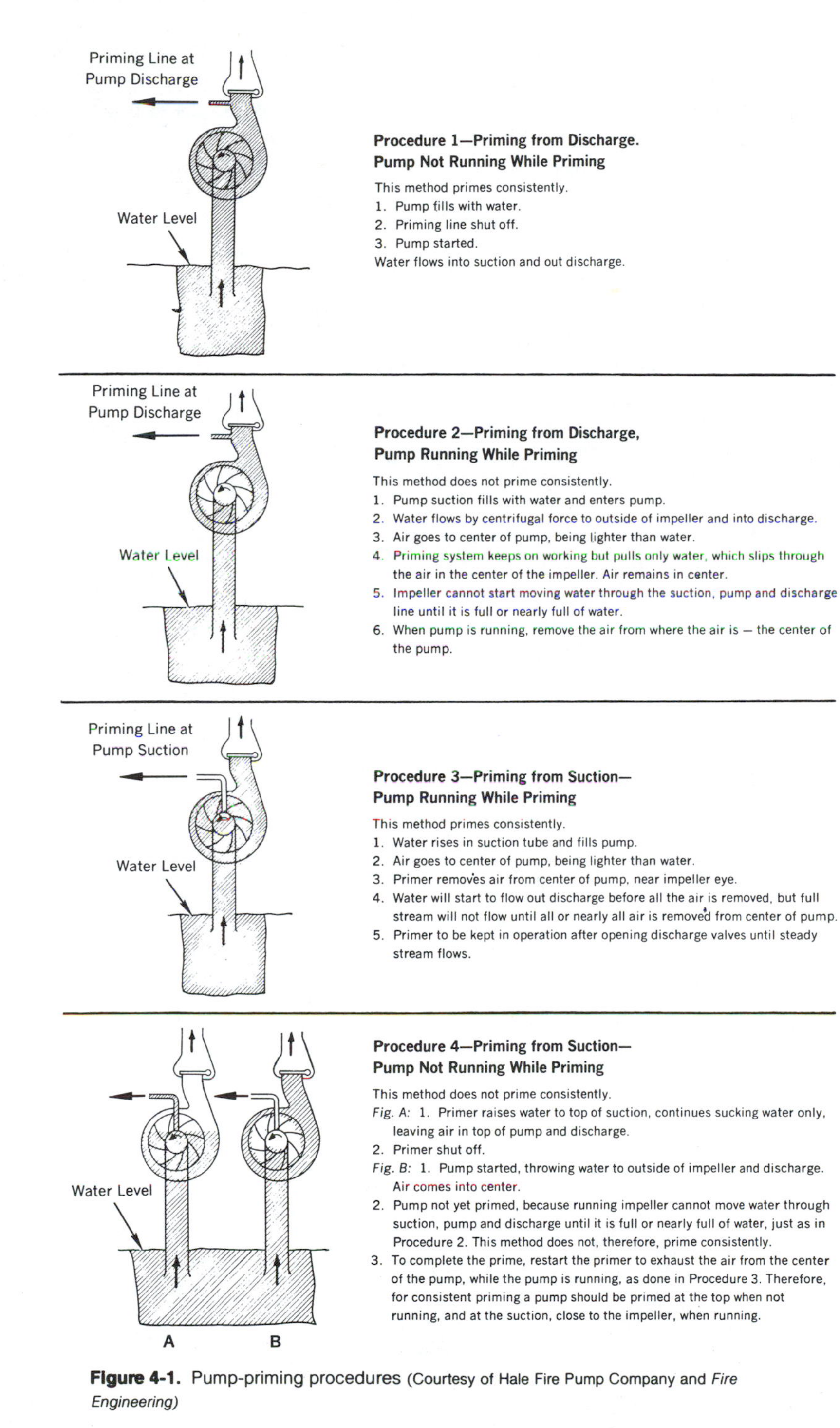

Figure 4-1. Pump-priming procedures (Courtesy of Hale Fire Pump Company and *Fire Engineering*)

this permits the pump operator to see the stream of water that issues from the primer when the main pump is primed. At the same time as the pump is primed, pressure will be indicated on the discharge gage as the main pump impellers are engaged. Efficiency of the exhaust primer depends upon the velocity of exhaust gases passing through the exhaust ejector; this requires engine speeds of from 1500 to 2800 rpm.

The hydrant conejector is provided so that if the priming valve should be opened with pressure on the suction side of the main pump, water would be discharged to the ground through check valve B. The vacuum created by the hydrant conejector would close check valve A, thus preventing any water from getting into the exhaust conejector unit.

Periodic and thorough inspections must be made of the entire priming system to make sure that it is in proper operating condition. The exhaust pipe and the gaskets between the engine and the primer should be without leaks. The butterfly valve must close fully; an accumulation of carbon or corrosion on the butterfly valve or in the exhaust conejector will greatly affect efficiency. All check valves must operate freely and seat properly.

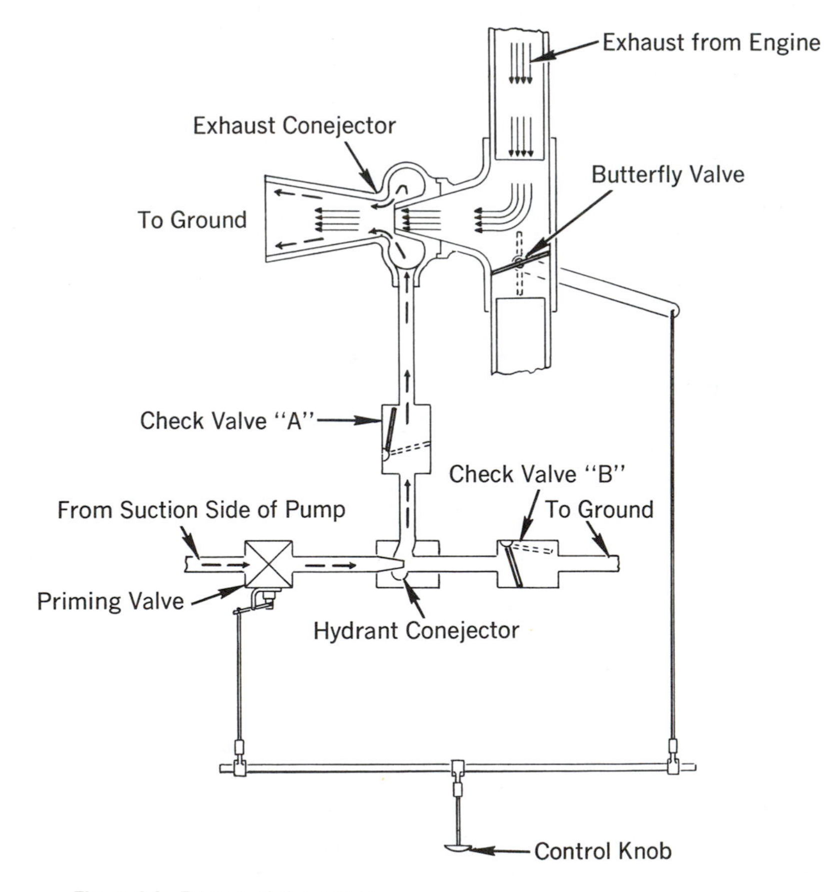

Figure 4-2. Exhaust-priming system

Intake Manifold System

An internal combustion engine in good condition develops between 20 and 22 in. of vacuum in the intake manifold while the engine is idling. This vacuum can be utilized to prime the main pump.

The primer assembly consists of a mechanism placed between the pump and the engine (Figure 4-3) to prevent water from being drawn into the intake manifold. When the prime handle is pulled out, air from the pump and the suction hose is drawn into the intake manifold. After the pump is primed, water rises into the wet chamber of the primer. The main float lifts sufficiently so that the lower ball valve is closed and the upper ball valve is raised, thus venting the prime to atmosphere and preventing further rise of water.

If, for any reason, the main float fails to rise, water will finally pass the open lower ball valve and enter the upper chamber on its way to the engine. When this does occur, water will cause the cork float to rise ahead of it and close off the entrance to the engine vacuum line. This eliminates any possible chance of water passing into the engine to cause damage.

The most effective vacuum is created at slightly above idling speed with little load on engine. Many difficulties experienced in priming can be related to excessive engine speed.

Engine Primer

A make of pumper utilizing a V-12 engine uses one bank of cylinders as a vacuum pump and the other bank for power.

The pump is placed in gear. When the priming control handles are pulled, valves close off fuel-air mixture from one carburetor to the intake manifold and permit only air within the pump to flow into the manifold, thus creating a vacuum. To prevent possible entrance of water from the pump into engine cylinders, a priming chamber equipped with an automatic shut-off valve is placed in line between the pump and the intake manifold.

Priming Pump Lubrication System

Most priming pumps are lubricated as shown in Figures 4-4 and 4-5. As rotors turn they create a vacuum on the suction side of the pump, pulling air from the main pump. To seal these positive displacement pumps so as to reduce leakage, engine oil is siphoned from a reservoir.

In the lubricating process (see Figure 4-5), as the rotor turns, it creates a vacuum on the suction side of the pump and pulls the air from the main pump. At the same time, oil is pulled in from the oil tank to lubricate the pump and bearings. Do not operate this priming pump unless oil is in the tank.

The air vent in the priming tank tube is used to break the siphon, thus preventing flow from the tank into the priming pump when it is not operating. This hole is made with a No. 60 drill and should not be enlarged; it, as well as the air vent in the filler cap, should be kept open at all times. A puddle of oil under the priming pump discharge may indicate a plugged air vent in the line.

Priming Valve

There is a spring loaded valve in the line between the priming pump and the main pump. The purpose of this valve, which is normally closed, is to control the flow from the centrifugal pump. Valve leakage may cause the main pump to lose its prime after use of the primer is discontinued; water running from the priming pump when the main pump is connected to a hydrant may indicate a bad seal.

Electric Driven Primers

Electric motors are often used to power positive displacement rotary pumps. An

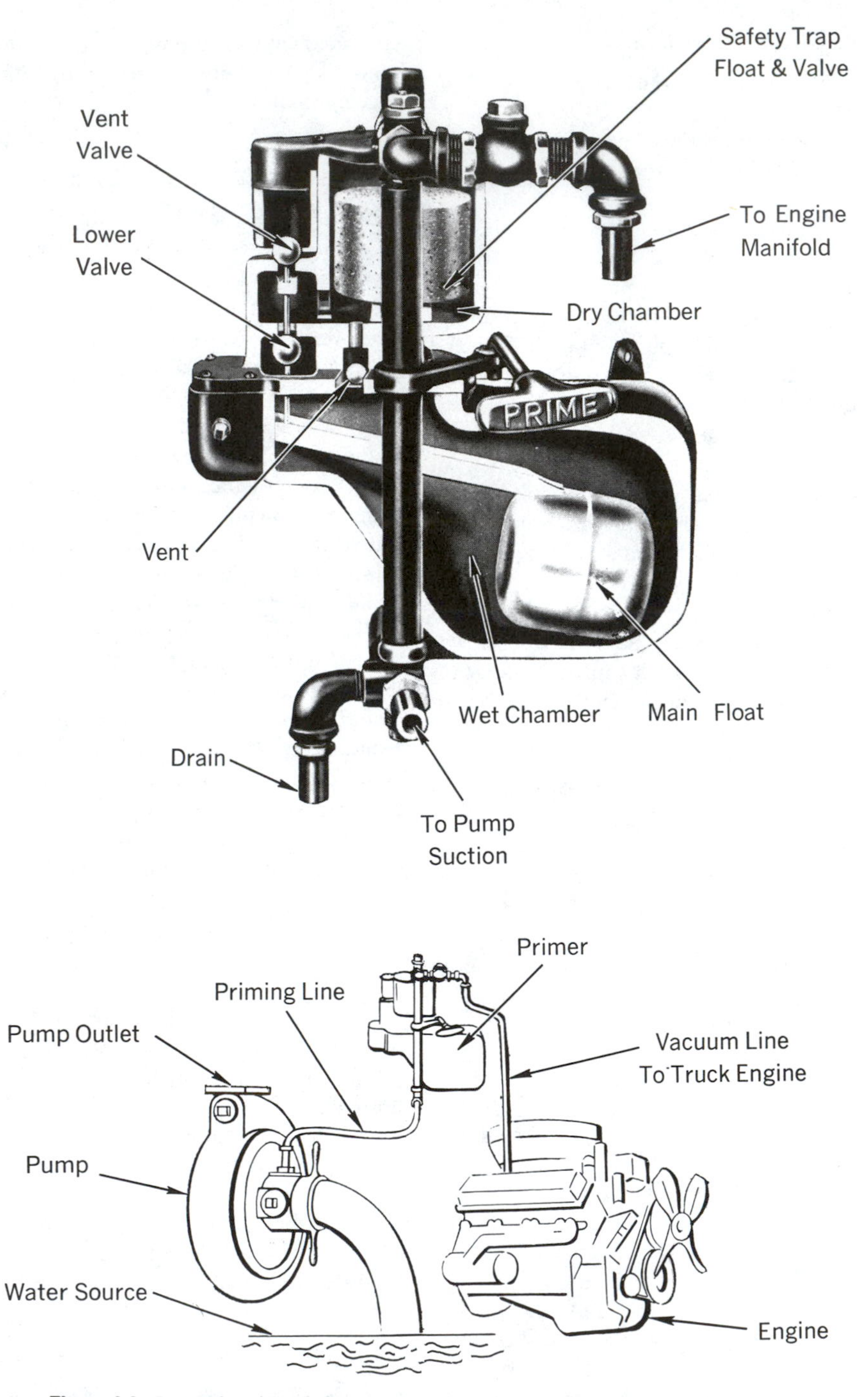

Figure 4-3. Barton-American engine vacuum primer (patented): *top,* cutaway view of primer; *bottom,* position of primer (Courtesy of American Fire Pump Company)

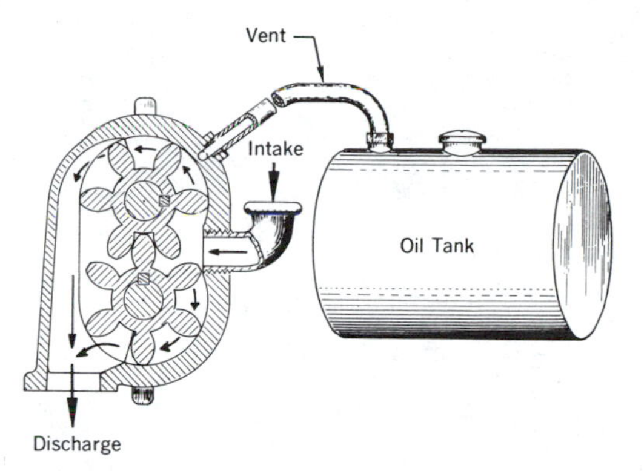

Figure 4-4. Rotary-gear priming pump

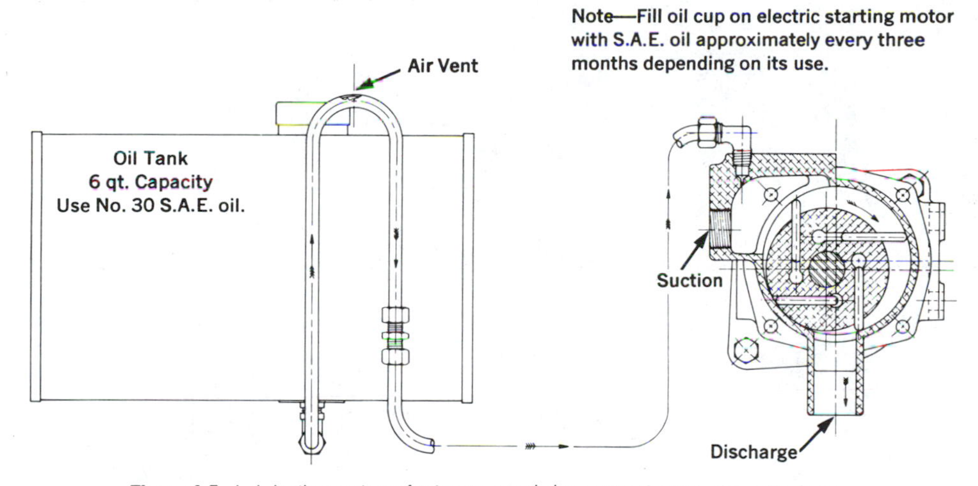

Figure 4-5. Lubricating system of rotary-vane priming pump (Courtesy of Hale Fire Pump Company)

electric primer is placed in operation by pulling out the primer control handle. This control actuates an electric motor by closing a switch; at the same time a valve between the priming pump and the main pump is opened. Releasing the handle disconnects the switch and allows the priming valve to close.

Unless the priming pump is automatically lubricated by siphoning oil from a supply tank, the lubricator on the pump panel should bė operated before attempting to prime the main pump.

Clutch Priming System

There are several variations of the clutch priming system.

One system (Figure 4-6) utilizes the pump transmission to drive the rotating clutch flange. When priming is desired, it is merely necessary to open a vacuum valve. When the priming button (see Figure 4-6) is pushed in, the inside of the priming valve diaphragm and the outside of the priming clutch diaphragm are subjected to vacuum from the engine intake manifold. In the priming valve, atmospheric pressure on the outside

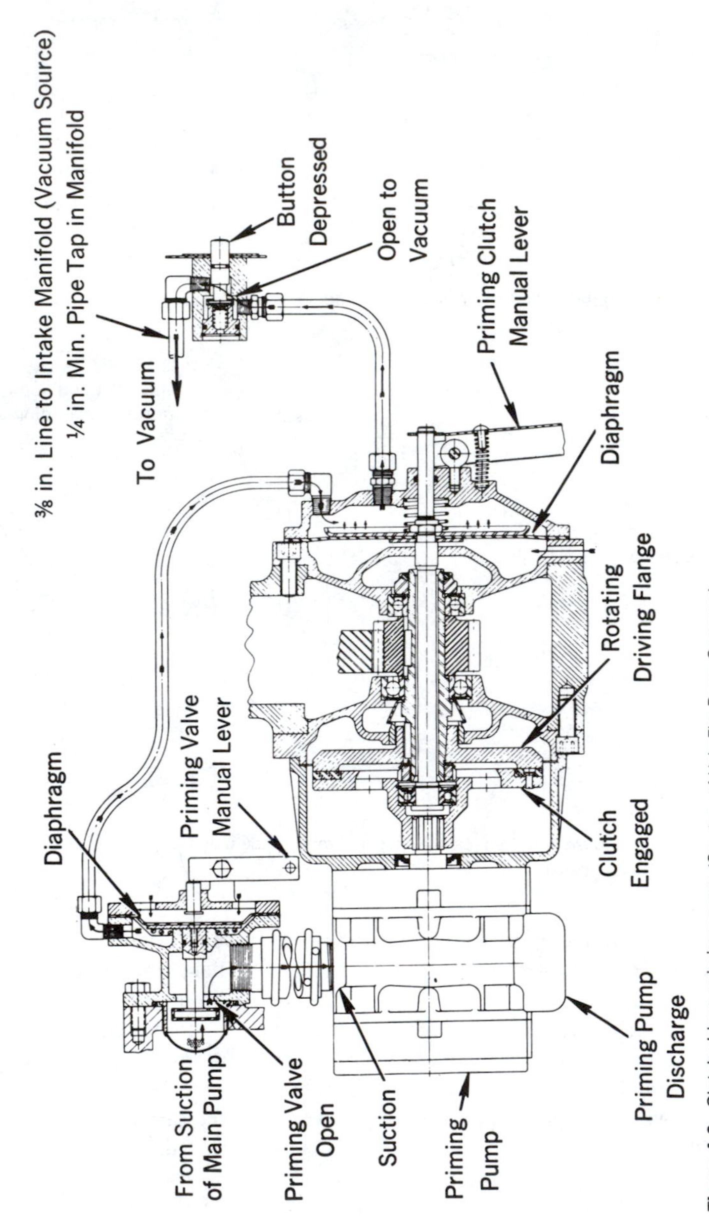

Figure 4-6. Clutch-driven priming pump (Courtesy of Hale Fire Pump Company)

of the diaphragm pushes the valve open and connects the priming pump suction to the main pump suction. The priming clutch is engaged by atmospheric pressure pushing on the inside of the priming clutch diaphragm, thus driving the priming pump.

Figure 4-6 shows the pump in the priming condition; when the pump is not priming, the priming button is released and the areas on both sides of the priming valve diaphragm and priming clutch diaphragm are subjected to atmospheric pressure. In this case, the priming valve is closed and the priming pump clutch is disengaged by spring force.

Another type operates similarly but is driven off the end of the impeller shaft. Obviously, this type will only operate when impellers are rotating.

Pressure Control Devices

The pump operator seldom knows in advance when a hose line is to be shut down; therefore, some method is needed to prevent excessive pump pressures from developing. The primary purpose of a pressure control device is to protect men at the nozzles from a dangerous pressure rise; a secondary purpose is to protect the hose and pumps.

If a single line being operated from a pump is suddenly shut down, there will be a momentary backing-up of pressure in the pump; this may be sufficient to rupture hose lines and damage pump casings, connections or the engine.

If more than one line is in operation and one of the lines is shut down, engine speed will increase as load is reduced; consequently, discharge pressure will also increase. Consequences of the rise in pressure may be serious, particularly if a line is being operated from a precarious position.

One way of explaining this pressure rise is to compare the engine reaction resulting from a decrease in horsepower requirements to events while driving a vehicle. If an engineer were driving a pumper up a steep hill so it required the accelerator pedal to be depressed three-fourths of the way to the floor to maintain a 25-mph steady speed, the truck would gain speed when reaching level ground at the top of the hill unless the accelerator were raised. Then if the accelerator were still maintained in the same position, the pumper would continue to gain speed when descending the hill on the far side because it requires far less power to descend a hill than it does to climb it. Therefore, if the throttle were adjusted to deliver an engine pressure of 200 psi while supplying two 250-gpm spray nozzles with 100 psi nozzle pressure, shutting down one of the tips would result in an engine speed increase because it would require less horsepower to pump 250 gpm than it would 500 gpm. As the engine speed increased, there would be a corresponding pump pressure escalation.

To maintain the desired discharge pressure, it is necessary to install some type of automatic pressure regulating device on fire fighting pumpers. Excessive discharge pressures are prevented by using one of the following devices:

1. An automatic relief valve that opens a bypass between the suction and discharge sides of the pump.
2. A pressure-operated governor to control the speed setting of the throttle on the engine.
3. Pressure-reducing valves on each discharge gate.

However, at very low volumes of discharge or minimum power settings, pressure-control devices are not really necessary. Under these conditions, when the discharge valves are closed, there would be an insufficient rise of pressure at the pump to cause damage.

Relief Valves

Relief valves prevent pressures from going above the desired setting by bypassing water from the discharge outlet back to the suction inlet. Figure 4-7 shows the flow of water through the relief valve when the pilot valve causes the relief valve to open and relieve the excess discharge pressure. In this way, a relief valve keeps both the pump and the engine under a constant load.

Principle of Relief Valves

To understand the function of a relief valve, one must first comprehend the basic principles of the internal combustion engine and the centrifugal pump.

An internal combustion engine develops power by burning fuel in its cylinders; power is equal to torque multiplied by engine speed. The pressure of the gases formed by the combustion pushes the pistons down and rotates the crankshaft. Thus, the torque developed depends on the amount of fuel burned during each cycle, and the horsepower depends on the torque and speed. At a given throttle setting, therefore, the engine will develop a certain torque.

A centrifugal pump develops pressure in the fluid being pumped by imparting kinetic (moving) energy to the liquid by means of the rotation of the impeller or impellers. The water enters the impeller near its center and is whirled by the impeller vanes; this action develops centrifugal force, which urges the water outward into the discharge manifold. At a specific impeller speed, the pressure at a given flow rate is always the same; as the flow rate is increased, the pressure decreases and the required amount of power increases. Vice versa, if the speed remains constant, when the pump flow is decreased the pressure is increased and the power requirement is decreased.

In the preceding discussion, the term *pressure* means net pump pressure, or the difference between discharge pressure and suction pressure. When a pump is operating from draft or a booster tank, the suction pressure is usually near or below zero; it will not vary appreciably with a change in flow rate. When the pump is operating from a hydrant or in a relay, however, the suction pressure ordinarily is at least 10 psi; the pump inlet pressure can vary appreciably with a change in flow.

The relief valve proper is merely a piston in a cylinder; one end of the piston normally closes a passageway between the pump dis-

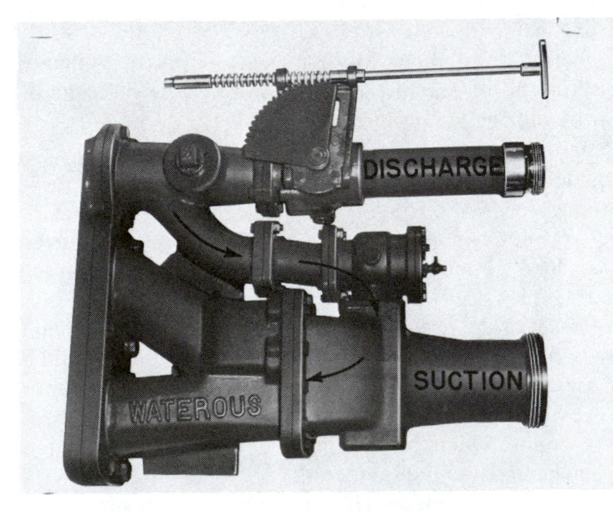

Figure 4-7. Relief-valve water flow (Courtesy of Waterous Company)

charge and the pump suction. The other end of the piston is subject to water pressure, which is automatically regulated by a pilot valve.

The pilot valve incorporates a spring-loaded valve that moves to allow the passage of water in response to changes in pump discharge pressure. The pilot valve is set at any desired pressure between 90 and 300 psi. By adjusting the compression on a coil spring that opposes the pump discharge pressure acting on a diaphragm or piston, some relief valve models have an even wider range. At any discharge pressure below the setting, the pilot valve will be closed; at any pressure higher than the setting, the pilot valve will be open.

To keep the discharge pressure the same when the flow through the discharge valves is decreased, the load on the engine must be regulated so that engine speed will not increase and cause a rise in net pump pressure. When operating from draft or from a booster tank, it is necessary only that the relief valve bypass the same flow that is shut off; thus, the torque load on the engine will remain constant. If the engine speed stays steady, so will the net pump pressure; because the pump suction pressure will not rise significantly, neither will the discharge pressure. In this case, as long as the flow that must be bypassed can pass through the relief valve without causing more friction loss than the net pump pressure, the relief valve will maintain the discharge pressure at the setting of the pilot valve. The larger the relief valve and the higher the net pump pressure, the more water can be bypassed before the friction loss becomes too high. The relief valve does not open fully until the maximum flow is being bypassed; at lesser flows the pilot valve modulates the pressure on the relief valve piston so that it opens just far enough to keep the friction loss through the relief valve and piping equal to the net pump pressure.

When operating from a hydrant connected to a high-capacity water distribution system, the suction inlet pressure may be quite high, but it will not change significantly when the flow rate changes. Usually the net pump pressure will not be as high as when operating from draft, so less flow can be bypassed without causing friction loss through the valve system equal to net pump pressure. With a suction gage pressure of 50 psi, the minimum discharge pressure that most relief valves will hold with a given reduction in flow is 50 psi higher than when operating at draft. As an example, if a relief valve will hold a 90-psi discharge pressure while operating at draft with a 750-gpm reduction in flow, it will hold a discharge pressure of 140 psi or higher when operating with a 50-psi suction gage pressure and the same 750-gpm decreased flow. The minimum regulated pressure refers to the net pump pressure. The system will stabilize when the flow through the relief valve and any discharge lines still open, and the friction loss through the relief valve and piping, create a pump torque load equal to the power being developed by the engine.

Therefore, because the relief valve holds a fairly constant engine speed by maintaining a uniform engine load, it cannot decrease the net pump pressure greatly. When operating from a hydrant or in a relay, if the suction pressure increases substantially because of a reduction in gpm flow, most models of relief valves will not be capable of controlling the pump discharge pressure to the selected setting. Also, most relief valves cannot under any circumstances control discharge pressures to an amount lower than suction gage pressure plus about 50 psi. When operating from draft or a booster tank, relief valves will not control at discharge pressures less than about 75 psi.

Waterous Relief Valve

The Waterous relief valve system (Figure 4-8) consists of two units: the relief valve itself and the pilot valve that controls it. The design of the system permits a rapid selection of any discharge pressure between 70 or

75 psi and 300 psi. After the pilot valve is adjusted for a desired pressure, it usually requires no further attention until another pressure setting is desired. The relief valve is available in two sizes. For pumps with rated capacities through 750 gpm, a 2-in. diameter is normally used; for rated capacities of 1000 gpm and higher, a 3-in. diameter relief valve is recommended. On the older large pumps, two of the smaller valves were often utilized instead of one large relief valve to allow the required quantity of water to be bypassed.

The schematic in Figure 4-8 illustrates a diaphragm attached to the pilot valve stem. On the newer models, an O-ring sealed piston actuates the pilot valve, instead of a diaphragm.

Hydraulic force is used to operate the bypass valve between the discharge and suction sides of the pump; this enables the relief valve to bypass large quantities of water with only a small pressure rise in the pump. Figure 4-9 illustrates the principle of the Waterous relief valve. A tube from the pump discharge is connected to the back of the large relief valve cylinder. This forces the valve to stay closed because water at the same pressure as that in the pump discharge is also pushing against the much larger area on the top side of the big relief valve. The tube from the pump to the relief valve has a constriction in it to limit the volume of flow. Another tube permits water to flow out of the system or to dump.

Water is metered through the constriction into the pressure chamber of the relief valve and out through the other pipe. The second pipe has a needle valve in it to control the rate of escape; in fact, this valve can be set so

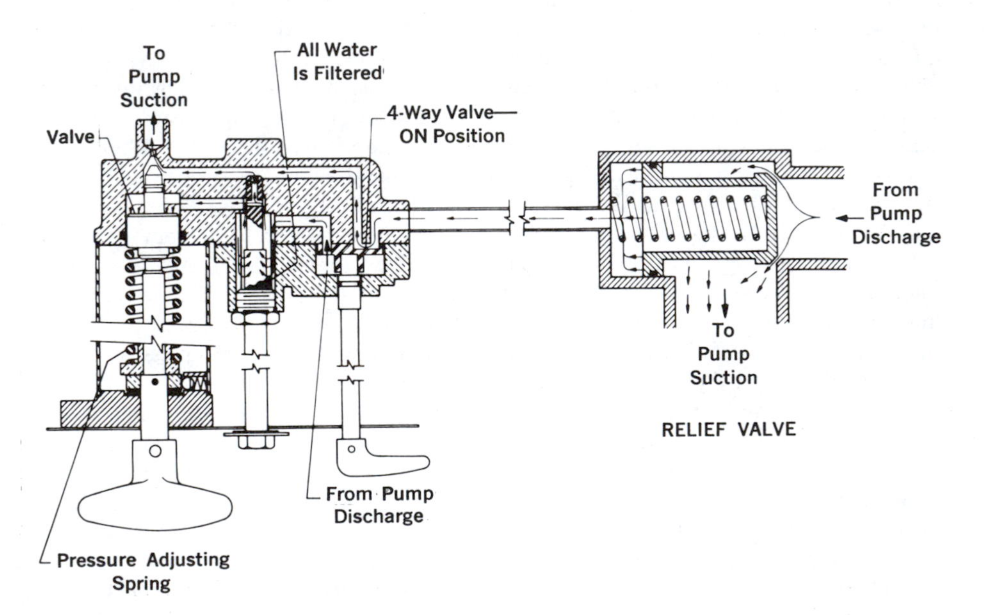

Figure 4-8. Schematic of Waterous relief-valve operation. In this view, the pump discharge pressure is higher than the pilot valve setting, so the relief valve is regulating the flow. (Courtesy of Waterous Company)

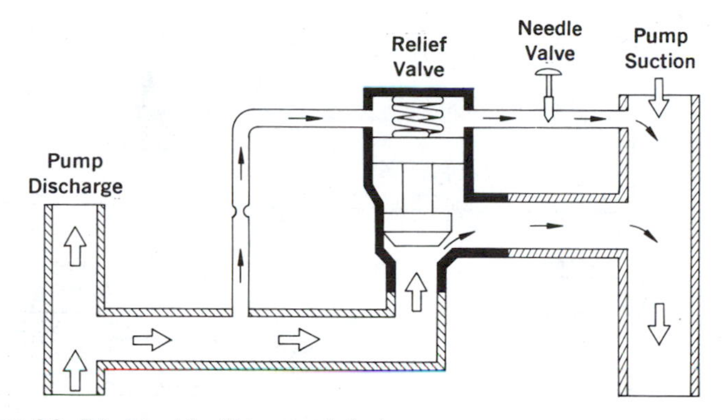

Figure 4-9. Principle of the Waterous relief valve

that water can escape from the cylinder more easily than it can enter through the constriction. The faster the water flows through the needle valve, the greater the pressure drop in the cylinder, and the lower the pressure behind the piston in the large relief valve. Adjustment of the needle valve setting (which changes the rate at which water is dumped) varies the hydraulic force behind the piston so that the relief valve can open part or all the way at any predetermined pressure in the pump discharge.

This type of system is manual, so the pump operator must keep a sharp eye on gages and keep adjusting the needle valve to compensate for pressure changes in pump discharge. In the pilot valve of an actual Waterous relief valve system, modulating of the needle valve, which dumps water to pump suction, is done automatically by changes in discharge pressure, but the principle of operation is roughly the same.

The spring in the main relief valve plays only a small part. Hydraulic force behind the main valve does most of the job, and this force is controlled by the amount of water dumped.

The relief valve proper is mounted on the pump between suction and discharge (see Figure 4-7). Arrows show the direction of water flow when the valve is open and bypassing. The pilot valve is mounted on the operator's panel of the apparatus. The pilot valve and its 4-way valve control operation of the relief valve.

Figure 4-10 shows the entire system. The pilot valve on the left is shown in two views to better illustrate the 4-way, or ON-OFF, valve. The relief valve is shown on the right. The needle valve in the pilot valve does double duty:

1. It meters the water from discharge through the pilot valve and tube to the back side of the main relief valve.
2. It also moves to the left to dump water back to pump suction.

Figure 4-10 shows what happens when the 4-way valve is turned to OFF position. This takes the entire relief valve out of operation, which is necessary to obtain pressures above 300 psi. Water from pump discharge goes directly through the 4-way valve to the main valve chamber, which is the right-hand side of the main valve in this illustration. Pressure is the same on both sides of the main valve, but the area is greater on the main valve chamber end (right), so the force is greater from that end. This greater force, plus force of the spring, keeps the valve closed so there is no flow back to pump suction.

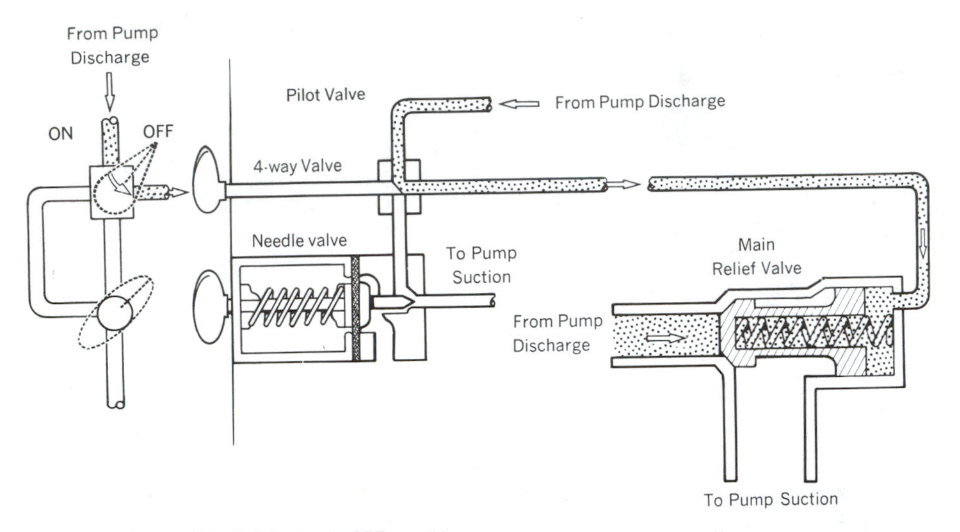

Figure 4-10. Relief valve in OFF condition

Figure 4-11 shows what happens when pump discharge pressure is lower than the pilot valve setting. In this illustration, the 4-way valve is ON, and the entire system is in operation. This time, the water flows from pump discharge to the pressure chamber of the pilot valve on the left. This pressure chamber is next to a diaphragm. Force of the spring against this diaphragm is regulated with the handle of the pilot valve; as long as the hydraulic force in this chamber is less than the force of the spring, the pilot valve will stay closed. Water pressure is then transmitted back through the 4-way valve to the main valve chamber (on right), and the main valve stays closed.

Figure 4-12 shows what happens when a pressure rise opens the pilot valve enough to overbalance the main valve. The restricted area around the needle valve in the pilot valve acts as a metering passage to control the rate of flow around it. When discharge pressure gets high enough to compress the diaphragm against the spring in the pilot valve, the needle valve moves to the left. This allows water to dump back to pump suction. This, in turn, reduces pressure in the tube and behind the main valve (right). Force on the small end of the main valve (left) is now greater. The valve opens and part of the water being pumped bypasses back to suction and reduces the discharge pressure. The main valve opens just enough to reduce the discharge pressure to the pilot valve setting. If the discharge pressure drops below the set pressure, the pilot valve reseats, and pressure builds up behind the large end of the main valve and closes it.

Operation. The 4-way valve allows the entire system to be placed in or out of operation without affecting the original pressure setting, which is determined by the spring tension on the diaphragm in the pilot valve.

The relief valve should normally be carried with the 4-way valve in the OFF position and the control handle set to a pressure somewhat higher than it is expected will be used. This relief valve will not control pressures higher than 300 psi. The lower draincock is kept closed.

As soon as the desired pump pressure is reached, and with at least one discharge line open, set the relief valve by turning the relief valve to ON position and rotate the control handle in a counterclockwise direction until the discharge gage indicates a pressure drop.

Then turn the handle clockwise until the gage indicates the desired pump pressure; water should not be continuously bypassing when the relief valve is properly set.

To raise the discharge pressure: screw in the control handle above anticipated pressure and increase the pump pressure; then readjust the relief valve.

To lower pressure: decrease the engine speed; then readjust the relief valve.

After shutting down, open the lower draincock to release water; then close. Because of the bypassing of water from the valve chamber through the pilot valve, the lower drain does not need to be opened to release air when the relief valve is being placed in operation. If the lower drain is opened when the relief valve is in control, pressure in the valve chamber is released and the relief valve opens.

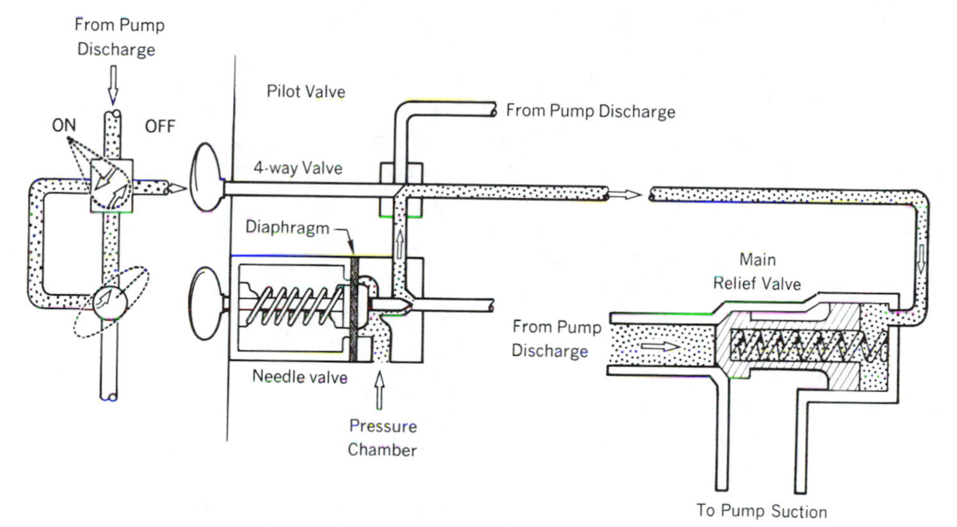

Figure 4-11. Relief valve in ON condition, not relieving

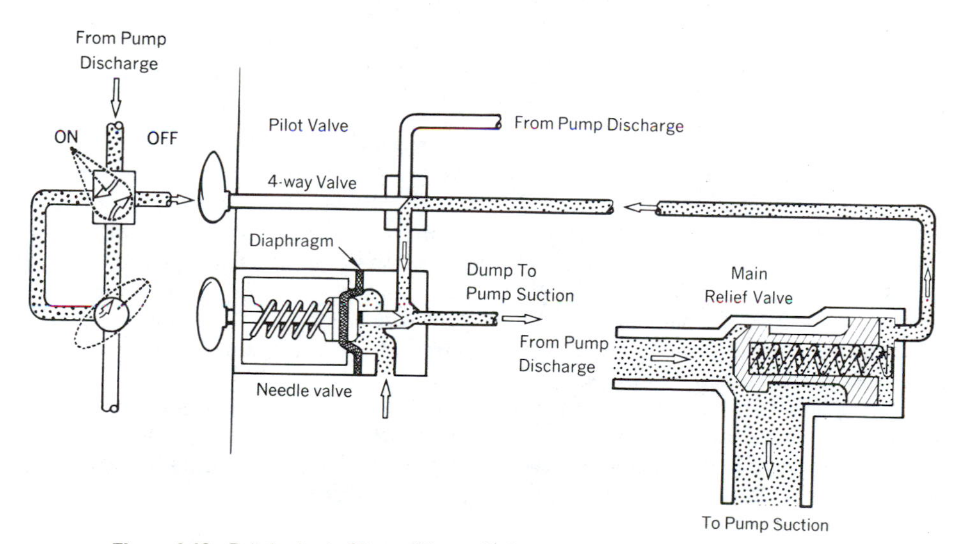

Figure 4-12. Relief valve in ON condition, relieving

Maintenance. Because of the design of the relief valve system, few disorders are apt to occur. If the relief valve operation is sluggish or erratic, the cause can usually be traced to fine sand, grit, or other foreign material clogging one of the valves or strainer. If a strainer is installed on the pilot valve, remove the strainer and clean it. With the 4-way valve OFF, the strainer can be removed with the pump operating. If the pilot valve has a flush line, opening the flush valve with the pump operating and the 4-way valve ON will usually clean the strainer and restore proper relief valve operation. Before reinstalling the strainer, check the condition of the gasket and replace if damaged. Tighten the strainer enough to seat the O-ring; high torque is not necessary.

If the relief valve operation is still erratic, operate the pump with the engine at about one-half throttle. Open one discharge valve enough to permit the discharge of a ½-in. stream of water to keep the pump cool. Observe the discharge pressure. Then with the 4-way valve ON, turn the pilot valve handle counterclockwise all the way to relieve spring pressure on the valve. The pilot valve should open, which will lower the pressure in the chamber behind the relief valve; the relief valve should open, which will cause the discharge pressure to decrease. If the pressure does not drop, either the pilot valve or the relief valve is stuck. To free either assembly, turn the 4-way valve to alternate between the ON and OFF positions rapidly several times to flush clean water through the pilot valve assembly and to reverse the forces acting on the relief valve. If this does not remedy the situation, unfasten the connection between the pilot valve and the relief valve (at the relief valve), or open the drain cock on the relief valve cover, if it is accessible. If the pressure then decreases, the relief valve is not stuck, and the trouble can be attributed to the pilot valve. If the pressure does not decrease, however, the relief valve is stuck. If either the pilot valve or the relief valve is stuck, disassemble and clean it according to the manufacturer's instructions.

Caution. The water in a centrifugal pump will heat up rapidly if the pump is operated with all of the discharge valves closed. Prolonged operation will cause damage from overheating. To prevent possible damage, allow a stream of water at least ¼-in. in diameter to discharge from the pump whenever it is running, unless a circulating line from pump to tank is provided.

After prolonged pumping from draft, especially dirty water, enough sand and dirt may accumulate to prevent the pilot valve from seating properly. Once a year, or as often as necessary, disassemble and clean the pilot valve assembly.

Hale Relief Valve, Latest Model

The modern Hale automatic relief valve (Figure 4-13) offers full-range pressure control and has ample capacity to prevent undue pressure rise when discharge lines are shut down. An easy-turning control handwheel sets the pressure quickly and accurately; a bright pilot light glows when the relief valve is open for flow. The principles of operation are similar to the older models of the Hale relief valve, which is discussed in the next section.

Operation. To set the relief valve, increase the pump discharge pressure to the required amount, using the discharge pressure gage. Then turn the handwheel slowly counterclockwise until the relief valve opens and the pilot light is lit. Then turn the wheel clockwise slightly until the light is out. The relief valve will now operate at the set pressure.

When the relief valve is not in operation, the adjusting handwheel should be turned clockwise so that the control is set above the normal operating pressure.

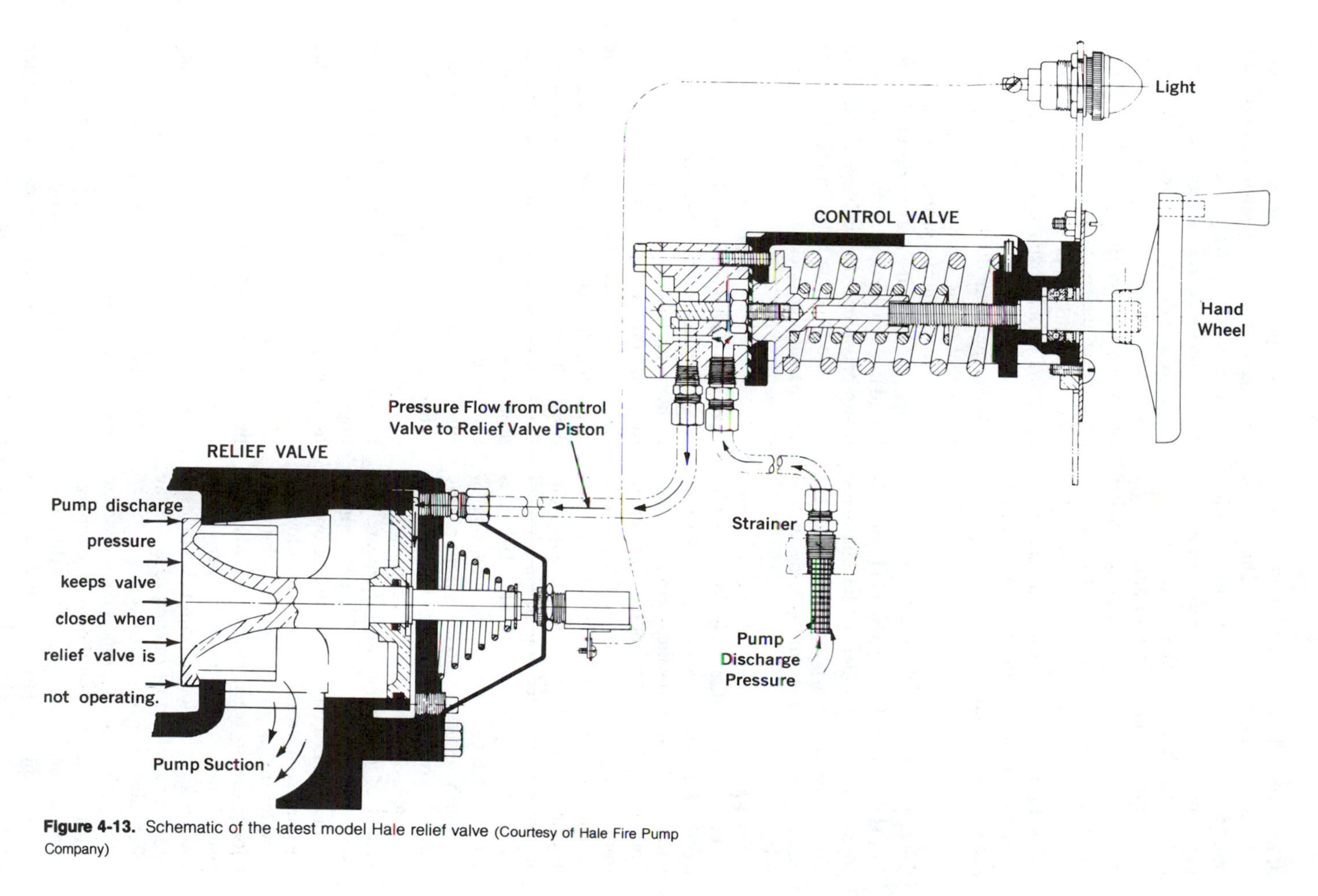

Figure 4-13. Schematic of the latest model Hale relief valve (Courtesy of Hale Fire Pump Company)

Maintenance. Test the relief valve frequently to make sure it is free to move. This can be done by turning the control valve adjusting handwheel clockwise as far as possible, thus locking out the control valve. Then bring the pump pressure up to 150 psi and turn the handwheel counterclockwise until the relief valve opens. Turning the control valve handle in and out with a pump pressure of 150 psi will cause the relief valve and control to operate; this action frees the valve and ensures proper operation.

Hale Relief Valve, Older Model

The Hale relief valve illustrated in Figures 4-14 and 4-15 is an older model automatic churn valve which bypasses water from the discharge side of the pump to the suction side.

The identifying letters of the components in the following discussion are keyed to figures 4-14 and 4-15 to make it easy to understand and follow the operation. Water from the discharge of the pump is prevented from flowing into the suction side by valve G. The valve is held closed by pressure of the discharge water. Water under pressure from discharge is prevented from flowing through the control line by pilot piston C, which is held closed by compression in spring S. Resistance of piston C to the water pressure may be varied by turning the knurled control cap A. Some models use a control wheel instead of a knurled cap. There are two drains on this relief valve. The upper drain, which is located on the back side of operating panel, is open at all times. This allows air or water that might be trapped behind the pilot piston to bleed away; otherwise, the piston could not operate properly. The draincock on the lower drain is kept closed and needs only to be opened to drain the relief valve to prevent freezing.

When the control spring is compressed to a greater amount than the pump discharge pressure, the relief valve remains closed.

When the discharge pressure becomes greater than the relief valve setting, pilot piston C opens and permits water to flow into chamber H of the relief valve proper. Water pressure applies force to piston J, which is on the shaft as valve G between the discharge and the suction sides of the pump. The same pressure is acting on both valve G and piston J but, as the area of the piston is greater than the area of the valve, the larger total force on the piston forces the valve open and permits bypassing of water in the pump.

When the pump discharge pressure drops below that of the compression pressure of the governor control spring, pilot piston C closes, and pressure against piston J is released through the upper drain. This permits pump pressure in the discharge chamber to close valve G. Flow between the discharge and the suction sides of the pump is then stopped.

Operation. The relief valve should normally be set and carried at a pressure somewhat higher than it is expected will be used. As soon as the desired pump pressure is reached, and with at least one discharge line open, set the relief valve by turning the control handle in a counterclockwise direction until the pump pressure gage begins to drop; then turn the control handle in a clockwise direction until the original pressure is reached. Water is about to bypass when the sleeve nut (located between the control handle and the upper chamber housing) starts to move away from the valve housing; the water may be heard when it is bypassing. Another method of setting the relief valve is to screw sleeve nut A back slowly until groove B on piston C comes in line with groove D on the valve casing E. Lock the sleeve nut in place with lock nut F. Water should not be continuously bypassing when the relief valve is properly set.

To raise the discharge pressure: screw in the governor-control handle above anticipated pressure, increase pump pressure, then readjust the relief valve.

To lower pressure: decrease the engine speed, then readjust the relief valve.

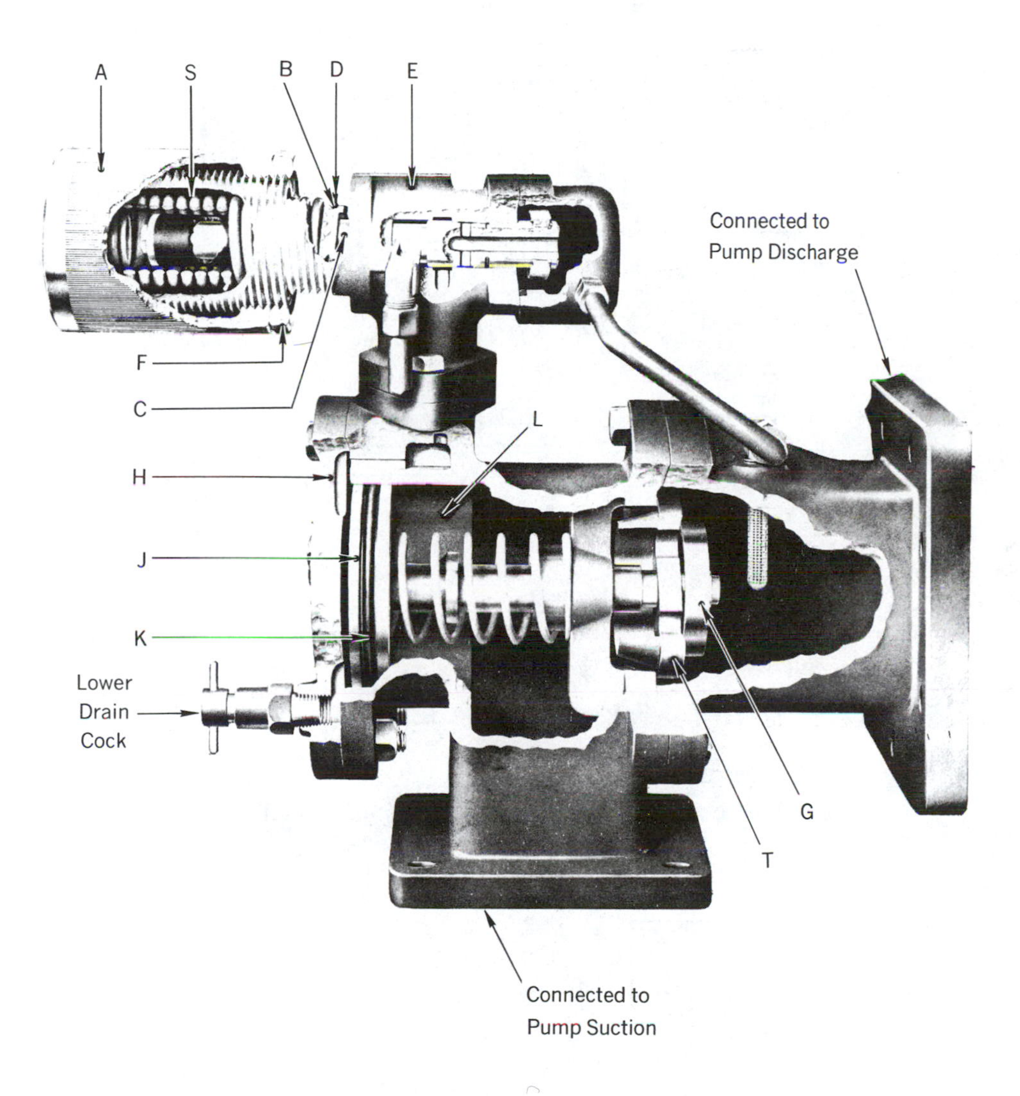

Figure 4-14. Hale relief valve (Courtesy of Hale Fire Pump Company)

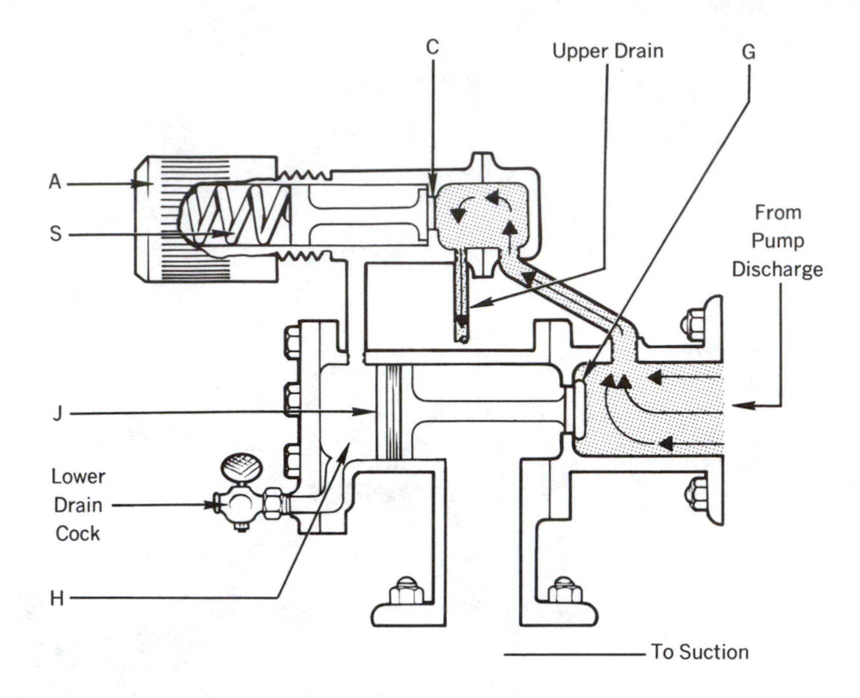

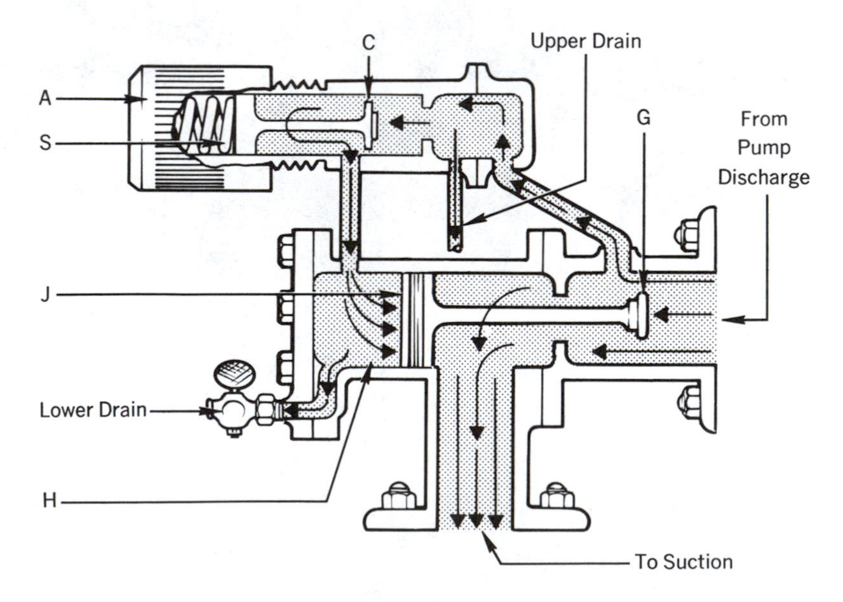

Figure 4-15. Schematic of Hale relief valve

Maintenance. At least once a month, set the control valve at approximately 150 psi and operate the pump at this pressure. Then turn the control handle counterclockwise until pressure drops. The main relief valve is now operating. Turn the control handle clockwise to raise pressure. Repeat this sequence several times to be sure the main relief valve is operating freely.

After use, remove the control handle and spring. With the fingers, thin-nosed pliers, or screw driver, rotate and move the plunger in and out to be sure it is free. Apply a thin film of oil and kerosene mixture to the exposed control valve through the opening in the upper chamber housing.

American LaFrance Relief Valve

The American LaFrance pressure-regulating valve (Figure 4-16) consists of a pilot valve assembly and a hydraulically (pump water) operated relief valve assembly.

The pilot valve assembly is mounted on the pump panel and consists of a pilot valve housing 6 (see key numbers in Figure 4-16), which contains the pilot valve 5, compensating spring 7, and four internal passages leading to tubing fittings and then through tubing to a drain, pump suction chamber, pump discharge chamber and relief valve. Between the pilot valve housing and pilot valve spring seat is a diaphragm 4, which allows the spring to act on the pilot valve, but prevents water from entering the spring housing. The load on the pilot valve spring is varied by means of an adjusting screw 2, which increases spring pressure as the handwheel 1 is rotated clockwise.

The relief valve is an integral part of the pump discharge header and controls bypassing of water from the pressure to suction chamber of the pump. The assembly consists of a churn valve 8 mounted in the discharge header casting in such a way that the outer end 9 forms a shutoff valve.

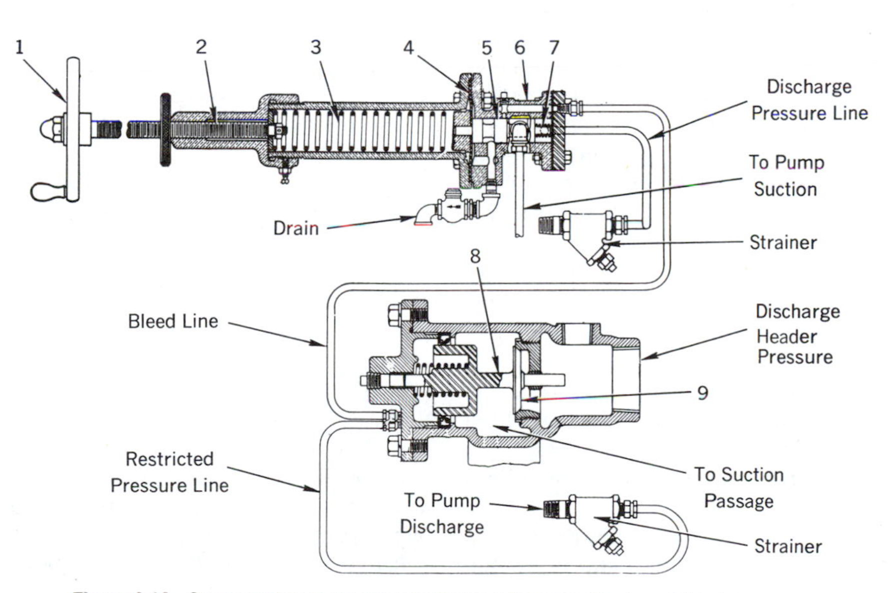

Figure 4-16. Cross section of American LaFrance relief valve (Courtesy of American LaFrance)

In normal operation, with the pump discharging through the fire hose, all of the interconnecting lines are full of water. The pilot valve and the churn valve are closed, blocking any flow of water from the pump discharge to suction. When a hose line is shut off, the resulting increase in pressure in the line from the pump discharges to the pilot valve will cause the pilot valve to move, compressing the pilot valve spring 3 until the port in the pilot valve housing is uncovered, thus allowing water at the cover of the relief valve to flow through the line between the relief valve and the pilot valve and on through the pilot valve to the pump section. This reduces pressure on the left side of the churn valve below the discharge pressure on the right side of the valve, allowing it to move to the left and to bypass water from the discharge side of the pump to the suction side; this, in effect, increases the flow within the pump and reduces its discharge pressure. The pilot valve and churn valve will settle at the required equilibrium position to maintain a pump pressure not exceeding 20 to 30 psi above the set pressure.

When a nozzle is opened, the momentary drop in pressure causes the pilot valve to move to the closed position, shutting off flow from the relief valve to the pump suction. Thus, pressure on the left side of the churn valve will increase due to water flowing in from the pump discharge. When pressure in the chamber equals the discharge pressure, the churn valve will move to the closed position due to action of the churn valve spring; this eliminates the bypass from discharge to suction within the pump and increases pressure.

To place the relief valve in operation:

Turn the handwheel clockwise to the limit of its travel and lock.

Open the throttle until desired pressure is reached.

Unlock the relief valve handwheel and turn it counterclockwise until the pump pressure gage indicates a slight drop. (On pumpers equipped for high-pressure operation the relief valve hydraulic lock must be open.)

Turn the handwheel clockwise slightly beyond the point where the desired discharge pressure is regained. Lock the handwheel in position. With the valve set in this manner, hose lines may be shut off without serious pressure rise.

After the relief valve has been set, the setting may be reduced by simply unlocking the relief valve handwheel and turning it counterclockwise until the new, lower pressure is reached.

Once set, a relief valve pressure setting can be increased only by unlocking the handwheel and turning it clockwise as far as it will go. Then the throttle can be reset to the new higher pump pressure after which, the relief valve can be reset.

Governors

The function of governors is to control the pump discharge pressure by regulating the speed of the engine.

Crown Pressure Governor

The Crown pressure governor shown in Figures 4-17 and 4-18 controls the pump discharge pressure by regulating the speed of the engine.

Governor assembly. The governor assembly is placed near the carburetor and is an integral part of the throttle linkage. It operates on the principle of balancing pump pressure against a spring-loaded piston within a cylinder. As pump pressure is increased, the governor spring is compressed, thus lengthening the governor assembly.

As long as the pump discharge pressure and the governor spring pressure are balanced, the throttle butterfly valve is held at throttle setting.

If a nozzle is shut down, pump pressure will increase. This greater pressure will further compress the governor spring, which

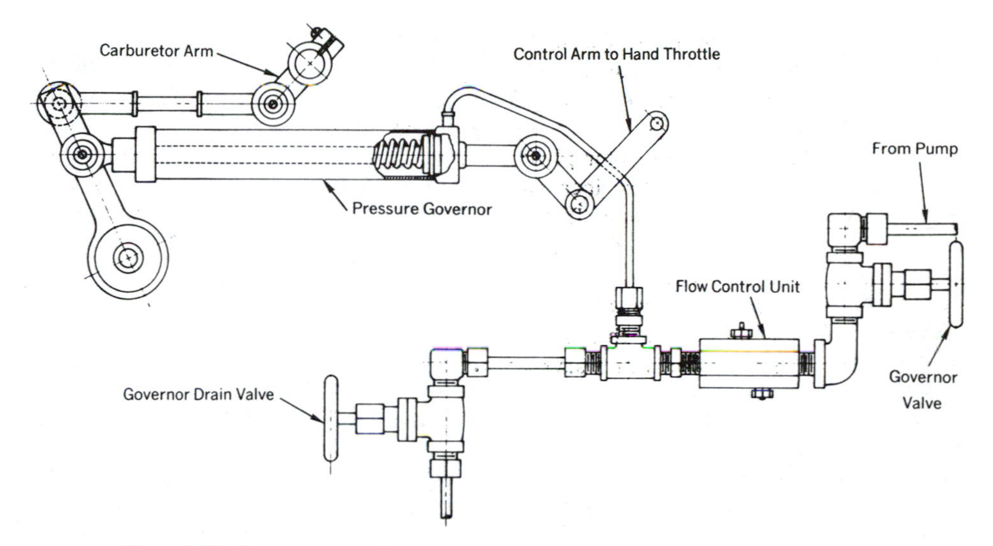

Figure 4-17. Crown governor (Courtesy of Crown Fire Coach Company)

will lengthen the governor assembly and close the throttle.

If the pump pressure decreases, the governor spring pressure will be greater than the water pressure; this will shorten the governor assembly and open the throttle.

Flow control unit. The flow control unit is placed in the line between the pump discharge and the governor assembly (see Figure 4-17). It has two passages for the flow of water; one is controlled by a check valve and the other by a needle valve. The flow control unit allows an unrestricted flow of water from the pump to the governor assembly through a check valve. A rise in the pump discharge pressure will open the check valve and close the throttle *rapidly*.

A drop in pump pressure causes the governor assembly spring to force water back toward the pump. The check valve closes, and the water flows through the orifice controlled by a needle valve. This allows the governor assembly to shorten gradually and to open the throttle *slowly*. For safety reasons and to prevent fluctuating engine speeds, or hunting, it is desirable to open the throttle slowly and close it rapidly.

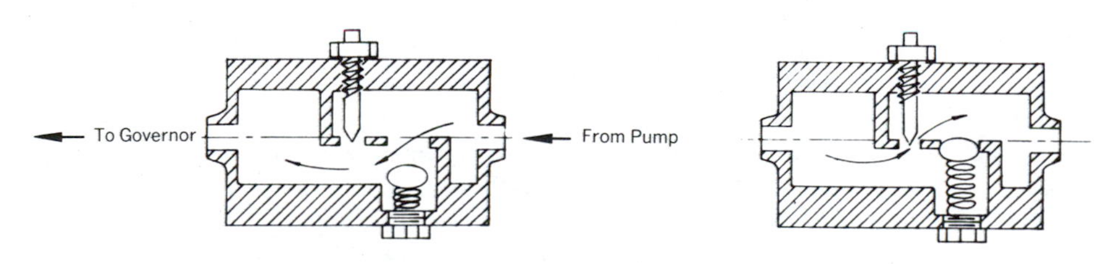

Figure 4-18. Crown governor flow-control unit (Courtesy of Crown Fire Coach Company)

Valves. Two valves control operation of the governor. The governor valve, when opened, allows water to flow from the discharge side of pump through the flow control unit to the governor assembly.

The governor drain valve, which is in the line past the flow control unit, is utilized to relieve any pressure that may be locked between the flow control unit and the governor assembly, thus keeping the governor assembly extended. This can occur if the flow control unit should become clogged or if the governor valve is closed with pump pressure in the governor assembly. Opening the governor drain valve will release pressure and allow the governor assembly to return to its normal position.

Operation. To place the governor in operation, close the governor drain valve; if both hands are utilized to operate these valves simultaneously, there will be a small flow of water through the drain line to flush out any possible sediment. If there is no discharge of water when the governor valve and the governor drain valve are both open, and there is pressure in the pump, it probably indicates that there is a blockage in the flow control unit or tubing.

With water flowing and the required discharge outlets open, advance the throttle to the desired pump discharge pressure. The governor will automatically set at this pump pressure. If working pressure requirements increase or decrease, readjustment of the throttle is all that is necessary; it is not necessary to open or close any valves.

Once the governor has been placed in operation and is in control, *the drain valve should not be opened for any reason.* Opening the governor drain valve when the governor is in control reduces water pressure being applied against the spring-loaded piston. This reduction in pressure takes the governor out of operation and causes the piston to return to its near normal position; the carburetor butterfly valve will advance to an almost full throttle position, which results in dangerously high pressures being supplied to men on the hose lines. This condition will also occur if the water supply should be exhausted, such as when pumping from the tank.

It is important that an operator place the governor in operation *before* increasing the pump discharge pressure. If the desired pressure is obtained first, and the governor is then placed in operation, the sudden increase in pressure causes the governor assembly to extend; this closes the throttle valve and results in a severe drop in pressure.

To take the governor out of operation, retard the hand throttle to its normal position, close the governor valve, and open the governor drain valve. *Always retard the hand throttle before taking the governor out of operation.*

Flow control unit adjustment. Occasionally the flow control unit may need a minor adjustment; if it is not correct, it can cause an excessive amount of engine rpm fluctuation during pumping operations. The flow control unit may be adjusted by the following procedure:

1. *Adjustment is to be made with the engine shut off.*

2. Loosen the lock nut that holds the needle valve in position.

3. Close the needle valve until it seats. *Caution*—do not force the needle valve against the seat.

4. Open the needle valve three-quarters to one full turn.

5. Tighten the lock nut. *Caution*—do not use excessive pressure when tightening the lock nut.

Maintenance. Throttle linkage should be maintained active and free at all times to insure that the governor will be operating at maximum efficiency and to facilitate driving operation.

Use chassis lubricant in the grease fitting and on the sliding rod at the frame cross members. Apply engine oil on the front and rear bell cranks and on all other joints.

Seagrave Pressure Governor

The Seagrave pressure governor (Figure 4-19) controls the pump discharge pressure by regulating the speed of the engine. This governor is of the spring-loaded diaphragm type with hydraulic remote control.

Governor assembly. The governor proper is mounted on the engine under the carburetor and consists of a throttle box and shaft, the shaft having a gear on its outer end which is engaged by teeth turned on the governor piston.

On some models, the governor is to the rear of the carburetor and operates the carburetor throttle butterfly by linkage; there is no separate butterfly valve between the carburetor and intake manifold.

The governor piston is contained in a cylinder and is held against the diaphragm by the governor spring, which normally holds the butterfly valve in a wide-open position.

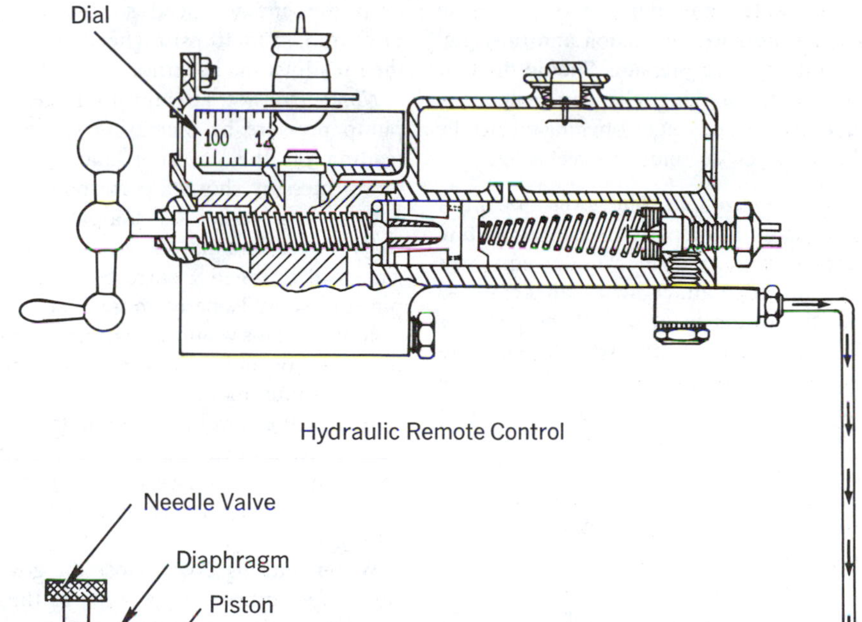

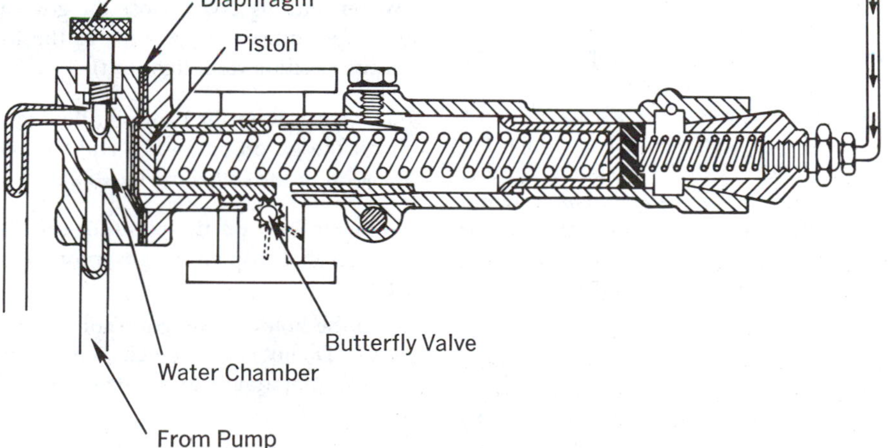

Figure 4-19. Seagrave governor (Courtesy of Seagrave Fire Apparatus, Inc.)

The governor spring is mounted at the end opposite the diaphragm in a hydraulic piston with a rubber cup similar to that in a conventional automobile wheel brake cylinder. This cylinder (called the slave cylinder), in which the hydraulic piston works, is connected by copper tubing to the remote control unit mounted on the control panel.

A water chamber on the opposite side of the diaphragm receives water from the fire pump. This water chamber is equipped with a needle valve; the point of the needle valve is ground off to permit a slight, continuous flow of water; the purpose of this is to stabilize the governor action and thus prevent a fluctuating pressure. Should the flow stop due to clogging, the obstruction can usually be flushed out by opening the needle valve for a few seconds and reclosing.

Governor control unit. The governor control unit consists principally of an automobile hydraulic brake master cylinder equipped with a threaded attachment with a crank to move the piston in or out. If the piston is moved in, it forces the hydraulic fluid out through the tubing line and into an hydraulic cylinder on the engine unit. This forces the slave cylinder forward, compressing the governor spring and thereby raising the pump pressure to the level necessary to close the governor throttle. When the control piston is moved outward, compression of the governor spring is lessened, thus lowering the regulated pump pressure. The governor control crank screw and nut attachment, which operates the master cylinder, also engages a vertical shaft on which the pressure indicating dial is mounted. The dial is calibrated to show the *approximate* pump pressure that corresponds with the governor setting. The dial carrier also acts as a commutator to control the dial light.

Operation. If pressure regulation is to be turned over to the governor, have the pump operating at 5 to 10 psi higher pressure than desired. Set the governor control dial by means of the crank to about 25 psi above operating pressure, then open the governor valve. The governor should then be adjusted by turning the crank handle so that the desired pump pressure shows on the pressure gage. (Turning the handle clockwise increases pressure; turning counterclockwise decreases pressure.)

When a change of discharge pressure is desired, it is not necessary to close the governor valve. To raise pressure, set the governor dial to about 25 psi above desired pressure, raise the pump pressure until it is 5 to 10 psi *higher* than desired, then readjust the governor. To lower governed pressure, drop the engine speed so the discharge pressure is 5 to 10 psi higher than desired, then readjust the governor.

Note: Do not attempt to increase the pump pressure by increasing the pressure reading on the governor dial unless the pump pressure shows a corresponding rise. Setting the governor dial at a pressure materially higher than that shown on the pressure gage will result in a corresponding rise of pressure should one or more hose lines be shut down. This would defeat the purpose of the governor and endanger those operators on the remaining fire streams.

To test the governor, momentarily close the governor valve and see that the engine speeds up and the pressure rises; upon opening the valve, pressure returns to its original setting.

When shutting down, close the governor valve and adjust the governor to the low-pressure position (dial light out).

Maintenance. Keep the hydraulic fluid reservoir on the governor control unit full. Use engine oil in the following locations:

1. In the hole on the governor set shaft below the dial strip on the governor control unit.

2. In the holes on the governor assembly cylinder. Do not use too much oil in the hole near the diaphragm as it may cause the diaphragm to rot.

3. At the point where the governor butterfly shaft passes through the manifold.

4. The gear on the butterfly shaft.

5. At points to keep linkage, bell cranks, and joints working freely.

Hale Governor

The Hale governor (Figures 4-20 and 4-21) is a basic design using a "balanced piston-air spring" principle to maintain relatively constant pump pressure by regulating the engine throttle.

The principle of operation can be explained by reference to Figure 4-20: trapped air pressure in the tank exerts a force A on the piston which tends to open the throttle and raise the pump pressure until it exerts a balancing force B on the opposite side of the piston. If the pump pressure tends to rise (from closing a nozzle) it will overcome force A and close the throttle. Conversely, when the pressure drops (from opening a nozzle), force A will overcome force B and open the throttle.

Following is an explanation of the function of the governor controls on the panel:

Actuator. An actuator valve controls the ON-OFF condition of the governor. If this actuator is pushed in (Figure 4-21), the governor is OFF because it allows the pump pressure to act on both sides of the piston. Consequently, the return spring will hold the throttle closed.

When this actuator is pulled out (see Figure 4-20), the governor is ON. This position of the actuator engages an O-ring which seals off the pump discharge pressure so that it can only push against one side of the piston. At the same time, air pressure is trapped in the tank which results in a constant "reference pressure" pushing on the other side of the piston.

Throttle. With the actuator pushed in, the throttle knob may be used as a conventional panel-mounted, hand throttle control (Figure 4-21). That is, backing out on the knob (turning it counterclockwise) will compress the return spring and open the throttle; turning it in (clockwise) closes the throttle.

Dampener needle. The purpose of the check valve and adjustable dampener needle valve is to minimize the tendency of the piston to override or go into a hunting or surging condition in which the engine speed will fluctuate. The check valve opens on rising pump pressure to permit a fast response of the piston in closing the throttle, thus protecting operators on the hose lines. When the pump pressure drops, the check valve closes, and water leaving the cylinder is diverted through (and around) the needle valve. The needle valve contains a short length of capillary tubing. This slows down the rate of throttle opening and prevents the piston from overriding or surging. For faster pressure recovery (or rate of throttle opening), the needle valve may be backed out (screwed counterclockwise) a few turns; however, backing out too far may cause a tendency to hunt. Do not unscrew this needle valve all the way while the pump is operating. Usually, satisfactory operation can be obtained with the dampener needle screwed all the way in.

Operating the governor for automatic control. Always start pumping with the actuator knob pushed in. With the required discharge valve open, raise the pump pressure by backing out on the throttle knob to obtain about 5 or 10 psi above the desired operating pressure. Wait approximately 3 seconds (to allow the air tank to fill up to operating pressure), then pull the actuator knob. Be sure to pull the actuator all the way out to engage the O-ring seal. Turn the throttle knob all the way in; the governor is now in operation. To change the pressure setting while pumping, back out on the throttle knob until it abuts the acorn nut. Push the actuator in and reset as described previously. To shut down, push the actuator in.

There is an alternate method of setting the governor which may be used for pumping conditions that have a slight tendency to hunt when all valves are closed. This is most likely at low power settings, e.g., with the pump in series, at low flow and pressure.

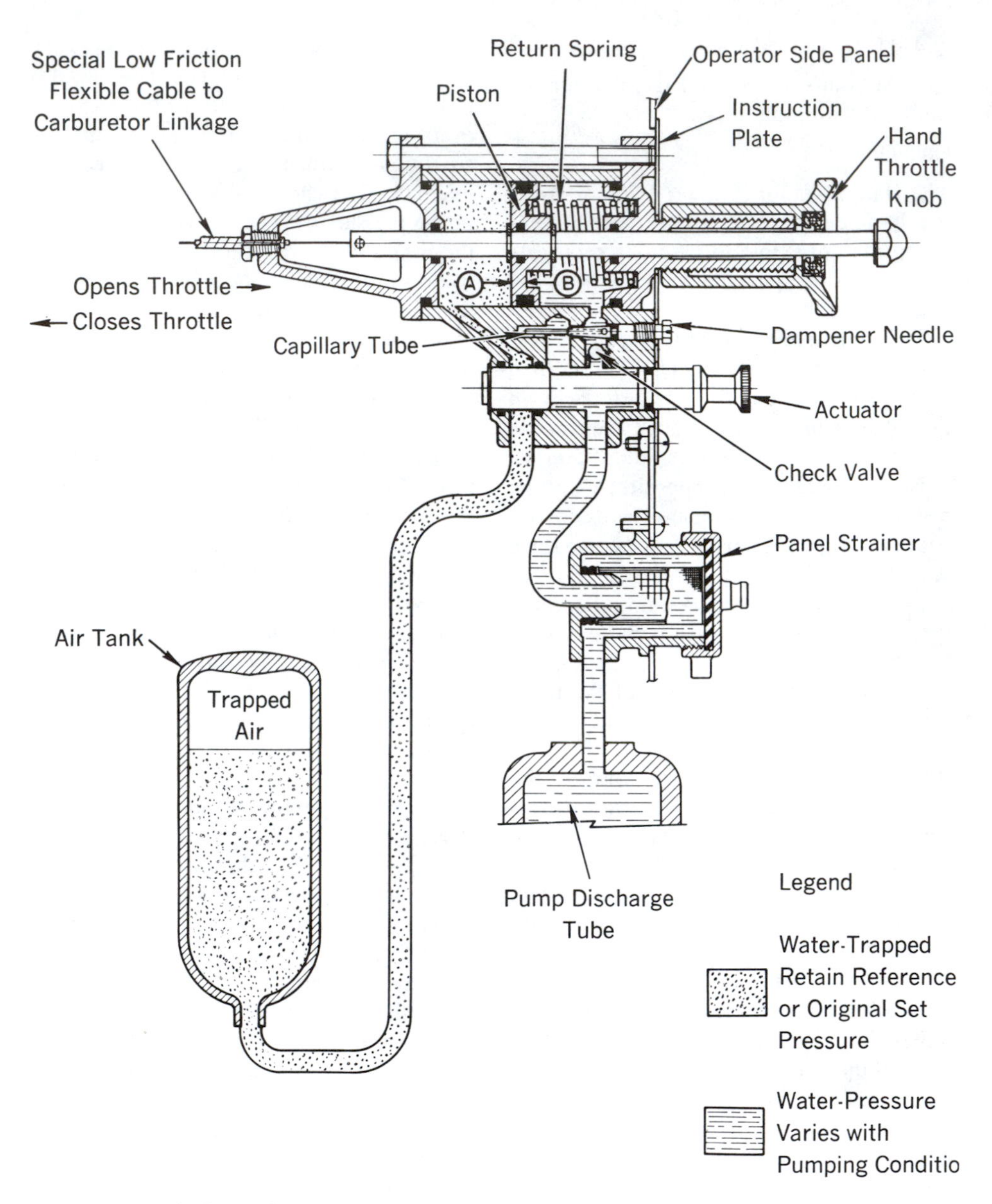

For Automatic Control

With required discharge valves open, turn out throttle knob (counterclockwise) to about 10 psi above desired operating pressure. Wait approximately 3 seconds, then pull out actuator knob. Turn throttle knob in. Governor is now operating on automatic control.

To shut down, push actuator knob in.

Figure 4-20. Hale governor in operation (automatic control) (Courtesy of Hale Fire Pump Company)

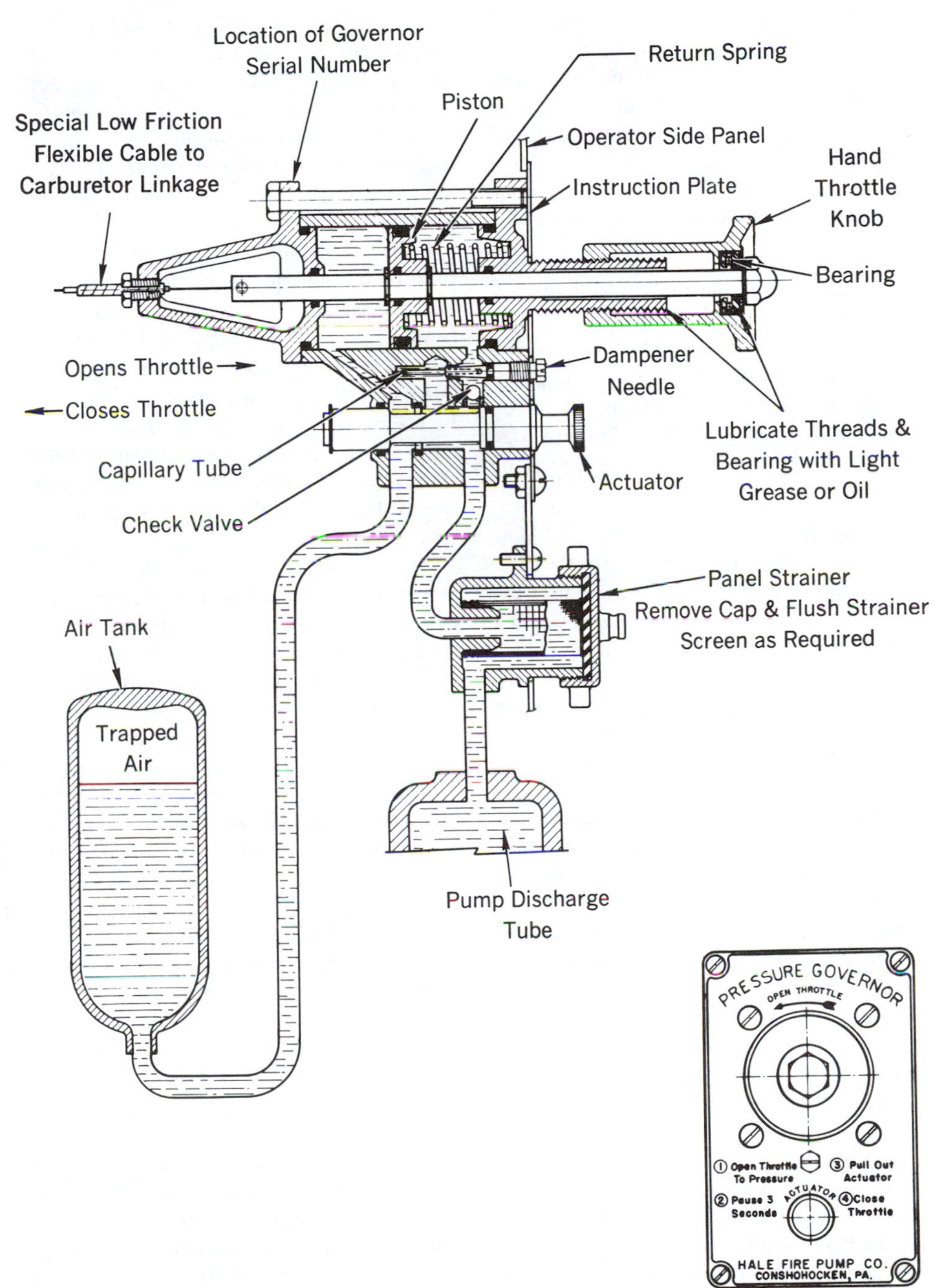

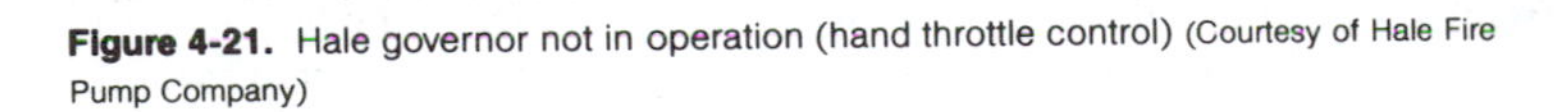

Figure 4-21. Hale governor not in operation (hand throttle control) (Courtesy of Hale Fire Pump Company)

With all the discharge valves closed, and the actuator in, open the throttle to about 10 or 15 psi above the desired operating pressure. Pause 3 sec and pull out the actuator (do not change the hand throttle setting); open the discharge valves. The governor is now engaged and will automatically open the throttle to maintain pump pressure.

Pressure rise. There is no minimum or maximum pressure rating on this governor; it will operate over the entire normal working range of the fire pump. The pressure rise should be from about 5 to 30 psi, depending on the pumping conditions. Generally speaking, high operating pressures and large changes in the throttle setting will give the highest pressure rises.

Servicing. Dirt blockage causes very slow pressure recovery. If the hand throttle knob should become difficult to turn, remove the dampener needle (after shutting down the pump and draining systems to relieve pressure) and check for dirt blockings by running a 5 gage (0.014 in.) or smaller wire through the capillary tube. (This size wire can usually be obtained from a wire brush.) Care should be taken to relieve the pressure completely before removing the dampener needle. If this is not done, the check valve ball may be blown out of the dampener needle hole by pump pressure when the dampener needle is removed for cleaning. Lubricate threads on the hand throttle knob and ball bearing under the acorn nut occasionally with light grease.

When starting to pump (with the governor locked out), the hand throttle knob may be difficult to turn rapidly at low pressure. This is a natural condition caused by the time delay required to fill the air tank and to bring the piston into a balanced pressure condition. Therefore, in opening the throttle, the first few turns should be taken gradually; as the tank pressure increases, the hand throttle knob will turn more freely.

If it requires considerably longer than three seconds for the air tank to fill to operating pressure, the governor panel strainer may be clogged. Remove the panel strainer cap, pull out the strainer screen and flush until this screen is clear. To replace, make sure two O-rings are on the spool in the strainer body, then slide the strainer screen over the spool and replace the cap. *Do not operate the governor without this 60-mesh screen strainer in the circuit.*

If there is a gradual drop in pressure while running the governor, check for a leak in the air tank circuit. This situation could result from a worn or damaged O-ring on the piston or in the actuator stem bore.

An excessive pressure rise could be caused by an excessive amount of water in the air tank. When the pump is not operating, there should be no water in the tank. An occasional draining of the tank may be necessary.

During freezing weather, the governor and its piping system will require draining. Open the draincock on the bottom of the air tank to drain the tank. With the main pump drains open, moving the hand throttle knob to its extreme position will help drain the governor body. Be sure to close all drains when finished.

When pumping with the *governor operating*, pushing the actuator knob in should close the engine throttle to idle. If this does not happen, there is excessive friction in the mechanism. Clean and oil the throttle linkage and check for kinks in the throttle cable. It may be necessary to disassemble the governor and clean the cylinder bore. A light film of oil (SAE 10 or 20) on the bore and on the piston stem will help reduce friction.

American LaFrance Governor

The American LaFrance governor (Figure 4-22) consists of a water chamber 3 connected to the discharge side of pump 1, a piston 4, a threaded rod connected to the piston 6 and a spring 5. The governor actuates the carburetor's auxiliary butterfly valve 9 on the engine by means of plunger 11 acting on the threaded rod through the governor adjusting link 7.

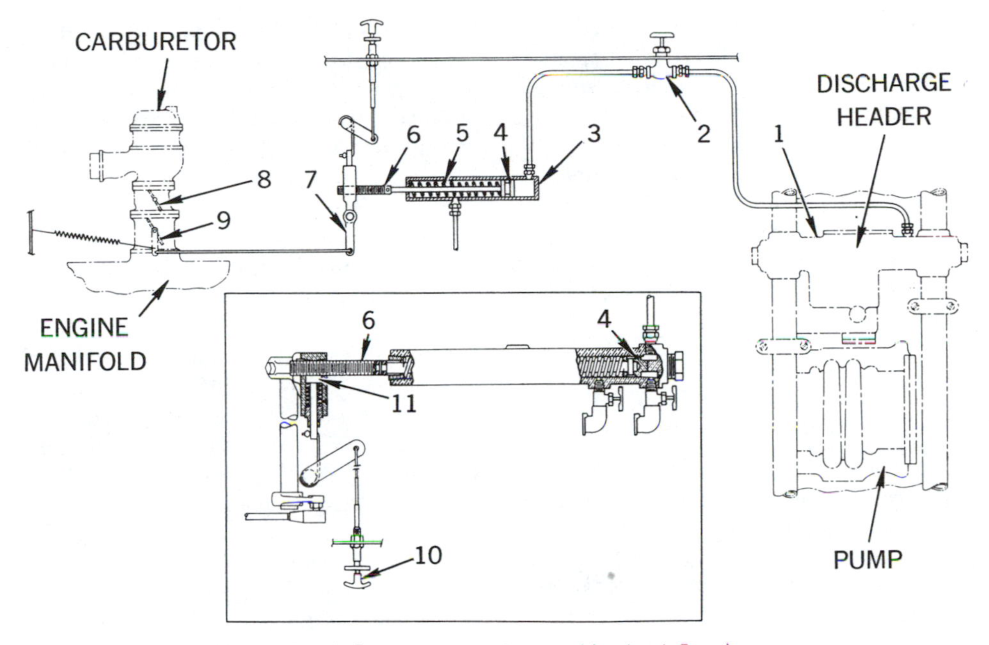

Figure 4-22. American LaFrance governor (Courtesy of American LaFrance)

To place this governor in operation, disengage the governor from the auxiliary butterfly valve by pulling out on the governor control handle 10; this disengages the plunger and threaded rod. Open the governor shutoff valve 2. With the discharge gates open and the pump discharging water, set the pump at the desired discharge pressure by gradually adjusting the engine throttle 8. Water pressure moves the piston, compresses the governor spring, and causes the threaded rod attached to the piston to be extended. The connecting link to the carburetor butterfly valve is engaged by pushing in the governor control handle at the panel. This releases the plunger and reengages it with the threaded rod so that any movement of the governor piston makes a corresponding movement of the connecting lever. If the discharge pressure is to exceed 300 psi, close the governor shutoff valve; the governor will not control pressure in excess of 300 psi.

When a hose line is shut off and the governor is set, there will be a slight pressure rise in the pump; this will cause the governor piston 4 to further compress the governor spring 5 and extend the governor shaft further out of its housing. By this time, the governor shaft is connected to the governor throttle linkage so that any movement of the piston is transmitted to the throttle. Thus a pump pressure increase will be translated into a movement of the throttle linkage; this will result in a slight closing of the throttle 9 until stabilized pressures are reached, which are slightly above the governor's original set pressure.

Conversely, if an additional discharge gate is opened, a momentary pressure drop occurs which causes the governor spring 5 to move the piston to the left; this opens the throttle and increases the pump pressure back to its set value.

Upon completion of pumping, lock the governor control handle in the out position before locking the throttle. This will prevent damage to the threaded rod and plunger due to their being disengaged while the plunger is holding the threaded rod extended with no pressure in the water chamber.

Following every pumping operation, lubricate the governor by opening the oil cup pressure cocks and filling them with light oil. After the oil has run from the cup into the governor, close the oil cup pressure cocks tightly.

Waterous Automatic Pump Pressure Controller

The Waterous pressure controller is a mechanical-hydraulic engine governor that will maintain a constant pump pressure at any desired setting between 75 and 500 psi, thus protecting nozzle operators from sudden pressure surges. The controller works equally well with gasoline and diesel engines. Since it is connected to the engine speed control (throttle) linkage, it reduces engine and pump wear because the engine runs just fast enough to deliver the desired pressure. When a high-pressure surge occurs, the controller cuts engine and pump speed down to maintain the desired pressure. A simple ON-OFF valve places the system in or out of operation.

An important feature of this pressure controller is the provision for automatic shutdown of the engine if the pump should lose pressure. This is accomplished by the directional flow valve that automatically reverses its position when the pump pressure falls below 30 psi. This action diverts the control pressure stored in the accumulator to the opposite side of the piston, thus closing the throttle to the idle position. The system consists of three basic components or assemblies: the control assembly, the actuating cylinder assembly, and an accumulator.

Control assembly. The control assembly mounts on the pumper panel; it includes the automatic directional control valve assembly, an ON-OFF valve, a sensitivity control valve mounted behind the panel, which normally needs adjustment only when first installed, and a fine mesh Y-type strainer to prevent sand or other contamination from entering the system.

Actuating cylinder assembly. The actuating cylinder is mounted in the engine compartment; it contains a movable double-acting piston and rod assembly that connects to the throttle linkage. The piston has accumulator control pressure on one side, which tends to open the throttle, and pump pressure on the other side, which tends to close the throttle. The cylinder is Teflon-lined to reduce friction and increase response sensitivity.

Accumulator. The accumulator reservoir, which is a 1-gal tank, contains a rubber diaphragm to separate the water from the gas. It is precharged at the factory with nitrogen to a pressure of 75 psi (minimum control pressure) and, unless a leak develops, it never requires any maintenance; if recharging is needed, it can be accomplished by using compressed air. Because a bladder separates the gas from the water in the accumulator, the gas cannot become dissolved in the water; this would result in a slowly drifting discharge pressure and, ultimately, a waterlogged accumulator.

Operating instructions. All the pump operator has to do to place this governor in operation is to accelerate the engine with the hand throttle to obtain the desired discharge pressure, turn on the pressure controller, and then pull the hand throttle out all the way. For operating at all pressures and capacities, use the following procedure:

1. Turn the ON-OFF valve to OFF.
2. Prime the pump and begin discharging water.
3. Adjust the discharge pressure to the desired setting with the hand throttle.

4. Wait 5 to 10 sec for the pressure to stabilize in the accumulator.

5. Turn ON-OFF valve to ON.

6. Slowly open the manual throttle control all the way.

If the pump loses pressure during operation and the engine returns to idle speed, use the following procedure to return the pump to operation:

1. Close the hand throttle fully.

2. Follow the above operating procedure, starting with Step 1.

To disengage the pump or to reset the discharge pressure:

1. Close the hand throttle until the pressure stabilizes at a pressure below the set pressure.

2. To disengage, turn automatic control OFF, and adjust the throttle control to the desired pressure.

3. To reset, disengage and repeat steps 4, 5, and 6 above.

Operating principles. In the following discussion of the operation of the Waterous automatic pump pressure controller, the components are keyed to the numbers in Figures 4-23 through 4-27. The numbers in all these figures are keyed to the legend in Figure 4-23.

After the pump has been primed and is operating (Figure 4-23), water flows from the pump through line 6 and strainer 16 to the bottom port of the directional flow valve (DFV) 1, through the lower cross holes in spool 15, and out lines 7 and 8. Water flows through line 7 to the throttle end of the cylinder, through line 8 and the ON-OFF valve 5 to the side port of the DFV. Through internal porting in the DFV, the water then flows to needle valve 4, and through line 11 to the spring end of the cylinder. With the ON-OFF valve 5 in the OFF position (open), all parts of the controller are equally pressured, and the piston in the cylinder remains stationary relative to the cylinder.

When the pump pressure reaches 75 psi, it moves the spool in the DFV upward against the spring force, thus changing the flow pattern. Because the accumulator is precharged to 75 psi, water flows into it when the pump pressure exceeds that figure.

After spool 15 shifts upward, water flows from the pump through line 6 to the bottom port of the DFV, through the lower cross holes in the spool, out line 9, through the ON-OFF valve 5. The water then flows through line 8 back into the top side port of the spool, out the top of the spool, through line 10, and into the accumulator.

When the desired pressure is reached, the pressure on the water side of the accumulator will equal the discharge pressure. Turning the ON-OFF valve 5 to ON will activate the system for automatic control; it actually closes the valve, thus isolating the accumulator 2 and the throttle side of the piston 13 from discharge pressure on the opposite side of the piston. With the throttle opened fully, piston 13 moves in the closed direction relative to the cylinder assembly; this is due to the momentary increase in pressure when opening the manual throttle control. The piston will come to rest at some intermediate point between the two heads when the hand throttle is fully opened. The discharge pressure at this time will equal the pressure that has been trapped in the accumulator (Figure 4-24).

If a discharge gate is closed, or discharge pressure rises for any reason, the pressure will rise on the tube side of the piston; it will remain the same as that in the accumulator on the throttle side. The higher pressure will force the piston toward the closed position, and will consequently throttle the engine down to keep the set pressure (Figure 4-25).

If a discharge gate is opened or if the pump pressure drops for any reason, the pressure on the tube side of piston 13 will drop below the accumulator pressure. The

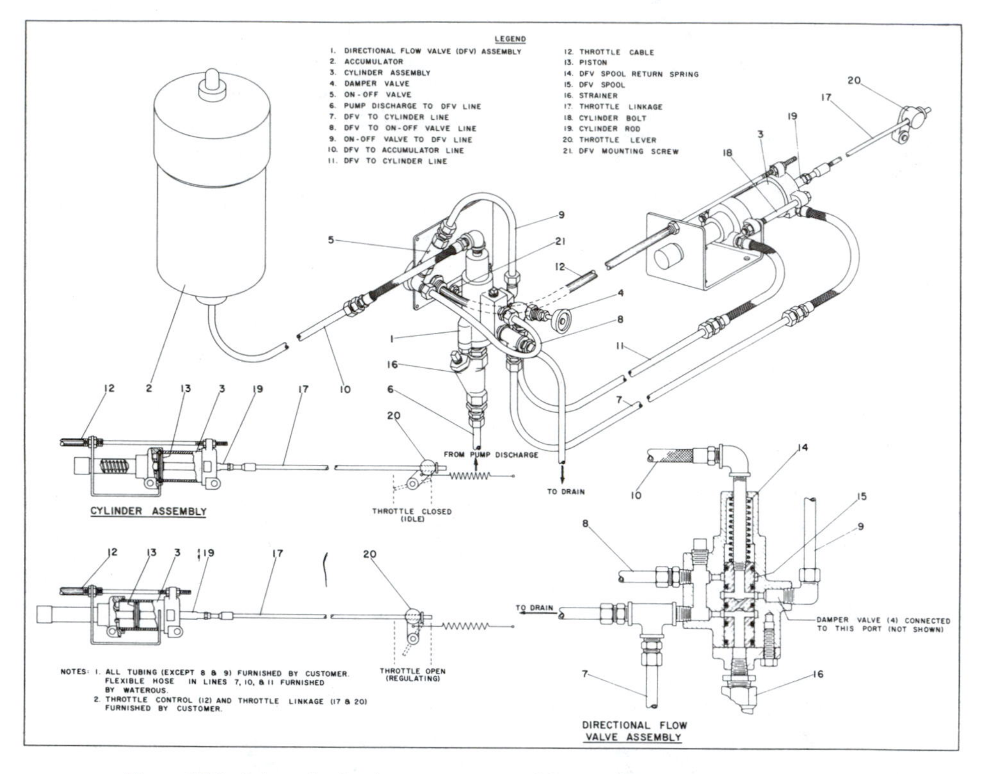

Figure 4-23. Schematic showing pressure controller operation (Courtesy of Waterous Company)

higher force on the opposite side of the piston will then move it toward the tube, and will open the engine throttle until the discharge pressure again equals the preset accumulator pressure (Figure 4-26).

If the pump runs out of water during operation, or loses its prime, or for some other reason the discharge pressure drops below 30 psi, spring 14 will force spool 15 down to the deactivated position. In this position, the passageways through the DFV connect the accumulator 2 with the tube side of piston 13 and the throttle side of the piston with line 6 from the discharge. Since the discharge pressure is zero and the accumulator is still pressurized, the piston and throttle will move toward the closed position, and the engine will return to idle speed (Figure 4-27).

Maintenance. If the governor does not operate properly, the cause can usually be traced to fine sand, grit, or other foreign material in the cylinder assembly or the DFV. Remove the strainer and clean it. If the system still appears to stick or does not react properly, pump *clean* water through the mechanism. With the control valve OFF, pressurize the pump to 200 psi, wait for the accumulator to become pressurized, then decrease the pressure to 50 psi. Do this a number of times to flush out the system.

If the governor still does not react, pressurize the system to 125 psi, and turn the control valve to ON; then open the throttle fully. If the pressure does not remain constant when the throttle is opened fully, then the piston is probably stuck in the cylinder assembly and must be disassembled.

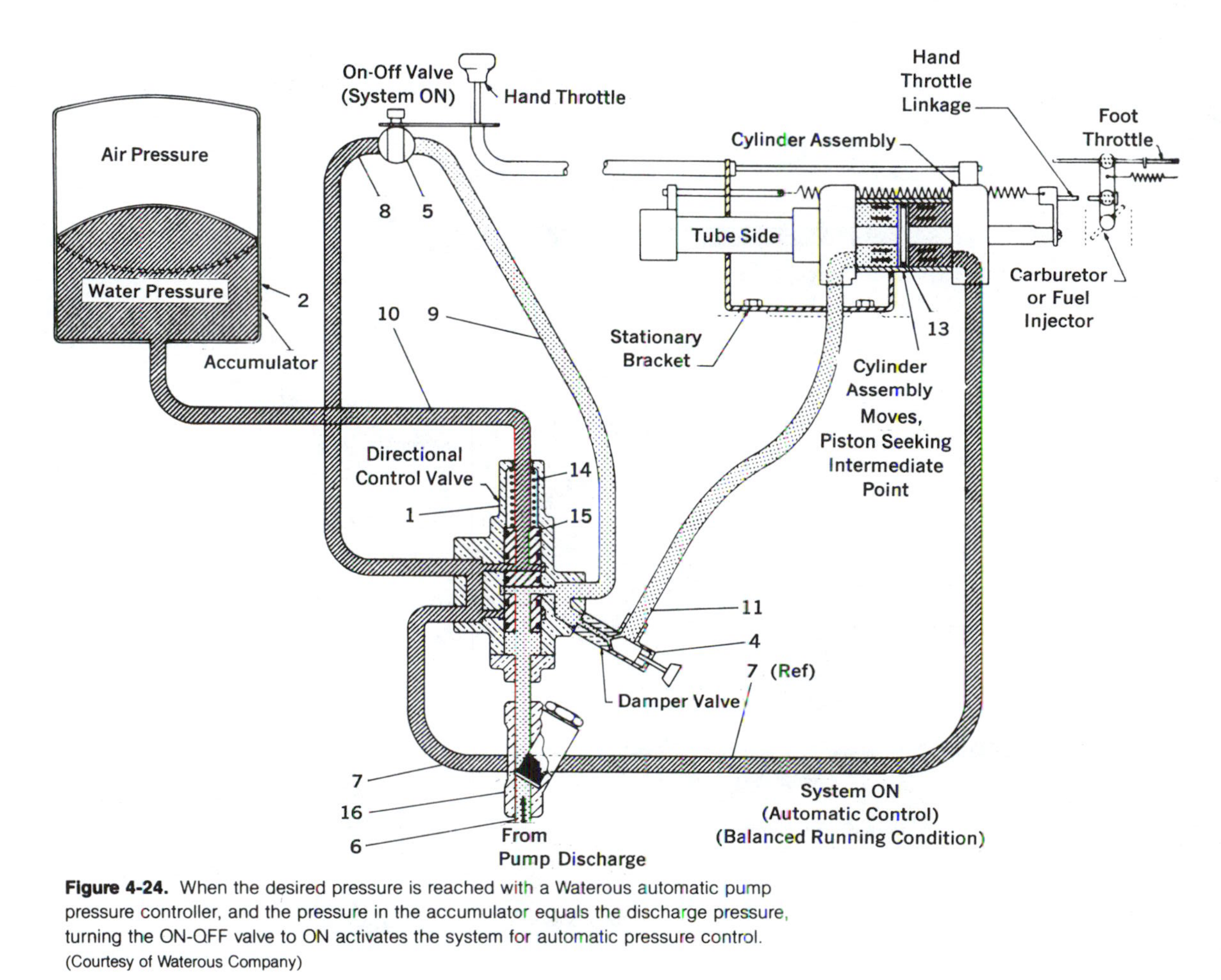

Figure 4-24. When the desired pressure is reached with a Waterous automatic pump pressure controller, and the pressure in the accumulator equals the discharge pressure, turning the ON-OFF valve to ON activates the system for automatic pressure control. (Courtesy of Waterous Company)

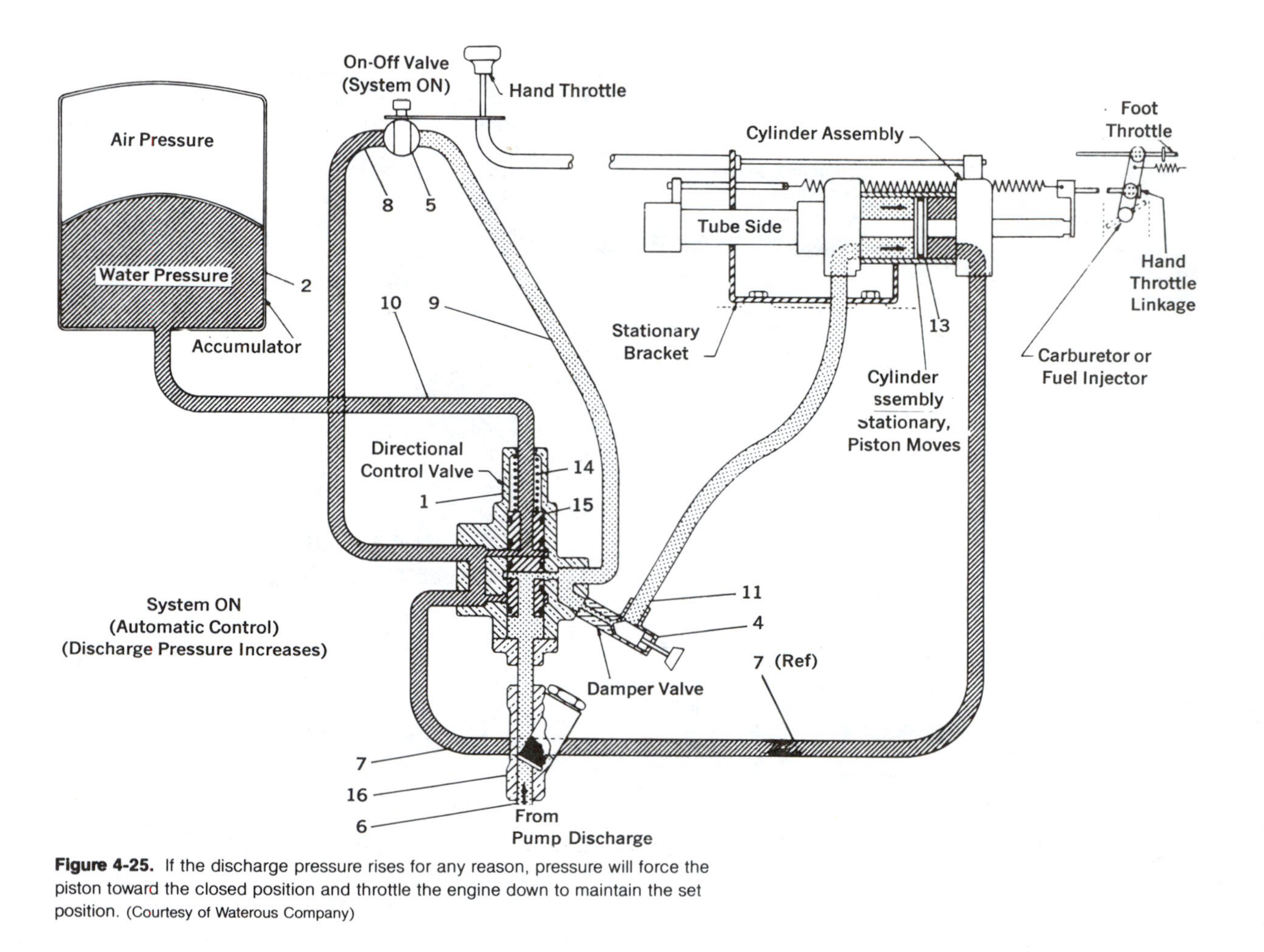

Figure 4-25. If the discharge pressure rises for any reason, pressure will force the piston toward the closed position and throttle the engine down to maintain the set position. (Courtesy of Waterous Company)

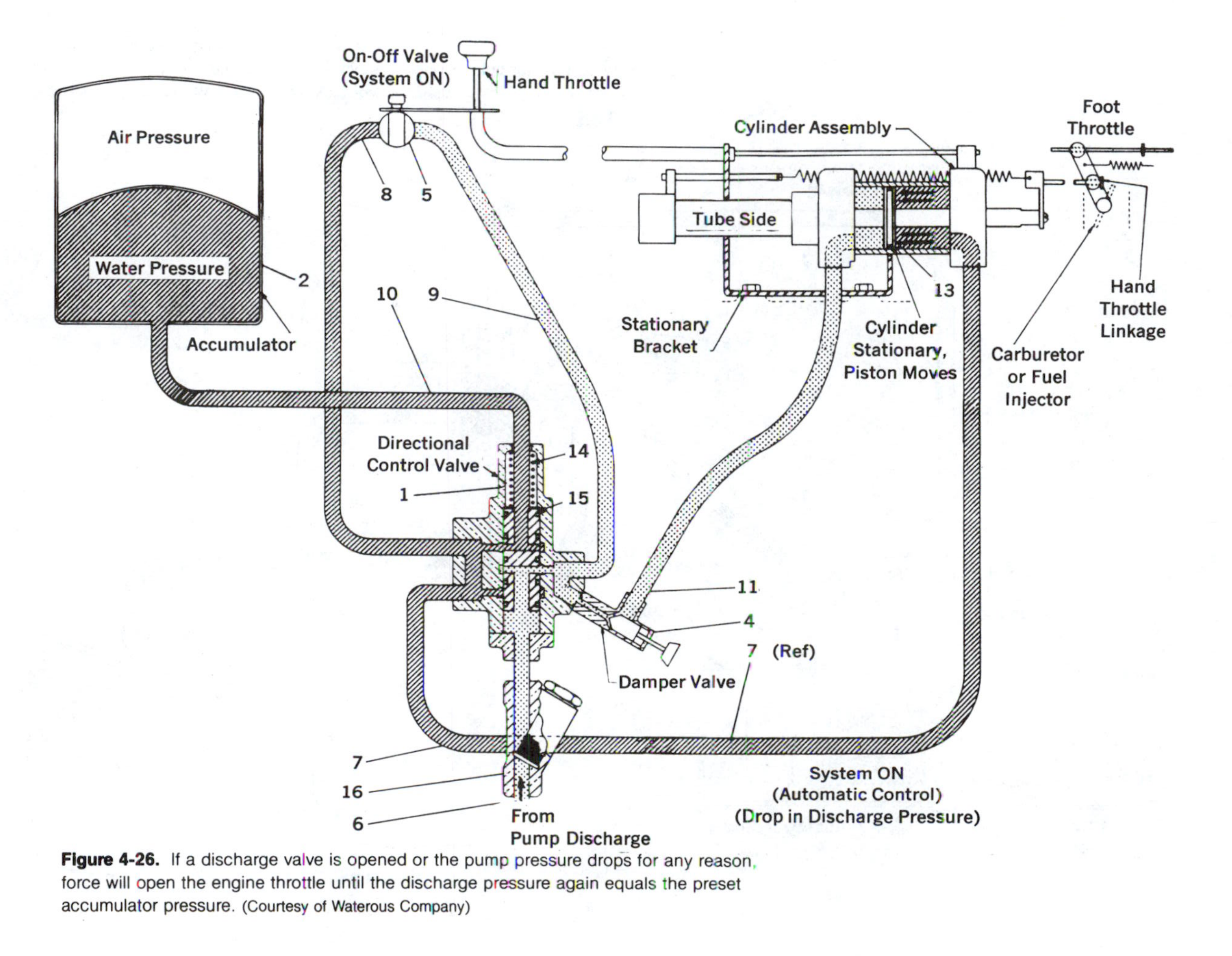

Figure 4-26. If a discharge valve is opened or the pump pressure drops for any reason, force will open the engine throttle until the discharge pressure again equals the preset accumulator pressure. (Courtesy of Waterous Company)

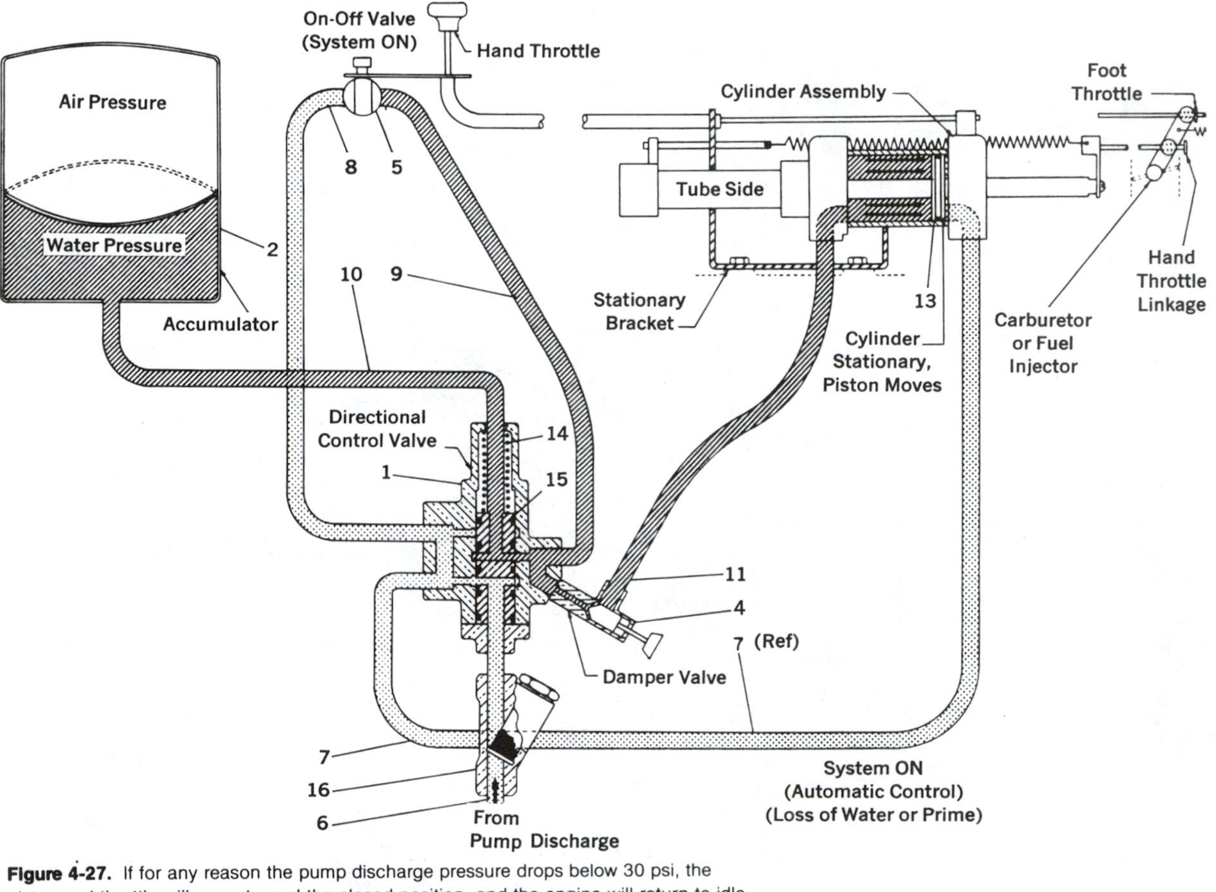

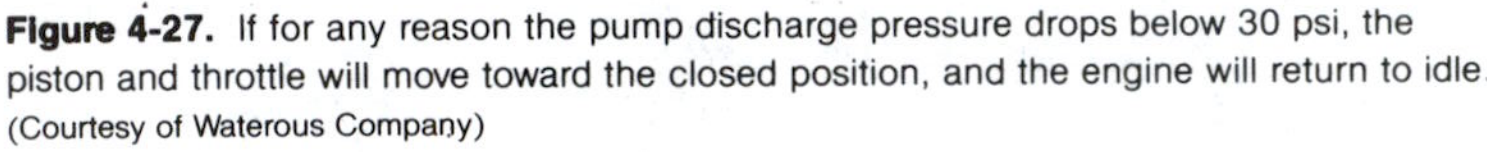
Figure 4-27. If for any reason the pump discharge pressure drops below 30 psi, the piston and throttle will move toward the closed position, and the engine will return to idle. (Courtesy of Waterous Company)

If the pressure slowly rises or drops during operation, there is probably an internal leak past the piston in the cylinder assembly; this is not readily apparent as it cannot be seen. This kind of leak can also cause an excessive pressure rise or drop when opening or closing discharge valves.

Cleaning. After prolonged pumping from draft, especially with dirty water, enough sand and dirt may accumulate to prevent movement of the DFV spool or cylinder assembly piston. Once a year, or as often as necessary, disassemble and clean the DFV and cylinder assembly. Lubricate the parts and O-rings with grease when reassembling.

American LaFrance "Pressurmatic" Governor

The "Pressurmatic" governor throttles the engine to vary the speed of the fire pump impellers to maintain a constant pump pressure from 90 to 300 psi, as preset by the pump operator. It will work equally well on both gasoline and diesel powered pumpers.

The engine and pump are protected against damage caused by a runaway condition if the water supply is reduced or lost. A hydraulic clutch connects the governor control to the throttle linkage; it is designed to hold engagement, once set, for the entire pressure range of 90 to 300 psi. If the water supply should be interrupted by cavitation or any other reason and drop the pump discharge pressure to 50 psi, even momentarily, the governor clutch will disengage and the engine will slow down to idle speed. The clutch is spring loaded to disengage completely when it is not pressurized; this allows free movement of the engine throttle arm during vehicle or pump operation prior to placing the governor in control.

System components. The governor consists of three basic components: the control valve on the pump operator's panel, the balancing cylinder and clutch assembly, and the reference pressure cylinder. The reference cylinder stores water from the fire pump discharge at the desired pressure; this water is trapped between air at the top of the reservoir and the reference side of the governor cylinder. The air in the reservoir acts as a spring or cushion to impart movement to the balancing piston.

Operating principles. Figure 4-28 shows the governor before it is placed in operation by opening the shutoff valve.

Figure 4-29 shows the governor shutoff valve open. The control valve on the pump operator's panel is pressurized and water under pressure moves the piston in the balancing cylinder to the lowest throttle position.

Figure 4-30 shows the control valve has been moved from the OFF to the SET position. This setting allows water flow to the reference pressure reservoir, pressure gage, and the reference end of the cylinder.

Figure 4-31 shows the governor in full control. As soon as the pump discharge gage and the governor gage show the same pressure, move the control valve to the RUN position and close the engine throttle. Moving the control valve to the RUN position cuts off the pump discharge pressure to the reservoir and to the reference end of the balancing cylinder. It opens the line for pump pressure to act through the control valve and lock the governor clutch; the engine throttle linkage is now controlled by the movement of the governor cylinder piston. The piston now has pump discharge pressure on one side and reservoir reference pressure on the other side. Any change in the pump discharge pressure, up or down, moves the piston to balance the pressure; this movement is transmitted to the engine throttle.

Placing governor in operation. The entire governor system must be drained after use and maintained dry. The control handle should be carried in the OFF position and the governor shutoff valve and the drain valves should be closed.

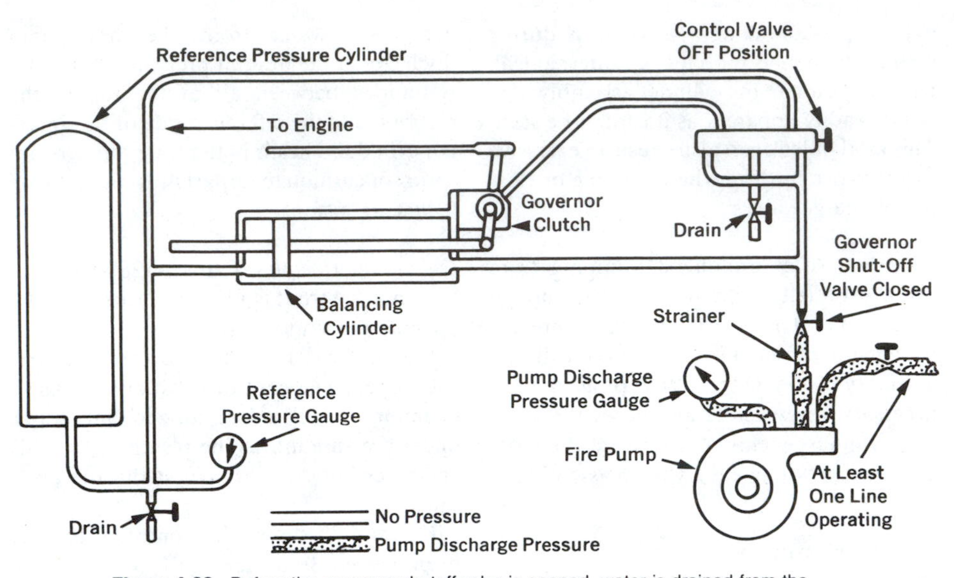

Figure 4-28. Before the governor shutoff valve is opened, water is drained from the entire governor system and the clutch is disengaged. (Courtesy of American LaFrance)

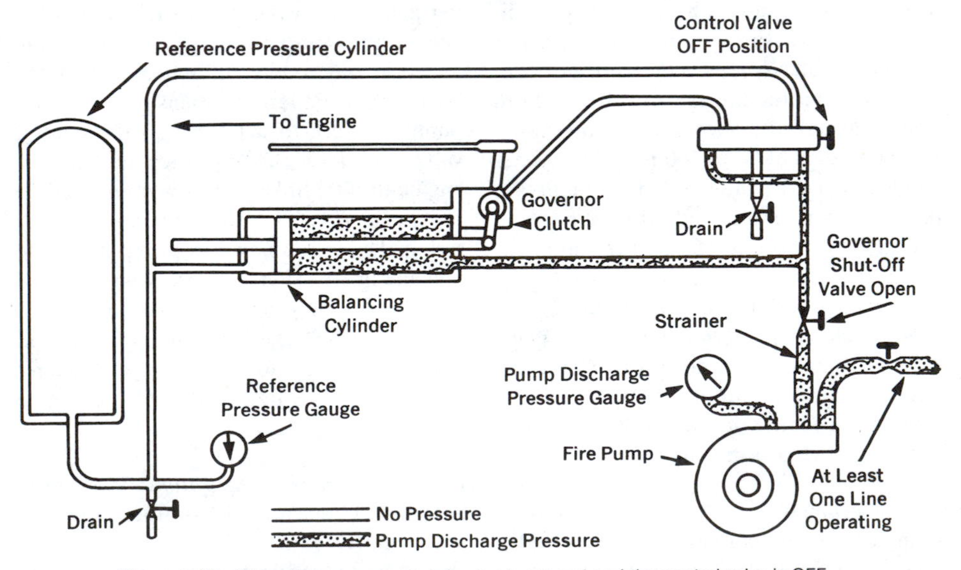

Figure 4-29. When the governor shutoff valve is opened and the control valve is OFF, pump discharge pressure flows to the balancing cylinder. (Courtesy of American LaFrance)

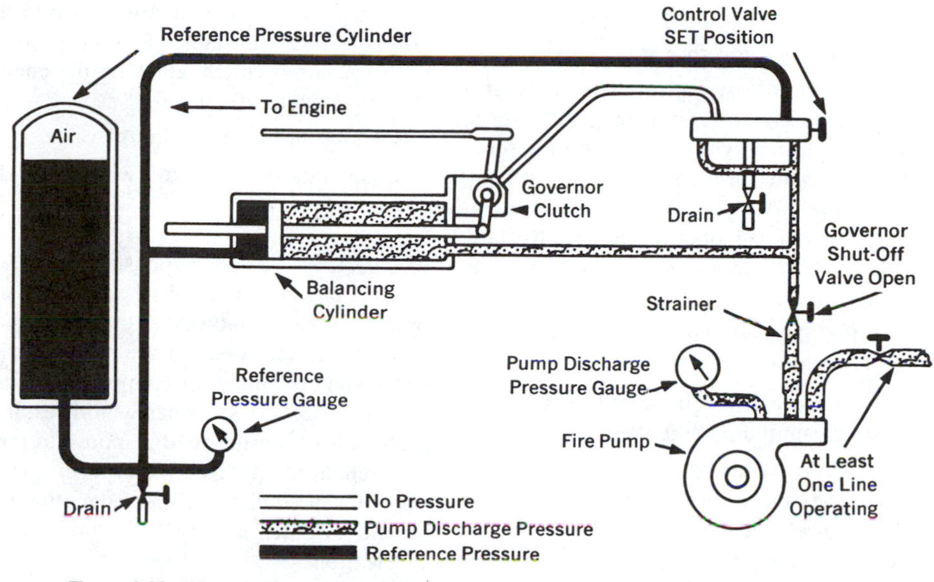

Figure 4-30. When the control valve is placed in the SET position, pressure in the reference cylinder becomes identical to the pump discharge pressure. (Courtesy of American LaFrance)

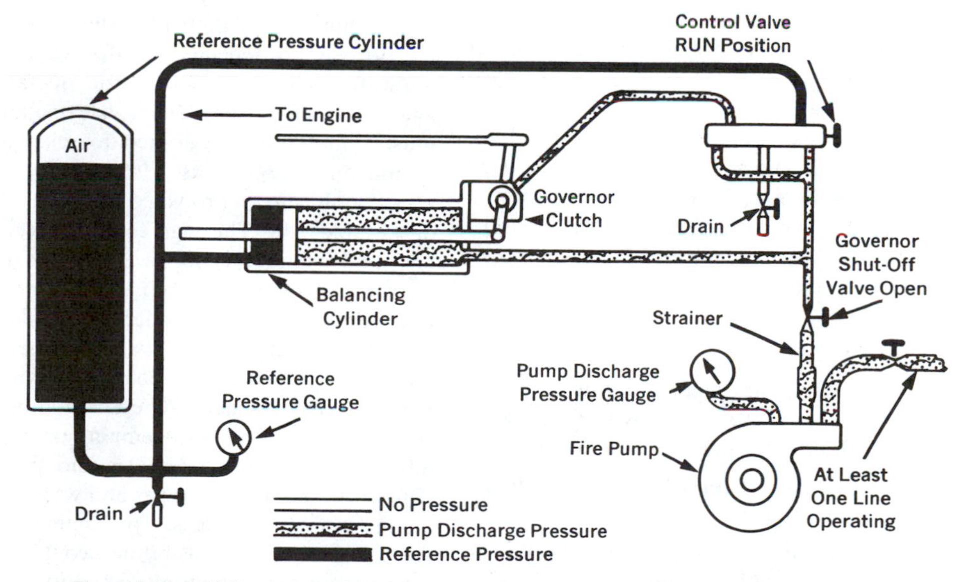

Figure 4-31. With the control valve in the RUN position, the governor clutch is engaged, and the cylinder piston movement is transmitted to the throttle linkage. (Courtesy of American LaFrance)

1. Place the fire pump in operation.

2. Open the governor shutoff valve.

3. Open at least one discharge gate. It is important that at least a small quantity of water is flowing when the governor is set to keep the water in the pump from overheating. If the nozzles are closed, and no water is being discharged, open a bleeder valve, tank fill, or other valve to cause some water to flow.

4. Pull the control handle to the middle, or SET, position.

5. Increase the engine speed to obtain the desired pump discharge pressure.

6. When the reference pressure gage reading is the same as the pressure shown on the pump discharge gage, pull the control valve handle to the RUN position.

7. Push the engine throttle control in. This closes the hand throttle so that if the pump discharge pressure drops below 50 psi, the governor clutch will disengage and the engine speed will drop to idle.

Changing the pump discharge pressure. The pressure may be increased or decreased at any time.

1. Increase the engine speed until resistance is felt or governor pressure gage shows a change. This will hold the throttle setting when the governor clutch is disengaged, holding the engine and pump speed at the present operating position.

2. Push the governor control valve in to the middle, or SET, position.

3. Adjust the engine throttle to either increase or decrease the discharge pressure until the desired pressure is attained.

4. Pull the control handle fully out to the RUN position.

5. Push the engine throttle fully in. The governor is again in control.

Shutdown procedure. When through pumping:

1. Push the governor control valve to the OFF position. This cuts off pump pressure to the governor clutch, allowing the engine to slow down to idle speed.

2. Close the governor shutoff valve.

3. Open drains to remove water from the governor system.

Maintenance. A slow response to maintain the set pressure when discharge lines are opened or closed indicates dirt or corrosion is present in the system. Care must be exercised when cleaning any components of the governor because scratches would result in some loss of sensitivity of response to pressure changes. A light waterproof grease should be used in the control valve; the seals in the cylinder assembly do not require lubrication.

The strainer in the line from the fire pump to the governor control should be cleaned monthly.

Automated Flow Control

Conventional regulation of water flow on the fire ground depends on effective communications between the nozzle operator and the pump operator. The nozzle operator must communicate the desired flow changes to the engineer who is permanently stationed at the pumper by voice, hand signals, or portable radio. The engineer must receive and understand the nozzle operator's request and then manually adjust the pumper to the desired flow rate. Under conditions of high background noise, distance, cluttered radio traffic, and darkness, effective communications can be difficult at best. In addition, the pump operator must remain stationed adjacent to the radio receiver and pump controls to monitor messages and watch the engine and pump gages. The engineer, if freed from these responsibilities, could perform other vital fire fighting tasks during the critical initial attack, such as manning a second attack line, aid in stretching hose lines, securing an additional water supply, assist in raising ladders, or any other of the multitude of duties which are required on the fire ground.

Nozzle Pump Operator System

The Grumman nozzle pump operator (NPO) system allows the nozzle operator on two of the hose lines to control the flow. This device consists of four major subsystems: the governor; flow control valves; communication system; and display and control panel.

The governor is an air-operated unit to control the engine speed for maintaining the pump discharge pressure between 0 and 300 psi.

Flow control valves on two of the pumper outlets control the water flow rate and pressure from 0 to 310 gpm in seven increments of approximately 45 gpm as directed by the communication system.

The communication system consists of two parts, the transmitter and the receiver/decoder. The transmitter is located between the last section of hose and the nozzle; it is operated by turning a spring-loaded plastic sleeve left (counterclockwise) about 35° and then releasing the sleeve to request a 45-gpm increase in flow; turning the sleeve right (clockwise) will send a decrease flow command. The sleeve must be allowed to spring back to the center position each time before transmitting another signal. The transmitter has a manual safety lock which, when engaged, prevents accidental flow commands. The receiver/decoder picks up the transmitter signal and relays the correct command to the flow control valves.

The display and control panel monitors the status of all subsystems with the use of both audio and visual displays. Lights indicate the amount of flow to each hose line and register malfunctions in four areas: overspeed of the engine, low oil pressure, low electrical output, and excessive engine temperature. Any malfunction in these areas causes the warning light on the panel to glow, thus designating the problem area. It will also sound the audible alarm and cause the radio-controlled hose streams to pulsate. In addition to the above alarms, when the engine overspeed warning is operating and the condition is not corrected, the system will automatically reduce the flow to each radio-controlled hose line by 45 gpm every 6 sec until the condition is corrected. The most common causes of engine overspeed are pump cavitation occurring when the fire stream demand exceeds the water supply, an air bound pump, or an empty booster tank. Therefore, the NPO system allows the nozzle operators to control the flow in two of the hose lines, and it will alert them or anyone else in the immediate area of any malfunction in the water supply or the pumper operation.

Automatic Pumper Radio Nozzle Control

The automatic pumper radio nozzle control system, manufactured by the Fire Research Corporation, consists of a governor, computerized monitoring system, radio receiver, nozzle control transmitters, and electrically controlled discharge valves. The complete modular system is quite simple. The nozzle operator has a 2-mi range transmitter mounted in his helmet. This helmet has two switches on it: press the left one and the pressure decreases; press the right switch and the pressure increases.

Back at the pumper there is a radio to receive the signals and operate the ball discharge valves as desired by the nozzle operator. Using the panel control, the engineer can electrically control the engine speed and the position of each discharge valve or transfer control over to each nozzle operator.

The engine is monitored constantly for problems with the cooling, electrical, or lubrication system. If there is an engine malfunction, an 8-in. bell will commence to ring with a distinctive sound to warn everyone of the problem.

The Fire Research Company also manufactures a radio-controlled hydrant valve which allows the pump operator to open and close the valve from a position at the pumper.

(Courtesy of the Chicago Fire Department)

5 Fire Fighting Hydraulics

Every fire fighter, whether the highest ranking chief officer or a new recruit, should possess a good practical working knowledge of hydraulics so that fire fighting tactics will have a sound basis and fire streams will be both safe and effective. Although many industries have installed fire fighting appliances that can be put into service at a moment's notice, most fires are ultimately controlled and extinguished by fire fighters with hose lines. In Figure 5-1, for example, a monitor supplied by a large water main could not be used after a tank ship exploded.

Principles of Hydraulics

The most effective and practical method of fire extinguishment is to cool the involved and exposed combustibles to a degree lower than their ignition temperatures. The fire temperatures of most materials usually reach 900° to 1800° F (dull-red to cherry-red heat of iron), and rarely exceed 2500°F (white heat of iron). The ignition and fire temperatures of all ordinary combustibles are higher than 300°F; therefore, if the involved and exposed materials are cooled to an approximate temperature of 300°F, the process of combustion ceases and extinguishment is accomplished.

Because of its high heat-absorption capability, water is the most widely used extinguishing agent. Successful fire fighting depends upon an adequate delivery of water on the burning substance. Therefore, every fire fighter should have a thorough knowledge of techniques used in producing effective fire streams. A clear understanding of hydraulic principles is necessary before any mathematical computations are attempted.

Fire Streams

A fire stream is considered to be a stream of water from a fire hose after it leaves the nozzle and until it reaches the fire. During the time the fire stream passes through air, it is influenced by gravity, wind velocity, heat currents, and friction with the air. The condition of the fire stream when it leaves the nozzle is affected by operating pressures, nozzle design, nozzle adjustment, and the condition of the nozzle tip.

Figure 5-1. A monitor supplied by a large water main could not be used after this tank ship exploded.

Pressure Loss or Gain Because of Elevation

Since one pound per square inch pressure is required to raise water in a fire hose or pipe 2.304 ft, 5 psi pressure will raise water 5 × 2.3 ft, or 11.5 ft. The average height of one story in a building is 11.5 or 12 ft; approximately 5 psi is required to raise the water for every story. The principle applies equally for ground elevations since a hose line that is laid up a hill will lose 5 psi for each 11.5 ft of vertical rise; if the nozzle is lower than the pump, the nozzle pressure will be 5 psi greater for every 11.5 ft of vertical drop in elevation. When there are no distinct floor levels, the correction for elevation can be estimated as 4½ psi for each 10 ft of elevation.

Static, Flow, and Residual Pressures

Static pressure is the pressure of the water when motionless. Water in a fire hydrant is under static pressure as long as the valve is not opened. No hindering forces are affecting it as long as it is not moving and, whether the pressure is produced by gravity or a pump, the line is under static pressure.

Flow pressure cannot be as great as static pressure because after the fire stream is set in motion, it is subjected to all the hindrances the waterway possesses, such as friction, short bends, etc.

Residual pressure is the pressure remaining when the valve is open and the water is flowing.

Suction

Normal atmospheric pressure at sea level may be considered to be 14.7 lb to the square inch. For practical purposes it may be assumed that water is incompressible; that is, it has a constant volume. Therefore, it may be said that a given weight of water will always occupy the same space, and a given volume of water will always have the same weight. A column of water 1 ft high with a cross-sectional area of 1 sq. in. will weigh 0.434 lb, and 1 lb of water will fill the same column to a height of 2.304 ft. Pressure exerted by a column of water is known as *head pressure.*

Since 1 lb of water will fill the column to a height of 2.304 ft, 14.7 lb will fill the column to a height of 14.7 × 2.304. This equals approximately 33.9 ft. Therefore, 33.9 ft is the maximum theoretical height to which it would be possible to lift water by suction at sea level.

In actual practice it is impossible to create a perfect vacuum even with the most delicate scientific apparatus, much less with machinery for practical use. Taking into consideration this inability to create a perfect vacuum, the effect of minute air leakage through connections and close fitting parts, and the frictional resistance through suction hose, the common practice is to use two-thirds of the theoretical distance, or 22.6 ft at sea level, for the practical limit of draft, or *lift.*

Atmospheric pressure decreases as elevation above sea level increases, at a rate of approximately ½ lb for every 1000 ft of elevation. Consequently, the theoretical height to which water may be lifted by suction decreases proportionally as the elevation increases. Therefore, the theoretical height to which a pumper can raise water decreases about 1 ft for every 1000 ft of altitude. Thus, at a point 2000 ft above sea level the maximum theoretical height of draft is 31.6 ft, and the practical limit is 21.2 ft.

Vacuum is calibrated from 0 to 30 in. on the vacuum scale of the compound gage; this is comparable to a vertical glass tube closed at the upper end, with the open end immersed in a vessel of mercury, the surface of which is subject to atmospheric pressure. With a vacuum in the tube, the mercury rises to the height where the weight of mercury in the tube is just balanced by the atmospheric pressure on the surface of mercury in the vessel. Mercury weighs 0.49 lb per cubic inch; thus, if the mercury stands at a height of 30 in., it indicates that a pressure of 14.7 lb (30 × 0.49) is present on the surface of mercury in the vessel.

To determine the height to which water will rise in the suction hose, the vacuum reading in inches should be multiplied by 0.49, thus obtaining the reduction per square inch of atmospheric pressure within the suction. For practical purposes, 1 in. of mercury (Hg) equals ½ psi of pressure; water will rise approximately 1 ft in the suction hose for each inch of vacuum within the suction hose (1 ft of water weighs 0.434 lb).

Drafting

When the source of water supply to a pumper is at a level below the inlet side of the pump, the pump operator must draft water by suction to obtain a discharge from the pump.

Atmospheric pressure acts uniformly on the surface of an open body of water. However, if this pressure is removed by vacuum from a portion of the surface of the water, a condition of unequal pressure is created. Water will not compress into a denser mass to resist this disturbance of pressure balance. Instead, that portion from which the pressure of the atmosphere has been reduced will rise to a height such that the quantity of water above the surface has the weight which will establish a pressure equal to that of the atmosphere and thereby restore the balance.

Figure 5-2 illustrates the effect of atmospheric pressure on water under vacuum. The identifying letters in the following discussion are keyed to those in the figure.

Tube A is open at both ends; when it is placed vertically into a body of water so as to submerge its lower end, atmospheric pressure acting on the surface of the water causes it to flow up into the tube until it reaches the same level as the water outside the tube. This is because the pressure acting on the surface of the water inside the tube through the open top end is the same as that on the surface of the water outside the tube.

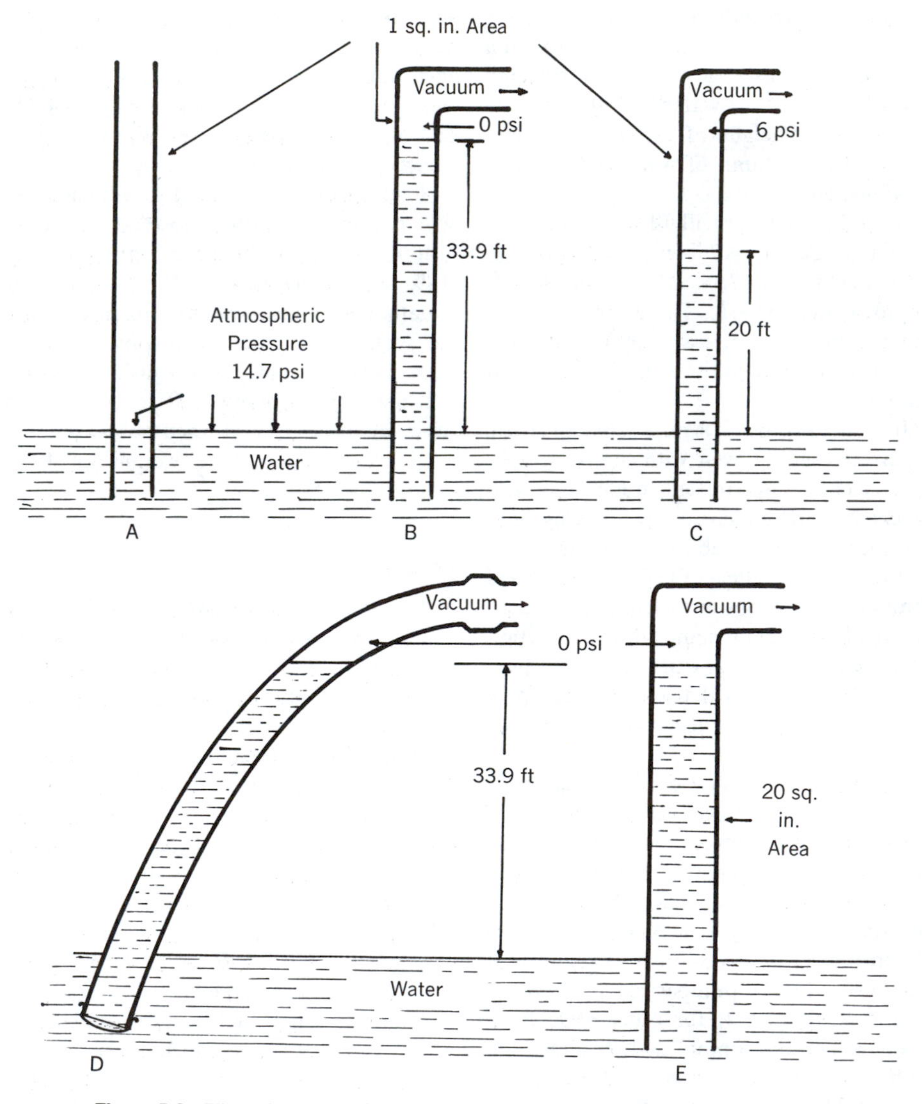

Figure 5-2. Effect of atmospheric pressure on water under vacuum

The upper end of tube B is sealed and connected to a pump which can create a complete vacuum. When the open end of tube B is placed in a body of water and pressure within the tube is gradually reduced, water flows up the tube until the pressure exerted at its base is equal to the pressure of the surrounding atmosphere. When air is completely exhausted from within the tube, an absolute vacuum has been formed, and water rises to a vertical distance of 33.9 ft, since at this height the weight of the 1-sq.-in. column of water will exert exactly the same pressure at its base as the atmospheric pressure on the surrounding surface of water, 14.7 psi.

From this explanation it can be seen that when a vacuum is formed in a long tube, the water is not pulled or sucked up by the formation of the vacuum, but it is forced up by the external pressure of the atmosphere acting on the exposed surface of water. If air is not completely exhausted from tube C, water will only rise to a height sufficient to balance the pressure differential between the inside and outside of the tube. For example, if the air pressure remaining in tube C is 6 psi, then the water will rise approximately 20 ft. (14.7 psi minus 6 psi equals 8.7 psi at atmospheric pressure remaining to force water up the tube. Each pound of air pressure will push water up the tube 2.3 ft; 8.7 psi multiplied by 2.3 ft equals 20 ft of lift.)

Tube D is a suction hose bent in a semicircle. Water within the suction hose would still rise to a height of 33.9 ft under an absolute vacuum. The pressure of water within the suction hose is exerted from the surface downward, in direct proportion to the vertical depth only.

Under conditions of a perfect vacuum, water will rise to a height of 33.9 ft regardless of the size of the conductor. Tube E has a cross-sectional area of 20 sq. in., but an atmospheric pressure of 14.7 psi is acting on the full area.

Drafting Hot Water

Where a pumper is drafting warm or hot water, its ability to lift water by suction is very much reduced, depending on the temperature of the water. This is because hot water tends to give off water vapor; this destroys or greatly reduces the ability of a pump to create a vacuum necessary for drafting water.

Friction Loss

Friction loss is a common term in the fire service, especially to those responsible for producing fire streams. It is loss in stream pressure in the entire waterway, pump to nozzle, and even through the pumps. Since fire stream velocity is measured in terms of pressure at the nozzle, frictional loss can be measured by the difference in pressure at the pump and at the nozzle. Figure 5-3 shows the effects of friction loss at various distances. Friction losses in a fire stream are due to the friction of the water stream upon different parts of the waterway, such as upon the interior surface of the hose, the couplings, the valves, the gaskets in the couplings, the hose appliances, etc. It is an ever-present hindering force and must be overcome.

Losses tend to be less as pressures increase because the hose expands to a somewhat larger diameter. Hose that has a jacket made of synthetic fibers does not expand under pressure as readily as cotton-jacketed hose; therefore, at a given pressure, the losses will be somewhat greater than with cotton-jacketed hose of the same nominal internal diameter. Some manufacturers compensate for this by increasing slightly the actual inside diameter of the hose; this can be done very readily because synthetic fibers provide a jacket of less thickness than cotton. Therefore, a larger diameter is readily accommodated by the coupling bowls. Good quality fire hose has a smoother inner surface that will create less friction than hose of a lower

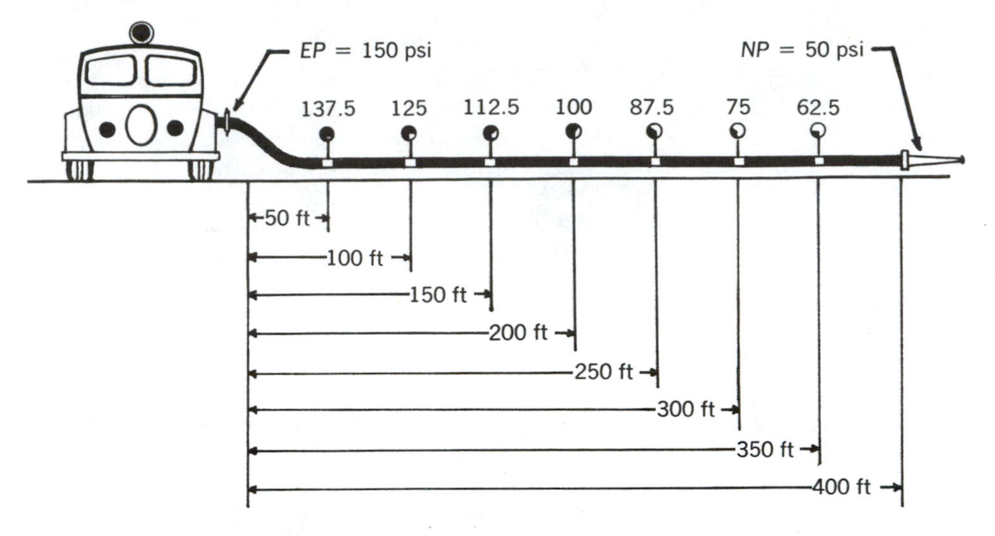

Figure 5-3. Effects of friction loss

grade. Hose lines laid in a zig-zag pattern have 5 to 6% greater loss than hose laid in a straight line. Friction loss in old hose may be 50% greater than in new hose.

Friction loss may be minimized by the waterway, but it will still be present to a noticeable extent. Consequently, the pump operator must count on adding pressure to overcome it. Since friction loss is a measure of the amount of energy or pressure loss, it is expressed in terms of pounds per square inch (psi). However, since under the same conditions there will be twice as much energy lost in passing through 2 ft of hose as is lost by the same amount of water passing through 1 ft of hose, the amount of energy lost must be measured in terms of loss over a given unit length. Friction loss in fire hose is measured in terms of pounds per square inch for a 50-ft section, or for each 100-ft of hose. Some methods of calculation are based on the length of the hose layout in 100-ft lengths; other formulas utilize the number of 50-ft lengths of hose.

The four fundamental rules that govern friction loss in hose lines and pipe are:

1. *All other conditions being equal, the loss by friction varies directly with the length of the line.* If friction loss for a given flow is 10 psi per 100 ft of 2½-in. hose, then as long as the conditions remain the same, there will be an additional 10 psi loss in each additional 100 ft of 2½-in. line. If the length of line is doubled, friction loss will be doubled.

2. *In the same size hose, friction loss varies directly as the square of the velocity flow.* As the flow increases, the friction loss also increases, but at a much greater rate. If the velocity of the flow is doubled, friction loss will be 2 × 2, or 4 times as great; if trebled, 3 × 3, or 9 times as great.

3. *For the same flow, friction loss varies inversely as the fifth power of the diameter of the hose.* This means that the larger the hose diameter, the less the friction loss, or the smaller the hose, the greater the friction loss. This may be shown by the following formula:

$$\text{Increase in loss} = \frac{(\text{diameter of larger hose})^5}{(\text{diameter of smaller hose})^5}$$

4. *For a given velocity of flow, friction loss is independent of the pressure.* As long as velocity of flow does not change, friction loss will remain the same regardless of pressure changes. The velocity of the flow, not pressure, is the determining factor in friction loss.

Comparing Hose Sizes

The relative friction loss in hoses, and their relative capacities, using 2½-in. hose as a basis, are:

A ¾-in. hose has a friction loss about 340 times that of 2½-in., or it would take 24 lines of ¾-in. to deliver the same quantity of water the same distance with the same loss of pressure.

A 1-in. hose has 86 times the friction loss and would require 11 lines to equal a 2½-in. line.

A 1½-in. hose has 13 times the friction loss and 3.9 lines would equal a 2½-in. line.

A 3-in. hose has the capacity of 1.7 lines of 2½-in. hose; its friction loss for any flow is equal to the friction in 2½-in. hose divided by 2.6.

A 3½-in. hose has the capacity of 2.6 lines of 2½-in., or the capacity of 1.5 lines of 3-in.; its friction loss for any flow is equal to the friction loss in 2½-in. hose divided by 5.8.

Advantage of Multiple Lines

The large quantity of water required by master streams produces a problem in overcoming friction loss, which can be solved by the use of multiple lines.

By using the formula $FL = 2Q^2 + Q$ (where Q = rate of flow, divided by 100), it may be determined that 600 gpm of water flowing through 100 ft of 2½-in. hose produces a friction loss of 78 psi; the same quantity of water flowing through 500 ft produces 390 psi friction loss. Such excessive friction loss can be overcome by using larger size hose or by using multiple lines.

If two parallel lines of the same size are used to carry the water, the friction loss may be divided by 3.6; thus, the friction loss in two lines of 2½-in. hose represents about 28% of the 78 psi loss in a single line of 2½-in. hose, or 21 psi.

A pump operator at a fire can figure roughly that two siamesed lines of 2½-in. hose will have only about one-quarter of the friction loss of a single line of 2½-in. hose of the same length and carrying the same quantity (gpm) of water. This large reduction is due to the fact that the 600 gpm is divided equally between the two hose lines and there is now 300 gpm flowing through each.

By the same reasoning, three parallel lines will cause 200 gpm to flow through each, reducing the friction loss to 10 psi for each 100 ft of hose layout.

Discharge Capacities of Nozzles on Hose Lines

An important factor in obtaining good fire fighting streams is to relate the flow of the nozzle selected to the practical water-carrying capacity of the size of fire hose employed at an acceptable friction loss in the hose. For practical purposes, fire hose can be identified with the flow that it is capable of carrying without excessive friction loss:

Nominal Hose Diameter	*Maximum Efficient Water Capability*
1 in.	30 gpm
1½ in.	100 gpm
2½ in.	250 gpm
3 in.	500 gpm
3½ in.	750 gpm

Larger flows can be handled by hose lines, but only with considerable added friction loss.

Nozzle Reaction

Forces developed by a fire stream will greatly affect the ease with which fire fighters can hold and maneuver a nozzle. Injuries may result when a nozzle escapes the nozzle operator's grasp. Reaction has overturned fire apparatus and may have a greater effect on fire boat movements than the rudder and propelling engines.

Nozzle reaction is based on the principle, "For every action there is an equal and opposite reaction." As water leaves a nozzle under pressure, it causes a force in the opposite direction. The amount of this reaction depends upon the size of the nozzle tip and the pressure, and is due to the acceleration of water in the nozzle. It may be compared to the acceleration of an automobile on a level road. A substantial force must be exerted to accelerate the car, and the more quickly the car accelerates to high speed, the greater the force must be. The situation in the nozzle of a hose differs from that of a car because water in the nozzle is continuously accelerated; the force acts continuously as additional water comes through. Water passing through a 1⅛-in. nozzle under a pressure of 50 psi accelerates from about 10 mph in a 2½-in. hose to about 50 mph leaving the nozzle. The 265 gal of water discharged in 1 min represents a weight of 2200 lb. A 1⅛-in. tip at 50 psi nozzle pressure has approximately 95 lb nozzle reaction. Experience has shown that this is about the maximum amount that a crew of three can handle for any appreciable time.

If a hose line were perfectly straight from the pumper to the nozzle, a reaction would be exerted against the pumper. In actual practice, however, there are always bends, and part of the reaction is exerted as a pull on the hose which is felt by fire fighters holding the nozzle. The actual forces which fire fighters must withstand when handling a nozzle will be less than the theoretical reactions because part of the reaction is taken up by the hose line, which usually is in contact with the ground or is otherwise supported.

Full theoretical reaction applies in the case of a nozzle directed horizontally at a right angle from a ladder, tending to push the ladder out from the building.

The reaction from a straight-bore tip is calculated with the formula

$$NR = 1.5\, D^2\, NP$$

where D = straight-bore nozzle diameter

NP = nozzle pressure in pounds per square inch (psi)

Example: What reaction is created by a 1⅛-in. straight-bore tip at 50 psi nozzle pressure?

$$\begin{aligned} NR &= 1.5\, D^2\, NP \\ &= 1.5 \times (1\tfrac{1}{8})^2 \times 50 \\ &= 1.5 \times 1.2656 \times 50 \\ &= 94.92 \text{ lb} \end{aligned}$$

The approximate reaction force generated by a fire stream flowing from a straight, spray, or fog nozzle is calculated with the formula

$$NR = 0.0505\, gpm\sqrt{P}$$

where NR = nozzle reaction in pounds of force

gpm = flow from nozzle in gallons per minute

P = nozzle pressure at base of nozzle

Most fog and spray nozzles are designed to discharge a definite quantity of water at a specific tip pressure, usually 100 psi. Thus, the formula can be simplified to

$$NR = \tfrac{1}{2}\, gpm$$

Example: What nozzle reaction will be created by a fog nozzle that discharges 300 gpm at 100 psi nozzle pressure?

$$\begin{aligned} NR &= \tfrac{1}{2}\, gpm \\ &= \tfrac{1}{2} \times 300 \\ &= 150 \text{ lb} \end{aligned}$$

The above reaction figures for fog and spray nozzles are the maximum that will be experienced when the nozzle is being operated at the designed 100 psi pressure. As the fog pattern is increased to larger than 30°, the reaction forces decrease.

Reaction of Heavy Streams

It should be kept in mind that a heavy stream at the top of an aerial ladder produces a leverage effect of considerable magnitude which may damage the apparatus. Tests have shown that nozzle reaction may exceed 400 lb (Figures 5-4 and 5-5) and may affect stability of the ladder and vehicle, particularly if pressure is reduced suddenly, as from a burst line, which may cause the ladder to whip. The maximum load capacity of elevating platforms when used as water towers decreases in direct proportion to the nozzle reaction force (measured in pounds); the nozzle reaction must be subtracted from the basket's rated carrying capacity.

Also of importance is the fact that a ladder pipe stream or other large discharge from a ladder, including 2½-in. hand lines, should always be operated in line with the main beam or trusses. Horizontal movement of ladder pipe should not exceed 15° either way from center because ladders have little resistance to torsional effects. Rotation of turntable is the correct way to rotate the stream from ladder pipes. Care should also be used in elevating and lowering the stream because this changes the direction of thrust on the ladder mechanism.

Opening Shutoffs on Nozzles

Shutoff nozzles on charged lines under pressure should be opened slowly. When a line is shut off at the nozzle, the flow ceases and full engine pressure may travel right up to the nozzle. Under such conditions, if the nozzle is opened quickly, fire fighters at the nozzle will, for a moment, have to handle a line with abnormally high nozzle pressure, and the line may get away from them.

Water Hammer

When the flow of water in a hose line is suddenly stopped, the line is subjected to a high-pressure surge that travels rapidly through the line. This pressure wave may be several times the normal pressure in the hose line and is known as a water hammer or ram. The instant closing of the valves or nozzles converts the kinetic energy possessed by the moving water into pressure energy. The surge starts at the point where flow is first stopped and travels back through the line. Because of the elasticity of fire hose, much of the shock of water hammer is deadened, but it remains a serious factor because it still may be sufficient to burst a hose.

The extent of the shock and pressure developed depends upon the velocity of the water and the suddenness with which the flow is stopped. Many burst sections of hose and suctions, as well as damaged pumps and ruptured water mains, can be attributed to water hammer caused by snapping nozzles or discharge gates closed. To prevent water hammer, nozzles, and valves should be closed slowly.

Pressure Differential

The entire pumping operation may be based on one principle, that of creating pressure differentials. A differential is a difference in quantity.

Water, a liquid, seeks its own level; if a pressure differential is established, the liquid seeks to overcome it. Likewise, drafting water is a process of creating a pressure differential that will be overcome by atmospheric pressure. When a suction hose is dropped into a water supply, atmospheric pressure is equal on the water surface both inside and outside the hose, but if the pressure on the water surface in the hose is decreased by suction, a pressure differential is established and the water level in the hose rises. If continued–within the limit of atmospheric pressure–water will be forced up from the source to the pump. On the other

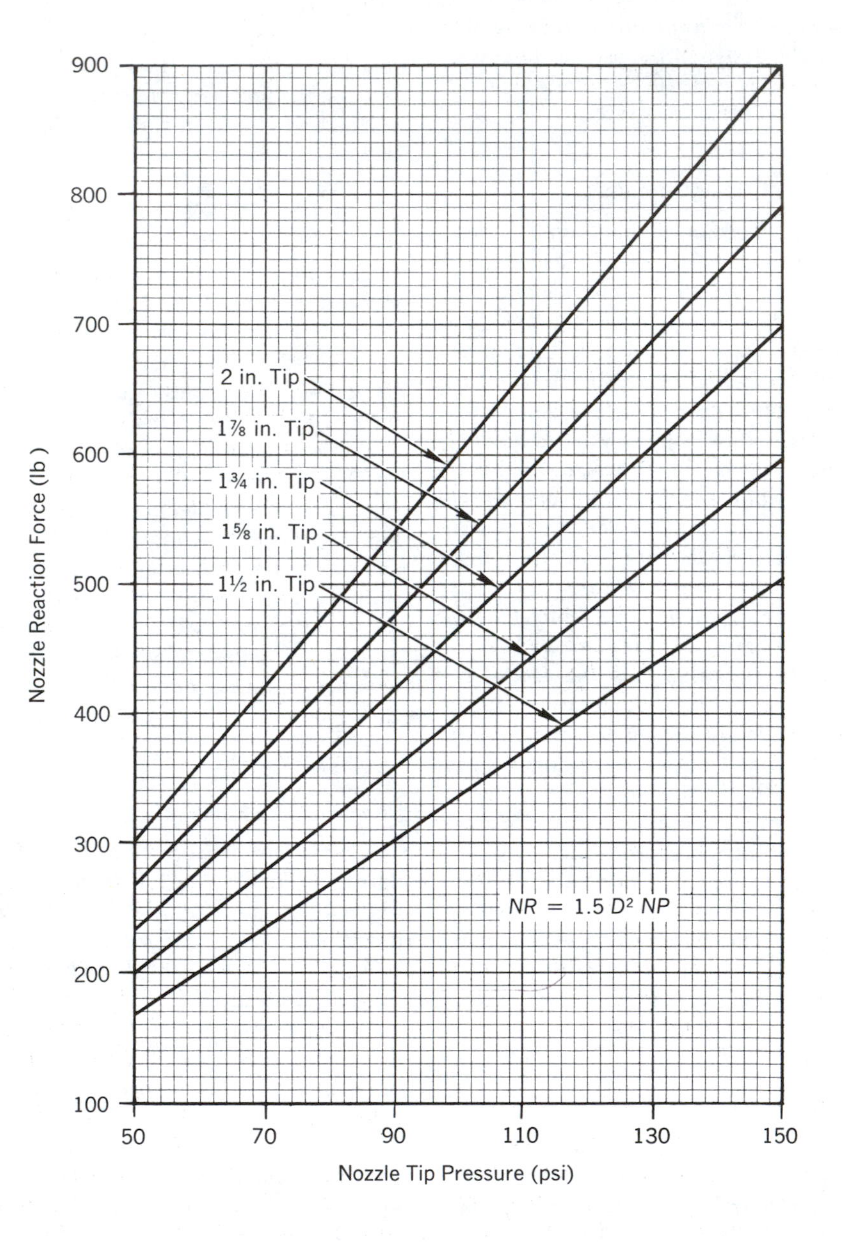

Figure 5-4. Nozzle tip reaction force vs. nozzle tip pressure

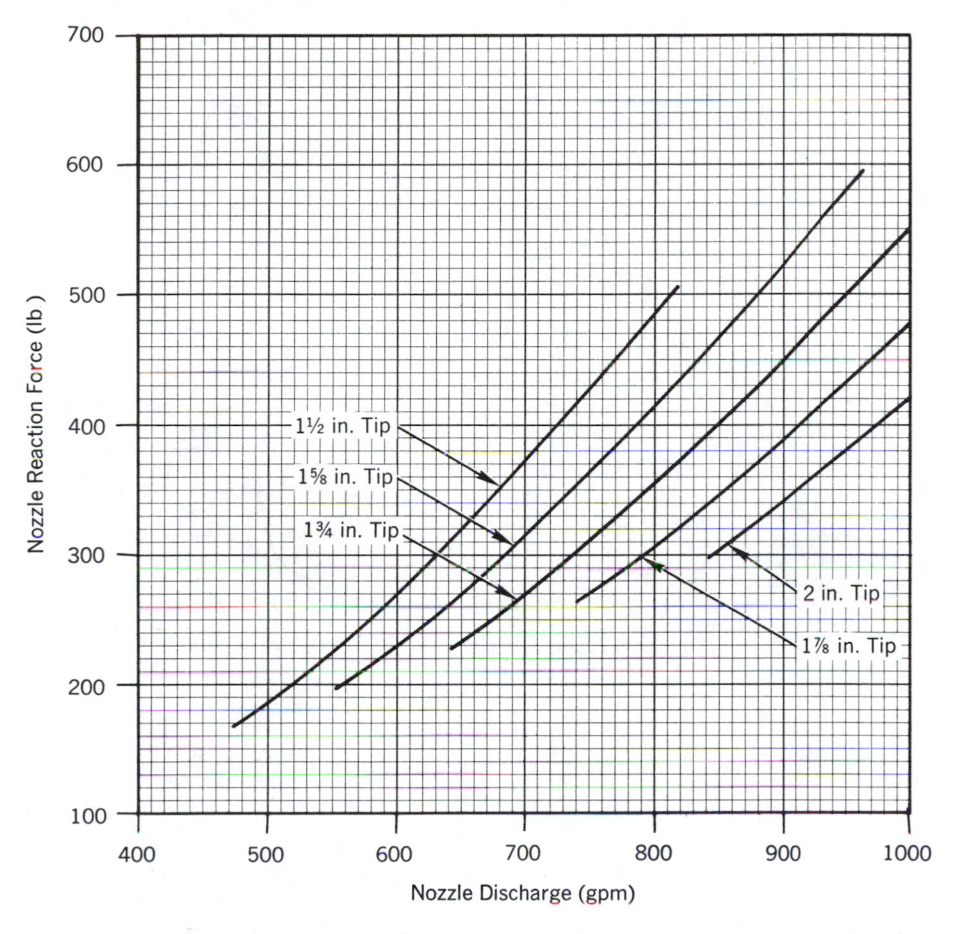

Figure 5-5. Nozzle discharge vs. reaction force

hand, to produce a fire stream, a pressure differential must be created by the pump. Pumping water when the pump is on the same level as an open line requires only a slight pressure differential. But since extra pressure is required at the nozzle to give the fire stream sufficient velocity to span the space between nozzle and fire, and since there is a certain psi loss developed for each 100 ft of hose, together with loss for other factors, a large pressure differential must be produced.

Just to flow water from the nozzle (Figure 5-6) requires only a few psi pump pressure because there are only a few gallons per minute flowing and friction loss is nil. To produce a fire stream that is effective, a nozzle pressure of about 50 psi is required. Because flow and friction loss increase when pressure increases (Figure 5-7), a pump pressure of 21 psi (friction loss) plus 50 psi (nozzle pressure), or 71 psi, must be maintained.

Additional pressure of 0.434 psi must be added to the engine pressure for each foot of elevation the nozzle is above the pump (5 psi per floor). This is to overcome the force of gravity (Figure 5-8). An engine pressure of

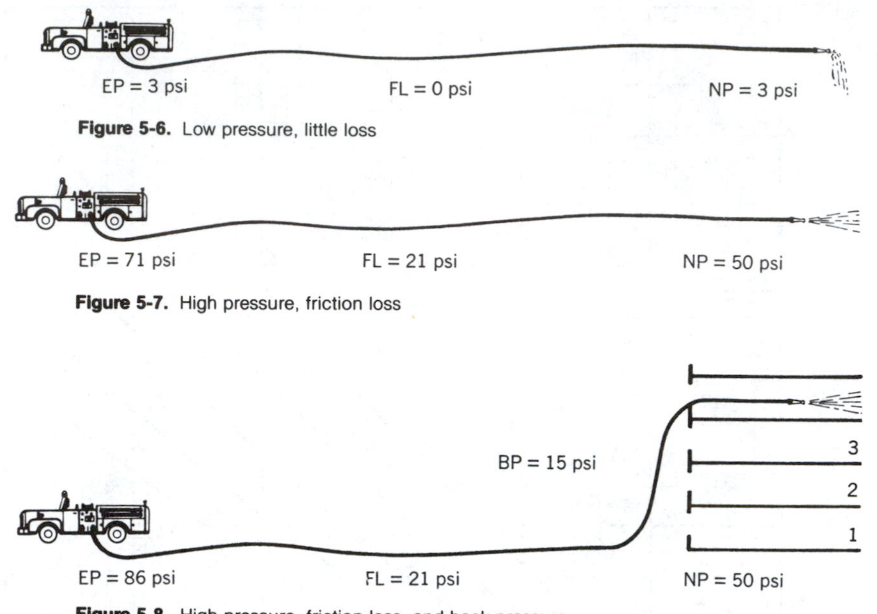

Figure 5-6. Low pressure, little loss

Figure 5-7. High pressure, friction loss

Figure 5-8. High pressure, friction loss, and back pressure

86 psi is required to supply 50 psi at the nozzle when 21 psi is needed to overcome friction loss in the hose, and there is 15 psi back pressure due to the nozzle being higher than the pump. When the nozzle is lower than the pump, gravity helps the pump supply the pressure, and the necessary engine pressure is correspondingly lower.

Fire Ground Hydraulics

This section is concerned only with producing effective hose streams at fires. A thorough knowledge of basic hydraulics is necessary to understand and visualize all the factors concerned with fire stream development, but theoretical aspects must often be subordinated to the practical.

On the fire ground, ability to quickly determine the engine pressure necessary to produce an efficient fire stream is mandatory. Exact nozzle pressures are not important; in fact, a moderate variation would not be readily discernible at the nozzle without use of a gage. The important object is to provide a nozzle pressure that has sufficient velocity to reach the fire, and yet will not risk injuring fire fighters with excessive nozzle reaction. An excessive amount of pump discharge pressure would not result in an equally excessive nozzle pressure because increased flow would produce a greater amount of friction loss.

Proper operation of the pumper depends to a great extent upon information the officer conveys to the pump operator. Each change that takes place at the nozzle will affect engine pressures. Replacing nozzles, placing a line into a heavy stream appliance, extending or reducing the length of hose lines, and taking a line above or below the ground level into a building are variations of activities that may easily occur without knowledge of the pump operator. Officers are responsible for keeping their crews informed so that engine pressures can be adjusted to fit conditions.

Friction Loss

Friction loss is the most important variable to be considered in fire ground hydraulics. (Each appliance, fitting, coupling, section of hose, and everything else through which water flows will impede the flow through friction.) All of these factors vary with condition, design, manufacturer, and age of hose and appliances; therefore, no exact allowances can be made for friction loss. However, tables of standard nozzle pressures and friction loss allowances are adequate for practical computations. (See Tables 5-7, 5-8, 5-9, 5-10, and 5-11 at the end of this chapter.)

Friction loss is real and cannot be avoided, but it can be kept within reasonable limits if judgment is used in layouts of hose and sizes of nozzles employed.

Use of parallel lines of hose reduces friction loss to approximately 28% of what it is with a single line, for the same flow of water.

Friction loss in 3-in. hose with 2½-in. couplings is about 40% that of 2½-in. hose. Thus, with the same engine pressure, it is possible to use a line of 3-in. hose two and one-half times as long as one of 2½-in. hose, to get the same nozzle pressure with the same size nozzle. Having 2½-in. couplings on 3-in. hose only increases friction loss by about 5%; using the same size couplings on both 2½-in. and 3-in. hose makes it possible to use these sizes interchangeably.

Provision of an additional hose line to supply a heavy stream device beyond the minimum required allows a greater proportion of the pump's energy to reach the nozzle. Thus, it is preferable to increase the water-carrying capacity of the hose lines rather than to increase the pressure at the pumper beyond the point at which the pump is designed to give its best performance. Frequently, the use of an additional line will allow the pump to supply the desired nozzle pressure while operating in the capacity position rather than in the pressure position.

Pumps provide rated capacity at 150 psi net pump pressure and only 70% of rated capacity at 200 psi net pump pressure. At 250 psi net pump pressure, the pump will discharge 50% of rated pump capacity. Excessive pressures, used in lieu of sufficient supply lines, merely reduce potential output of pumping apparatus; they can also result in serious overloads that can cause apparatus failures.

The following formulas are used to determine flow and friction loss:

$$Dis = 30D^2\sqrt{P}$$

$$FL = 2Q^2 + Q$$

where Dis = discharge measured in gallons per minute (gpm)

D = nozzle diameter

P = nozzle pressure

FL = friction loss in 100 ft of 2½-in. hose

Q = gpm divided by 100

For less than 100 gpm, pressure loss $= 2Q^2 + \frac{1}{2}Q$.

For ease of mental calculations on the fire ground, when the friction loss for a nozzle discharge cannot be remembered, the formula $FL = 2Q^2 + Q$ may be simplified to $FL = 2Q^2$. Omitting the $+ Q$ will greatly simplify mental computations without seriously affecting the nozzle pressure.

By the use of these formulas, it is evident that the friction loss in a hose is based on the quantity of water flowing. (See Table 5-1.)

If a pumper were pumping into four hose lines of the same nominal diameter and length, with similar size nozzles on the ends, a pitot gage would show a variation of nozzle pressures due to varying conditions and manufacturers of hose, gaskets, and fittings. Thus, the discharge in gallons per minute could vary considerably. Therefore, each department should establish standard nozzle pressures for all its equipment and should round off the theoretical discharge quantities for more practical usage.

TABLE 5-1. FRICTION LOSS

NOZZLE SIZE, in.	NOZZLE PRESSURE, psi	FLOW, gpm	FRICTION LOSS PER 100 ft OF 2½-in. HOSE, psi
1	71	250	15
1⅛	45	250	15
1¼	29	250	15
2½	1¾	250	15

Standard Nozzle Pressures

The usual standard nozzle pressures adopted in standard practice are as follows:

50 psi for smooth-bore tips on hand lines.

70, 75, 80, or 100 psi (depending upon department policy) for fog or spray tips on hand lines.

80 psi for smooth-bore nozzles on master stream appliances.

100 psi for spray tips on master stream appliances.

Most fog and spray nozzles are designed to discharge their rated capacity at 100 psi nozzle pressure; many fire departments set a lesser standard to reduce nozzle reaction on the hose crew. This reduced tip pressure will, except in constant-flow nozzles, discharge a corresponding lesser quantity of water. Comparison of the square roots of the pressures ($\sqrt{P}$) will allow a relative estimation of the reduced flow.

Friction loss will remain constant as long as the gpm remain the same. Regardless of the pressure, the friction loss in a given flow is approximately constant. A spray nozzle delivering 250 gpm at a nozzle pressure of 100 psi has the same related 15 psi friction loss in each 100 ft of 2½-in. hose as 50 psi nozzle pressure on a 1⅛-in. tip since both nozzles discharge approximately 250 gpm (Figure 5-9).

Advantages of GPM Method

This technique utilizes the same formulas and factors as the theoretical methods but is simplified for easier use on the fire ground. This relieves officers and pump operators of the necessity for learning one set of formulas, factors, and constants in order to understand the theoretical basis of hydraulics, and then learning an entirely different set to use for mental computations.

In this technique, all amounts are rounded off to aid memory and simplify calculation; this does not significantly affect estimates of nozzle pressures. The method requires fewer operations when figuring engine pressures, so there is less chance for error.

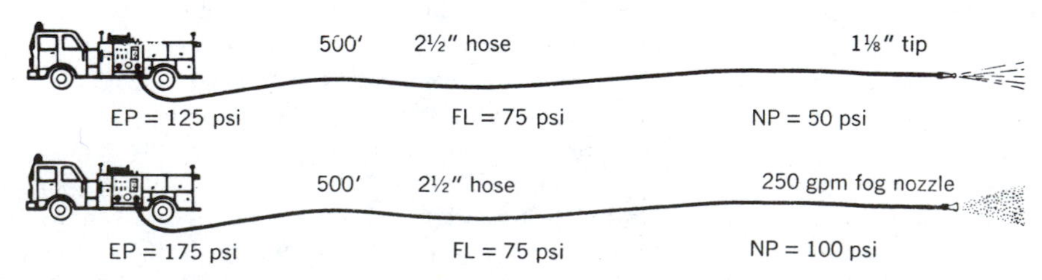

Figure 5-9. Friction loss depends on the volume of flow.

The method is as accurate as any. The theoretical answer does not always conform to the actual conditions because there are many variables in actual fire fighting operations which are difficult to determine precisely; for example: the exact length of the hose line, the elevation at which the nozzle is being operated, the condition of the hose and gaskets, and the efficiency of the nozzle and orifices. In fact, many spray nozzles have an adjustable flow; it would be impossible for the pump operator to know in what position this type of nozzle is being used.

Communications

The correct application of water on the fire can be determined only by the fire fighters at the nozzle; an adequate nozzle pressure can be maintained only by the pump operator. Close coordination between the two ends of the hose lines is necessary so that the pump operator can deliver the proper pressure. Arm signals are occasionally resorted to, but the widespread use of portable radios in the fire service greatly simplifies the necessary liaison at a fire.

Practical Application of the GPM Method

This method was checked by placing a gage at each 100 ft of hose line and at the nozzle. It was found that this method of determining engine pressures is far more accurate than its simplicity might indicate.

Every fire department uses different sizes of nozzles, variations of ready lines, preconnects, wye assemblies, appliances, and fittings. This section includes standards that may be varied to accommodate each department.

Straight-bore nozzles are designated by tip diameter; spray nozzles by coupling size and rate of flow.

Standard nozzle pressures are used.

When a department has a variety of spray nozzles with varying flows, it is suggested that an average gpm flow be determined and this amount be used at all times unless it is definitely known that a specific nozzle is being operated.

Hand lines. Unlike fog nozzles, the flow of water from solid-stream nozzles can be calculated accurately, given nozzle size and pressure. Table 5-2 gives the approximate gpm capacities of the smooth-bore nozzles and the corresponding amount of friction loss for each 100 ft of single 2½-in. line. The figures have been rounded off.

A standard nozzle pressure of 50 psi is recommended for all smooth-bore nozzles on hand lines. By keeping nozzle pressure constant, a specified nozzle size will always be related to a definite amount of friction loss; i.e., for 50 psi nozzle pressure always figure that there is 10 psi friction loss in each 100 ft of 2½-in. hose when using a 1-in. nozzle, 15 psi when using a 1⅛-in. tip, 25 psi when using a 1¼-in. nozzle, etc.

TABLE 5-2. FRICTION LOSS

SMOOTH BORE NOZZLE SIZE, in.	NOZZLE PRESSURE, psi	FLOW, gpm	FRICTION LOSS PER 100 ft OF 2½-in. HOSE, psi
⅝	50	80	2
¾	50	120	4
⅞	50	160	7
1	50	200	10
1⅛	50	250	15
1¼	50	325	25
1⅜	50	400	36

Spray and fog nozzles. Most fog and spray nozzles are engineered to discharge their rated capacities at 100 psi nozzle pressure. However, many fire departments standardize on a lower nozzle pressure to reduce the nozzle reaction on the crew. If a department has many different sizes of fog nozzles, it is useful to decide on an average gpm discharge for all of them; i.e., when nozzle discharges vary from 200 gpm to 300 gpm, figure all of them as flowing 250 gpm and allow a friction loss of 15 psi per 100 ft.

The friction loss in each 100 ft of 2½-in. hose is determined solely by the gpm flow through the line. Nozzles discharging a flow that is not listed in Table 5-3 will cause a friction loss in each 100 ft of 2½-in. hose which may be calculated by using the formula: $FL = 2Q^2 + Q$. For nozzles discharging less than 100 gpm, use the formula: $FL = 2Q^2 + \frac{1}{2}Q$. By closely associating nozzle sizes with a standard gpm flow, the friction loss in a hose line may be quickly calculated.

To use this method, determine the gpm flowing from the nozzle and the length of the hose line. Multiply the friction loss figure by the number of 100-ft lengths of 2½-in. hose; then add the nozzle pressure; this determines the recommended engine pressure.

Example: Find the engine pressure required to provide a good fire stream from a 1¼-in. nozzle supplied by a single 2½-in. line 500 ft in length (see Figure 5-10). Use the formula

Engine pressure = friction loss + nozzle pressure

The discharge from a 1¼-in. nozzle at 50 psi nozzle pressure is 325 gpm; friction loss is 25 psi per 100 ft of single 2½-in. hose.

$$\begin{aligned} 5 \times 25 = {} & 125 \text{ psi friction loss} \\ & \underline{50} \text{ psi nozzle pressure} \\ & 175 \text{ psi engine pressure} \end{aligned}$$

TABLE 5-3. FRICTION LOSS

FLOW, gpm	FRICTION LOSS PER 100 ft OF 2½-in. HOSE, psi	FLOW, gpm	FRICTION LOSS PER 100 ft of 2½-in. HOSE, psi
46	1	336	26
75	2	342	27
97	3	349	28
116	4	356	29
133	5	362	30
148	6	368	31
162	7	375	32
175	8	381	33
187	9	387	34
199	10	393	35
210	11	399	36
221	12	405	37
230	13	411	38
240	14	417	39
249	15	422	40
258	16	428	41
267	17	433	42
275	18	439	43
283	19	444	44
291	20	449	45
299	21	454	46
307	22	460	47
314	23	465	48
321	24	470	49
329	25	475	50

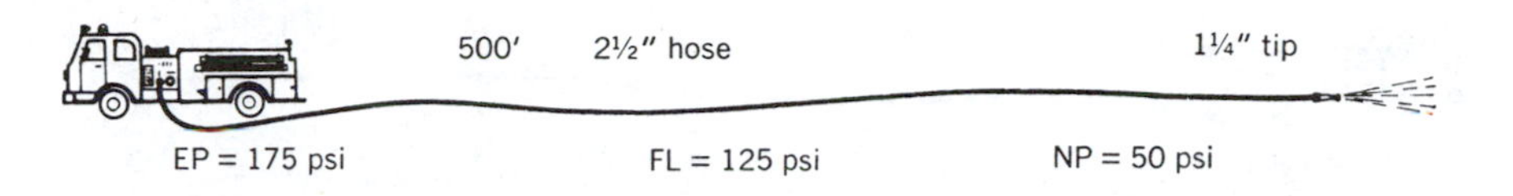

Figure 5-10. Engine pressure equals nozzle pressure plus friction loss.

Example: Determine the engine pressure of a company pumping through 1000 feet of 2½-in. hose and a 1-in. nozzle at 50 psi nozzle pressure (Figure 5-11).

$$EP = FL + NP$$

A 1-in. tip at 50 psi discharges 200 gpm; this produces 10 psi friction loss in each 100 ft of single 2½-in. hose.

10 × 10 = 100 psi friction loss
50 psi nozzle pressure
150 psi engine pressure

Example: Determine the engine pressure for a company pumping through 500 ft of 2½-in. hose. A 300-gpm fog nozzle is being used with 100 psi nozzle pressure (Figure 5-12).

$$EP = FL + NP$$

A flow of 300 gpm passing through a single line of 2½-in. hose creates 21 psi friction loss in every 100 ft.

5 × 21 = 105 psi friction loss
100 psi nozzle pressure
205 psi engine pressure

Example: Determine the pressure of an engine company pumping through 800 ft of 2½-in. hose to a 1⅛-in. tip. Nozzle pressure is 50 psi (Figure 5-13).

$$EP = FL + NP$$

The discharge from a 1⅛-in. nozzle at 50 psi is 250 gpm. When this quantity is flowing through a single length of 2½-in. hose, there is 15 psi friction loss in every 100 ft.

8 × 15 = 120 psi friction loss
50 psi nozzle pressure
170 psi engine pressure

Multiple hose lines. Multiple hose lines are often needed to bring friction losses within practical limits when supplying master stream appliances or relaying large volumes of water (Figure 5-14).

When two or more hose lines are used to supply water to a desired point or appliance, calculations are simplified by considering only the friction loss in each 100 ft of one line. Each hose line will deliver its equal share of water because the pressure applied by the pump will equalize itself in each hose line. Average the lengths of parallel hose lines when they are not equal.

When a 2-in. smooth-bore tip is being used at a nozzle pressure of 80 psi, 1000 gpm is being discharged. If only one hose line is used to supply this water, the entire 1000 gpm must pass through it. However, if another hose line of the same size and length is laid, it will carry approximately one-half of the water and each hose line would then handle 500 gpm. If a third line were laid, each would carry about 333 gpm, while only 250 gpm would flow through each line when four parallel lines are connected between the pump and the appliance.

Example: A single 2½-in. hose line is supplying a nozzle with 600 gpm. The total amount of water must pass through one 2½-in. hose line. Using the formula $FL = 2Q^2 + Q$, the friction loss in each 100 ft will be 78 psi (Figure 5-15).

If a second hose line is laid, only 300 gpm will be flowing through each line, and the friction loss will be reduced to about one-fourth of the original, or 21 psi (Figure 5-16).

When a third hose line is stretched, the 600 gpm will be flowing through three parallel lines; each line will be carrying 200 gpm

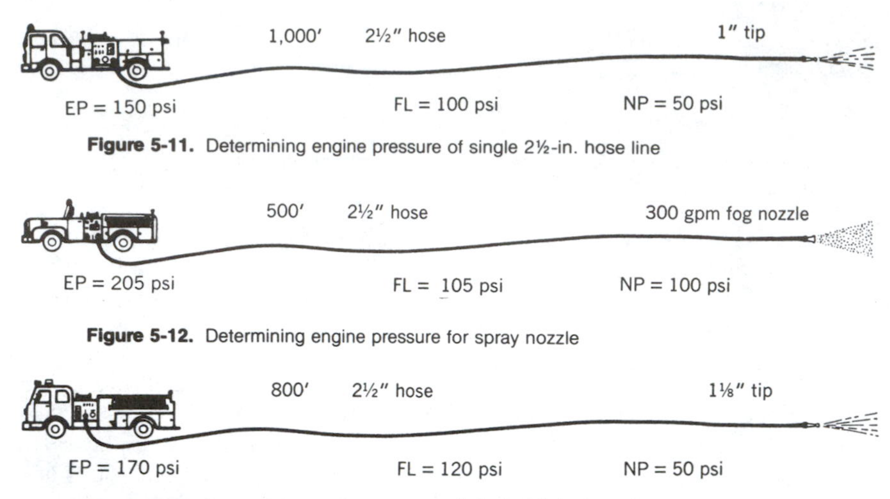

Figure 5-11. Determining engine pressure of single 2½-in. hose line

Figure 5-12. Determining engine pressure for spray nozzle

Figure 5-13. Determining engine pressure of single 2½-in. hose line

Figure 5-14. Because hose lines on the fire ground must supply a variety of nozzles and appliances, with nozzle pressures ranging from 50 to 100 psi, reasonably precise calculations are necessary to figure the correct engine pressures. (Courtesy of the Los Angeles City Fire Department)

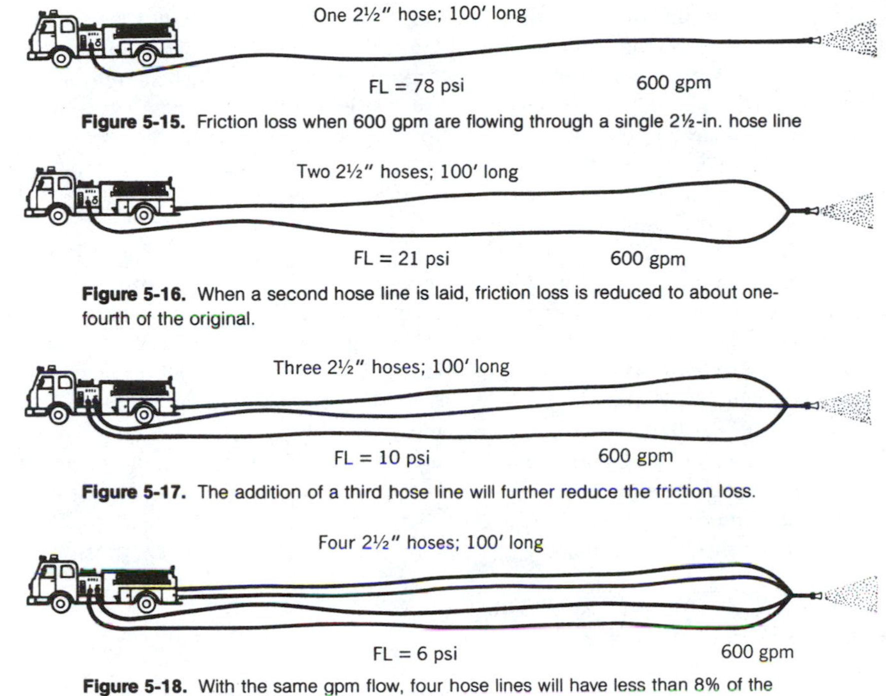

Figure 5-15. Friction loss when 600 gpm are flowing through a single 2½-in. hose line

Figure 5-16. When a second hose line is laid, friction loss is reduced to about one-fourth of the original.

Figure 5-17. The addition of a third hose line will further reduce the friction loss.

Figure 5-18. With the same gpm flow, four hose lines will have less than 8% of the friction loss in a single line.

with a friction loss of 10 psi for each 100 ft (Figure 5-17). The addition of the third hose line will reduce the friction loss in the hose layout by approximately one-half. This allows the pumper to supply the same quantity of water for twice the distance without increasing the engine pressure.

It would take 10 psi per 100 ft to push 200 gpm through one line of 2½-in. hose, and it would also require 10 psi per 100 ft for the same pumper to force 200 gpm each through three or five or more 2½-in. lines. If these lines were siamesed together before the nozzle, the nozzle *volume* would be proportionately greater depending on the number of lines, but nozzle *pressure* would be determined by subtracting 10 psi per 100 ft from the engine pressure *one time,* and not once for each 2½-in. hose line.

If a fourth hose line is stretched, the friction loss will be reduced to 6 psi as only 150 gpm will be flowing through each hose line (Figure 5-18).

Note: For fast mental comparisons, assume that the laying of a second hose line will reduce the friction loss to about one-fourth of the original line; laying a third hose line will further reduce the friction loss by one-half, or one-eighth of the original single line; and the addition of a fourth hose line of equal length and diameter will again reduce the friction loss by 40%. This allows the pumper to supply the same quantity of water for the same distance with correspondingly lower engine pressures.

Master stream appliances. Master stream appliances, because of nozzle pressure and nozzle tip size, require multiple hose lines or large-diameter hose to supply the necessary large volumes of water. The pump operator must know the total number of lines supplying the appliance, length of the lines being pumped into, type of appliance, nozzle pressure, and the size of the nozzle being pumped into.

A standard nozzle pressure of 80 psi is recommended for all smooth-bore nozzles on heavy stream appliances (Table 5-4). Large-spray nozzles should be operated at their designed pressure of 100 psi.

Standpipes and appliances. There is a considerable loss of pressure in all standpipe connections, elevating platforms, portable monitors, wagon batteries, and ladder pipes. The most accurate method of determining friction loss is to use pressure gages at the inlet and at the nozzle tip: the difference in gage readings will be friction loss. Some monitors and batteries have a pressure gage built into the appliance. Where an accurate determination cannot be made, the following allowances should be made to compensate for friction loss and back pressure:

Standpipe: 25 psi friction loss plus the back pressure from pump to nozzle.

Portable monitor: 25 psi friction loss plus back pressure from pump to nozzle.

Wagon battery: 25 psi for both friction loss and back pressure from pump to nozzle.

Ladder pipe or elevating platform (irrespective of height or flow): 80 psi for both friction loss and back pressure.

If this set of standard friction loss allowances proves to be incorrect for a specific apparatus because of larger or smaller piping, monitor or battery design, or other restrictions, then drilling and experimentation will allow a more practical figure to be adopted.

It is neither practical nor necessary to figure individually the friction loss and back pressure allowances which should be made when using ladder pipes and elevating platforms, because the elevation of the nozzle is changed quite often; each 10 ft the appliance is raised or lowered will affect the nozzle pressure about 4½ psi.

Use of the figure 80 psi to compensate for friction loss and back pressure in ladder pipes and aerial platforms is based on the following: the tip size most often used on a ladder pipe is 1½-in.; at 80 psi nozzle pressure, the flow is 600 gpm. A typical ladder pipe assembly consists of a clappered siamese inlet, 100 ft of 3-in. hose, and the ladder pipe appliance. A typical pressure allowance for back pressure is 25 psi. Therefore, friction loss and back pressure will average:

in clappered siamese = 10 psi
in 100 feet of 3-inch hose = 30 psi
in ladder pipe = 15 psi
Back pressure = 25 psi
Total for ladder pipe assembly = 80 psi

Arbitrary values cannot be assigned to the various components of an aerial platform. However, the manufacturer will often give these figures, or a gage may be attached to the appliance. Where an estimate must be made, assume 80 psi.

An intercom or a radio is usually furnished to the operator of the nozzle on an aerial ladder or aerial platform. The fire stream emerging from a nozzle should be watched. When the reach of the fire stream is

TABLE 5-4

NOZZLE SIZE, in.	NOZZLE PRESSURE, psi	FLOW, gpm
1¼	80	400
1⅜	80	500
1½	80	600
1⅝	80	700
1¾	80	800
2	80	1000

insufficient, or the fire stream is breaking up, the pump operator should be notified.

In the following examples, to find the engine pressure use the formula $EP = FL + NP$, plus allowances for portable monitor, wagon battery, ladder pipe assembly, or aerial platform, as the case may be.

Example: Two parallel lines of 2½-in. hose, each 500 ft in length, are supplying a 1½-in. tip on a portable monitor. Nozzle pressure is 80 psi. What is the engine pressure (Figure 5-19)?

A 1½-in. tip at 80 psi nozzle pressure discharges 600 gpm. Each line is carrying its equal share of water; this amounts to 300 gpm. Friction loss for 300 gpm flowing through 100 ft of 2½-in. hose amounts to 21 psi per 100 feet ($FL = 2Q^2 + Q$).

A friction loss and back pressure allowance of 25 psi should be made for the portable monitor. Pump and appliance are on the same elevation, so back pressure is not a factor.

5 × 21 = 105 psi friction loss
80 psi nozzle pressure
25 psi allowance for portable monitor
210 psi engine pressure

Example: Three parallel lines of 2½-in. hose, each 300 ft in length, are supplying a 1¾-in. tip on a portable monitor. Both the pump and the heavy stream appliance are on the ground level. Nozzle pressure is 80 psi. What is the engine pressure (Figure 5-20)?

A 1¾-in. tip at 80 psi nozzle pressure discharges 800 gpm. Each line is carrying its equal share of water; this amounts to about 270 gpm. Friction loss for 270 gpm flowing through 100 ft of 2½-in. hose amounts to 17 psi.

3 × 17 = 51 psi friction loss
80 psi nozzle pressure
25 psi allowance for portable monitor
156 psi engine pressure

Example: Determine the engine pressure of a pumper delivering water through two parallel lines of 2½-in. hose, each 300 ft long, to a wagon battery equipped with a 1¾-in. tip. Nozzle pressure is 80 psi (Figure 5-21).

A 1¾-in. tip at 80 psi nozzle pressure discharges 800 gpm. Each line is carrying 400 gpm, so friction loss amounts to 36 psi per 100 ft of 2½-in. hose.

3 × 36 = 108 psi friction loss
80 psi nozzle pressure
25 psi allowance for wagon battery
213 psi engine pressure

Note: The minimum pump capacity that can be used for this situation is 1250 gpm, unless the flowing pressure at the hydrant is high. A 1000-gpm pumper can only supply 700 gpm at a net pump pressure of 200 psi.

Example: Determine the engine pressure of a pumper delivering water through three parallel lines of 2½-in. hose to a 1¾-in. tip on a wagon battery. Each line is 300 ft long and the nozzle pressure is 80 psi (Figure 5-22).

A 1¾-in. tip at 80 psi nozzle pressure discharges 800 gpm. Each line is carrying approximately 267 gpm. It is necessary to interpolate odd gallonages into the standard friction loss figures. Knowing that 250 gpm results in 15 psi friction loss and 300 gpm causes 21 psi friction loss, it may be estimated that 17 psi is the approximate friction loss in 100 feet of 2½-in. hose when the flow rate is 267 gpm.

3 × 17 = 51 psi friction loss
80 psi nozzle pressure
25 psi wagon battery allowance
156 psi engine pressure

Note: The addition of the third line reduces the friction loss by one-half. Now a 1000 gpm pumper could easily supply this wagon battery.

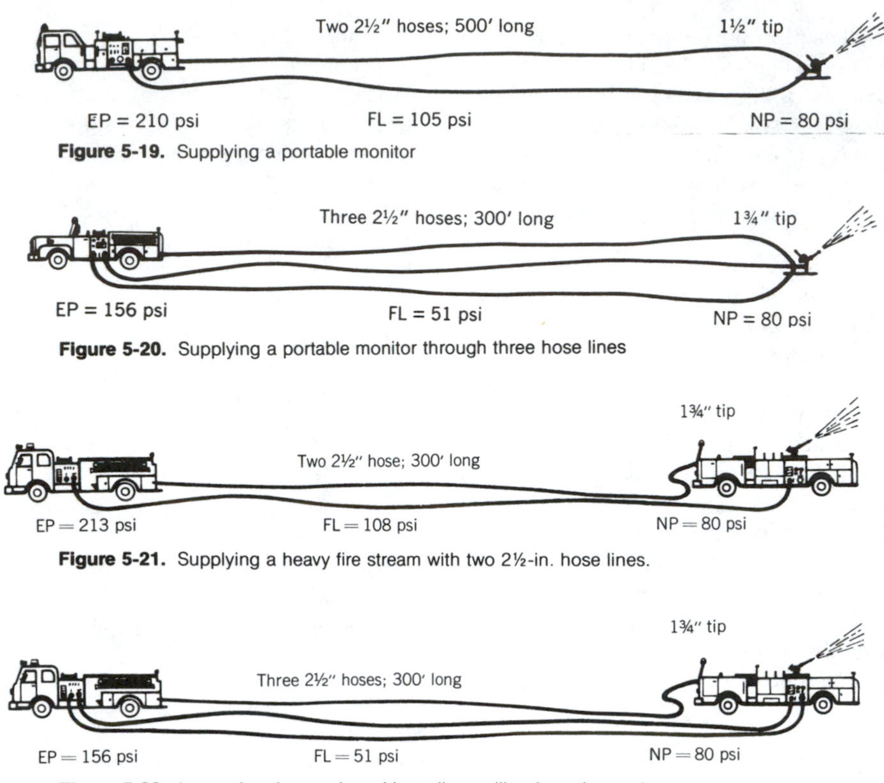

Figure 5-19. Supplying a portable monitor

Figure 5-20. Supplying a portable monitor through three hose lines

Figure 5-21. Supplying a heavy fire stream with two 2½-in. hose lines.

Figure 5-22. Increasing the number of hose lines will reduce the engine pressure.

Example: Find the engine pressure required to supply a ladder pipe equipped with a 500-gpm spray nozzle through three lines of 2½-in. hose which average 500 ft in length. The nozzle pressure is 100 psi (Figure 5-23).

An 80 psi allowance (irrespective of ladder height or volume of flow) is made for friction loss and back pressure in the entire ladder pipe assembly.

When 500 gpm are flowing through three lines of 2½-in. hose, each line is carrying about 167 gpm. Knowing that there is a 10 psi friction loss when the flow rate is 200 gpm, mental interpolation will establish that there is approximately 7 psi friction loss in each 100 ft of 2½-in. hose when 167 gpm are flowing.

$5 \times 7 =$ 35 psi friction loss
80 psi for ladder pipe assembly
100 psi nozzle pressure
215 psi engine pressure

Example: Find the engine pressure required to supply a ladder pipe equipped with a 1½-in. tip through three 2½-in. hose lines which average 400 ft in length (Figure 5-24).

The standard nozzle pressure for smoothbore tips is 80 psi. A 1½-in. nozzle at 80 psi nozzle pressure is discharging 600 gpm. Each hose line is carrying about 200 gpm. Friction loss when 200 gpm are flowing through 100 ft of single 2½-in. hose equals 10 psi. Because there is 10 psi friction loss in

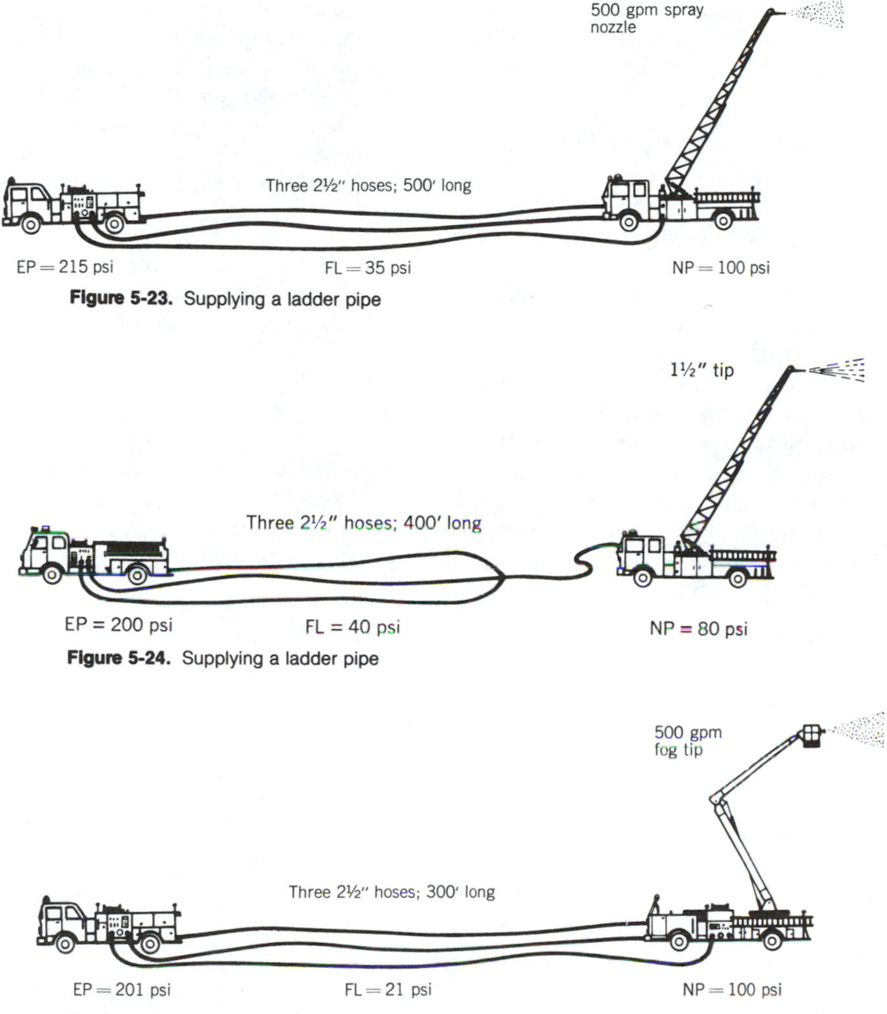

Figure 5-23. Supplying a ladder pipe

Figure 5-24. Supplying a ladder pipe

Figure 5-25. Supplying an aerial platform

each 100 ft of single 2½-in. hose when 200 gpm are flowing, there is 10 psi friction loss in each 100 ft of three 2½-in. lines when the flow rate is 600 gpm.

4 × 10 =	40 psi friction loss
	80 psi for ladder pipe assembly
	80 psi nozzle pressure
	200 psi engine pressure

Example: Find the engine pressure required to supply an aerial platform equipped with a 500-gpm spray nozzle through three lines of 2½-in. hose which average 300 ft in length. The nozzle pressure is 100 psi (Figure 5-25).

An 80 psi allowance, regardless of height or flow, is allowed for friction loss and back pressure in the entire aerial platform piping.

When 500 gpm are flowing through three lines of 2½-in. hose, each line is carrying about 167 gpm. Knowing that there is a 10 psi friction loss when 200 gpm are flowing, then mental interpolation will establish that there is approximately 7 psi friction loss in each 100 ft of the hose layout when 167 gpm are flowing through each of the three hose lines.

$3 \times 7 =$ 21 psi friction loss
80 psi for aerial platform
100 psi nozzle pressure
201 psi engine pressure

Conversion of various hose sizes. When the hose layout includes other than 2½-in. lines, to figure friction loss there must be some method of converting the various hose sizes to an equivalent amount of 2½-in. hose or of comparing the amounts of water that may flow through them. The following method converts the gpm being delivered through the various sizes of hose into equivalent gpm through a single 2½-in. hose line by means of a conversion factor:

Hose sizes	*Factor*
3½″ to 2½″	0.4
3″ to 2½″	0.6
1½″ to 2½″	4.0
1″ to 2½″	11.0
¾″ to 2½″	24.0

The equivalent gpm flow in a single 2½-in. hose is determined by multiplying the quantity delivered through the various sized hoses by the conversion factor. The friction loss can then be obtained by using the formula $FL = 2Q^2 + Q$ (for flows less than 100 gpm, use $FL = 2Q^2 + \frac{1}{2}Q$) or by referring to a friction loss table or chart. The commonly used friction loss amounts may be easily memorized. For ease of mental calculations on the fire ground, the formula $FL = 2Q^2 + Q$ may be simplified to $FL = 2Q^2$ without great loss of accuracy.

When parallel lines of other than 2½-in. hose are used, average the length of the lines and divide the flow between the parallel lines before converting; i.e., if there are 600 gpm flowing through two parallel lines of 3-in. hose, one 600 ft long and the other 400 ft long, figure the problem for 300 gpm flowing through 500 ft of 3-in. hose.

Example: 800 gpm are being delivered through a single 3½-in. hose line. The equivalent gpm in a single 2½-in. line would be $0.4 \times 800 = 320$ gpm. By using the formula $FL = 2Q^2 + Q$, or by referring to a friction loss table, it will be found that the friction loss amounts to 24 psi in each 100 ft of hose.

Example: Find the engine pressure necessary to supply 80 psi nozzle pressure to the 1½-in. tip on an aerial platform that is being supplied through a single 3½-in. hose line which is 450 ft long (Figure 5-26).

A 1½-in. tip at 80 psi nozzle pressure discharges 600 gpm. The equivalent flow in a single 2½-in. hose will be $0.4 \times 600 = 240$ gpm. When 240 gpm are flowing through a single 2½-in. hose line, the line loses about 14 psi in friction loss for every 100 ft in length.

$4\frac{1}{2} \times 14 =$ 63 psi friction loss
80 psi nozzle pressure
80 psi for aerial platform
223 psi engine pressure

Adding a parallel 3½-in. hose line causes each hose to carry 300 gpm (Figure 5-27). This reduces the engine pressure; 300 gpm in a single 3½-in. hose line is equivalent to 120 gpm flowing through a single 2½-in. hose line. $FL = 2Q^2 + Q$, so $FL = 4$ psi.

$4\frac{1}{2} \times 4 =$ 18 psi friction loss
80 psi nozzle pressure
80 psi for aerial platform
178 psi engine pressure

Example: Find the engine pressure necessary to supply 100 psi nozzle pressure to the 600 gpm fog tip on a wagon battery through one 3½-in. hose line 600 ft in length (Figure 5-28).

The 600 gpm flow through 3½-in. hose is equivalent to 240 gpm passing through a single 2½-in. line ($0.4 \times 600 = 240$ gpm). When 240 gpm is flowing through a single 2½-in. line, it loses 14 psi in friction loss for every 100 ft in length.

A 25 psi allowance should be made for friction loss and back pressure in the wagon battery.

6 × 14 =	84 psi friction loss
	25 psi for wagon battery
	100 psi nozzle pressure
	209 psi engine pressure

Example: Find the engine pressure required to supply 500 ft of 3½-in. line which is wyed into two 600 ft lengths of 2½-in. hose. There is a 1⅛-in. smooth-bore nozzle on each line (Figure 5-29).

Each 1⅛-in. tip, with the standard nozzle pressure of 50 psi, discharges 250 gpm. Therefore, each 2½-in. line is carrying 250 gpm, and there are 500 gpm flowing through the 3½-in. hose.

The same pressure that will produce a 500 gpm flow in 3½-in. hose will produce 0.4 × 500 = 200 gpm in 2½-in. hose. When 200

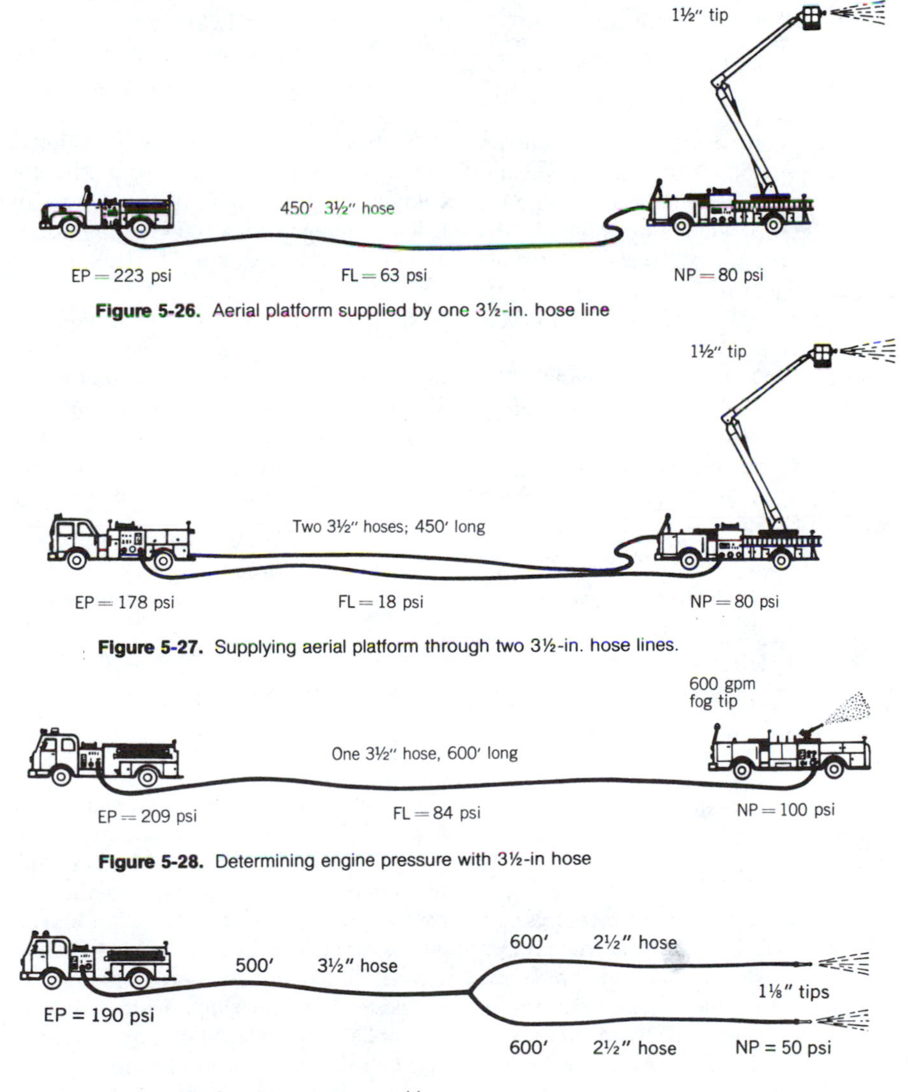

Figure 5-26. Aerial platform supplied by one 3½-in. hose line

Figure 5-27. Supplying aerial platform through two 3½-in. hose lines.

Figure 5-28. Determining engine pressure with 3½-in hose

Figure 5-29. Supplying a wye assembly

gpm are flowing through a single 2½-in. hose, 10 psi friction loss is created. Likewise, 10 psi friction loss is created in each 100 ft of 3½-in. hose when 500 gpm are flowing.

When 250 gpm are flowing through a single 2½-in. line, 15 psi friction loss is created in every 100 ft.

5 × 10 =	50 psi friction loss in 3½-in. hose
6 × 15 =	90 psi friction loss in wyed 2½-in. hose
	50 psi nozzle pressure
	190 psi engine pressure

Example: Find the engine pressure required to supply 300 ft of 3-in. line which is wyed into two lengths of 2½-in. hose. One length of 2½-in. is 400 ft long and has a 1⅛-in. tip on it; the other 2½-in. is 200 ft long and is equipped with a 2½-in. spray nozzle which discharges 325 gpm at 75 psi nozzle pressure (Figure 5-30).

If the friction losses in the individual 2½-in. lines vary considerably, then an average loss must be taken; when the variation is small, pump to the higher pressure.

A 1⅛-in. tip at the standard 50 psi nozzle pressure discharges 250 gpm; this causes a friction loss of 15 psi per 100 feet.

4 × 15 =	60 psi friction loss
	50 psi nozzle pressure
	110 psi engine pressure

A 325-gpm spray nozzle at 75 psi nozzle pressure creates a friction loss of 25 psi per 100 ft.

2 × 25 =	50 psi friction loss
	75 psi nozzle pressure
	125 psi engine pressure

In this case, average the two pressures and use the figure of 118 psi for the friction loss in the 2½-in. hose and the nozzle pressures.

In actual use at a fire, the pump operator would use the length of 2½-in. hose and nozzle size which from experience he believed necessary. It would be neither practical nor necessary to compute both lengths of 2½-in. unless there was an extremely wide variation.

When 575 gpm are flowing through a single length of 3-in. hose (250 and 325), its friction loss is equal to that of 345 gpm (0.6 × 575) flowing through a single 2½-in. line. This amounts to 27 psi friction loss in each 100 ft of 3-in. hose

118 psi for nozzle pressure and friction loss in 2½-in.
81 psi friction loss in 3-in. hose
199 psi engine pressure

Example: Find the engine pressure required to supply a 1½-in. wye assembly composed of two lengths of 1½-in. hose 100 ft long equipped with 100-gpm fog nozzles at 75 psi nozzle pressure. The length of the 2½-in. line is 800 ft (Figure 5-31).

Multiply the flow passing through one 1½-in. line (100 gpm) by the conversion factor (4.0) to obtain the equivalent flow in 2½-in. hose; 4 × 100 = 400 gpm. This produces a friction loss of 36 psi per 100 ft.

The friction loss caused by 200 gpm passing through 100 ft of 2½-in. line is 10 psi.

8 × 10 =	80 psi friction loss in 2½-in. hose
	36 psi friction loss in 1½-in. hose
	75 psi nozzle pressure
	191 psi engine pressure

Under normal conditions, 1½-in. wye assemblies, preconnected lines, ready lines, and other standard hose layouts should have a precalculated allowance to simplify fire ground operations. In the above example, the pump operator would always allow 110 psi for nozzle pressure and friction loss in a wye assembly consisting of two 1½-in. lines, 100 ft in length, and two 100-gpm fog nozzles at 75 psi nozzle pressure. Other wye assemblies or standard hose layouts should also be calculated ahead of time.

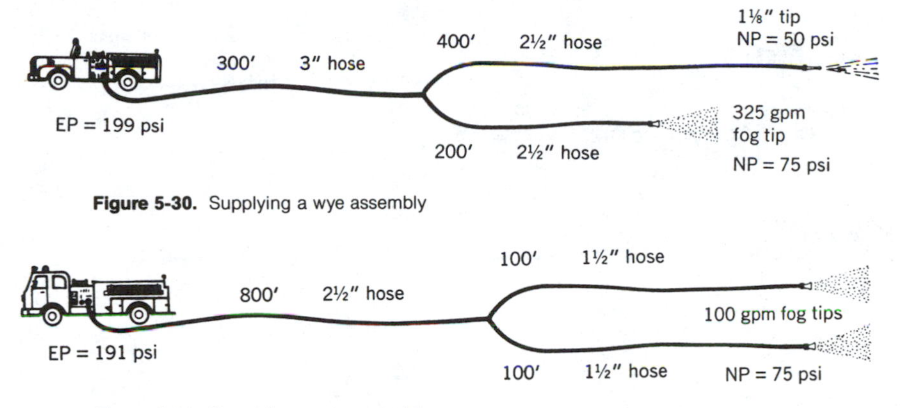

Figure 5-30. Supplying a wye assembly

Figure 5-31. Supplying a wye assembly

TABLE 5-5. FRICTION LOSS

HOSE SIZE, in.	NOZZLE SIZE, in.	NOZZLE PRESSURE, psi	FLOW, gpm	FRICTION LOSS PER 100 ft, psi
1	1/4	50	13	5
1	5/16	50	20	12
1	3/8	50	29	25
1 1/2	1/4	50	13	1
1 1/2	5/16	50	20	2
1 1/2	3/8	50	29	3
1 1/2	1/2	50	52	10
1 1/2	5/8	50	81	25

NOTE—Friction loss figures rounded off to nearest whole number.

Small lines. Friction loss in small lines may be determined by using the conversion factors. However, with standard nozzle sizes and pressures, the friction loss can be precalculated; the calculations will depend upon hose size and gpm flow of the nozzle. By remembering a few related amounts, fire ground calculations are greatly simplified. It is not necessary to memorize entire charts or tables, just the friction losses related to nozzles and hoses carried on the apparatus. The friction loss figures in Table 5-5 are rounded off.

Fog and spray nozzles have a large variety of nozzle pressures and discharge flows. An operator should refer to a friction loss conversion chart, and then round off the corresponding friction loss, to determine quickly and easily the proper pressure loss for the various flows.

Example: What engine pressure is necessary to supply a 100-gpm spray nozzle, at 75 psi pressure, through 500 ft of 1½-in. hose (Figure 5-32)?

The factor to convert the flow of water through 1½-in. hose to 2½-in. hose is 4.0. Therefore, 100 gpm passing through 1½-in. hose is equivalent to 400 gpm flowing through an equal length of 2½-in. line (4.0 × 100 = 400 gpm). The friction loss is 36 psi per 100 feet ($FL = 2Q^2 + Q$).

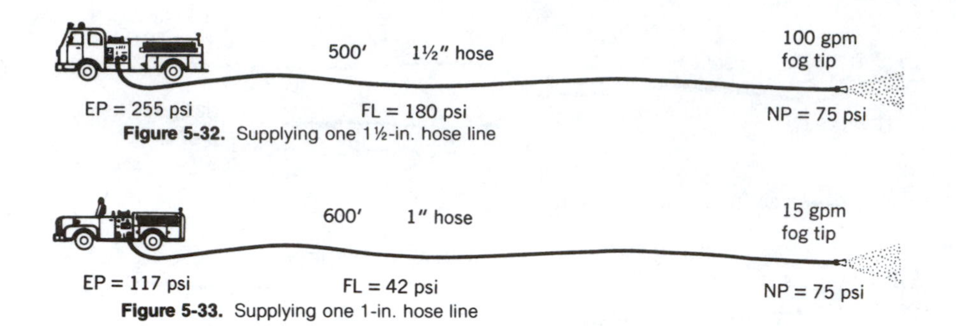

Figure 5-32. Supplying one 1½-in. hose line

Figure 5-33. Supplying one 1-in. hose line

5 × 36 = 180 psi friction loss
75 psi nozzle pressure
255 psi engine pressure

Example: What engine pressure is necessary to supply a 15-gpm spray nozzle, at 75 psi pressure, through 600 ft of 1-in. hose (Figure 5-33)?

The factor to convert the flow of water through 1-in. hose to 2½-in. hose is 11.0. Therefore, 15 gpm passing through 1-in. hose is equivalent to 165 gpm flowing through an equal length of 2½-in. line (11.0 × 15 = 165 gpm). The friction loss amounts to 7 psi per 100 ft ($FL = 2Q^2 + Q$).

6 × 7 = 42 psi friction loss
75 psi nozzle pressure
117 psi engine pressure

Unequal discharge pressure. When a pumper must deliver water into two or more hose lines, each of which requires a different engine pressure, the operator should pump to the highest needed pressure and then close down the discharge gates to reduce pressures.

Example: Find the engine pressure required to supply the following hose lays. The left front discharge is connected to 600 ft of 2½-in. line with a 1⅛-in. tip. The left rear discharge is delivering water to 400 ft of 2½-in. hose equipped with a wye assembly composed of two 100-ft lengths of 1½-in. hose and two 100-gpm fog nozzles at 75 psi nozzle pressure. The right rear discharge is connected to 500 ft of 2½-in. hose which is equipped with a 2½-in. fog nozzle that discharges 300 gpm at 75 psi nozzle pressure. What will be the required engine pressure? What pressures will be shown on the individual gages which are connected to the discharge gates (Figure 5-34)?

The left front discharge gate requires a pressure of 140 psi.

The left rear discharge gate requires 150 psi pressure.

The right rear discharge gate requires 180 psi pressure.

Therefore, an engine pressure of 180 psi is necessary to supply these hose layouts. The pump discharge gage and the individual gage for the right rear discharge gate will register a pressure of 180 psi. The discharge gates on the left front and the left rear outlets will be closed down until the individual gages indicate pressures of 140 psi and 150 psi respectively.

Example: A 1000-gpm pumper at draft is supplying four 1⅛-in. tips through four 2½-in. lines, each 600 ft long. What is the engine pressure (Figure 5-35)?

When more than one hose line is supplied by the same pumper, hose lines are usually near equal length. If the nozzles are the

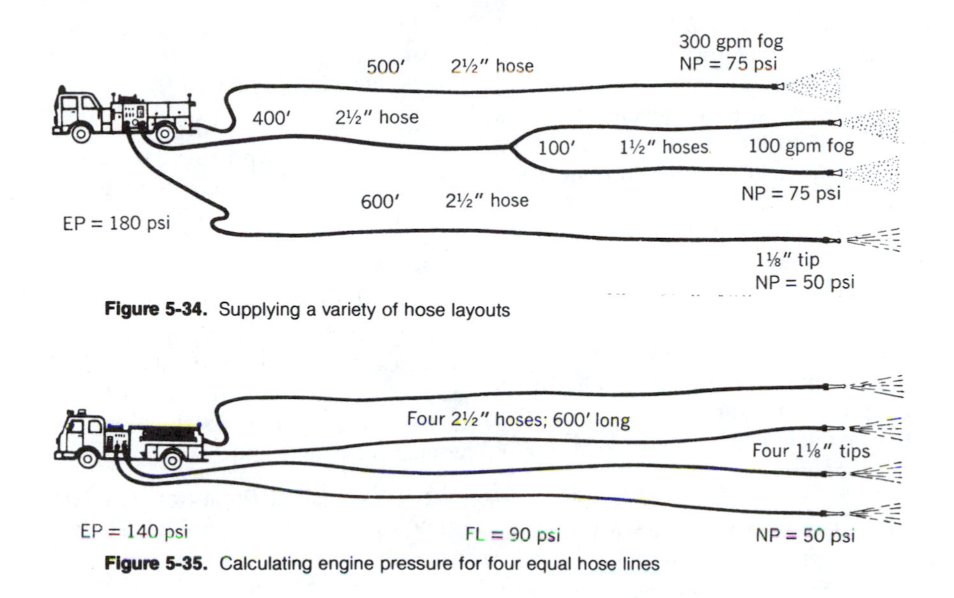

Figure 5-34. Supplying a variety of hose layouts

Figure 5-35. Calculating engine pressure for four equal hose lines

same, the flow is equal and the resistance to be overcome by the pumper is the same. The pump speed must be increased to deliver additional water, but the discharge pressure at the pump will remain the same. When hose lines are about the same length, and gpm flows are similar, figures for only one line need be computed because the other lines will have approximately the same pressure.

A 1⅛-in. tip at 50 psi nozzle pressure delivers 250 gpm. A friction loss of 15 psi per 100 feet occurs when 250 gpm flows through a single 2½-in. line.

$6 \times 15 =$ 90 psi friction loss
50 psi nozzle pressure
140 psi engine pressure

A 1000-gpm pumper delivers its full capacity at 150 psi net pump pressure; this is just about all the water that this pumper can handle when it is at draft. ("Net Pump Pressure" is the pressure actually developed in the pump; it is determined by subtracting intake pressure from discharge pressure.)

Back pressure. When pumping up or down hills, or to upper floors of a building, an allowance of 0.434 psi must be made for each difference of a foot in elevation. When pumping up hills or to an upper floor of a building, engine pressure *(EP)* = friction loss *(FL)* + nozzle pressure *(NP)* + back pressure *(BP)*, or

$$EP = FL + NP + BP$$

When pumping down hills or to a basement, gravity will help the pump to provide pressure:

$$EP = FL + NP - BP$$

so the back pressure is subtracted (in this case, back pressure may be referred to as forward pressure).

For ease of computation, allow 4½ psi for each 10 ft. A floor in a building is considered to be 12 ft in height, so a standard back pressure of 5 psi per story above the first floor should be allowed.

When pumping up or down hills with unknown grades, assume a 10% grade. The rise or drop will be 10 ft for each 100 ft of hose laid out, or 4½ psi.

Example: What back pressure would be developed if 1000 ft of hose were laid up a mountainous road and the slope were not known?

Figure it as a 10% grade, or 4½ psi loss for each 100 ft of hose; for 1000 ft, this amounts to 45 psi back pressure.

Example: Find the back pressure that should be overcome when pumping to the fifth floor of a building.

The first floor is not counted, so 4 × 5 = 20 psi back pressure.

Example: What is the engine pressure necessary to supply a portable monitor equipped with a 600-gpm fog nozzle placed on the roof of a one-story building? Nozzle pressure is 100 psi. The pump is positioned 20 ft lower than the base of the building and is pumping through two parallel lines of 2½-in. hose, each 300 ft long (Figure 5-36).

The friction loss produced by 600 gpm flowing through two parallel lines of 2½-in. hose is the same as that produced by 300 gpm passing through a single line, or 21 psi per 100 ft.

A 25 psi allowance should be made for friction losses developed in the portable monitor.

The roof of a one-story building gives the same back pressure as the second floor or 5 psi.

3 × 21 =	63 psi friction loss
2 × 4½ =	9 psi back pressure for elevation
	25 psi allowance for portable monitor
	5 psi back pressure for building
	100 psi nozzle pressure
	202 psi engine pressure

Example: A 200-gpm fog nozzle working on the roof of a two-story building is supplied by 450 ft of single 2½-in. hose. Nozzle pressure is 75 psi; what is the required engine pressure (Figure 5-37)?

10 × 4½ =	45 psi friction loss
2 × 5 =	10 psi back pressure
	75 psi nozzle pressure
	130 psi engine pressure

Example: A pumper is supplying a standpipe through two parallel lines of 2½-in. hose, each 200 ft long. To the tenth-floor outlet is connected 100 ft of 2½-in. hose with a 1-in. smooth bore nozzle. What is the engine pressure (Figure 5-38)?

2 × 3 =	6 psi *FL* in two 2½-in. lines
1 × 10 =	10 psi *FL* in single 2½-in. line
9 × 5 =	45 psi back pressure
	25 psi allowance for standpipe
	50 psi nozzle pressure
	136 psi engine pressure

Example: Determine the discharge pressure of an engine company pumping through 800 ft of 2½-in. hose down an incline into a 1-in. nozzle placed 50 ft below the pump (Figure 5-39).

A 1-in. nozzle at 50 psi pressure discharges 200 gpm; this amount flowing through a single 2½-in. hose develops 10 psi friction loss in every 100 ft.

8 × 10 =	80 psi friction loss
	50 psi nozzle pressure
	130 psi engine pressure

For each 10 ft of elevation, there is 4½ psi back pressure.

The nozzle is below the pumper, and gravity aids the pump; this effect of elevation is sometimes referred to as forward pressure.

	130 psi engine pressure
−	23 psi forward pressure
	107 psi adjusted engine pressure

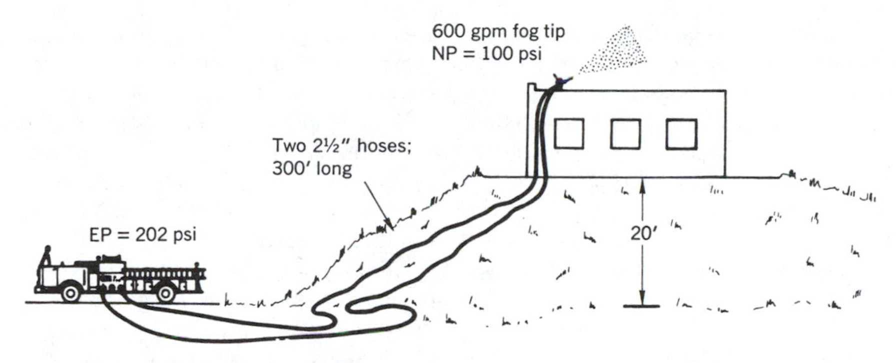

Figure 5-36. Supplying a portable monitor on a building

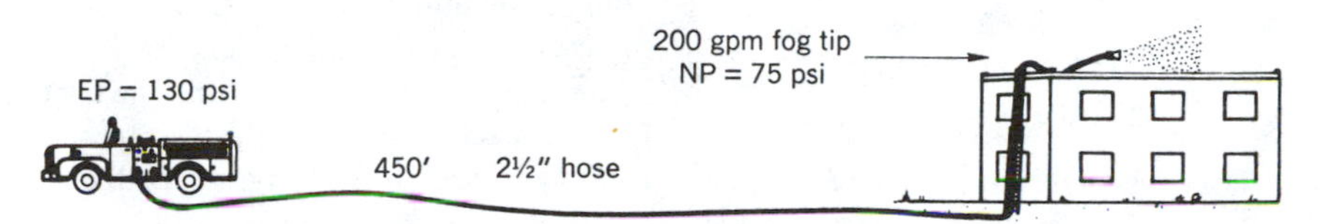

Figure 5-37. Determining engine pressure, two-story building

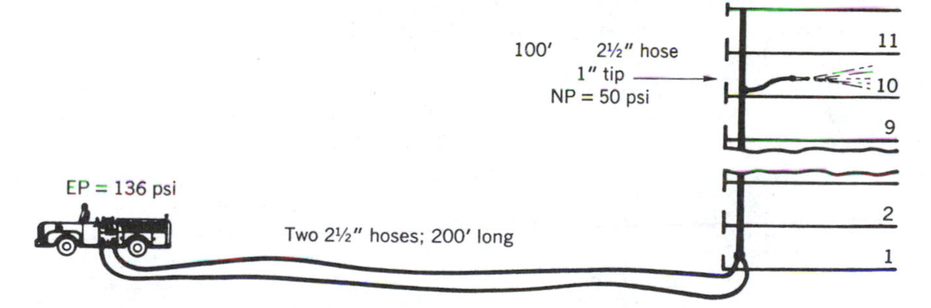

Figure 5-38. Determining the engine pressure for a fire on the tenth floor of a building

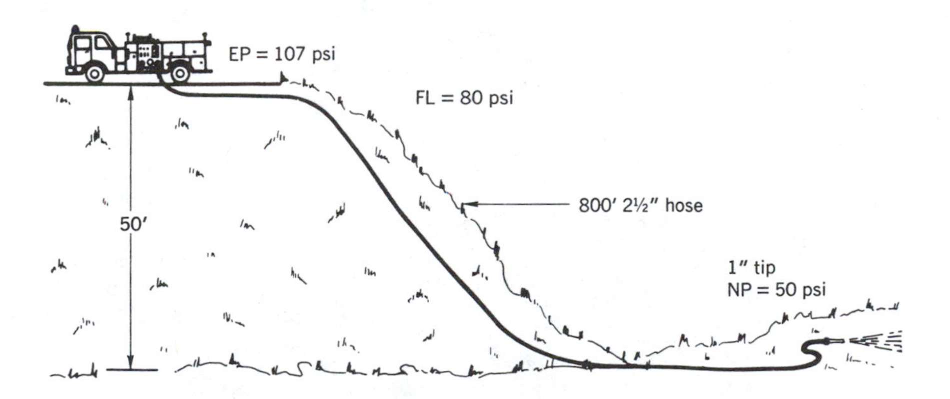

Figure 5-39. Calculating engine pressure when pumping downhill

Example: A fire is on the fourth floor of a building which is located on top of a 100-ft hill. A pumper at the bottom of the hill must supply 100 psi nozzle pressure to a 300-gpm spray tip through 800 ft of 2½-in. hose. What engine pressure is required (Figure 5-40)?

The hill back pressure amounts to 0.434 psi for each foot of elevation; for convenience, allow 4½ psi for each 10 ft of rise. Building back pressure averages 5 psi for each floor above the first.

10 × 4½ = 45 psi hill back pressure
3 × 5 = 15 psi building back pressure
8 × 21 = 168 psi friction loss
100 psi nozzle pressure
328 psi engine pressure

For most pumpers, this pressure would be difficult to attain; also, fire hose may burst at the higher pressures. The fastest and most practical method of lowering the required pump discharge pressure to a practical amount is to replace the tip with a smaller one and lower the nozzle pressure. A 200-gpm spray tip at 70 psi nozzle pressure would require:

10 × 4½ = 45 psi hill back pressure
3 × 5 = 15 psi building back pressure
8 × 10 = 80 psi friction loss
70 psi nozzle pressure
210 psi engine pressure

If the extinguishing effect of a 300-gpm spray tip at 100 psi nozzle pressure is required for fire fighting, and a relay pumper cannot be placed on top of the hill by the building, then use parallel lines of 2½-in. hose or larger diameter hose to reduce the friction loss.

Unequal hose diameters. When parallel hose lines have different diameters, there are additional complications. For each line to carry its proportional share of the water, the friction loss in each hose line must be identical. This would require computing what percentage of the total flow would be carried by each line, individually calculating the friction loss in each line, adding on the other pressure requirements (nozzle pressure, back pressure, appliance allowance, etc.), and then regulating the pressure at each discharge gate so that the far ends of the hose lines would have identical pressures. Obviously, this would not be practical on the fire ground.

However, a practical solution may be obtained by calculating an average engine pressure for the lines; the flow through each hose would stabilize at a point where the friction losses are the same. Divide the total gpm by the number of hose lines (regardless of sizes), change to equivalent 2½-in. flows using the conversion factors, find the friction loss in each line, and then average the friction losses. To this figure, add on the other pressure requirements.

Example: Find the engine pressure necessary to supply 10 psi pressure to the suction inlets of a second pumper which is operating a 750-gpm spray tip on a wagon battery at 100 psi pressure. The hose lay consists of two parallel lines: one is 400 ft of 2½-in. hose and the other is 500 ft of 3½-in. hose (Figure 5-41). Also, what engine pressure would the second pumper require?

Divide the total gpm flow by the number of hose lines (750 gpm ÷ 2 = 375 gpm.) Using the conversion factor of 0.4, 375 gpm in a single 3½-in. hose is equivalent to 150 gpm in a single 2½-in. line (375 × 0.4 = 150).

Friction loss of 150 gpm in a single 2½-in. = 6 psi per 100 feet. Friction loss of 375 gpm in a single 2½-in. = 32 psi per 100 ft.

5 × 6 = 30 psi *FL* in 500 ft of 3½-in.
4 × 32 = 128 psi *FL* in 400 ft of 2½-in.
128 + 30 = 158 psi
158 ÷ 2 = 79 psi average friction loss in the two parallel lines. An inlet pressure of 10 psi is required.

The operator of the second pumper is not concerned with the supplying hose layout, as

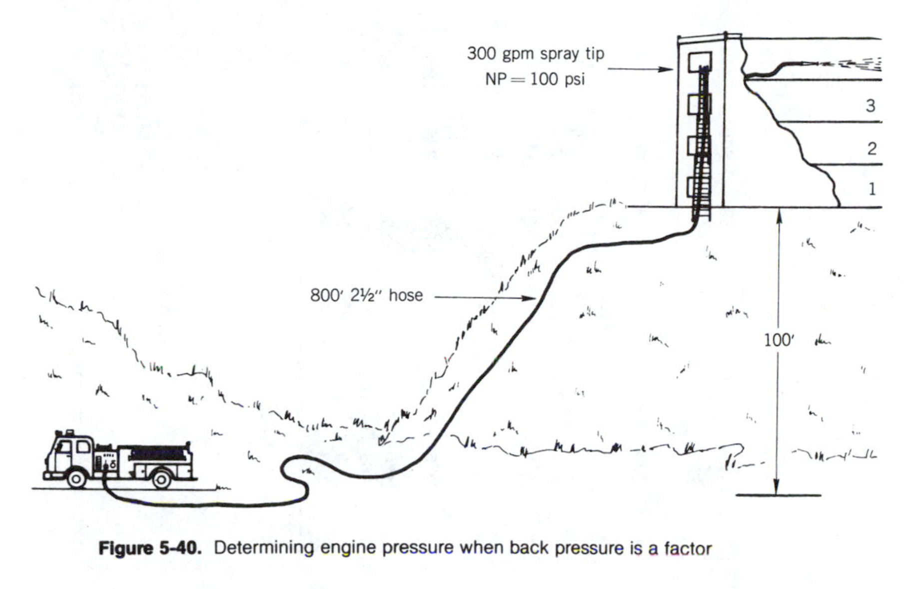

Figure 5-40. Determining engine pressure when back pressure is a factor

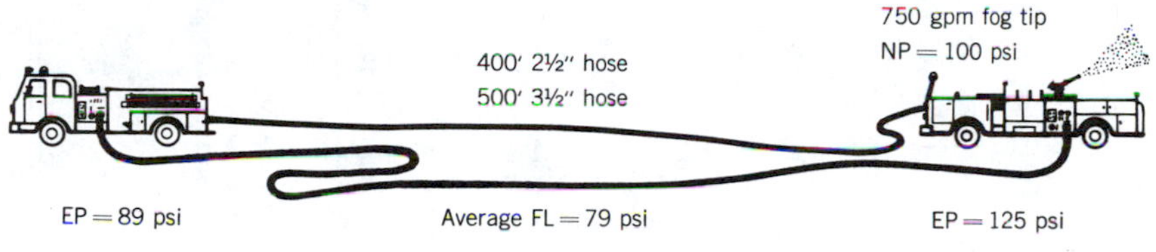

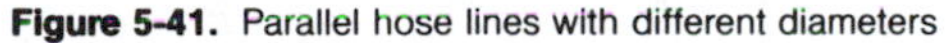

Figure 5-41. Parallel hose lines with different diameters

long as an adequate quantity of water is received. The operator's only concerns are the friction loss between the pump and the wagon battery, and the nozzle pressure:

25 psi wagon battery allowance
100 psi nozzle pressure
125 psi engine pressure

Sprinkler systems. On the fire ground, it is difficult to determine the pressure that will be required to supply a sprinkler system. Adequate working pressure depends upon the number of sprinkler heads open and the size and location of the heads. It is a good rule to pump at 150 psi, whether the fire is apparent or not. A change in pressure can be made after information is received about the amount of water which is flowing.

Hydraulics Calculator

A hydraulics calculator or chart (see Table 5-6) does not relieve an officer or pump operator of the responsibility for having the knowledge and ability to compute mentally the proper engine pressures, but it does serve the following purposes:

1. It will enable experienced pump operators to check their mental calculations.

2. It will be a guide for acting engineers until they become proficient in hydraulic computations.

TABLE 5-6. HYDRAULICS CHART

HOSE LAYOUT	TIP SIZE	NOZZLE PRESSURE, psi	FLOW, gpm	FRICTION LOSS, psi PER 100′	ENGINE PRESSURE, psi, PER LENGTH OF HOSE LAYOUT									
					100′	200′	300′	400′	500′	600′	700′	800′	900′	1000′
One	¼″	50	13	5	55	60	65	70	75	80	85	90	95	100
1″ Line	1″ Fog	75	15	7	80	90	95	105	110	115	125	130	140	145
One	⅝″	50	80	25	75	100	125	150	175	200	225	250	275	300
1½″	1½″ Fog	75	80	25	100	125	150	175	200	225	250	275	300	*
Line	1½″ Fog	75	100	35	110	145	180	215	250	285	*	*	*	*
	1½″ Wye Assembly	75	200	10	120	130	140	150	160	170	180	190	200	210
One	1″	50	200	10	60	70	80	90	100	110	120	130	140	150
2½″	1⅛″	50	250	15	65	80	95	110	125	140	155	170	185	200
Line	1¼″	50	325	25	75	100	125	150	175	200	225	250	275	300
	2½″ Fog	75	300	21	95	115	140	160	180	200	220	245	265	285
Two	1½″	80	600	21	100	120	145	165	185	205	225	250	270	290
2½″	1¾″	80	800	36	115	150	190	225	260	295	*	*	*	*
Lines	2″	80	1000	55	135	190	245	300	*	*	*	*	*	*
	2½″ Fog	100	500	15	115	130	145	160	175	190	205	220	235	250
Three	1½″	80	600	10	90	100	110	120	130	140	150	160	170	180
2½″	1¾″	80	800	17	95	115	130	150	165	180	200	215	235	250
Lines	2″	80	1000	25	105	130	155	180	205	230	255	280	*	*
	2½″ Fog	100	500	7	105	115	120	130	135	140	150	155	165	170

FRICTION LOSS AND BACK PRESSURE ALLOWANCES

Portable Monitor .... add 25 psi
Wagon Battery .... add 25 psi
Ladder Pipe .... add 80 psi
Standpipe .... add back pressure plus 25 psi

SPRINKLER INLET

Whether or not fire is showing .... pump at 150 psi

EQUIVALENT NOZZLE DIAMETERS

Cellar Nozzle .... 1¼″
Bresnan, Small .... 1½″
Bresnan, Large .... 1¾″

BACK PRESSURE ALLOWANCES

5 psi per floor .... 4½ psi per 10 ft
Mountain Slopes .... 4½ psi for each 100 ft of hose laid out.

*Annual hose test pressure (300 psi) should not be exceeded.

3. It will be a source of information so officers and pump operators may refresh their memories.

4. It will quickly show the benefits of laying additional hose lines to lower the friction loss.

Hydraulic charts should be individually designed for each fire department and should reflect only the hose layouts, tip sizes, and nozzle pressures that are standard with that organization. The calculator, Table 5-6, was designed for a department that uses only 1-in., 1½-in. and 2½-in. hose. This department has standardized nozzle pressures of:

50 psi for smooth-bore tips on hand lines;

75 psi for fog and spray tips on hand lines;

80 psi for smooth-bore tips on master stream appliances; and

100 psi for fog and spray tips on master stream appliances.

Fire departments using other sizes of hose, nozzles, or nozzle pressures should design a calculator that will indicate the conditions which their officers and pump operators are likely to encounter; other information that may be needed on the fire ground may be included. These charts should be affixed near the operator's panel on the apparatus and others should be readily available to all officers on the fire ground.

It should be realized that a centrifugal pump, unless aided by a good hydrant pressure, will not be able to exceed its rated capacity. Pumps are rated to deliver 100% of capacity at a net discharge pressure of 150 psi, 70% of capacity at 200 psi, and 50% of capacity at 250 psi. Therefore, unless aided by incoming pressure, a 1000-gpm pumper would discharge 1000 gpm at 150 psi, 700 gpm at 200 psi, and 500 gpm at 250 psi. Asterisks, instead of pressures, should be inserted in spaces where it would not be possible for the pumper, because of rated capacity, to supply the desired quantity at the required net pump pressure.

The extreme left-hand column of Table 5-6 describes the hose layout, while the right half is concerned with the length of the hose layout.

The second column lists the various tip sizes in common use by that department. The listed wye assembly consists of a gated wye fitting, two 100-ft lines of 1½-in. hose, and two 100-gpm fog nozzles. For this assembly, 110 psi is allowed ($NP = 75$ psi + $FL = 35$ psi).

The third column shows the standard nozzle pressures.

The fourth column gives the gallons per minute. These figures have been rounded off for ease of computing.

The fifth column shows the friction loss in 100 ft of the hose layout. These figures have been rounded off.

The figures under "Length of Hose Layout" indicate the correct engine pressures for ground lays. These figures have been rounded off to the nearest 5 psi and show the nozzle pressure and the friction loss ($EP = FL + NP$).

Example: What pump pressure is necessary to supply a 1⅛-in. tip through one 2½-in. line 500 ft long?

In the "One 2½-In. Line" bracket, follow the 1⅛-in. tip size column across the page until it intersects the 500-ft "Length of Hose Layout" column. This figure (125 psi) indicates the correct pressure to be pumped.

Example: The wagon battery of a pumper is being supplied by three 2½-in. lines into the pump suction through a relay. The battery is using a 2-in. tip, and the pump inlet pressure is 10 psi. What is the correct engine pressure?

The friction loss and back pressure in the supplying lines will not be of any consequence to the pump operator of the apparatus with the wagon battery, so long as an adequate volume of water is received. The operator's only concerns are the friction loss between the pump and the wagon battery and the nozzle pressure:

25 psi allowance for wagon battery
80 psi nozzle pressure
105 psi engine pressure

Example: A pumper is supplying a ladder pipe equipped with a 500-gpm fog nozzle through two 2½-in. lines which average 300 ft in length. What pump pressure is required?

In the "Two 2½-Inch Lines" bracket, follow the 500-gpm fog nozzle size across the chart until it intersects the 300-ft "Length of Hose Layout" column. To this figure (145), add the ladder pipe allowance (80 psi).

145 psi friction loss in two 2½-in. lines
80 psi allowance for ladder pipe
225 psi engine pressure

Example: In the previous example, how would the engine pressure be affected if a third 2½-in. line 300 ft long were laid between the pumper and the base of the ladder?

By comparing the corresponding figures on Table 5-6, it may be quickly ascertained that the pump pressure would be lowered 25 psi (145 psi − 120 psi = 25 psi).

Example: Two pumpers are relaying water to a wagon battery equipped with a 500-gpm fog nozzle. There are two 2½-in. hose lines that average 1000 ft in length between the first and the second pumpers. If 10 psi inlet pressure is desired on the second pumper, what engine pressure should be maintained on the first pumper?

When 500 gpm are flowing through two 2½-in. hose lines, 15 psi friction loss is developed in each 100 ft.

$10 \times 15 =$ 150 psi friction loss
10 psi inlet pressure
160 psi engine pressure

TABLE 5-7. THEORETICAL DISCHARGE OF NOZZLES IN U.S. GALLONS PER MINUTE

HEAD		VELOCITY OF DISCHARGE,	DIAMETER OF NOZZLE, in.												
lb	ft	ft PER sec	1/16	1/8	3/16	1/4	3/8	1/2	5/8	3/4	7/8	1	1 1/8	1 1/4	1 3/8
10	23.1	38.6	0.37	1.48	3.32	5.91	13.3	23.6	36.9	53.1	72.4	94.5	120	148	179
15	34.6	47.25	0.45	1.81	4.06	7.24	16.3	28.9	45.2	65.0	88.5	116.	147	181	219
20	46.2	54.55	0.52	2.09	4.69	8.35	18.8	33.4	52.2	75.1	102.	134.	169	209	253
25	57.7	61.0	0.58	2.34	5.25	9.34	21.0	37.3	58.3	84.0	114.	149.	189	234	283
30	69.3	66.85	0.64	2.56	5.75	10.2	23.0	40.9	63.9	92.0	125.	164.	207	256	309
35	80.8	72.2	0.69	2.77	6.21	11.1	24.8	44.2	69.0	99.5	135.	177.	224	277	334
40	92.4	77.2	0.74	2.96	6.64	11.8	26.6	47.3	73.8	106.	145.	189.	239	296	357
45	103.9	81.8	0.78	3.13	7.03	12.5	28.2	50.1	78.2	113.	153.	200.	253	313	379
50	115.5	86.25	0.83	3.30	7.41	13.2	29.7	52.8	82.5	119.	162.	211.	267	330	399
55	127.0	90.4	0.87	3.46	7.77	13.8	31.1	55.3	86.4	125.	169.	221.	280	342	418
60	138.6	94.5	0.90	3.62	8.12	14.5	32.5	57.8	90.4	130.	177.	231.	293	366	438
65	150.1	98.3	0.94	3.77	8.45	15.1	33.8	60.2	94.0	136.	184.	241.	305	376	455
70	161.7	102.1	0.98	3.91	8.78	15.7	35.2	62.5	97.7	141.	191.	250.	317	391	473
75	173.2	105.7	1.01	4.05	9.08	16.2	36.4	64.7	101.	146.	198.	259.	327	404	489
80	184.8	109.1	1.05	4.18	9.39	16.7	37.6	66.8	104.	150.	205.	267.	338	418	505
85	196.3	112.5	1.08	4.31	9.67	17.3	38.8	68.9	108.	155.	211.	276.	349	431	521
90	207.9	115.8	1.11	4.43	9.95	17.7	39.9	70.8	111.	160.	217.	284.	359	443	536
95	219.4	119.0	1.14	4.56	10.2	18.2	41.0	72.8	114.	164.	223.	292.	369	456	551
100	230.9	122.0	1.17	4.67	10.5	18.7	42.1	74.7	117.	168.	229.	299.	378	467	565
105	242.4	125.0	1.20	4.79	10.8	19.2	43.1	76.5	120.	172.	234.	306.	388	479	579
110	254.0	128.0	1.23	4.90	11.0	19.6	44.1	78.4	122.	176.	240.	314.	397	490	593
115	265.5	130.9	1.25	5.01	11.2	20.0	45.1	80.1	125.	180.	245.	320.	406	501	606
120	277.1	133.7	1.28	5.12	11.5	20.5	46.0	81.8	128.	184.	251.	327.	414	512	619
125	288.6	136.4	1.31	5.22	11.7	20.9	47.0	83.5	130.	188.	256.	334.	423	522	632
130	300.2	139.1	1.33	5.33	12.0	21.3	48.0	85.2	133.	192.	261.	341.	432	533	645

(Continued)

TABLE 5-7 (Cont.). THEORETICAL DISCHARGE OF NOZZLES IN U.S. GALLONS PER MINUTE

HEAD		VELOCITY OF DISCHARGE	DIAMETER OF NOZZLE, in.												
lb	ft	ft PER sec	1/16	1/8	3/16	1/4	3/8	1/2	5/8	3/4	7/8	1	1 1/8	1 1/4	1 3/8
135	311.7	141.8	1.36	5.43	12.2	21.7	48.9	86.7	136.	195.	266.	347.	439	543	656
140	323.3	144.3	1.38	5.53	12.4	22.1	49.8	88.4	138.	199.	271.	354.	448	553	668
145	334.8	146.9	1.41	5.62	12.6	22.5	50.6	89.9	140.	202.	275.	360.	455	562	680
150	346.4	149.5	1.43	5.72	12.9	22.9	51.5	91.5	143.	206.	280.	366.	463	572	692
175	404.1	161.4	1.55	6.18	13.9	24.7	55.6	98.8	154.	222.	302.	395.	500	618	747
200	461.9	172.6	1.65	6.61	14.8	26.4	59.5	106.	165.	238.	323.	423.	535	660	799
250	577.4	193.0	1.85	7.39	16.6	29.6	66.5	118.	185.	266.	362.	473.	598	739	894
300	692.8	211.2	2.02	8.08	18.2	32.4	72.8	129.	202.	291.	396.	517.	655	808	977

HEAD		VELOCITY OF DISCHARGE	DIAMETER OF NOZZLE, in.												
lb	ft	ft PER sec	1 1/2	1 3/4	2	2 1/4	2 1/2	2 3/4	3	3 1/2	4	4 1/2	5	5 1/2	6
10	23.1	38.6	213	289	378	479	591	714	851	1158	1510	1915	2365	2855	3405
15	34.6	47.25	260	354	463	585	723	874	1041	1418	1850	2345	2890	3490	4165
20	46.2	54.55	301	409	535	676	835	1009	1203	1638	2135	2710	3340	4040	4810
25	57.7	61.0	336	458	598	756	934	1128	1345	1830	2385	3025	3730	4510	5380
30	69.3	66.85	368	501	655	828	1023	1236	1473	2005	2615	3315	4090	4940	5895
35	80.8	72.2	398	541	708	895	1106	1335	1591	2168	2825	3580	4415	5340	6370
40	92.4	77.2	425	578	756	957	1182	1428	1701	2315	3020	3830	4725	5710	6810
45	103.9	81.8	451	613	810	1015	1252	1512	1802	2455	3200	4055	5000	6050	7210
50	115.5	86.25	475	647	845	1070	1320	1595	1900	2590	3375	4275	5280	6380	7600
55	127.0	90.4	498	678	886	1121	1385	1671	1991	2710	3540	4480	5530	6690	7970
60	138.6	94.5	521	708	926	1172	1447	1748	2085	2835	3700	4685	5790	6980	8330
65	150.1	98.3	542	737	964	1220	1506	1819	2165	2950	3850	4875	6020	7270	8670
70	161.7	102.1	563	765	1001	1267	1565	1888	2250	3065	4000	5060	6250	7560	9000
75	173.2	105.7	582	792	1037	1310	1619	1955	2330	3170	4135	5240	6475	7820	9320

(Continued)

TABLE 5-7 (Cont.). THEORETICAL DISCHARGE OF NOZZLES IN U.S. GALLONS PER MINUTE

HEAD		VELOCITY OF DISCHARGE,	DIAMETER OF NOZZLE, in.												
lb	ft	ft PER sec	1 1/2	1 3/4	2	2 1/4	2 1/2	2 3/4	3	3 1/2	4	4 1/2	5	5 1/2	6
80	184.8	109.1	602	818	1070	1354	1672	2020	2405	'3280	4270	5410	6690	8080	9630
85	196.3	112.5	620	844	1103	1395	1723	2080	2480	3375	4400	5575	6890	8320	9920
90	207.9	115.8	638	868	1136	1436	1773	2140	2550	3475	4530	5740	7090	8560	10210
95	219.4	119.0	656	892	1168	1476	1824	2200	2625	3570	4655	5900	7290	8800	10500
100	230.9	122.0	672	915	1196	1512	1870	2255	2690	3663	4775	6050	7470	9030	10770
105	242.4	125.0	689	937	1226	1550	1916	2312	2755	3750	4890	6200	7650	9250	11020
110	254.0	128.0	705	960	1255	1588	1961	2366	2820	3840	5010	6350	7840	9470	11300
115	265.5	130.9	720	980	1282	1621	2005	2420	2885	3930	5120	6490	8010	9680	11550
120	277.1	133.7	736	1002	1310	1659	2050	2470	2945	4015	5225	6630	8180	9900	11800
125	288.6	136.4	751	1022	1338	1690	2090	2520	3005	4090	5340	6760	8350	10100	12030
130	300.2	139.1	767	1043	1365	1726	2132	2575	3070	4175	5450	6900	8530	10300	12290
135	311.7	141.8	780	1063	1390	1759	2173	2620	3125	4250	5550	7030	8680	10490	12710
140	323.3	144.3	795	1082	1415	1790	2212	2670	3180	4330	5650	7160	8850	10690	12930
145	334.8	146.9	809	1100	1440	1820	2250	2715	3235	4410	5740	7280	8990	10880	12960
150	346.4	149.5	824	1120	1466	1853	2290	2760	3295	4485	5850	7410	9150	11070	13200
175	404.1	161.4	890	1210	1582	2000	2473	2985	3560	4840	6310	8000	9890	11940	14250
200	461.9	172.6	950	1294	1691	2140	2645	3190	3800	5175	6750	8550	10580	12770	15220
250	577.4	193.0	1063	1447	1891	2392	2955	3570	4250	5795	7550	9570	11820	14290	17020
300	692.8	211.2	1163	1582	2070	2615	3235	3900	4650	6330	8260	10480	12940	15620	18610

NOTE—The actual quantities will vary from these figures, the amount of variation depending on the shape of nozzle and size of pipe at the point where the pressure is determined.

TABLE 5-8. FRICTION LOSS IN RUBBER OR RUBBER-LINED FIRE HOSE

	PRESSURE LOSS, psi PER 100 ft HOSE									
FLOW IN U.S. gpm	¾″	1″	1½″	2″	2½″	TWO 2½″ LINES SIAMESED	THREE 2½″ LINES SIAMESED	3″	TWO 3″ LINES SIAMESED	3½″
10	13.5	3.5	0.5	0.1	....	....	....	....	....	....
15	29.0	7.2	1.0	0.3	....	....	....	....	....	....
20	50.0	12.3	1.7	0.4	....	....	....	....	....	....
25	75.0	18.5	2.6	0.6	....	....	....	....	....	....
30	105.0	26.0	3.6	0.9	....	....	....	....	....	....
35	....	35.0	4.8	1.2	....	....	....	....	....	....
40	....	44.0	6.1	1.5	....	....	....	....	....	....
45	....	55.0	7.6	1.9	....	....	....	....	....	....
50	....	67.0	9.3	2.3	....	....	....	....	....	....
60	....	....	13.2	3.2	....	....	....	....	....	....
70	....	....	17.0	4.3	....	....	....	....	....	....
80	....	....	23.2	5.4	....	....	....	....	....	....
90	....	....	27.7	6.7	....	....	....	....	....	....
100	....	....	33.0	8.4	2.5	....	....	1.2	....	....
110	....	....	40.0	10.0	3.2	....	....	1.4	....	....
120	....	....	47.0	11.7	3.9	....	....	1.6	....	....
130	....	....	55.0	13.6	4.5	....	....	1.8	....	....
140	....	....	....	15.5	5.2	....	....	2.0	....	....
160	....	....	....	19.7	6.6	....	....	2.6	....	....
180	....	....	....	25.0	8.3	....	....	3.2	....	....
200	....	....	....	30.6	10.1	2.5	....	3.8	1.2	....
220	....	....	....	36.0	12.0	3.2	....	4.6	1.4	....
240	....	....	....	42.0	14.1	3.9	....	5.4	1.6	....
260	....	....	....	....	16.4	4.5	....	6.3	1.8	....
280	....	....	....	....	18.7	5.2	....	7.2	2.0	....
300	....	....	....	....	21.2	5.9	2.5	8.2	2.3	....
320	....	....	....	....	23.8	6.6	3.2	9.3	2.6	....
340	....	....	....	....	26.9	7.4	3.6	10.5	2.9	....
360	....	....	....	....	30.0	8.3	3.9	11.5	3.2	....
380	....	....	....	....	33.0	9.2	4.3	12.8	3.5	....
400	....	....	....	....	36.2	10.1	4.7	14.1	3.8	7.5
420	....	....	....	....	....	11.1	5.2	15.4	4.2	6.9
440	....	....	....	....	....	12.0	5.6	16.8	4.6	7.5
460	....	....	....	....	....	13.0	6.1	18.2	5.0	8.1
480	....	....	....	....	....	14.1	6.6	19.7	5.4	8.8
500	....	....	....	....	....	15.2	7.1	21.2	5.9	9.5
520	....	....	....	....	....	16.4	7.7	22.7	6.3	10.3
540	....	....	....	....	....	17.5	8.3	24.3	6.7	11.1
560	....	....	....	....	....	18.7	8.9	26.0	7.2	11.9

(Continued)

TABLE 5-8 (Cont.). FRICTION LOSS IN RUBBER OR RUBBER-LINED FIRE HOSE

	PRESSURE LOSS, psi PER 100 ft HOSE									
FLOW IN U.S. gpm	¾"	1"	1½"	2"	2½"	TWO 2½" LINES SIAMESED	THREE 2½" LINES SIAMESED	3"	TWO 3" LINES SIAMESED	3½"
580	....	....	....	....	....	19.9	9.5	28.0	7.7	12.7
600	....	....	....	....	....	21.2	10.1	29.9	8.2	13.4
620	....	....	....	....	....	22.5	10.7	31.6	8.7	14.2
640	....	....	....	....	....	23.8	11.4	33.5	9.3	15.0
660	....	....	....	....	....	25.3	12.0	35.5	9.9	15.9
680	....	....	....	....	....	26.9	12.7	37.5	10.5	16.8
700	....	....	....	....	....	28.3	13.4	39.5	11.0	17.7
720	....	....	....	....	....	29.9	14.1	....	11.6	18.7
740	....	....	....	....	....	31.5	14.8	....	12.2	19.7
760	....	....	....	....	....	33.0	15.6	....	12.8	20.7
780	....	....	....	....	....	34.6	16.4	....	13.4	21.7
800	....	....	....	....	....	36.2	17.2	....	14.1	22.7
820	....	....	....	....	....	38.0	18.0	....	14.8	23.8
840	....	....	....	....	....	39.9	18.7	....	15.5	24.9
860	....	....	....	....	....	....	19.5	....	16.1	26.0
880	....	....	....	....	....	....	20.4	....	16.8	27.1
900	....	....	....	....	....	....	21.2	....	17.5	28.2
925	....	....	....	....	....	....	22.3	....	18.4	29.7
950	....	....	....	....	....	....	23.4	....	19.3	31.2
975	....	....	....	....	....	....	24.5	....	20.2	32.7
1000	....	....	....	....	....	....	25.8	....	21.2	34.3
1100	....	....	....	....	....	....	31.0	....	25.2	41.0
1500	....	....	....	....	....	....	53.0	....	47.0	....

NOTE—Rough rubber lining may increase the losses as much as 50%.

TABLE 5-9. FRICTION LOSS IN COTTON RUBBER-LINED HOSE (For Heavy Streams from Large Hose or Siamesed Lines)

	PRESSURE LOSS, psi PER 100 ft					
FLOW, gpm	2 LINES OF 2½" SIAMESED	2½" AND 3" SIAMESED	2 LINES OF 3" SIAMESED	3 LINES 2½" SIAMESED	3" HOSE	3½" HOSE
400	10.1	5.9	3.9	4.7	14.1	6.3
420	11.1	6.5	4.2	5.2	15.4	6.9
440	12.0	7.1	4.6	5.6	16.8	7.5
460	13.0	7.7	5.0	6.1	18.2	8.1
480	14.1	8.3	5.4	6.6	19.7	8.8
500	15.2	9.0	5.9	7.1	21.2	9.5
520	16.4	9.6	6.3	7.7	22.7	10.3
540	17.5	10.4	6.7	8.3	24.3	11.1
560	18.7	11.1	7.2	8.9	26.0	11.9
580	19.9	11.9	7.7	9.5	28.0	12.7
600	21.2	12.7	8.2	10.1	29.9	13.4
620	22.5	13.5	8.7	10.7	31.6	14.2
640	23.8	14.2	9.3	11.4	33.5	15.0
660	25.3	15.1	9.9	12.0	35.5	15.9
680	26.9	15.9	10.5	12.7	37.5	16.8
700	28.3	16.8	11.0	13.4	39.5	17.7
720	29.9	17.7	11.6	14.1		18.7
740	31.5	18.6	12.2	14.8		19.7
760	33.0	19.5	12.8	15.6		20.7
780	34.6	20.4	13.4	16.4		21.7
800	36.2	21.5	14.1	17.2		22.7
820	38.0	22.5	14.8	18.0		23.8
840	39.9	23.6	15.5	18.7		24.9
860		24.5	16.1	19.5		26.0
880		25.6	16.8	20.4		27.1
900		26.7	17.5	21.2		28.2

TABLE 5-10. FRICTION LOSS IN COTTON RUBBER-LINED HOSE—Single Hand Lines

	PRESSURE LOSS, psi PER 100 ft						
FLOW gpm	2½" HOSE	2¾" HOSE	3" HOSE	FLOW gpm	2½" HOSE	2¾" HOSE	3" HOSE
100	2.5	1.7	1.2	260	16.4	9.9	6.3
110	3.2	2.1	1.4	270	17.5	10.5	6.7
120	3.9	2.4	1.6	280	18.7	11.2	7.2
130	4.5	2.8	1.8	290	19.9	11.9	7.7
140	5.2	3.1	2.0	300	21.2	12.7	8.2
150	5.8	3.6	2.3	310	22.5	13.5	8.7
160	6.6	4.0	2.6	320	23.8	14.3	9.3
170	7.4	4.5	2.9	330	25.3	15.2	9.9
180	8.3	5.0	3.2	340	26.9	16.2	10.5
190	9.2	5.6	3.5	350	28.4	17.1	11.0
200	10.1	6.1	3.8	360	30.0	18.0	11.5
210	11.1	6.7	4.2	370	31.5	18.9	12.2
220	12.0	7.2	4.6	380	33.0	19.8	12.8
230	13.0	7.8	5.0	390	34.6	20.7	13.4
240	14.1	8.5	5.4	400	36.2	21.7	14.1
250	15.3	9.2	5.9				

TABLE 5-11. INCHES OF MERCURY WITH EQUIVALENT HEAD OF WATER

INCHES OF MERCURY	FEET HEAD OF WATER	INCHES OF MERCURY	FEET HEAD OF WATER
1	1.13	16	18.08
2	2.26	17	19.21
3	3.39	18	20.34
4	4.52	19	21.47
5	5.65	20	22.60
6	6.78	21	23.73
7	7.91	22	24.86
8	9.04	23	25.99
9	10.17	24	27.12
10	11.30	25	28.25
11	12.43	26	29.38
12	13.56	27	30.51
13	14.69	28	31.64
14	15.82	29	32.77
15	16.95	30	33.90

(Courtesy of American LaFrance)

6 Pumping Procedures

The pump operator's job is a responsible one and calls for alertness and good judgment. Because many of the early fire department pump operators held steam engineer licenses, they are still referred to as engineers. Many fire departments use a variety of makes and models of pumping apparatus. The engineer is expected to be equally adept at operating every one of them, and the only way to do this is to become thoroughly acquainted with, and experiment with, the capabilities of the apparatus.

Pumping Apparatus

Because all pumpers fulfill the same purpose, all pumping equipment is basically the same. Water enters the pump suction inlet and is discharged at a sufficiently high pressure to form an effective fire stream. Different types of pumping equipment are differentiated mainly by their methods of transmitting power to the pump and by the type and position of their various controls.

Control Panel

The pump operator's position in most pumpers is on the left side of the vehicle for midship-mounted pumps and in front of the truck for front mounts. Some manufacturers have placed the control panel midship (Figure 6-1), just in front of the hose bed. Regardless of the location, the pump controls, gages, and other instruments must be convenient for quickly and easily placing the pump in operation and monitoring its performance.

Auxiliary Cooling

The burning of a vapor/air mixture in the cylinders of an internal combustion engine produces heat; a large amount of the surplus heat must be controlled by the engine's cooling system to avoid an excessively high temperature. The cooling system is engineered so that the flow of air through the radiator and around the engine maintains the proper engine temperature while the vehicle is moving. When the pumper is expected to produce a great amount of work at a fire, such as pumping while hooked up to a hydrant, the engine could overheat.

To compensate for this cooling limitation, some manufacturers install an extra-large radiator in the vehicle. Older pumpers are cooled by allowing water, controlled by a

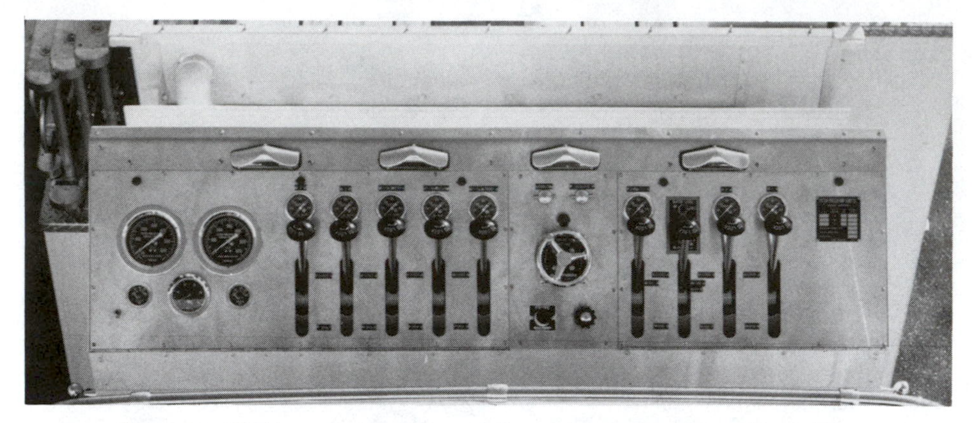

Figure 6-1. Midship-mounted control panel allows the pump operator's position to be situated where visibility is not impaired. (Courtesy of Peter Pirsch & Sons Company)

valve, to flow directly from the fire pump into the cooling system. This valve must be used with care, and the radiator cap should be removed, because when the high pump pressure is directed into the engine it may cool it too rapidly or burst the radiator or hoses by exceeding the overflow capacity of the cooling system.

Other pumpers are equipped with heat exchangers, which use pump water in one way or another to maintain an even temperature in the engine coolant. One method of doing this is to circulate pump water around tubes through which the coolant is passing; another method is to pass pump water through coils in the bottom tank of the engine's radiator.

The engine temperature in the older apparatus is usually maintained between 170 and 190°F as they have an open type cooling system. The newer apparatus with pressurized cooling systems may be operated at higher temperatures to obtain more efficiency from the engine. The manufacturer's recommendations must not be exceeded.

Series-Parallel Pump Capacities

The two-stage series-parallel fire apparatus pump, long the standard of the fire service, has a wide range of capacities over an expanse of pressures roughly twice the range possible with a single-stage pump. A 1000-gpm series-parallel pump will deliver 500 gpm at 300 psi net pump pressure at essentially the same speed and power demand required for 1000 gpm at 150 psi, while a typical 1000-gpm single-stage pump must be driven at a speed 30% to 35% higher to deliver 500 gpm at 300 psi than for 1000 gpm at 150 psi, and it will require 80% to 85% more power. More important, in many cases, is the ability of the two-stage pump to develop net pressures up to 400 to 500 psi when it is necessary or desirable to do so, while the single-stage pump is usually limited to a net pressure of about 300 psi or less.

To enable a pump operator to use a series-parallel pump to the best advantage, however, it is essential to know how to use each position, series (pressure) or parallel (volume).

The best transfer valve position for any particular pumping condition will depend on the characteristics of the particular pump and engine conbination used, and also on whether the pump is drafting, taking water from a hydrant, or in a relay. Generally, if a positive suction pressure can be maintained, as in a hydrant or relay operation, the pump can usually handle up to 75% of its rated capacity while in the series position without any serious loss of efficiency; while drafting, there is danger of cavitation if the pump is expected to deliver over 50% of rated capacity in the series position. Because some pumps can best perform their service tests of delivering 70% of their rated capacity at 200 psi net pump pressure in the series position, while others perform better if the transfer valve is in the parallel position, it is obvious that the best way for an engineer to really become acquainted with the capabilities of his apparatus is through dedicated experimentation and practice.

Hose Connections and Layout Considerations

Whenever an option exists as to which side of the apparatus the suction hose and the hose lines should be attached, it is recommended that they be connected to the side of the pumper on which the controls are located. The pump operator can thus notice pulsation and flabbiness in the suction hose if the pump begins to run away from the water, and a glance will show which valve controls which hose line. An exception would be warranted if the hose lines would thus protrude sufficiently into the street so they would create a traffic hazard.

Pump discharge pressure. Adjustment of the pump controls so the correct pressure will be produced for any particular hose layout is one of the most important responsibilities of an engineer. Because a pumper is expected to adequately supply fire streams that range from a small booster line to a heavy stream appliance requiring the full capacity of the apparatus, a thorough knowledge of the engine is essential. Pump pressures should be reasonably exact. Too little pressure will result in inadequate and ineffective fire streams; pressures that are too high are difficult for the nozzle operators to control and could result in burst hose or failure of a critical part of the equipment.

The hose lines should be charged slowly and the pressure gradually increased to the desired amount; this will allow the nozzle operators to get set before the maximum working pressure reaches the nozzle. A good practice when operating from hydrants having a flow pressure of less than 100 psi is first to charge the line by allowing the hydrant pressure to flow through the pump and hose when water is called for. This will allow the air to escape and will provide a stream of water at the nozzle without the usual jolting nozzle reaction. The engineer should then slowly advance the throttle as necessary to develop the necessary engine pressure.

Never operate a dry pump at a high speed for a long period of time because it will result in serious damage to the seal rings, impellers, and shaft packing. If the pump has been drained because of freezing weather, and pump operation is necessary for engineer instruction or for drilling, prime the pump with a garden hose or from the booster tank. However, it is always best to draft or connect up to a hydrant anytime adequate practice is desired. If the pumper is operating from a hydrant or drafting, and pump pressure is no longer needed on the hose line, the operator should be notified so that pressure can be reduced or the pump taken out of gear. When a pump is operating without water flowing, the pump will heat up. The higher the pressure at which a pump is operating, the more heat it will generate.

An alert engineer will prevent any excessive overheating; a bleeder cock or an unused discharge gate can be partly opened to allow a small flow of water to be discharged to keep the pump cool. When the fire has been knocked down, the discharge pressure should be lowered or the pump taken out of gear and hydrant pressure used for the necessary wetting down and overhaul. The lower pressures will reduce the nozzle reaction and make the hose lines more maneuverable.

Charging hose lines. Where hose lays are long or lines must be taken aloft, pump connections will normally be completed before the nozzles are in position. Pump operators must make allowances when these or similar conditions occur. It is better for a crew to be in position and to wait a few seconds for water than to be prematurely pinned down by a loaded line. However, undue delay in getting water on a fire will increase property damage and possibly endanger lives. This part of the pump operator's job calls for good judgment and heads-up operation.

Other considerations in charging lines are: be sure that the proper discharge gate is opened; observe the line to be charged and see that it actually fills with water; see that the line is sufficiently straightened out near the pump to remove any kinks that would impede the flow of water.

Obvious hose layouts. Quite often hose layouts will call for pressure or capacity operation and present little or no problem to the pump operator.

A typical example of a hose layout obviously calling for pressure operation is a single 2½-in. line 1000 ft long equipped with a 1⅛-in. tip. The quantity is relatively small, 250 gpm at 50 psi nozzle pressure, and the engine pressure requirement quite high—approximately 200 psi.

On the other hand, a typical example of a definite capacity layout for a 1000 gpm capacity pumper is three 2½-in. lines, each 300 ft long, all equipped with 1¼-in. tips. The quantity is fairly large—975 gpm—if each tip is operating at 50 psi nozzle pressure, while the engine pressure requirement is moderate—approximately 125 psi.

Suggested rules for determining proper operating position. Somewhere between the two extremes given above are hose layouts that will place inexperienced pump operators in doubt as to which position (pressure or capacity) should be used.

A suggested rule of thumb for determining the proper position is to operate in pressure position for volumes not exceeding the rated capacity of the pump at 200 psi. For a 1250-gpm pump, this is 70% of the rated capacity, or 875 gpm. By applying this rule, a pump operator will be able to quickly decide which hose layouts have capacities beyond the capabilities of this pump in pressure position. For example, suppose a pump is supplying a wagon battery equipped with a 1¾-in. tip at 80 psi nozzle pressure through four lines of 2½-inch hose, each 600 ft long. The flow is 800 gpm, and the required engine pressure is approximately 170 psi. A 1000 gpm pump operating from a good hydrant could handle this layout in the pressure position. A 1250-gpm pump supplying a wagon battery in the example above could suuply the desired amount of water at the required pressure in either pressure or capacity position. The general rule in cases of this kind is to choose the pump position that will give the desired results at the lowest engine speed. In following this rule, there will be occasions when the engine will be operating below the minimum speed recommended. However, under these conditions, the engine will be operating with little or no load, and, since the minimum rpm's listed are for working under a fairly heavy load, it will not be detrimental to the engine.

Influence of pump design on operating position. Some pumps are designed so that they may be changed from one pump position to the other at any pressure; others should be changed only when the differential between the pump intake and the discharge pressure is 40 psi or less. This is a factor pump operators should consider, under certain conditions, in selecting pressure or capacity operation. For example, a pump is supplying two 2½-in. hand lines at a large fire, and the pump operator has every reason to believe that a third or possibly a fourth line will be connected to the pump. If the pump is the type that can be changed from pressure to capacity at any time, the operator can supply the two original lines in the pressure position and change to the capacity operation when additional lines are connected without affecting the lines already in operation. However, if the pump is of the type that should be changed from one position to the other with a 40 psi or less differential between the pump intake and the discharge pressures, probably the best procedure would be to start the pumping operations in the capacity position. This would make it unnecessary to interrupt the flow of water to lines already in operation.

Safety considerations. When pumps are designed so that the change from one position to the other can be made at any pressure—that is, without altering the engine speed—one important safety factor should be remembered. When the pump is changed from capacity to pressure, a considerable increase in the discharge pressure can occur. This sudden increase in pressure and the increased reaction it causes at the nozzle might be dangerous to operators of the hose line. When changing from a capacity to a pressure operation, pump operators should watch the pump gages closely and be prepared to make any necessary throttle adjustments.

Discharge Pressure Regulation vs. Pressure Control Devices

When several lines are in operation, the relief valve or governor setting must be high enough to permit adequate pressure for the line requiring the greatest pump pressure. This will mean that the automatic pressure regulation for the other lines may not be as effective as might be desired. In this connection, it is well to observe that although modern pressure controlling devices, properly maintained and set, are usually extremely efficient, they do not relieve the engineers and the nozzle operators of the responsibility of opening and closing discharge valves and nozzles very slowly. As with any safety device, it is best not to expect pressure controls to be completely effective for preventing excessive pressure increases and surges.

Set pressure control devices as soon as required engine pressure is obtained. When the flow of water from a centrifugal pump is stopped or reduced, the pressure produced by the pump will increase sharply. When one line is in operation, increased pressure built up by the pump when the line is shut down may be sufficient to burst the hose. If two or more lines are in operation and one line is suddenly closed, the pressure buildup may be sufficient to affect the remaining hose lines; this could possibly endanger the lives of fire fighters on the nozzle or burst the hose lines. Pressure control devices provided on each pump are designed to control this excessive increase in pump pressure *when properly set.*

Do not use pressure control devices to provide changes in pressure requirements. Any change in pump pressure requirements should be accomplished by using the engine throttle. Relief valves and governors are not intended for this purpose but rather to control pressure increases. If a control device is used to raise or lower pressure requirements, the throttle setting will usually be much higher than required. Under this condition, pump pressures could be increased dangerously should the control device become inoperative.

Resetting pressure control devices. When a new pump pressure is required during pumping operations and a control device is in operation, it will be necessary on most pumps to reset it for the new pressure. Should the new pressure requirement be higher, the setting must be increased to a point above the new pressure requirement before advancing the throttle. The governor or relief valve should be reset immediately after the new pump pressure is established. Should the new pressure requirement be less, the procedure is to decrease the pump pressure and then reset the governor or relief valve.

Supplying lines with different pressure requirements. When supplying lines of different lengths or with different size nozzle tips, the discharge pressure of the pump is set to take care of the line having the highest pressure requirement with the discharge gate for that line fully opened. Other lines with lesser pressure requirements are controlled by partially closing or feathering the discharge valves to which they are connected.

The task of properly feathering the discharge gates is greatly simplified when the pumps are equipped with individual pressure gages for each outlet. In such a case, pump operators can determine the proper degree of feathering by observing the gage for the discharge gate to which the line with the lesser pressure requirement is connected. For pump operators who have pumps not equipped with individual pressure gages, proper control of flow is a matter of judgment based on experience.

There are three conditions under which a pump operator could be required to provide proper nozzle pressures for lines with different pump discharge pressure requirements: first, when both lines are placed in operation at the same time; second, when the line with a higher pressure requirement is placed in service after other lines are already in operation; third, when the line with a lesser pressure requirement is placed in service after other lines are already in operation.

Placing both lines in operation at the same time. For example, suppose pump pressure requirements are 125 psi for the first line and 150 psi for the second line. The pump operator should open both discharge gates fully when charging the lines, then advance the throttle until the pump pressure reaches 125 psi, the pressure requirement for the first line. At this point, the discharge gate to which the first line is connected should be slowly feathered to prevent excessive flow and nozzle pressure, while, at the same time, the throttle is being advanced to meet the pressure requirements of the second line.

Placing a second line with a higher pressure requirement in operation. For example, suppose the pump is supplying one line with a pressure requirement of 125 psi at the pump. A second line is laid which has a pressure requirement of 150 psi at the pump. In this case, the pump operator should slowly open the discharge gate to which the second line is connected, advancing the throttle at the same time to maintain the 125 psi pump pressure required for the original working line until the discharge valve is fully opened. At this point, the discharge gate of the original working line should be slowly feathered to prevent excessive flow and nozzle pressure, advancing the throttle simultaneously to meet the pressure requirements of the second line.

Placing a second line with a lower pressure requirement in operation. For example, suppose the pump is supplying one line with a pressure requirement of 150 psi at the pump. A second line is laid which has a pressure requirement of 125 at the pump. In this case, the pump operator should slowly open the discharge valve to which the second line is connected until the proper flow and nozzle pressure for this line are established, advancing the throttle at the same time to maintain the original 150 psi pump pressure.

Booster Tank Operation

The majority of fire departments extinguish over 80% of their fires with small hand lines of ¾, 1, or 1½-in. diameter. The water supply for these lines is carried in a tank on the apparatus, and pump operation is essentially an operation at draft. The engineer has no problem with connecting a suction hose to an outside source because the tank and pump are permanently connected with pipe, but the use of the pump primer may be required to properly evacuate air from the pump and piping. Pressure may be derived from a small booster pump, the main fire pump, or for extremely high pressures, from the main pump in series with a booster pump.

When the officer decides that the booster tank and small hand lines will be used, the apparatus is spotted in the most convenient location for the minimum length of hose line. The brakes are set, the road transmission gear shift is placed in the proper position, and the pump drive is engaged. It must be ascertained that all drains are closed and the pilot valve on the relief valve is properly set or the governor supply valve is closed. Open the tank to pump dump valve, and a short engagement of the primer will assure evacuation of all air.

While the pump operator is engaged in the above operations, the lines have been laid and connected. The discharge gate valves for the operating lines can now be opened and the discharge pressure built up to the desired amount by slowly advancing the throttle. For this service, the parallel-series centrifugal pump should be operated with the transfer valve set in the pressure position.

When operating small ¾ and 1-in. booster lines, the relief valve or governor is seldom used. The basic purpose of a pressure control device is to protect the fire fighters and the hose against an undue rise in discharge pressure if one or more lines are shut off. The volume of flow for booster lines is relatively small, and shutting off one or two lines makes little difference in pump discharge pressure. With 1½-in. hose, the relief valve or governor is generally set to operate for any pressure rise above the preset pressure, as is done for 2½-in. and larger lines.

Most pumpers and tankers have a small bypass line from the pump discharge to the water tank; there is seldom a valve in this line as it is important that it be left open whenever using water from the tank. Sometimes a small hole in the clapper of the dump line check valve accomplishes the same purpose. Small lines are frequently shut off for short periods while they are working on the extinguishment of a fire. The engine and pump are still working although no water is being discharged; water in the pump is heated by the churning. Without this circulation bypass, the heat buildup in the pump is sufficient to produce boiling water; in high-pressure operation, it could produce superheated steam. This hot water or steam could damage the pump and has severely burned fire fighters when they opened a nozzle after water had been churning in the pump for awhile. When a pump is operating from a hydrant or other source of water under pressure, water will flow from the pump into the tank in a steady stream; the higher the discharge pressure, the greater the flow. Unless there is a means of shutting off this bypass line, the tank will fill and water will be discharged from the tank overflow. As long as the engineer knows what this discharge of water means, there is no reason to be concerned as this is a natural occurrence.

When no water is being discharged from the hose lines while pumping from a hydrant or drafting, water overheating can occur. It is suggested that the pump operator open a pump drain, bleeder valve, or discharge gate if the pump is left in gear. For lengthy shutdowns, the pump should be taken out of gear.

Air Locks

When utilizing the tank to supply water at a fire, difficulty in developing an effective fire stream is sometimes encountered. Pressure may be insufficient, or it may be nonexistent. This predicament may be encountered whether the apparatus utilizes a small auxiliary booster pump or the main pump for small hose line operations.

Some departments carry their pumps and piping full of water at all times, others drain out the water from the hose and pumps and leave them dry. Carrying the pumps fully primed will give the companies a few more gallons of water, which could be critically needed at a fire, and carrying the pump wet tends to keep the packing moist and it will last longer. However, particularly in areas with extremely low temperatures, water left in the hose, pump, and plumbing may freeze during cold weather and it does tend to get the fire fighters' feet wet when they remove the suction caps to connect up suction hoses.

Upon arrival at a fire scene where the decision has been made to use the booster tank and small lines, procedures normally state that the tank to pump dump valve should be opened and the pump placed in gear, though not necessarily in that order, and then the engine throttle is used to increase the pump impeller speed. The operator now has every reason to expect the pressure shown on the discharge gage to rise as the engine speed increases. And it will, if there is no air in the pump or piping between the tank and the pump.

The problem is that when the impellers rotate, and the air in the pump gathers at the eye of the impeller or impellers. This happens because water is heavier than air; the centrifugal action of the pump hurls the water outward to the casing and the air is left in the impeller eye area. There is only one solution, prime the pump. If a discharge valve is opened before the pump has been completely primed, the pump capacity will be partially reduced because of some remaining air.

There are two methods of priming the pump. The fastest is to expel the air with the priming device, similar to drafting from a pond. The priming device should be operated long enough to eliminate all of the entrapped air; the common error at this point is to be in too great a hurry and disengage the priming device at the first indication of water spraying from the primer outlet. The length of time for a complete prime will vary according to whether the priming pump is connected to the center of the pump, to the top of the pump, or to the high point of the piping. Theoretically, a centrifugal pump with spinning impellers is primed more quickly and effectively when the air is drawn from the eye of the impellers. Remember, prime long enough to eliminate all of the entrapped air. On apparatus where this dilemma occurs often, routinely operate the priming pump every time the tank is utilized; this will not excessively increase the time it takes to supply the nozzle operator with adequate fire streams, and it will prevent embarrassing delays. Pumps with transfer valves may develop an air lock in only one stage when the valve is in the parallel (volume) position. This could reduce the pump output by one-half. When this occurs, place the transfer valve in the series (pressure) position until the pump is fully primed. Generally, pumpers operate most efficiently in the pressure position when discharging small quantities of water.

A second, though more lengthy solution, is to prime the pump by gravity. Open the tank to pump dump valve and the highest pump discharge outlet. Any air in the system will escape from the highest point; close the valves when the water flows freely from the discharge gate. A better method is to carry the pump and piping full at all times. After a fire or wet drill, do not drain the pump unless there is a danger of freezing. If the pump leaks a puddle onto the fire station floor, the solution is to properly adjust the packing.

Pumping Procedures

Pumping procedures are a crucial element in fireground operations. It is important that personnel operating the pumper apparatus understand the capabilities and proper use of the equipment.

Following are the recommended pumping procedures of four major pump manufacturers. There are similarities and differences. When available, the manufacturer's recommendations should be followed explicitly. However, when one fire department has many makes and models of fire apparatus, and personnel are expected to be equally adept at operating every one of them, a simplified pumping procedure may be devised that is equally applicable to all of that department's pumpers.

Hale Pump Procedures

The Hale pump illustrated in Figure 6-2A is mounted on a Ward LaFrance pumper. The accompanying legend in Figure 6-2B identifies the parts and instrument dials shown in Figure 6-2A. Following are the Hale pump operating procedures.

Transfer Valve

With the transfer valve in *volume* (parallel) position, each of the two impellers acts as a separate single-stage pump working in parallel. Each impeller takes suction from outside the pump and discharges its water to the pump discharge. Hence, in parallel the impellers pump large volumes of water.

With the transfer valve in *pressure* (series) position, the impellers act as a two-stage pump. The discharge of one impeller is directed into the suction of the second impeller, therefore producing half the volume produced when the valve is in the parallel position but doubling the pressure. Hence, in series the impellers pump high pressure.

The transfer valves of various pumps are operated manually by push-pull rods or hand wheels. They are also operated by vacuum or hydraulic pressure. In the latter cases, means are provided for transferring manually in case of power failure.

Pumps should be operated in the range (*pressure* or *volume*) that gives the desired flow and pressure at the lowest engine speed. To change from one range to the other, slow the engine until there is not over 30 psi difference between the suction and the discharge gage pressures, then shift the transfer valve to the other position. When shifting the transfer valve, a metallic click, or two clicks, may be heard. This will be the check valves closing. If this click is too severe, it is an indication that the change-over was made while the discharge pressure was too high.

Hale pumps are designed to pump up to 200 psi net pump pressure in *volume* position at reasonable engine speeds. In general, *volume* position should be used at any *net* pump pressure under 150 psi, especially when pumping from a hydrant. When pumping from a booster tank or draft, *pressure* position may be used when the volume is less than one-half pump capacity and desired pressure is over 100 psi.

Gearshift Operation

The gearshift lever moves a sliding gear in the drive unit. This gear has three or four positions, depending on the type of priming system.

Three-position (clutch or electric priming):

1. Road: all the way back to drive truck.
2. Neutral.
3. Pump: all the way forward to drive pump.

Four-position (gearshift priming):

1. Road: all the way back to drive truck.
2. Neutral.
3. Prime: to drive priming pump.
4. Pump: all the way forward to drive pump.

Note: These positions apply to the sliding gear and not necessarily to the position of the gearshift lever.

Fireground Procedures

The following procedures are used when the pump is to be put into operation immediately after arrival at the fire. If standing by without pumping, the pump should not be engaged.

Working from Hydrant

1. Engage pump.
2. Place the road transmission into the proper pumping gear (usually high gear).

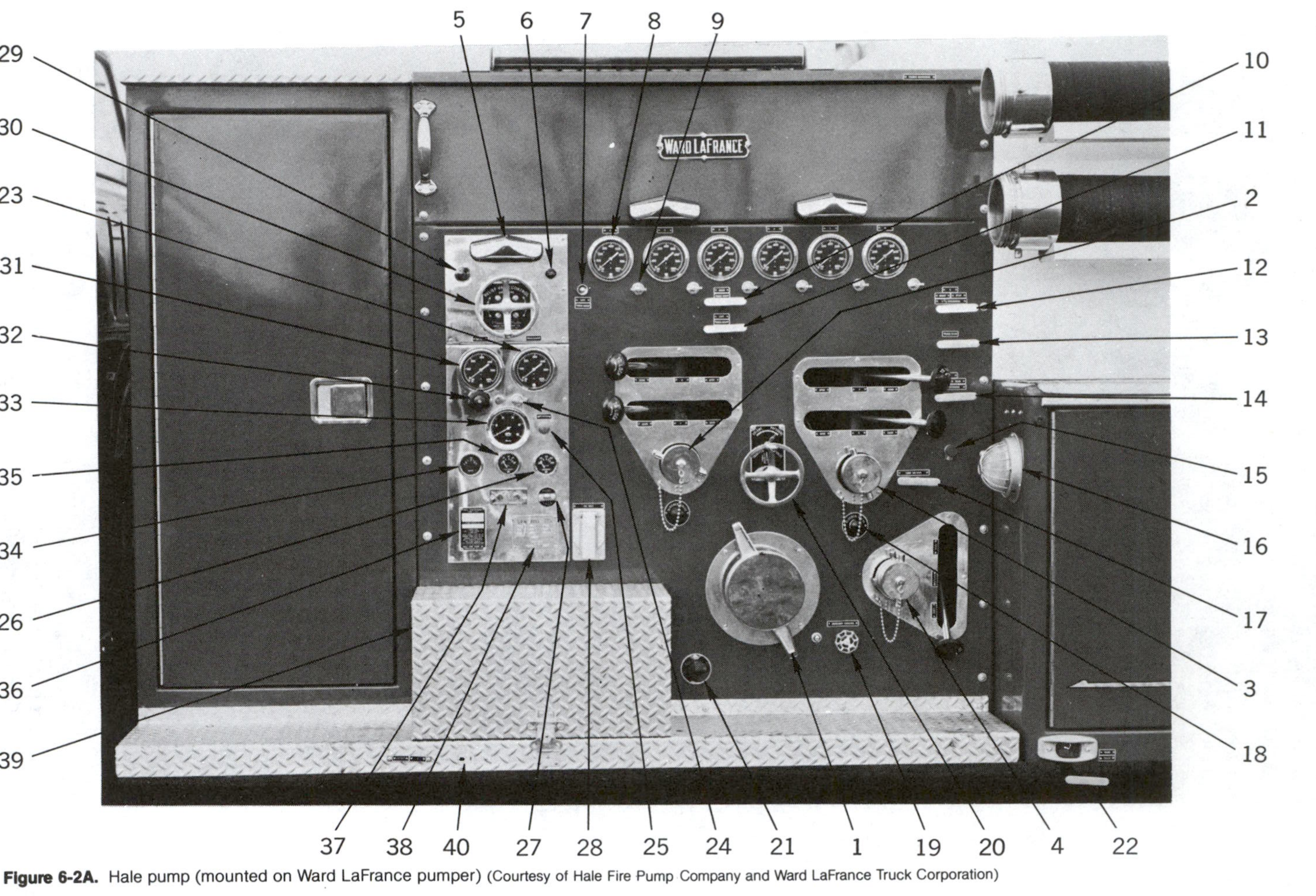

Figure 6-2A. Hale pump (mounted on Ward LaFrance pumper) (Courtesy of Hale Fire Pump Company and Ward LaFrance Truck Corporation)

1 Suction inlet
2 Front left discharge gate and remote control for front right discharge gate (upper lever controls right side)
3 Rear left discharge gate and remote control for rear right discharge gate
4 2½ in. auxiliary suction
5 Light, operator's panel
6 Switch, operator's panel light
7 Switch, electric hose reel rewind
8 Gage, Discharge pressure (one for each outlet)
9 Valve, steady (to keep needle on pressure gage steady)
10 Valve, right hand hose reel discharge
11 Valve, left hand hose reel discharge
12 Valve, right rear 1½ in. discharge
13 Valve, tank filler (to fill tank while pumping or can be used to bypass water when nozzles are closed to prevent heating water in pump)
14 Valve, left rear 1½ in. discharge
15 Counter, rpm (hand counter outlet to check tachometer)
16 Light, operator's station
17 Valve, tank suction (controls flow of water from booster tank to pump)
18 Valve, bleeder (to drain hose line attached to that discharge gate)
19 Valve, auxiliary cooling (to control water from pump flowing to cooling coils in engine heat exchanger)
20 Valve, transfer (manual type illustrated)
21 Valve, master drain (drains all chambers of pump at same time)
22 Valve, booster tank drain
23 Gage, main pump pressure
24 Valve, steady (to keep needle on pressure gage steady)
25 Switch, starter (to restart stalled engine)
26 Gage, engine temperature
27 Valve, primer
28 Outlet, 110 volt electrical
29 Light, relief valve (to show if relief valve is in operation)
30 Valve, relief (to prevent excessive pressure when nozzles are closed)
31 Gage, vacuum (indicates pressure or vacuum on suction side of pump)
32 Throttle (turn knob for fine adjustment, push center knob for rapid changes)
33 Tachometer, engine rpm
34 Gage, booster tank water level
35 Gage, engine oil pressure
36 Plate, name (shows pump serial number)
37 Connection, plugs for checking main pump gages (left side is vacuum, right side is pressure)
38 Plate, pump data (shows gpm, psi, and rpm of pump tests)
39 Box, battery
40 Plug, polarized battery charging

Figure 6-2B. Legend for Hale pump in Figure 6-2A (Courtesy of Hale Fire Pump Company and Ward LaFrance Truck Corporation)

Release the clutch and open the throttle to a fast idle.

3. Attach one end of the suction hose to the hydrant and the other end to the suction tube of the pump. If possible, flush dirt from the hydrant first.

4. Close the discharge valves and lock the relief valve or the governor out of control.

5. Open the hydrant.

6. Open the discharge valve or valves.

7. Open the engine throttle gradually until the desired pressure is reached. If the compound gage shows a vacuum before the desired pressure is reached, it is a definite indication that the hydrant will not supply any more water. In that case, the only way to get more pressure is to use smaller nozzle tips.

8. As soon as the desired pressure is reached, regulate the engine cooling valve.

9. Set the relief valve or governor control. When the pump pressure is changed, this control must be reset in the same manner.

Working from Draft

Get as close to the water as is safely possible. The pump will do better than its rated capacity with a 10-ft vertical lift, but as the lift increases above 10 ft, pump capacity will fall off. This applies to any type or make of pump.

The pump should normally be primed with the transfer valve in *volume* position.

1. *Gearshift-type priming:* place in "neutral" position.

2. *Clutch-type or electric priming:* place in "pump" position.

For both types, attach the suction hose to the pump, put a strainer on the opposite end, and submerge the strainer in the water. Close all the discharge valves, drain valves and draincocks. Engage the priming pump.

On a *gearshift type priming,* the pump impellers remain stationary while priming. Operation is as follows:

1. Open the priming valve.

2. With the engine idling, disengage the clutch and place the pump gearshift in prime position.

3. Engage the clutch and open the throttle to 1000 to 1200 engine rpm.

4. When the water first reaches the priming pump, it will come out mixed with air. Wait a few seconds until the discharge from the priming pump is uniform, then close the priming valve, shift into the main pump, open the throttle until a pressure of about 20 psi is built up.

5. Open the discharge valve slowly.

On a *clutch type or electric priming,* the pump impellers rotate while priming. Operation is as follows:

1. Set the throttle to 1300 to 1500 engine rpm.

2. Push the priming button.

3. In 10 to 30 seconds water will enter the main pump and pressure will rise. Open the discharge valve slowly and lock it in position. *Do not release the priming button until a full, steady stream is flowing through the discharge hose.* Follow this procedure exactly. *Caution:* If the priming pump does not discharge water in 30 seconds, do not continue to run; stop and look for air leaks.

Nothing can be gained by running the engine at high speed while priming; it is much better to take it deliberately and be sure.

4. Open the throttle gradually until the desired pressure is reached.

5. Regulate the valve that cools the engine.

6. Set the relief valve or the governor.

As the throttle is opened, pressure should build up as the engine speed increases.

Should the engine accelerate without a corresponding increase in pressure, the pump would cavitate or run away from the water.

If a temporary shutdown is desired when working from draft, simply slow down to about a 30 psi discharge pressure and close the discharge valves; this will prevent the pump from losing its water if there are no air leaks. To resume pumping, open the discharge valves and the throttle.

If the pump gets hot from continued churning without flow, open a discharge valve or the pump drain periodically to release hot water.

Working from Booster Tank

1. Place the pump sliding gear into "pump" position.
2. Place the transfer valve in *volume* position.
3. Open the valve between the tank and the pump suction.
4. Close the discharge valves. Close the valve from the pump discharge to the booster hose.
5. Prime exactly as when working from draft.
6. After priming, if a lower engine speed is desired, shift the transfer valve to *pressure* position.

When pumping a small volume of water through a fog nozzle or small booster hose tip, it is advisable to switch from *volume* to *pressure* two or three times to clear the pump of entrapped air.

Waterous Pump Procedures

Basic pumping procedures with Waterous pumps (Figures 6-3 and 6-4) will be identical, regardless of apparatus manufacturer. Pump controls will vary widely between different models as these pumps may be obtained with a variety of pump transmissions, but the principles are all the same.

Pump Transmission

The Crown Fire Coach uses two different types of pump transmissions, both of which are described to explain general procedures.

Waterous transmission. The main pump is driven through the pump transmission, which is located behind the road transmission. At all times when the main pump is in operation, the road transmission must be in direct drive.

When the main pump is in operation, a small amount of cooling water from the pump discharge is circulated through a manifold in the pump transmission; it is then returned to the suction side of the pump. This cooling is automatic, but only occurs when the main pump is in operation.

The pump gearshift handle for a Waterous pump transmission is located in the cab. An electrically operated pump gearshift is an optional item, but when used it is located on the driver's dash panel. The gearshift handle has two positions:

1. All the way back: road position.
2. All the way forward: pump position.

Crown transmission. The main pump is driven by means of a separate pump transmission not connected to the road transmission. It consists of a gearbox installed between the engine clutch housing and the road transmission housing. At all times, while the main pump is in operation, the road transmission remains locked in neutral.

The pump transmission is equipped with an internal oil-cooling unit of the heat-exchanger type; when the main pump is in operation, a small amount of water from the pump discharge is circulated through a manifold at the bottom of the pump transmission, and is then returned to the suction side of the pump. This cooling is automatic, but only occurs when the pump is in operation.

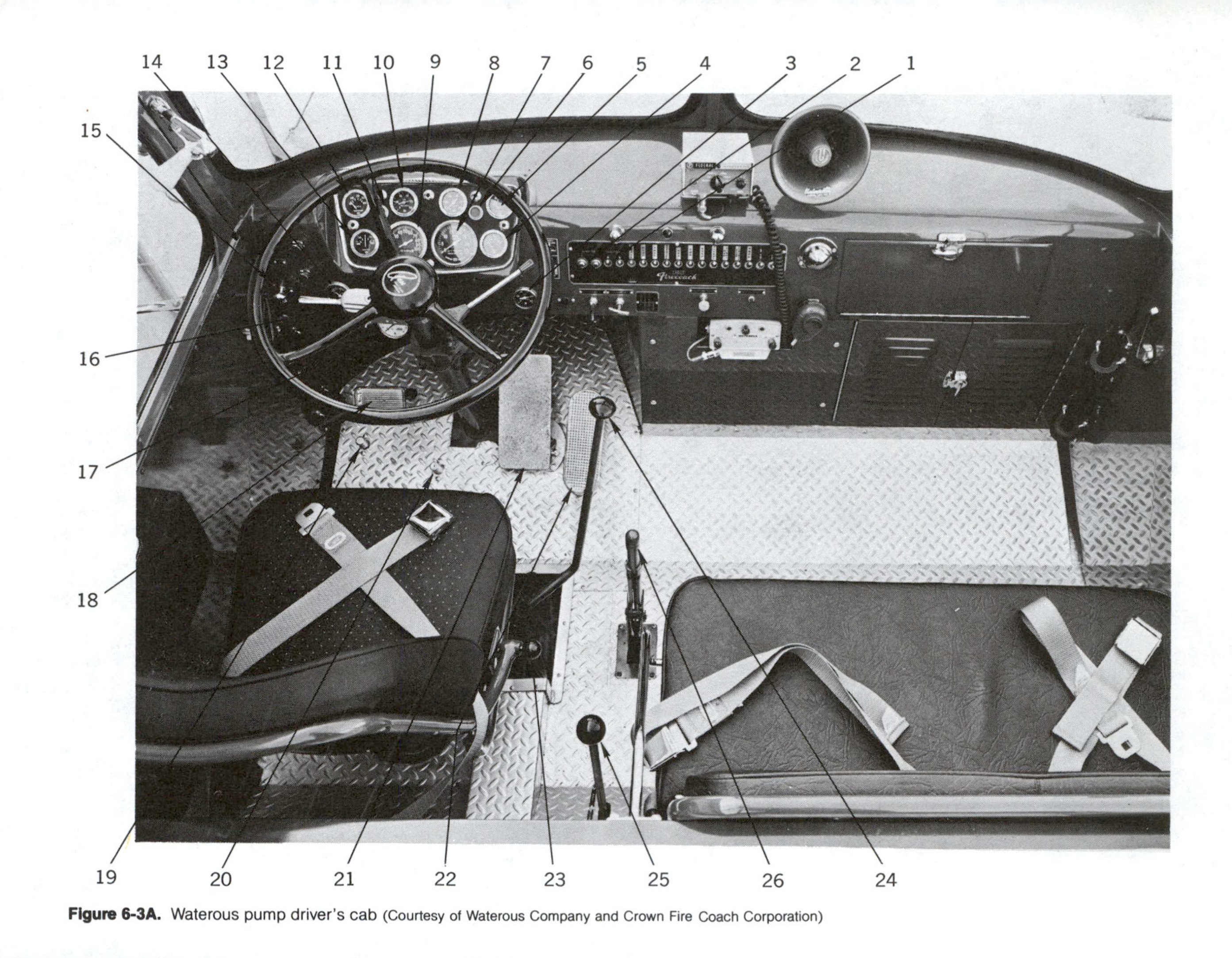

Figure 6-3A. Waterous pump driver's cab (Courtesy of Waterous Company and Crown Fire Coach Corporation)

1 Switches, ''A'' & ''B'' Ignition
2 Switches, ''A'' & ''B'' Starter
3 Gage, Air pressure (indicates amount of air in emergency reservoir)
4 Gage, Fuel
5 Gage, Engine oil pressure
6 Indicator light, Low oil pressure
7 Tachometer, Engine speed
8 Gage, Air pressure (indicates air pressure in primary and main reservoirs)
9 Speedometer and Odometer
10 Gage, Applied air pressure (indicates pressure delivered to brake diaphragms when brakes are applied)
11 Indicator light, low air pressure
12 Gage, Engine water temperature
13 Ammeter
14 Gage, Engine vacuum (indicates vacuum at the engine intake manifold)
15 Switch, Battery change over
16 Switch, Directional signal
17 Gage, Booster pump pressure
18 Clutch pedal
19 Switch, Headlight dimmer
20 Switch, Foot siren control
21 Foot brake pedal
22 Accelerator pedal
23 Seat adjustments
24 Gear shift lever
25 Auxiliary booster pump lever
26 Parking brake

Figure 6-3B. Legend for Waterous pump in Figure 6-3A (Courtesy of Waterous Company and Crown Fire Coach Corporation)

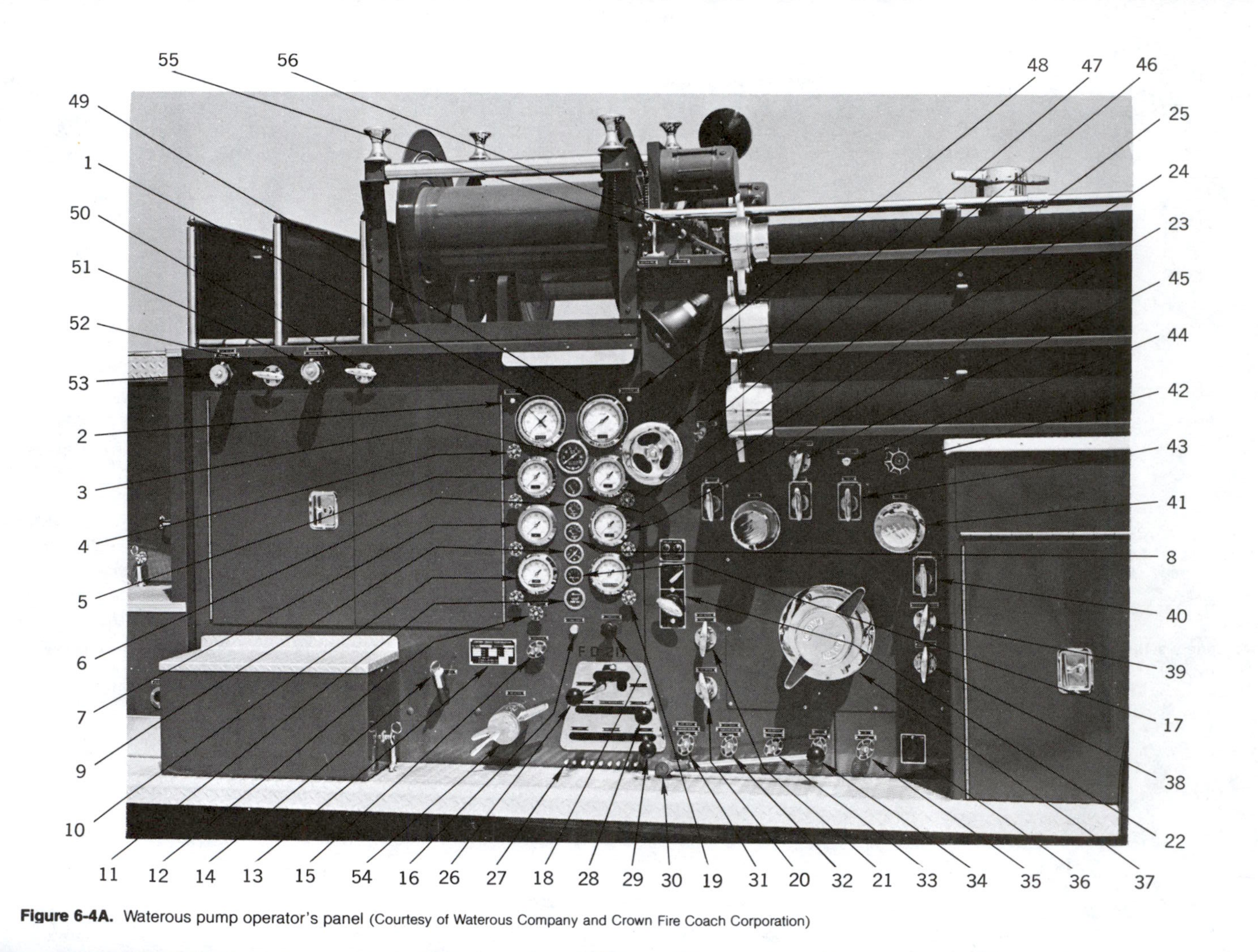

Figure 6-4A. Waterous pump operator's panel (Courtesy of Waterous Company and Crown Fire Coach Corporation)

1 Gage, Main pump pressure
2 Check outlet, Main pump pressure gage
3 Tachometer, Engine rpm (dial is marked with stationary red lines which show 80%, 90%, and 100% of peak engine rpm)
4 Valve, Vibration dampening (pump gages and discharge gate gages are equipped with individual adjusting valves to control water flow and prevent needle vibration)
5 Gage, Discharge gate No. 1
6 Gage, Engine oil pressure
7 Gage, Discharge gate No. 4
8 Gage, Engine water temperature
9 Gage, Pump transmission oil flow indicator
10 Gage, Left rear discharge outlet
11 Gage, Pump log (records hours of pump operation)
12 Valve, Radiator fill (auxiliary cooler valve must be open when used)
13 Plate, Pump data
14 Priming tank dip stick
15 Valve, Auxiliary cooler
16 Outlet, Rpm counter (to check tachometer)
17 Gage, Booster tank water level
18 Engine throttle
19 Gage, Booster pump pressure
20 Valve, Auxiliary suction
21 Valve, Tank fill (from auxiliary suction inlet)
22 Control, Relief valve
23 Gage, Discharge gate No. 3
24 Gage, Fuel
25 Gage, Discharge gate No. 2
26 Lever, Main pump transmission control (operation of this lever allows the selection of either main pump or road gear)
27 Fittings, Pump lubrication (these nine fittings allow lubricating pump and associated valving from a central location)
28 Valve, Booster tank dump (opening this valve allows water to flow from booster tank to booster pump)
29 Valve, Booster tank dump (opening this valve allows water to flow from booster tank to main pump)
30 Valve, Front suction (this lever controls the 4 in. front suction inlet)
31 Valve, Left front transverse hose bed drain
32 Valve, Left rear transverse hose bed drain
33 Valve, Auxiliary suction drain
34 Valve, No. 2 discharge outlet drain
35 Valve, No. 3 discharge outlet drain
36 Valve, Main pump manifold drain
37 Inlet, Main pump suction
38 Transfer valve control
39 Valve, Right rear 1½ in. discharge
40 Discharge gate, Right side No. 4
41 Discharge outlet No. 3
42 Valve, Tank fill (to fill booster tank from discharge side of main pump)
43 Discharge gate, Left side No. 3
44 Control, Primer
45 Discharge gate, Left rear 2½ in.
46 Valve, Fuel
47 Relief valve control, Booster pump
48 Check outlet, Compound gage
49 Gage, Main pump compound
50 Valve, Left rear transverse hose bed discharge
51 Discharge outlet, 1½ in.
52 Valve, Left front transverse hose bed discharge
53 Discharge outlet, 1½ in.
54 Inlet, Auxiliary suction (controlled by No. 20)
55 Valve, Left hose reel
56 Valve, Right hose reel

Figure 6-4B. Legend for operator's panel in Figure 6-4A (Courtesy of Waterous Company and Crown Fire Coach Corporation)

The Crown pump transmission is equipped with an individual lubricating oil pump system that is automatic; it operates only when the main pump is in operation. The oil pump picks up cool oil from the bottom of the transmission case and pumps it to the top, allowing oil to flow over the gears.

The pump gearshift handle is located at the operator's panel. Its movements are air actuated, and it has three positions:

1. All the way back, pump position.
2. All the way forward, road position.
3. In a raised position, which disengages the clutch.

Transfer Valve

As a general rule, keep the transfer valve in *pressure* (series) position when using half or fewer of the pump discharge valves and in *volume* (parallel) when using more than half of the valves. If pump is operating at a high lift, or with a large amount of water through one or two valves, using the *volume* position may be necessary to avoid cavitation.

If high pressure is required (more than 200 psi), operate pump in *pressure* position even if it means closing one or more valves to avoid cavitation.

If changing the transfer valve from one position to the other is desirable while the pump is operating, reducing the discharge pressure will decrease the effort required. If a pump has a manually operated transfer valve, reduce the discharge pressure to 75 psi or less. With electric transfer valves, reducing the discharge pressure is necessary only if it exceeds 250 psi.

Before Pumping

1. Put the road transmission in neutral and set the parking brake, unless the brake is in the drive line between the transmission or transfer case and the pump. Block the wheels if the brake cannot be used, no auxiliary braking device is installed, or the truck is parked on a grade.
2. Close all the discharge valves and drain openings.
3. Unless pumping from a booster tank, connect the suction hose to the pump.
4. When connecting the discharge hoses, make sure they are free of kinks and sharp bends.
5. Make sure all hose connections are tight. For proper operation, the suction and discharge gaskets must be clean and in good condition.

Pumping from Hydrant or in Relay

1. Close the valves between the water tank and the pump.
2. Connect the suction hose between the hydrant or relaying pumper and the pump suction.
3. Open the suction, hydrant, or other valves as necessary to allow water to enter the pump.
4. Switch the transfer valve to the desired position.
5. Engage the pump drive.
6. Open the discharge valves and accelerate the engine to obtain the desired discharge pressure and capacity. Do not attempt to pump more water than is available from the hydrant or relaying pumper.

It is best to operate at a minimum suction pressure of 10 psi to prevent the hose or soft suction from collapsing.

Pumping from Draft

1. Close the valves between the water tank and the pump.
2. Connect a hard suction hose to the pump fitting and attach a suction strainer to

the intake end of hose.

3. Switch the transfer valve to desired position.

4. Prime the pump.

5. Engage the pump drive.

6. Open the discharge valves, and accelerate the engine to obtain the desired discharge pressure and capacity.

Pumping from Water Tank

1. Make sure the suction caps are tight on pump.

2. Switch the transfer valve to the desired position.

3. Open the valves in piping between the water tank and the pump suction, and at least one discharge valve.

4. Allow about 30 sec for water to flow to the pump, engage the pump drive, and accelerate the engine to obtain the desired discharge pressure and capacity. Priming the pump may be necessary under some conditions because of air trapped in the piping.

After Pumping

1. Slow the engine to idling speed.

2. Disengage the pump drive.

3. Disconnect the suction and the discharge hoses. Leave the caps off until the pump is completely drained.

4. Install the suction and the discharge caps.

5. If the transfer valve was not used during the pumping operation, switch it back and forth several times to keep it working freely.

Seagrave Pump Procedures

The Seagrave pump is illustrated in Figure 6-5. Following are pumping operation procedures.

Transfer Valve

The changeover valve is operated by an air cylinder connected to an arm on the transfer valve. The air cylinder is actuated by a four-way valve on the operator's panel. It is usually not necessary to close the throttle or reduce pressure when changing the pump position at pressures below 300 psi. Use caution, particularly when changing from parallel to series, because theoretically the pressure would double.

A red pointer on the discharge gage indicates the point at which to change the pump from volume to pressure when attempting to get the maximum amount of water. Pressure indicated by this stationary pointer depends on the model of the pump. Pressure below the red hand should be in *volume* position and above the red hand in *pressure* position. When pumping from a hydrant, add the hydrant flowing pressure to the red hand reading to determine the changeover point.

The pump is provided with the *volume* (parallel) feature to obtain a large volume of water. When only a small quantity of water is required, as for a single line of 2½-in. hose or a booster line, operating in *pressure* (series) position will give slower engine speeds and will be easier on the engine and the pump. On the other hand, if large volumes of water are required, operating in *pressure* position will reduce the capacity of the pump and increase the chance of cavitation.

Pumper Operation at Draft

Locate the pumper as near the source of water as is safely possible. Set the parking brake.

Before leaving the driver's seat, depress the clutch and shift the road transmission into direct drive. While holding the clutch pedal down, engage the clutch control switch on the driver's instrument panel to hold the clutch pedal depressed. Leave the cab and shift the pump control level into *out* position. Releasing the shift lever handle

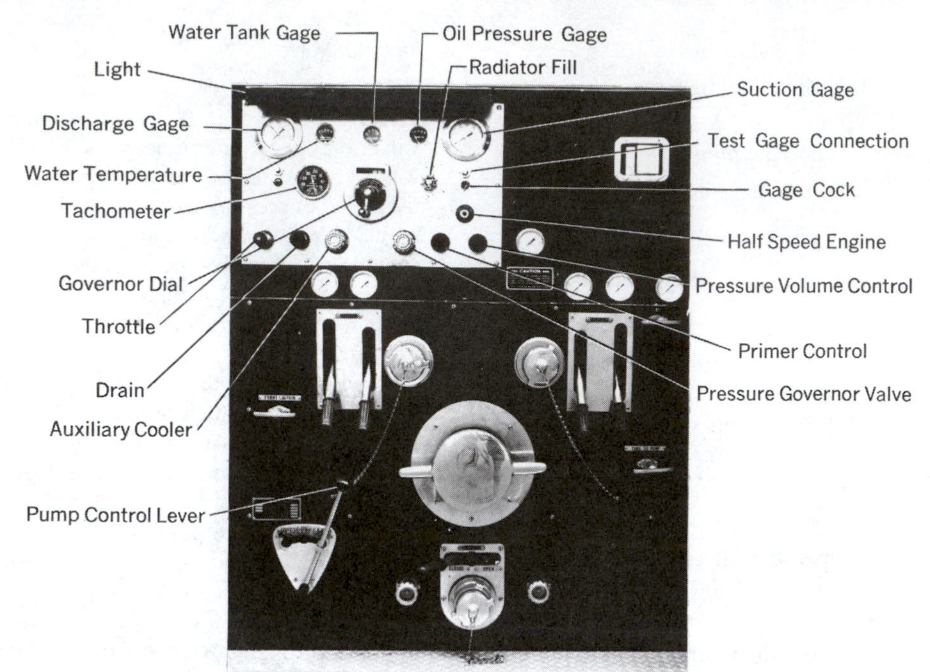

Figure 6-5. Seagrave pump (Courtesy of Seagrave Fire Apparatus, Inc.)

engages the engine clutch with the pump transmission in neutral. It can be left in that position while the pump suction hoses and discharge lines are being attached.

Connect the suction hoses, being sure that the suction strainer is attached and all gaskets are in place and in good condition.

Be certain that all discharge valves and drains are closed.

Connect the discharge lines.

Pull out on the pump shift lever handle to disengage the engine clutch. Count to about 5 to allow the transmission and the drive shaft time to almost stop rotating. Shift the pump lever to *pump* position and release the handle.

Open throttle control to give an engine speed of 700 to 800 rpm. Pull out the primer button and hold until water is discharging from the vacuum pump, or until pressure shows on the pressure gage, then release the button.

Open the discharge gate slowly, and at the same time advance the throttle until a steady flow is established. If for some reason the pressure drops as the discharge gate is opened, reengage the primer for a few seconds to start a water flow and pressure buildup again. The throttle should be left advanced and the discharge gate left open when the primer is operated the second time.

As soon as pumping has started, open the cooler valve about one full turn and adjust it later to hold the engine temperature at 170 to 190°.

Open the pressure governor valve and adjust the governor.

Pumper Operation at Hydrant

Operation at a hydrant is exactly the same as at draft, with two exceptions:

First, all priming operations are eliminated.

Second, the hydrant pressure shown on the suction gage should be added to the pressure indicated by the red pointer on the discharge gage. The sum of these two pressures represents the pressure at which a changeover from volume to pressure, or vice versa, should be accomplished; for example, if the red pointer indicates 190 psi and the suction gage shows 40 psi, 190 plus 40 equals 230 psi. Therefore, the transfer valve should be operated at 230 psi instead of 190 psi. The pump will function satisfactorily without this precaution, but if maximum performance is desired, this is the correct procedure.

The suction hose may be attached and the hydrant turned on at any time, but *do not open the discharge gates* until the pump control lever is in *pump* position; otherwise, with water flowing through the pump, the pump shaft will spin and make it difficult to engage the pump gears.

Pumper Operation for Booster Lines

1. After the pumper is spotted, proceed as with the drafting operation. Place the road transmission in direct drive, with the clutch pedal depressed and clutch switch on.

2. Open the booster tank valve (or the hydrant connection if a tank is not used).

3. When ready to pump, shift the pump control lever to *out* position and hold the control lever out for about the count of five, then shift to *pump*.

4. Open the throttle to give an engine speed of 700 to 800 rpm and open the valve to the booster hose.

5. If the pump was dry, engage the priming pump for about 5 sec to expel air from the pump; if the pump was carried full of water, this will not be necessary.

6. Advance the throttle only enough to give the desired pressure.

7. Change the transfer valve to *pressure* position and pump in series, regardless of the pressure required.

8. Readjust the throttle to the desired pressure.

It is not necessary to use the auxiliary cooler when operating a booster line. If, however, 2½-in. or larger lines are used in addition to the booster line, use of an auxiliary cooler and a pressure governor will be the same as when pumping from draft or hydrant.

Shutting Down

1. When pumping is finished, close the throttle and move the pump control lever to *out* position.

2. Close the cooler and the governor valves; adjust the governor control to low pressure position (dial light out).

3. Close the hydrant (if at hydrant); open hose-line drains and pump drains.

4. Pull out on the pump shift lever, hold for a count of 5 to allow the gears to stop rotating, then shift to *road.*

5. Enter the cab, place the road transmission in neutral, and operate the dash control switch to release the clutch. The pumper is now ready to drive.

American LaFrance Pump Procedures

The American LaFrance pump is illustrated in Figure 6-6A, with accompanying legend in Figure 6-6B. Following are the operating procedures for this pump.

Transfer Valve

The pump should ordinarily be operated in *capacity* position for pressures up to 210 psi. This is the proper setting for this pressure range regardless of the size of the stream to be discharged. If discharge pressure in excess of 210 psi is required, change the pump from *capacity* to *pressure* position. If it is necessary to change from *capacity* to *pressure,* or vice versa, while pumping, it is desirable to reduce pressure by closing the throttle; this will reduce the turning effort

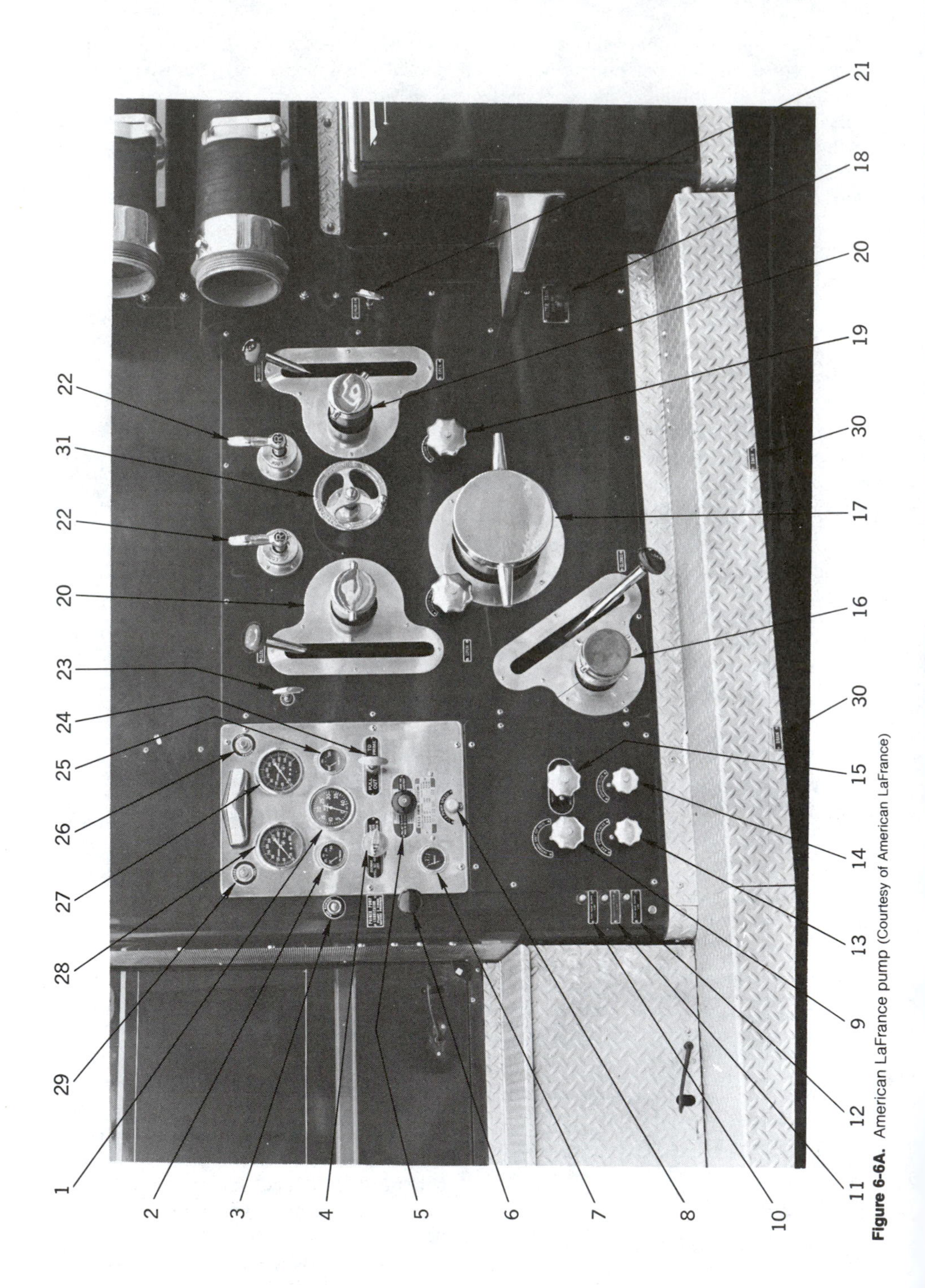

Figure 6-6A. American LaFrance pump (Courtesy of American LaFrance)

1 Gage, Engine tachometer
2 Gage, Oil pressure
3 Hose reel rewind
4 Governor control
5 Engine throttle
6 Primer lubricator
7 Gage, Booster tank
8 Tachometer take off
9 Valve, Governor shut off
10 Lubrication fitting, Pump front bearing
11 Lubrication fitting, Change over valve
12 Lubrication fitting, Pump packing
13 Valve, Auxiliary cooler
14 Valve, Radiator filler
15 Valve, Booster cooling
16 Auxiliary 2½ in. suction inlet
17 Main pump suction inlet
18 Valve, Water tank
19 Valve, Bleeder
20 Gates, 2½ in. discharge
21 Valve, 1½ in. live line
22 Controls for right hand discharge gates
23 Valve, Hose reel control
24 Primer control
25 Gage, Engine temperature
26 Gage test
27 Gage, Discharge pressure
28 Gage, Suction
29 Gage test
30 Valves, Drain
31 Valve, Transfer

Figure 6-6B. Legend for American LaFrance pump in Figure 6-6B (Courtesy of American LaFrance)

on the changeover-valve (transfer valve) handwheel and will prevent a dangerous sudden increase in the discharge rate at the nozzle.

The *pressure* setting may be used for decreased volumes down to 175 psi and for any practical volumes at pressures above this figure.

Operating from Draft

1. Upon arriving at the water supply, place the pumper in the position requiring the shortest length of suction hose. Set the parking brake.

2. Disengage the clutch, shift the road transmission to direct-drive position, then engage the pump by moving the pump shift lever to the pumping position. Engage the clutch. Lock the road transmission in direct drive, and lock the pump transmission in the pumping position.

3. Check the position of the transfer valve.

4. Connect the discharge hose to the desired outlets.

5. Make certain that all the discharge gates, bleeder valves, and drains are closed.

6. Lock the relief valve or the governor out of operation.

7. Advance the throttle to a fast idle, approximately 900 to 1000 rpm.

8. Pull out the primer control and hold it until priming is complete. When water appears through the primer discharge line to the ground, release the primer control to shut off the primer.

9. Open the throttle until pump pressure is stabilized at 40 or more psi.

10. Open the discharge gate to which the hose line is attached.

11. Increase the throttle until the desired pressure is reached.

12. If the pump should lose prime during operation, it can be reprimed by closing the discharge gate and repriming as outlined above.

13. Set the relief valve or the governor.

14. To shut the pumper down, close the throttle and the discharge gates. With the discharge gates closed, the pump will hold its prime. To resume pumping, open the discharge gates and increase the throttle setting.

Operating from Hydrant

1. Place the pumper in a position where the shortest length of suction hose can be used to connect the hydrant to the pump suction. Set the parking brake.

2. Disengage the clutch, shift the road transmission to direct drive and engage the pump by moving the pump shift lever to the pumping position. Lock the road transmission in direct drive and the pump transmission in pumping gear.

3. Check the position of the transfer valve (changeover valve).

4. Connect the suction hose from the hydrant to the pump suction.

5. Connect the discharge hose to the desired outlets and open the discharge gates.

6. Make certain that all other discharge gates, bleeders, and drain valves are closed.

7. Lock the relief valve or the governor out of operation.

8. Open the hydrant and advance the throttle to obtain the desired discharge pressure.

9. Set the relief valve or the governor.

10. To shut the pumper down, close the throttle and the discharge gates. To resume pumping, open the discharge gates and increase the throttle setting.

Operating as a Booster

1. Place the pumper in position and set the hand brake.

2. Disengage the clutch, shift the road transmission to direct drive, and engage the pump by moving the pump shift lever to the pumping position. Engage the clutch. Lock the road transmission and pump transmission in gear.

3. Be sure the suction caps are tightly in place so that no air can leak into the pump around the gaskets.

4. Open the tank valve in the line between the water tank and the pump suction.

5. Check the position of the transfer valve. The pump should ordinarily be operated in *capacity* for pressures up to 210 psi, regardless of the size of stream to be discharged.

6. Pull out the primer control and hold until priming is complete. When water appears through the primer discharge line to the ground, release the primer control to shut off the primer.

7. Make certain that all discharge gates, drain, and bleeder valves are closed.

8. Lock the relief valve or the governor out of operation.

9. With engine idling at approximately 900 to 1000 rpm, pull out the primer control.

10. Hold the throttle in this position until a full prime is indicated by a steady stream of water discharging from the primer outlet to the ground. Adjust the throttle until a stabilized pressure of 40 psi is reached.

11. Open the booster-line discharge valve, close the primer control, and adjust the throttle to obtain the desired pump pressure.

12. It normally will not be necessary to set the governor during the booster operation, since the low flow through the booster hose can be shut off without a serious pump pressure rise.

13. Always open the booster-line cooling valve during this type of operation.

14. To shut off the flow of water, simply close the shut-off at nozzle.

15. To shut down permanently, close the throttle, close the booster discharge valve and close the water-tank valve.

Operating as a High-Pressure Booster

1. On pumps equipped to operate at high pressure (in excess of 330 psi), shift and lock the road transmission in overdrive, and proceed as outlined under "Operating as a Booster."

2. Before priming, turn the transfer valve to its *pressure* position. Always open the booster-line cooling valve when using a booster stream in order to avoid overheating the pump.

Simplified Pumping Procedure

Basically, all fire department pumpers are similar; the major differences are in the controls. Therefore, a logical sequence of pumping procedues, except for a few variations, will be equally applicable to all models. Deviations may be necessary because of minor differences in apparatus, accessories, hose lays, suction hook-up, or other conditions. For example, alterations in this procedure will be made to prevent the pump from running dry for a long period of time or to keep the clutch from being kept disengaged too long. When provided, the *out* position on pump shifts may be utilized, or the sequence can be changed.

A thorough knowledge of apparatus construction and pumping theory is necessary for the pump operator to intelligently operate the pump in the most efficient manner with the least confusion. Although some discretion is allowed at times, there are good reasons for completing some steps ahead of others.

Transfer Valves

Nearly all parallel-series centrifugal pumps are most efficient when their transfer valve is in the *pressure* position while pumping quantities of water less than 70% of the pump's capacity. In *volume* position, they are more efficient when discharging over 70%. Thus, a 1000 gpm pump would probably effectively discharge quantities of less than 700 gpm in the *pressure* position and over 700 gpm in the *volume* position. This rule varies moderately between pumps, but is generally accurate. The transfer valve should be carried in the position most likely to be used at the next fire; most departments keep the valve in the *pressure* position.

The transfer valve should be in the position that will give the desired flow and pressure at the lowest engine speed. Tentatively, position the valve according to the quantity of water to be pumped. While pumping, note the engine speed and check to ascertain which position of the valve will give the desired quantity and pressure at the lowest engine speed.

Some transfer valves are constructed so they may be changed over at high pressures, while others are only operable at low discharge pressures. However, when pumping into hose lines at high pressures, safety dictates that the pressure should be reduced. This is particularly critical when changing from parallel to series. In theory, switching from parallel to series with a two-stage pump would cause the pressure to double. Safety requires good judgment.

Pumping Procedures

The following procedures are equally adaptable for both hydrant and drafting operations:

Spot the apparatus and set the parking brake.

For apparatus with pump transmission ahead of road transmission, first place the road transmission in neutral (if the control is in the cab, also engage the pump), then leave the cab.

For apparatus with pump transmission behind road transmission, first place the road-to-pump shift lever in pump position. (This lever should always be first in and first out; this method will allow the gears in the pump transmission to be rotated in case of a difficulty in engaging or disengaging the gears.) Next, place the road transmission in proper gear (usually in direct). Then leave the cab.

If there is the slightest possibility that the parking brake will not hold the apparatus safely, chock blocks should be placed at this time.

Connect the suctions (hard or soft). It is good practice to connect a soft suction to the hydrant first in case the apparatus is incorrectly spotted and must be moved.

Hydrant operation. Open the hydrant. Take any kinks out of the soft suction before leaving the hydrant area.

Connect the hose lines. Do not open the discharge gates.

Operating at draft. Prime the pump. It will depend upon the apparatus whether the pump will be in or out of gear. Priming of some pumps is facilitated by having the transfer valve in *volume,* while others do equally well in either position.

Engage the pump.

SLOWLY open the discharge gates. Good judgment is necessary to prevent the fire fighters on the nozzle from waiting an excessively long period of time; yet, precautions must be taken so the hose lines are not loaded before they are in position.

Take any kinks out of the hose lines.

Position the transfer valve, if necessary. The transfer valve should be in the position that will give the desired flow and pressure at the lowest engine speed.

Some makes of governors must be placed in operation at this time, before raising the discharge pressure.

SLOWLY advance throttle to the desired pressure. Always advance and retard the throttle slowly and evenly by rotating the knob.

Adjust the governor or the relief valve, except when the governor is already set.

Maintain a close check on all gages. Operate the auxiliary coolers and pump transmission coolers when necessary. As long as the temperature is held below the boiling point of water, the engine is not considered to be running hot. Observe the manufacturer's recommendations.

Temporary Shutdown

SLOWLY retard the throttle. Use the throttle panic button only for emergencies.

SLOWLY close the discharge gates.

If pumping at draft, maintain a 30 to 35 psi discharge pressure to avoid losing the prime.

If shut down for an extended period of time, take the pump out of gear or open a discharge-gate bleeder valve to prevent churning of the pump from heating the water.

Permanent Shutdown

SLOWLY retard the throttle. Use the throttle panic button only for emergencies.

SLOWLY close the discharge gates.

Take the pump out of gear.

Open the discharge-gate bleeder valves to drain the hose lines.

Close the hydrant.

Reposition the transfer valve, if necessary.

Return the pressure regulating device and all other valves and controls to their normal position.

For apparatus with pump transmission behind road transmission, first place the road-to-pump shift lever in road position, then place the road transmission in neutral.

Stop the engine.

Disconnect the hose lines.

Disconnect the suction hoses.

Pick up the chock blocks before moving the apparatus.

Note: Any time that an engine is working hard, whether driving or pumping, a large amount of heat is generated. Therefore, the engine should be allowed to idle for a period of time after taking the pump out of gear in order to circulate the coolant and draw air through the radiator. Run the engine at a slow rate of speed for about 5 min to remove the excess heat.

(Courtesy of the Los Angeles City Fire Department)

7 Pumper Tests

Early tests of pumping apparatus often provided the opportunity for spectacular competition between manufacturers. For example, apparatus was purchased after comparing pumpers to see which projected a stream of water the highest vertical distance. Some cities invited all manufacturers to engage in an endurance contest and purchased the model that outlasted the others.

The current program of testing and listing fire department pumpers was an outgrowth of a practice started during the annual meetings of the International Association of Fire Engineers (now the International Association of Fire Chiefs) in the early 1900s. Apparatus manufacturers would display their new "automobile pumping engines" at the meeting, and engineers would check performance with 12-hr tests. The 12-hr test was later shortened to 6 hr as improvements in engines and pumps reduced the number of failures.

Early testing procedures consisted of a 100% delivery of rated capacity during a 6-hr run, 50% of capacity for 3 hr at 200 psi, and 33⅓% of capacity for 3 hr at 250 psi. This test procedure continued for many years and eventually became the procedure for testing what are now known as Class "B" pumpers.

In 1910, the National Board of Fire Underwriters, which performed tests for manufacturers and municipalities, published the first edition of *Fire Department Pumper Tests and Fire Stream Tables.* This pamphlet was prepared to assist fire department officials and others in determining the condition of pumpers and their compliance with pumping requirements of the purchase specifications. Test procedures incorporated in this publication have been adopted by most fire departments.

In 1950, a change in testing procedures was made. As buildings were becoming

The author wishes to thank the Insurance Services Office for information and excerpts adapted for this chapter from *Fire Department Pumper Tests and Fire Stream Tables,* 8th ed. This chapter includes copyrighted material of Insurance Services Office with its permission.

larger and taller, larger quantities of water were necessary at high pressure. New tests established Class "A" pumpers, which specified the delivery of rated capacity at 150 psi, 70% of capacity at 200 psi, and 50% of capacity at 250 psi. This is the test now in use. Class "B" pumpers are no longer built, though some are still found in departments that have not yet had a chance to replace them.

The American Insurance Association (successor to the National Board of Fire Underwriters) discontinued the pumper testing and listing program, having decided that this was the kind of service that Underwriters' Laboratories, Inc. was specially set up to perform. The Insurance Services Office, 160 Water Street, New York, has assumed the responsibilities of the American Insurance Association (AIA).

All pumpers should be tested annually, as well as after major repair work, as part of the regular fire department preventive maintenance program. Service tests similar to those described here are witnessed by engineers of the Insurance Services Office and other fire insurance rating organizations and bureaus when making surveys of the fire defenses of municipalities throughout the country.

Fire pump testing has two primary objectives:

1. To determine the present condition and capability of the pump and engine.

2. To provide records for comparison with previous and future test results.

In normal drills and fire fighting operations, little pumping is performed at the full rated capacity of the pump; it is because of this that members often remark about the noise and speed of the engine during a test and become unduly alarmed. Failure of a pumper to deliver the required quantity of water at a fire has often been the result of an inexperienced engineer throttling the engine back to an excessively slow speed because the normal vibration and noise of a hard-working engine caused undue apprehensiveness. It is at this time that the engineer can actually observe what might be expected of the apparatus under rigorous conditions; variance in operation from manufacturers' specifications can also be seen. Accuracy of gages and other instruments are often taken for granted in daily operation; the tests afford a good opportunity to check them with calibrated instruments.

Types of Tests

All pumpers should be subjected to three types of tests: certification, delivery, and service (Tables 7-1 and 7-2). During these tests, all pumpers are required to pass dry-vacuum, quick-lift, and pressure-control tests. They are required to pump with a lift that does not exceed 10 ft at the following capacities and pressures:

100% of rated capacity at 150 psi net pump pressure to check the condition of the engine and pump.

100% of rated capacity at 165 psi net pump pressure (overload test) to check the engine's reserve power.

70% of rated capacity at 200 psi net pump pressure to check the condition of the engine and pump.

50% of rated capacity at 250 psi net pump pressure to check the condition of the engine and pump.

All tests are to be conducted without the engine exceeding the manufacturer's maximum no-load governed speed.

The overload test for capacity at 165 psi is not part of the service test recommended in *Fire Department Pumper Tests and Fire Stream Tables,* published by the Insurance Services Office. However, many fire departments include this test annually to further check the pumper's performance.

TABLE 7-1. PUMPER TEST REQUIREMENTS

VOLUME*	PRESSURE**	CERTIFICATION TEST	DELIVERY TEST	SERVICE TEST
100% @	150 psi	2 hr	2 hr	20 min
100% @	165 psi	10 min	10 min	5 min
70% @	200 psi	30 min	30 min	10 min
50% @	250 psi	30 min	30 min	10 min

*Volume is the percentage of the rated capacity of the pump.
**Pressure is the net pump discharge pressure.

TABLE 7-2. PUMP TEST PRESSURES AND CAPACITIES

RATED CAPACITY AT 150 psi NET PRESSURE, gpm	70% OF RATED CAPACITY AT 200 psi NET PRESSURE, gpm	50% OF RATED CAPACITY AT 250 psi NET PRESSURE, gpm
500	350	250
750	525	375
1000	700	500
1250	875	625
1500	1050	750
1750	1225	875
2000	1400	1000

Certification Test

Underwriters' Laboratories, Inc. provides a Certification of Fire Department service (Figure 7-1) to verify inspection and testing of pumpers at the manufacturer's plant before delivery to the purchaser. The certificate confirms that the identified pumper has successfully completed a pumping capability test witnessed by a representative of Underwriters' Laboratories, Inc., and forms the basis for certifying performance at altitudes to 2000 ft. The certificate also covers certain other characteristics and equipment of the pumper that are considered essential to proper pumping performance. It does not cover characteristics relating to the operation of the pumper as a vehicle, such as the road tests.

Delivery Test

The delivery test, which is also referred to as the acceptance test, is essentially the same as the certification test. Every new pumper, on delivery and before acceptance by the fire department, should be subjected to a test that will determine its ability to meet contract specifications. The test should be conducted under the same conditions as the future service tests so there will be a reasonably exact comparison of results. Delivery tests should be made with all accessories and power-consuming appliances attached and using the grade of fuel recommended by the manufacturer.

When testing pumpers to determine if they meet their contract requirements, the discharge should be not more than about 15 gpm above the rated discharge and the net pump pressure should be not more than about 5 psi above the pressure rating at which the pump is being tested.

The acceptance test is not a useless duplication of the manufacturer's certification test, even though the pumping tests are similar. There have been occasions where the pumper was not able to satisfactorily pass the delivery test after it arrived at the

3000-375
5M 12-75 PTD. IN U.S.A.

UL

UNDERWRITERS LABORATORIES INC.

333 PFINGSTEN ROAD · NORTHBROOK, ILLINOIS 60062

an independent, not-for-profit organization testing for public safety

CERTIFICATE OF INSPECTION FOR FIRE DEPARTMENT PUMPER

RATED CAPACITY..........................gpm Date..

This certifies that the pumper described below has performed acceptably and is provided with items of equipment as shown.

Manufacturer: ....Model No..............................Serial No......................

For:..Location..

Chassis: Mfr..Model No..............................Serial No......................

Engine: Mfr..Model No..............................Serial No......................

Pump: Mfr..Model No..............................Serial No......................

Test Conditions: Barometric Pressure..................in. Hg. (corrected to Sea Level); Temp...............F; Elevation...............ft.;

Suction Hose: Size..................in.; Length..................ft.; Pump elev. above water source..ft.;

Performance Certified to..ft. Elevation.

TEST CONDITIONS	PUMP CONTROL POSITION	FLOW GPM	DISCHARGE PRESSURE PSI	SUCTION LIFT PSI (NEG.)	NET PUMP PRESSURE PSI	PUMP SPEED RPM	ENGINE SPEED RPM	GEAR RATIO ENGINE TO PUMP
Capacity 150 psi—2 Hrs.								
Overload 165 psi—10 Min.								
70% Capacity 200 psi—30 Min.								
50% Capacity 250 psi—30 Min.								

Automatic Pressure Control Test: Max. Increase.......................psi. Vacuum Test:...........................in. Hg. drop in 10 Min.

No Load Governor Speed...........................rpm; Specified...........................rpm. Pump Location..

EQUIPMENT CHECK

CONTROLS	Yes	No
Visible, Accessible & Illuminated	X	
Catches or locks	X	
Pump position indicated	X	
Hand throttle positively held	X	
Operating data plate #	X	
Gear Shift position latch	X	
Pump drive gear indicated at transmission shift lever	X	
PUMP CONNECTIONS		
Gated suction inlet		
Strainers	X	
Outlet area minimums	X	
Drains for outlets	X	
Priming device		
Tank to pump valve		
Tank-pump min. water flow		
Supply valve to heat exchanger	X	

FUEL SYSTEM	Yes	No
Tank Capacity, 2 hr. min.	X	
Tank max. liquid level below carburetor/injectors		
Solenoid in fuel line		
Fuel boost pump at tank		
INSTRUMENTS AND GAUGES		
Oil pressure #		
Engine coolant thermometer #	X	
Tachometer #	X	
Odometer/Hourmeter	X	
Hand counter (ratio..........to..........engine/pump)	X	

..........In. Suction gauge..........In. Hg. to..........psi

..........In. Discharge gauge..............to.............psi

— AT OPERATOR'S POSITION

Not Valid Unless Countersigned

UNDERWRITERS LABORATORIES INC.

SPECIMEN

Signed: D. L. Breting

D. L. BRETING
Vice President
Follow-Up Services

CERTIFICATE № 28864

Figure 7-1. Underwriters Laboratories, Inc., Certification Test (Courtesy of Underwriters Laboratories)

purchaser's location; the engine and pump may have been subjected to wear and abuse subsequent to the certification test from demonstrations and road mileage during delivery from the factory to the purchaser. The acceptance test is a realistic check of the pumper, because it is conducted under the conditions and at the elevation where the apparatus will be operated. Also, there are other requirements to be checked out that are not included in the certification test, such as driving performance, carrying capacity, cooling system, suspension, and braking system. For these tests, the apparatus should be loaded with its full complement of hose, booster tank water, equipment, and manpower so the rig will attain its full anticipated weight.

A delivery test should be conducted under the joint supervision of representatives from the manufacturer and the fire department to discover the condition of the apparatus and to determine whether it fully complies with the contract specifications. It will be extremely beneficial if the fire company that will be assigned the apparatus can be present at these tests. This will give the pump operators and drivers the opportunity to drive the vehicle and to operate the pump and other devices and equipment at full capacity and at high pressures under the direct observation of experts who have the knowledge to answer their questions and to correct their mistakes.

The delivery test is probably the most important of the three tests because it assures that the fire department is obtaining the vehicle it is paying for and provides the performance criteria against which the annual service tests will be compared.

Service Tests

Pumpers in service or in reserve should be tested annually and after any extensive repairs to the engine or pump. Tests made during the service life of the apparatus can be compared with previous service tests and the acceptance test to check how well the engine and pump are maintaining their original efficiency as a pumping combination (Figure 7-2). Investigation has shown that when regular and systematic tests of pumpers are not made, existing defects may continue undetected for considerable periods under the light demands of ordinary fires and drills; they may only become apparent at a large fire when the pumper is called upon to perform at or near its full capacity. A gradual lessening of efficiency is seldom apparent during routine and normal operations.

Furthermore, regular tests are a valuable drill for pump operators because in very few fire departments do they receive sufficient training in operating pumpers under a variety of conditions, such as at high pressure, at full capacity, and at draft. The failure of an apparatus to function effectively at a fire, or the inability of the crew to operate the pumper properly, may cause needless loss of life and property.

Service tests are based on the capacity specified in the original purchase specifications; they should be conducted under the same conditions as the delivery test so that a realistic comparison of results is possible. Tests of apparatus at pressures or quantities less than those specified by the original contract do not show the true condition of the apparatus and are of little value for other than operator training.

Data submitted at the time of the acceptance tests and all results of service tests should be maintained in a permanent file so that the condition of the pumper can be compared over years of operation.

Preparatory Procedures for Testing

Pumpers should be tested at draft. When possible, the vertical lift (the vertical distance from the surface of the water to the center of the pump suction inlet) should not

F-71—1M—12-65 (F-72)

CITY OF LOS ANGELES — DEPARTMENT OF FIRE
APPARATUS SERVICE TEST RECORD

Shop No. 60020

Date ______ Make of apparatus Mack Eng. Co. 37

Where tested Test Pit Make of engine Hall-Scott Engine No. 606093

Tested by Leephart & Philpott No. cylinders 6 Bore 5-3/4" Stroke 6"

Operated by Jack Wyan H.P. 300 at 2,400 R.P.M.

Road gear used 4th Pump make Hale Mod. C 21 F Ser. 21056 Type Cent.

Priming device type Gear Suction size 6" Where taken Left Side

Rated capacity 1,250 gpm at 150 lbs. pressure. Gear ratio, Engine to pump 1 to 1.94

TEST GAUGE NOS. ENG. 3 VAC. 1 PITOT 2 R.P.M.'s checked by Sun Tach.

Lift in feet to center of pump from ground 3' 7" Total lift 9' Water obtained in 28 seconds

FIRST TEST: (Dry vacuum)

16" obtained in 8 seconds. Shut down at $24\frac{1}{2}$" " and dropped $9\frac{1}{2}$" " in 30 seconds.

SECOND TEST: (CAPACITY) Layout 4-50' lines into 1-$3\frac{1}{2}$" lines gated **Noz. size** $2\frac{1}{4}$"

Time	Pitot	Pressure Eng.	Pressure Test	Vacuum Eng.	Vacuum Test	R.P.M. Tach	R.P.M. Check	Ign.	Temp.	Oil P.	Manifold Vacuum
1:38	70	150	150	13"	$12\frac{1}{2}$	1725	1900	B&B	185	55	
1:43	70	150	150	13"	$12\frac{1}{2}$	1725	1900	"	190	50	
1:48	70	150	150	13"	$12\frac{1}{2}$	1725	1900	"	185	48	
1:53	70	150	150	13"	$12\frac{1}{2}$	1725	1900	"	185	42	
1:58	70	150	150	13"	$12\frac{1}{2}$	1725	1900	"	185	40	

THIRD TEST: (PRESSURE) Layout 3-50' lines into one $3\frac{1}{2}$" line, gated **Noz. size** 1-3/4"

Time	Pitot	Pressure Eng.	Pressure Test	Vacuum Eng.	Vacuum Test	R.P.M. Tach	R.P.M. Check	Ign.	Temp.	Oil P.	Manifold Vacuum
2:35	95	208	200	10"	$11\frac{1}{4}$	1850	2025	"	180	50	
2:40	95	208	200	10"	$11\frac{1}{4}$	1850	2025	"	190	48	
2:45	95	208	200	10"	$11\frac{1}{4}$	1850	2025	"	185	46	

FOURTH TEST: (High Pressure) Layout 2-50' lines into one $3\frac{1}{2}$" line, gated **Noz. size** 1-5/8"

Time	Pitot	Pressure Eng.	Pressure Test	Vacuum Eng.	Vacuum Test	R.P.M. Tach	R.P.M. Check	Ign.	Temp.	Oil P.	Manifold Vacuum
2:46	64	258	250	10"	$11\frac{1}{2}$	1575	1750	"	185	42	
2:51	64	258	250	10"	$11\frac{1}{2}$	1575	1750	"	185	41	
2:56	64	258	250	10"	$11\frac{1}{2}$	1575	1750	"	185	41	

FIFTH TEST: (Relief valve or governor) Make Hale

Relief valve or governor set at 150 psi Shut off at 158 psi raised pressure 8 lbs.

SUMMARY

	2nd test	3rd test	4th test
Duration in minutes	20	10	10
Engine pressure	150	200	250
Gauge correction	---	---	---
Corrected E.P.	150	200	250
Vacuum, lbs. per sq. in.	6.9	5.9	4.9
Net pump pressure	156.9	205.9	254.9
Pitot reading	70	95	64
Gauge correction	--	--	--
Corrected Pitot reading	70	95	64
G.P.M.	1260	886	626
Engine R.P.M.'s	1900	2025	1750
Pump R.P.M.'s	3686	3928	3395

Mechanical work done during test

Figure 7-2. Example of service test form

exceed 10 ft. A well-designed drafting pit of sufficient size to prevent turbulence and the excessive heating of the water provides ideal conditions.

When a drafting site must be improvised from a river, lake, pond, or other body of water, the various possible sites should be carefully studied to select the best. The site should, if possible, be located along an improved roadway or on solid ground where the water level is from 4 to 8 ft below the grade. It should be possible to reach the water from the pumper suction inlet with not more than 20 ft of hard suction hose with the strainer submerged at least 2 ft and with no humps in the hose. The water should be at least 4 ft deep where the strainer is located to provide clearance below the strainer and sufficient depth above it. If drafting from shallow water is necessary, a special basket or container should be used to prevent the suction in the hose from drawing in particles from the stream bed. Clean, fresh water is desirable, but where salt water is drafted, the pump, pipe, fittings, and pressure regulating devices should be thoroughly flushed out after testing.

While tests should always be made at draft, if possible, it may sometimes be necessary to run service tests by connecting to a hydrant. However, this procedure does not test the ability of the priming system or of the pump to maintain a prime, and it involves the additional problem of getting rid of the large amount of water that is discharged. Preferably, the hydrant used should be one at which the initial pressure is below 40 psi, and the hydrant should be connected to a large-enough water main to assure a sufficient supply of water.

The pumper should be parked as close as possible to the water's edge; it is usually more convenient to have the pump control panel side of the pumper away from the water. Front or rear suction inlets should be avoided because the piping between the pump and the inlet is usually restricted.

The size of the suction hose to be used will depend on the altitude and the lift, as well as on the rated capacity of the pump to be tested. The suction hose used should be of the minimum sizes for the pumps indicated:

500 gpm pumpers–4-in.

750 gpm pumpers–4½-in.

1000 gpm pumpers–6-in. or, at low altitudes, 5-in.

1250 gpm pumpers–6-in.

1500 gpm or larger pumpers–6-in. or two suction hoses (5-in. minimum size) into a siamese

Test Hose Layouts

The discharge hose layout should consist of 2½-in. hose lines to one or more smoothbore nozzles of suitable size. The hose performs two functions: (1) to carry the water from the pumper to the nozzle and (2) to provide enough total friction loss, thus reducing the required pump discharge pressure to the desired nozzle pressure. If only a relatively short length of hose is required to perform the first function, the second function can be performed by increasing the friction loss by partly closing the discharge outlet gate or gates on the pumper or, preferably, by partly closing a valve inserted in the discharge lines for that purpose. Suggested hose and nozzle layouts are shown in Table 7-3.

The lengths of 2½-in. hose between the discharge gates and the nozzle or heavy stream appliance should not exceed 100 ft in length; the number of hose lines will depend on the required quantity of flow. If the friction loss is so excessive that the required flow cannot be obtained at the necessary pump discharge pressure, reduce the number of supplying hose lines or shorten one or more of the lines to 50 ft. Remember, the friction loss should be less than that actually needed to operate the pump at the required

TABLE 7-3. HOSE AND NOZZLE LAYOUT SUGGESTIONS*

DISCHARGE, gpm	120 OR 150 psi TEST
2000	Two 100′ lines, 2″ nozzle, in duplicate
1750	Two 100′ lines, 2″ or 1⅞″ nozzle, in duplicate
1500	Three 100′ lines, 2″ nozzle; and one 50′ line, 1⅜″ or 1½″ nozzle
1250	Three 100′ lines and one 50′ line; 2¼″ nozzle or one 50′ line, 1⅝″ nozzle, in duplicate
1000	Three 100′ lines; 2″ nozzle
750	Two 100′ lines; 1¾″ nozzle
500	One 50′ line; 1⅜″ or 1½″ nozzle

*Copyright, Insurance Services Office, 1975.

NOTE—Where two or more lines are indicated, they are to be siamesed into a heavy stream appliance.

discharge pressure and have the proper nozzle pressure. The additional friction loss is achieved by partly closing one or more of the pump discharge gates or line valves.

Observe all safety precautions at all times. Do not straddle or stand over any hose line that is under pressure; never stand in the vicinity of any charged hose line unless absolutely necessary because they are a constant threat of injury. Always stand to one side of a hose layout so that if a coupling pulls out or if the hose bursts, the danger of being hit will be reduced. If a hose breaks or a coupling pulls loose, the hose will tend to straighten out at first. Then it may start to whip back and forth if the pressure is sufficient. Therefore, it is safer to stand on the inside of a bend in the hose line, as compared to the outside, because the initial movement will be in the opposite direction. Hose lines that are being subjected to extremely high pressures should be lashed in place to prevent whipping and flailing if one should burst. Discharge lines should be securely fastened to the pumper outlets to avoid personnel injury if the hose should come loose from the coupling during the test.

The size of the nozzle is usually chosen to give the desired discharge quantity at a nozzle pressure between 60 and 70 psi, or reasonably close to it. This pressure is not so high that the pitot gage would be difficult to handle in the stream of water, nor is it so low that the normal inaccuracies of a gage at low pressures would occur. The nozzle should always be securely fastened; never allow a test to be made with dependence placed on any person or persons to hold the nozzle because pressure produces high nozzle reaction forces. Failure to abide by this precaution has caused serious injuries in several instances. A portable monitor or other heavy stream appliance will also provide safe stability.

The suggested nozzle sizes and pressures to be used are listed in Table 7-4. Other nozzle sizes and pressures can be utilized as long as the correct nozzle pressures for supplying the required quantities can be maintained. If large tips are needed, but are not available, two smaller tips that will deliver the required quantity may be substituted. For example, a 2¼-in. nozzle is required for the 150-psi capacity test of a 1250-gpm pumper, but the only sizes of large tips that the department owns are 1½-in., 1¾-in., and 2-in. By referring to Table 5-7 in Chapter 5, it can be seen that a 1½-in. tip at 64 psi will discharge about 533 gpm and a 1¾-in. tip at 64 psi will discharge about 727 gpm; the total of approximately 1260 gpm is sufficiently accurate for these tests.

The discharge quantities from smaller-size nozzles are given in Table 5-7 in Chapter 5 or may be obtained by taking the discharge from a nozzle of twice the given size operating at the same nozzle pressure

TABLE 7-4. NOZZLE DIAMETERS AND PRESSURES FOR PUMP TESTS

RATED gpm	NOZZLE DIAMETER, in.	NOZZLE PRESSURE, psi	ACTUAL gpm FLOW
250	1	70	250
	1⅛	45	253
350	1⅛	88	351
	1¼	58	351
375	1¼	65	376
500	1⅜	80	505
	1½	58	508
	1⅝	42	508
525	1½	62	525
	1⅝	46	531
625	1½	88	626
	1⅝	64	627
700	1⅝	80	700
	1¾	60	708
750	1¾	68	750
	1⅞	52	754
875	1¾	94	881
	1⅞	70	875
	2	55	886
1000	2	70	1001
	2¼	44	1000
1050	2	78	1050
	2¼	50	1065
1225	2¼	66	1224
	2½	43	1230
1250	2¼	70	1260
	2½	45	1252
1400	2¼	88	1412
	2½	56	1403
1500	2½	65	1506
1750	2½	87	1749
	2¾	60	1748
	3	42	1750
	1–1¾ & 1–2	70	1755
	2 & 2	54	1746
2000	2¾	78	2003
	3	56	2000
	2 & 2	70	1988

and dividing this discharge by four; for nozzles larger than those given in Table 5-7, the discharge may be obtained by taking the discharge for a nozzle of half the given size and multiplying this by four. This adjustment is possible because, for the same pitot pressure, the increase in discharge is very nearly proportional to the square of the ratio of the nozzle diameters.

Slowly open and close all nozzles, discharge gates, valves, hydrants, and engine throttles to prevent sudden pressure changes, water hammer, or pressure surges that could cause injury to personnel or damage to hose, apparatus, equipment, or water mains.

Net Pump Pressure

To calculate the net pump pressure while drafting, the pump inlet pressure must be subtracted from the pump discharge pressure. The net pump pressure is actually the sum of the pump discharge gage pressure (corrected for any gage error), plus the lift (the vertical distance from the water level to the suction gage), plus suction losses (friction loss in the suction hose and strainer). Allowances for lift and suction losses are usually given in feet and must be divided by 2.3 to convert the figures into pounds per square inch (psi). Table 7-5 gives the allowances in feet of lift for the suction hose friction loss.

The formula for calculating suction pressure is

$$\text{Suction pressure} = \frac{\text{lift} + \text{suction hose allowance}}{2.3}$$

For example, a 1000-gpm pumper is spotted so there is a 7-ft elevation from the water surface to the center of the side suction inlet; the test gages are placed 2 ft above the center of the suction inlet. For calculating the suction pressure, 9 ft of lift is used in the formula. Twenty feet of 5-in. suction hose is utilized. By referring to Table 7-5, 9½ ft was determined to be the equivalent suction hose friction loss. Divide the lift and the suction hose friction loss allowance in feet by 2.3 to convert the total to pounds per square inch (psi):

$$\text{Suction pressure} = \frac{9 + 9\frac{1}{2}}{2.3} = \frac{18.5}{2.3} = 8 \text{ psi}$$

The net pump pressures are the pressures that will be indicated on the discharge pressure test gage. All gage errors should be corrected when making calculations.

For the capacity test, the 8 psi is subtracted from the 150 psi test pressure to obtain a net pump discharge pressure of 142 psi. This would be the discharge gage pressure.

When making the 200 psi and the 250 psi pressure tests, an additional allowance is made. For the 70% of capacity at 200 psi, deduct 1 psi from the suction pressure to obtain the net pump pressure; for the 50% of capacity test at 250 psi, deduct 2 psi from the suction pressure to determine the net pump pressure. In the above example, the net pump pressure for the 200 psi test would be 200 – 7, or 193 psi and for the 250 psi test, it would be 250 – 6, or 244 psi.

When drawing water from a hydrant, the net pump pressure is equal to the pump discharge pressure less the pressure on the pump suction gage. If the pump suction gage shows a flowing pressure of 40 psi while pumping from a hydrant at a pump discharge pressure of 150 psi, the actual work being done by the pump is equivalent to a net pump pressure of 150 – 40, or 110 psi as the hydrant pressure is aiding the pump. If these gage pressures were obtained during a pumper test, 190 psi on the pump discharge pressure gage would indicate a net pump pressure of 150 psi.

Lift

The theoretical lift of the pump is the vertical distance from the center of the pump to

TABLE 7-5. ALLOWANCES FOR FRICTION LOSS IN SUCTION HOSE*

RATED CAPACITY OF PUMPER, gpm	DIAMETER OF SUCTION HOSE, in.	ALLOWANCE, ft	
		FOR 10 ft OF SUCTION HOSE	FOR EACH ADDITIONAL 10 ft OF SUCTION HOSE
500	4	6	plus 1
	4½	3½	plus ½
750	4½	7	plus 1½
	5	4½	plus 1
1000	4½	12	plus 2½
	5	8	plus 1½
	6	4	plus ½
1250	5	12½	plus 2
	6	6½	plus ½
1500	6	9	plus 1
	4½ (dual)	7	plus 1½
	5 "	4½	plus 1
	6 "	2	plus ½
1750	6	12½	plus 1½
	4½ (dual)	9½	plus 2
	5 "	6½	plus 1
	6 "	3	plus ½
2000	4½ "	12	plus 2½
	5 "	8	plus 1½
	6 "	4	plus ½

NOTE—The allowance computed above for the capacity test should be reduced by 1 psi for the allowance on the 200-psi test and by 2 psi for the allowance on the 250-psi test.

Example: 1000-gpm pumper, 9-ft lift, 20 ft of 5-in. suction.

$$\text{Pressure Correction} = \frac{9 + 8 + 1\frac{1}{2}}{2.3} = 8 \text{ psi (capacity test)}$$

the water level measured in feet. However, when testing pumps, the lift that is recorded should be the vertical distance from the pressure test gage to the water level because this is what is required to obtain the actual net pump pressure.

Maximum lift is the greatest difference in elevation at which the pumper can draft the required quantity of water under the established physical characteristics of operation; these include the design of the pump, the adequacy of the engine, the condition of the pump and engine, the size and condition of suction hose and strainers, the elevation of the pumping site above sea level, and the temperature of the water. The theoretical values of lift and maximum lift must be reduced by the entrance and friction losses in the suction hose equipment to obtain the actual or measurable lift.

An adequate-size suction hose should be used for testing in order to obtain a true measure of performance (Table 7-6). Too small a suction hose will restrict the flow of water and increase the friction loss in the suction hose; it may prevent the delivery of

TABLE 7-6. MINIMUM DISCHARGE TO BE EXPECTED OF A PUMPER IN GOOD CONDITION OPERATING AT DRAFT AT VARIOUS LIFTS*

CONDITIONS—Operating at net pump pressure of 150 psi; altitude of 1000 feet; water temperature of 60°F; barometric pressure of 28.94" Hg (poor weather conditions).

RATED CAPACITY, PUMP		500 gpm		750 gpm		1000 gpm		1250 gpm	1500 gpm		
SUCTION HOSE SIZE		4"	4½"	4½"	5"	5"	6"	6"	6"	DUAL 5"	DUAL 6"
Lift in Feet: 4	20' Suction Hose (Two Sections)	590	660	870	945	1160	1345	1435	1735	1990	2250
6		560	630	830	905	1110	1290	1375	1660	1990	2150
8		530	595	790	860	1055	1230	1310	1575	1810	2040
10		500	560	750	820	1000	1170	1250	1500	1720	1935
12		465	520	700	770	935	1105	1175	1410	1615	1820
14		430	480	650	720	870	1045	1100	1325	1520	1710
16		390	430	585	655	790	960	1020	1225	1405	1585
18	30' Suction Hose (Three Sections)	325	370	495	560	670	835	900	1085	1240	1420
20		270	310	425	480	590	725	790	955	1110	1270
22		195	225	340	375	485	590	660	800	950	1085
24		65	70	205	235	340	400	495	590	730	835

*Copyright, Insurance Services Office, 1975.

NOTES—Net pump pressure is 150 psi. Operation at a lower pressure will result in an increased discharge; operation at a higher pressure, a decreased discharge.

—Data based on a pumper with ability to discharge rated capacity when drafting at not more than a 10-ft lift. Many pumpers will exceed this performance and therefore will discharge greater quantities than shown at all lifts.

the rated capacity of the pumper at the available vertical lift. Other factors that affect operations at draft are the length and condition of the suction hose, altitude, water temperature, and barometric pressure.

Another element that greatly affects the performance is the difference between the actual discharge being pumped and the designed rated capacity. A 750-gpm pumper with a 30-ft length of 5-in. suction hose would probably have a maximum lift of about 15 ft, while a 1500-gpm pumper drafting only 750 gpm through 30 ft of 5-in. suction hose would probably have a maximum lift of about 20 ft or, through 30 ft of 6-in. suction hose, a maximum of 23 ft.

Testing Procedures

Pumper tests are no more accurate than the attention to detail and the precision with which nozzle pressures and engine pressures are maintained. If close watch is kept on the pitot gage reading, the line valve can be gradually adjusted as engine speed is increased, until pump pressure and pitot readings are both as desired. Care should be taken to make sure that the valve does not vibrate either open or closed; in either case, both the discharge and pump pressure will be affected.

Pumper Tests

When testing a pumper, three variable factors, which are interrelated in that a change in one factor will always produce a change in at least one of the others, must be considered: pump speed, net pump pressure, and pump discharge. For example, an increase in the speed of the pump will increase the discharge, or the pressure, or both. Adjustments of these variables through a change in the position of the engine throttle (modifying pump speed), a change in the hose layout or gate valve positions (modifying pump pressure), and changing the nozzle size (modifying discharge) are the only ways to reach the standard test condition desired.

The pumper should be operated at reduced capacity and pressure for several minutes to allow engine and transmission to warm up. The engine is gradually speeded up until the desired pump pressure is reached. If the pressure will not increase to the desired amount, a length or two of hose may have to be added, a smaller nozzle used or a discharge gate throttled. When the desired pressure at the pump is obtained, the pitot gage should be read to see if the required quantity of water is being delivered.

If the discharge is not as great as desired, and it is believed that the pump will deliver a greater quantity of water, the discharge may be increased by further speeding up the pump. If speeding up the pump increases the pump pressure more than 5 or 10 psi, a length of hose should be taken out of the layout, a discharge gate opened slightly, or a larger nozzle used.

When the pump pressure and discharge quantity are satisfactory, the test can be officially started and should be run for at least 20 min. Readings should be taken on the pitot gage and the pressure gage with sufficient frequency to obtain a good average. If the pressures vary, readings need to be taken with greater frequency than if the pressures hold steady.

The speed should be taken at frequent intervals, corresponding to the times the pressure readings were taken. Counting the revolutions for 1 min generally gives readings of sufficient accuracy.

After the engine has warmed up, there should be little change in the engine speed. It should be realized that any change in engine speed must of necessity produce a corresponding change in pump discharge and hence in pitot reading, and that, other things being equal, any change in pitot reading indicates a change in engine speed. A change in pump speed will also cause a change in pump pressure, so whenever pump speed, pump pressure, and pitot gage readings do not show corresponding changes, it is safe to say that some reading is in error or some condition has arisen that affects the readings and needs correction. Engine speeds can be changed by working the hand throttle at the operator's position. Automatic relief valves and pressure governors controlling the speed of the pump should be disengaged during the tests.

Gages and Instruments

All the time that a pumper is being tested, its operation should be observed carefully. All the gages and other instruments should be checked at frequent intervals. Compare the gages on both the pump panel and the cab dash board to ascertain if they are both registering and are reasonably accurate. Constantly check the oil and temperature gages, as well as the ammeter, pump log, engine hour meter, and any other instrument on the pump panel or the dash board.

Service tests are designed and conducted in such a manner that their principal purpose is to detect and avoid any potential malfunctions before they can affect a pumper's performance at a fire. Unusual noises and vibrations should be checked out and corrected immediately; if the cause cannot be quickly determined, shut down the test before damage is caused. Watch for water and oil leaks. This is the best opportunity to determine if everything is operating properly.

When operating any apparatus, it is important that the engine temperature be kept within the proper range; neither a cold nor an excessively hot engine will give as good service as one run at the proper temperature. If the engine temperature rises above the manufacturer's recommendation, crack the engine cooling valve slightly to maintain an even temperature.

The oil pressure of the engine should be watched to see that the engine lubrication is being properly maintained; pump transmissions should be frequently checked for overheating; and any leak in the pump casing, packings, or connections should be noted and taken care of. This is especially true of centrifugal pumps, which are not self-priming and could, therefore, lose their water supply from a leak in the suction line. A slight leakage at the pump packing gland is beneficial because it keeps the packing cool.

Other defects in the performance of the engine or pump should be recorded. Minor defects should be corrected immediately, if possible. Pumpers should be retested after any major repairs.

Pressure Gages

During the tests the pumper gages should not be relied upon for complete accuracy; they should be simultaneously checked by comparing their readings with those of reliable test gages. No tests can be more reliable than the accuracy of the measuring instruments employed.

Two plugged ¼-in. standard pipe thread connections for test gages, one on the pressure side and one on the suction side of the pump, are suitably identified and readily available on the pump panel. Connect the test gages to these fittings with flexible hoses so the gages may be placed in a position for convenient readings. When comparing the readings on the apparatus gages with those of the test gages, the two gages should be at the same level; a difference of 1 ft in elevation would cause an inaccuracy of almost ½ psi. While this slight difference would not cause any problems, a greater divergence in head pressures should be considered and allowances made accordingly.

It is a common but faulty practice to read a pressure gage at the highest point in the swing of its needle; the center of the needle swing should always be read, as this is the average pressure. A petcock in the line to the test gage may be throttled to prevent excessive vibration, but if throttled too much, the gage pointer will no longer indicate the pressure correctly; never attempt to entirely eliminate the needle vibration. Any leak in the line to the test gage will result in an incorrect gage reading.

Tachometer

All modern pumpers have a tachometer at the pump operator's position to indicate the speed at which the engine is operating when driving the pump. During every pump test, the accuracy of the tachometer should be checked during the four pumping tests; if there is any significant inaccuracy, record a true rpm count every time the pressure gage readings are made. The speed of the engine for each test should be compared with the acceptance test and the previous service tests to ascertain whether the engine speeds must be increased to meet the pumping requirements as the apparatus ages.

A revolution counter, with an extension if necessary, or an accurate tachometer should be used for obtaining true engine speed readings. Some portable commercial tachometers of an electric type are very accurate.

Any watch with a full sweep second hand that may be clearly read can be used. When using a stop watch, the best and most accurate method is to leave the stop watch running at all times, engaging the revolution counter at a chosen instant and disengaging it when the hand of the stop watch makes a passage to the same point on the dial 1 min later. The best method of performing this test is to utilize a hand counter for exactly 60 sec; taking a measurement for only 30 sec doubles the margin of error.

Pitot Gage

When conducting a pump test, the correct quantity of water must be pumped at the designated net pump discharge pressure. Unless a fire department has a flow meter, the best method of measuring the pump discharge gallonage is to utilize straight-stream nozzles of the correct size. When the tip size is known, measuring the velocity pressure of the nozzle stream and referring to a discharge table to determine the quantity of water being discharged are the only steps necessary.

A pitot gage assembly (Figure 7-3) equipped with a gage measures the velocity pressure of the fire stream flowing from a nozzle. To ensure accuracy, the pitot tube should be securely clamped in position at the nozzle so that it will remain in the center of the fire stream; the leading edge of the pitot tube should be one-half the nozzle diameter away from the end of the nozzle. If the pitot tube is brought up too close to the nozzle, the gage reading will be erroneously increased.

It is very beneficial if a length of flexible hose is inserted between the pitot assembly and the gage; this will allow the gage to be mounted close to the pump panel where the engineer can watch both the nozzle pressure and the pump discharge pressure gages simultaneously. This will speed up the tests considerably and the hose will also absorb vibrations, thus reducing gage needle flutter. Readings will be easier to take and will be more accurate.

Once the correct balance between pump and nozzle pressures has been achieved, the test may begin. Various readings should be taken often enough to ensure an accurate average for the entire test.

Calculating the Results

The quantity of water pumped can be calculated from discharge tables; use the average pitot gage reading, corrected for gage errors, and the size of nozzle used. If readings were taken at exact and equal intervals, the proper average is the arithmetic average of the recorded readings; if the readings were not taken at equal intervals of time, a judgment average of the readings taken should be used; give more weight to those readings that were maintained for longer periods of time. The same principle also applies to the readings of pump pressure and speed.

Knowing the true engine speed and the pump transmission gear ratio, the pump speed can be calculated. It is an indication of a loss of pump efficiency when the pump impeller must rotate at a higher rpm to discharge the rated quantities of water.

Trouble Shooting

Most tests are conducted without incident. Occasionally trouble does develop during some tests and an effort should be made to locate the source of trouble while the pumper remains at the test site.

Failure to prime a centrifugal pump is a frequent source of trouble and is usually caused by an air leak in the assembly. One method to trace this trouble is to remove all attached hose lines, cap all suction and discharge openings, and operate the priming mechanism in accordance with the manufacturer's recommendations. Study the suction gage to determine the maximum vacuum developed; this should be at least 20 in. of mercury (20 in. Hg). Stop the primer

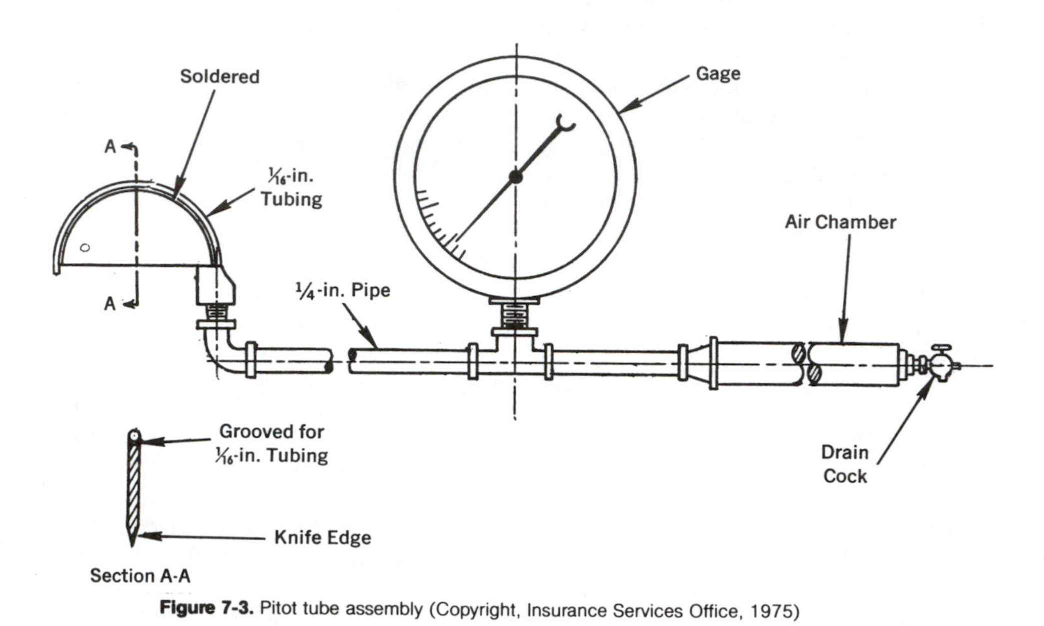

Figure 7-3. Pitot tube assembly (Copyright, Insurance Services Office, 1975)

and attempt to hold the vacuum in the pump. If the vacuum drops from the maximum to about 12 in. in less than 10 min, there is a leak in the pump assembly; it may be in a valve, drain cock, piping, casing, or pump packing. The leak may be located by listening for air movement. Another method is to connect the pumper to a convenient hydrant, cap the discharge outlets, open the hydrant, and watch for water leaks. A leak can usually be corrected at the test site.

Two other possible causes of an air leak are the wagon battery control valve and the booster tank dump valve. Remove the battery nozzle and replace it with a cap; check the booster tank to be assured that it does contain water. It has happened at a lengthy drafting operation that the pump suddenly lost its suction. The cause was determined to be that tank water was leaking past the closed dump valve and when the tank became empty, the consequent air leak was too great for the pump to maintain its high drafting lift.

If the pump cannot deliver its rated capacity at the various pressures, Chapter 8, on trouble shooting, should be referred to for a more complete discussion. The most likely cause of failure is the inability of the engine to develop sufficient horsepower. Insufficient power is the result of the engine not running at the proper speed for the desired conditions. It may be that the pump operator is hesitant about advancing the throttle to a fully opened position, or the transmission may be in the wrong position. If a sufficiently high engine speed cannot be attained from a gasoline engine, a tuneup with new spark plugs and distributor points may be indicated. A slipping clutch or an unlocked automatic transmission may be allowing a power loss.

Excessive engine speed may be the result of operating the pumper with the wrong transmission gear in use, stuck throttle control cable, restricted suction, or not having the suction hose strainer under a sufficient depth of water. Restricted suction is a common cause of a pumper's inability to deliver the rated capacity of water during a test and it may be the result of any one or a combination of the following conditions: suction hose too small; high altitude; suction lift too high; improper type of strainer; clogged suction strainer at the pump or at the end of the suction hose; aerated water; water too warm (over 95° F); leaks on the suction side of the pump at the intake or couplings; collapsed or defective suction hose; and foreign material in the pump.

Insufficient pressure when operating a centrifugal pump may be the result of pumping too large a volume of water for the power available and, in multistage pumps, pumping in the volume position instead of the required pressure position. This can be checked by partly closing off all discharge valves until only a small flow is observed, then opening the throttle until the desired pressure is reached, followed by slowly opening all discharge gates and increasing the throttle as necessary to maintain pressure until the desired volume is obtained. An improperly adjusted or inoperative transfer valve may prevent the building up of adequate pressure.

If the engine is operating efficiently, and the road transmission is in the correct gear, check the trouble shooting lists in Chapter 8, "Pump Trouble Shooting," and correct the deficiencies. If corrections cannot be made without extensive repairs or a complete overhaul, then record the volumes of water delivered to the nozzles at the correct net pump discharge pressures. This will allow a comparison of the reduction of pump capacity.

As an example, if a 1000-gpm pumper cannot discharge its full capacity at a net pump pressure of 150 psi, slowly reduce the quantity of water flowing to the nozzle. As the discharge gate or line valve is slowly closed, thus increasing the friction loss in the hose layout, the discharge pressure will rise correspondingly. When the correct net pump pressure is indicated on the test gage, take a pitot reading of the test nozzle pressure to discover what the true pump discharge quantity is. A 2-in. tip at 72 psi pressure will discharge 1008 gpm; if only 64 psi nozzle pressure can be attained, the pump is merely discharging 951 gpm at the net pump pressure of 150 psi. Thus, the pump is only delivering 94% of its rated capacity at 150 psi, instead of 100%.

Service Test

There is no exact sequence for performing the various tests described in this section, but they should be carried out with the least intervening time between them, just long enough to change the hose and nozzle layouts. Experience has shown that the following sequence will require the least changing and manipulation of hose and nozzles:

1. Perform the Dry Vacuum Test before the hard suctions are placed in the water; this test should indicate any suction problems that could affect the later tests.

2. The Drafting Test is next. Drop the hard suctions into the water and check to find out how long it takes to prime the pump.

3. Perform the Capacity Test, then the Overload Test, and then the Automatic Pump Pressure Control Test for volume operation as these three tests all require the same hose and nozzle layouts.

4. Check the pump for its ability to deliver 70% of capacity at 200 psi; although it is not required, it is a good idea to again check the relief valve or governor at this pressure.

5. Check the ability of the pump to deliver 50% of capacity at 250 psi; then perform the Automatic Pump Pressure Control Test.

Dry Vacuum Test

A pumper should have the ability to develop an adequate suction for drafting; the pump and piping should be free of excessive air leaks. To determine this proficiency, a vacuum test, using the pump priming device, shall develop vacuum of 22 in. of mercury at altitudes up to 1000 ft and hold the vacuum with a drop of not in excess of 10 in. in 10 min. For every 1000 ft of additional elevation, the required vacuum shall be reduced 1 in. The primer shall not be used after the 10-min test period has started; all discharge outlets should be uncapped. This is basically a test of the priming system and the tightness of the pump and fittings, not a check of the pump's ability to maintain a vacuum while drafting water.

Drain the pump, close all drain, discharge, and inlet valves. Inspect all gaskets in suction hoses and inlet caps. Check to be certain that the suction hose is of the same size as was utilized in the previous tests so an exact comparison of the test results may be made. Connect 20 ft of suction hose to the side suction inlet and cap the far end. Operate the priming device until the vacuum test gage indicates vacuum of 22 in. of mercury; if sufficient vacuum cannot be obtained within 1 min, discontinue the operation and check for leaks. After the 22 in. of vacuum is obtained, stop priming, disengage the pump if it is rotating, and time the pressure decrease. The gage should indicate a drop of not over 10 in. of mercury (10 in. Hg) in 10 min.

If air leaks are excessive, they should be located and corrected; failure to correct defects could adversely affect the subsequent tests. Leaks may be detected by pulling a high vacuum and then shutting off the engine; air leaks may cause a perceptible sound. The most common sources of air leaks are open valves and leaking suction hose gaskets.

Record the maximum vacuum attained and the amount it dropped on the test form.

Drafting Test

To determine the mechanical condition of the priming system and the absence of air leaks the pump should be capable of taking suction and discharging water with a lift of 10 ft in not more than 30 sec through 20 ft of suction hose of the appropriate size; pumpers of 1500-gpm or larger capacity shall perform the test in not over 45 sec. The pump should be dry and be capable of performing the test with the transfer valve in both the parallel and series position, when the pump is of a parallel-series type.

Spot the pumper close to the water source. Open the pump drains to be certain that the pump is dry, then close all drains, discharge gates, bleeder valves, and other openings into the pump. Make sure that suction gaskets are in place and free from foreign matter. Connect 20 ft of the correct size suction hose with an attached strainer to the side suction inlet. Lower the strainer end of the hose into the water; secure the strainer with a rope so it is held at least 2 ft below the water surface and 2 ft from the bottom. Connect the necessary 2½-in. hose lines to the discharge gates.

Start the priming mechanism, noting the time; after the prime is obtained, operate the pump controls and throttle as necessary to develop pressure; then open one discharge gate to permit the flow of water; note the time that water first issues from the tip. Water should be discharged within 30 or 45 sec, depending on the size of the pump.

The procedure is the same for positive displacement pumps, except that one discharge valve must be opened to permit the evacuation of air and the churn valve must be closed before starting to draft.

Another important benefit from this test is that it offers a practical method of determining a pump operator's skill and knowledge. The required pumping procedures for drafting are necessarily more complex and critical than those used during hydrant operations.

Capacity Test

This test of the pumper's ability to deliver 100% of its rated capacity at a net pump discharge pressure of 150 psi is conducted for 2 hr during the certification and the delivery tests; it occupies 20 min during the service tests. Before starting this test, operate the pump at about 100 psi engine pressure for a few minutes to allow the engine, road transmission, pump transmission, and other power train components to warm up. Check the instruments to be sure that everything is operating satisfactorily. When so equipped, the transfer valve should be in the parallel (volume) position.

Slowly increase the engine speed until the discharge test gage indicates a net pump pressure of 150 psi. Use the pitot gage to measure the velocity pressure at the nozzle to determine the quantity of water flowing. Open or close the water control valve to adjust the nozzle flow and attain the desired discharge. Closing the valve will increase the friction loss and decrease the nozzle pressure and the water flow; opening the valve will decrease the friction loss and increase the nozzle pressure and the water flow. Close cooperation between the pump operator, slowly adjusting the micrometer throttle control to attain the correct net pump pressure, and the helper adjusting the water control valve to attain the correct nozzle pressure will quickly and effectively allow the desired pump pressure and discharge to be attained.

After the correct balance of pump pressure and nozzle discharge has been achieved, the test may start. For the 2 hr tests, it is recommended that the gage readings be recorded every 10 min; for the annual service test of 20 min, readings at 5-min intervals are suggested. Record the time of starting and the gage readings, then take an engine speed reading with either a test tachometer or a hand counter. Repeat these recordings at the specified intervals. Minor adjustments of the engine speed or the water control valve may be necessary to maintain the correct pump pressure and nozzle discharge.

During the tests, constantly check the oil pressure, engine temperature, pump transmission temperature, ammeter, and all other instruments; constantly be on the alert to detect any signs of an incipient malfunction; check the pump packing to ascertain that a few drops per minute are leaking through to keep the packing cool; this is the best opportunity to observe the functioning of the pumper working under a maximum load.

Overload Test

The overload test, sometimes referred to as the spurt test or the excess power test, checks the engine's reserve power to ascertain if it has more than the bare amount of energy to perform the pumping operations. One reason for this test was that some pumpers were being manufactured with engines of the barest minimum power to pass the road and pumping tests when the apparatus were new. After the apparatus had aged somewhat, and the engines suffered the average loss of efficiency due to wear, the pumpers were not able to pass the annual tests. Incorporating this check into the annual service tests will allow an observation of how much of the excess power the engine retains as the apparatus grows older.

To pass this test, the pumper must deliver 100% of its rated capacity at 165 psi net pump pressure without exceeding the engine manufacturer's certified maximum no-load governed speed.

Utilize the same hose layout and nozzle arrangement as was used for the capacity test. Slowly increase the engine speed to raise the net pump discharge pressure to 165 psi; simultaneously, slowly close the valves regulating the water to the hose lines. When the discharge gage indicates a net pump pressure of 165 psi and the pitot gage shows that 100% of the pump's capacity is being discharged, record the readings on the test sheets. During the certification and delivery tests this overload test is conducted for ten minutes; during the service tests, five minutes will allow a sufficient opportunity to observe the results.

Automatic Pump Pressure Control Test

Pumpers are equipped with an automatic pump pressure control to prevent the pump discharge pressure from increasing a hazardous amount when a nozzle is shut off. Pumpers may be equipped with either a pressure governor, which controls the engine speed, or a relief valve, which allows water to bypass from the discharge side of the pump back to the suction side. Either type of device, when set in accordance with the manufacturer's instructions and properly maintained, should control the pump discharge pressure with very little increase.

With the pressure control set at 10 psi higher than the desired operating pressure and the pump discharging water through one or more hose lines, the pump pressure should not increase more than 30 psi when the discharge valves are slowly closed. The device should be capable of operation over a range of 90 to 300 psi discharge pressure.

This test is usually made immediately following the capacity test to utilize the same hose layout and nozzle. It checks the performance of the automatic pump pressure control while pumping from draft and discharging rated capacity at 150 psi pressure. While not a requirement, it is a good practice to slowly open the discharge gates again to ascertain how accurate the control is at returning the discharge pressure to the original setting.

After the 150 psi test has been completed, reestablish the controls so the pump is again discharging its rated capacity at 150 psi net pump pressure. Reduce the pump pressure to 90 psi by using the engine throttle, readjust the pressure control according to the manufacturer's instructions, slowly close the discharge gates, and the engine pressure should not increase over 30 psi.

When the pump is discharging 50% of its rated capacity at 250 psi, repeat the above test; the pressure should not increase over 30 psi.

Most engineers also check out the relief valves and governors by performing the pressure control test after the pump has been tested on its ability to provide 70% of its rated capacity at 200 psi. This is not a required test of the device, but it does allow another check of the ability of the automatic pump pressure control to regulate the pump discharge pressures; these tests also provide valuable practice for the pumper operators to adjust the mechanism. Another helpful benefit of this testing is that it makes the device operate over its full range of control, thus keeping the mechanism operating freely.

Use the same procedures in the 200 psi and the 250 psi relief valve and governor tests as were used in the 150 psi check.

Test of 70% at 200 psi

This test will check the ability of the pumper to deliver 70% of its rated capacity at a net pump pressure of 200 psi. Some parallel-series pumps will perform this test best if the transfer valve is in the series (pressure) position; others are more efficient in the parallel (volume) position. Check the previous tests for the correct position; if in doubt, experiment with each.

Change the nozzle to the correct size for the required quantity of flow at the 200 psi pressure. If a 1000 gpm pumper is being tested, 700 gpm will represent 70% of its rated capacity. Balance the pump discharge pressure and the nozzle discharge as suggested in the capacity test. Record the gage readings and the engine speeds at 5-min intervals during the service test and at 10-min intervals for the longer tests.

Test of 50% at 250 psi

This test will check the ability of the pumper to deliver 50% of its rated capacity at a net pump pressure of 250 psi. Transfer valves should be placed in the series (pressure) position.

Change the nozzle to the correct size for the required quantity of flow at 250 psi pressure. If a 1000 gpm pumper is being tested, 500 gpm will represent 50% of its rated capacity. Balance the pump discharge pressure and the nozzle discharge as suggested in the capacity test. Record the gage readings and the engine speeds at 10-min intervals for the longer tests and at 5-min intervals for the service test.

Testing Positive Displacement Pumps

Theoretically, a positive displacement pump will discharge a definite quantity of water each time the shaft rotates once. Slippage is the difference between the nominal (theoretical) displacement and the quantity of water that is actually pumped.

The actual displacement of positive displacement pumps should be calculated, and from this and the nominal displacement, the percent of slippage is calculated. The actual displacement is found by dividing the discharge in gallons per minute (gpm) by the speed of the pump in revolutions per minute (rpm); this gives the displacement in gallons per revolution. The nominal displacement is sometimes found stamped on the pump name plate, but if not, it can be obtained from the manufacturer, provided that the nominal displacement has not been changed by repairs to the pump. The difference between the nominal displacement and the actual displacement, divided by the nominal displacement, times 100, is the percent of slip.

Long usage and the accompanying wear serve to increase the slip so that the amount of slippage is an indication of the condition of the pump. The slip, of course, will be much higher for the pressure test than for the capacity test. Some pumps will have a much higher slippage than others even when new.

A comparison of pump slippage and engine speeds as obtained in the service test, with those obtained in tests made at previous times and especially the acceptance test made when the pumper was new, will give a good indication of the condition of both the pump and the engine. If the pumper will not deliver its rated capacity, it may be due to a much increased slip or may be due to the poor condition of the engine. A very high slip on a piston pump often indicates the valves are in poor condition.

Slippage, as the term is used in these tests, is not a factor in testing centrifugal pumps.

WATER TANK

8 Pump Trouble Shooting

Difficulties in pump operations may occur from time to time. There may be trouble in priming the pump; sufficient water may not be delivered to the nozzles; adequate fire streams may not be developed; or any number of other problems may be encountered. The ability to recognize the difficulty, diagnose the possible causes, and identify the remedy will largely determine the competence of operators.

The majority of difficulties will be readily identified and remedied by experienced and knowledgeable operators. Many mechanical problems will not occur if the apparatus is well maintained and is regularly operated at fires and at wet drills under realistic conditions. Operators and machines both become sluggish, inefficient, and ineffective when they are not occasionally exercised to their full capacity.

In addition, long periods of inactivity permit transfer valves, governors, relief valves, priming pumps, and other auxiliary equipment to become inoperative or inaccurate. These problems are very likely to occur when nonuse or lack of setting or adjustment allows parts to become corroded or clogged. An accumulation of rust particles, sand, and other debris will cause sticking, freezing, or inaccurate responses. The best method of guaranteeing that these devices will operate properly is to periodically use and adjust them while pumping water under pressure, and then lubricate them for correct maintenance.

At least once every month, all pumps and accessory equipment should be put through a comprehensive wet drill. This practice will develop and train competent operators; they will quickly learn the capabilities and the limitations of the apparatus; they will be able to correctly diagnose and correct any malfunction.

Performance Affected by Environment

Terrain, weather, and other environmental situations over which we have no control greatly affect the abilities of machines to develop their full potentials. These conditions may include any or all of the following: altitude, air temperature, barometric pressure of the atmosphere, relative humidity, and temperature of the water to be pumped. Internal combustion engines and pumps are designed and engineered to develop their full potentials at close to sea level and at a certain atmospheric temperature (commonly 85° F).

Atmospheric Pressure

The effects of atmospheric pressure on pump efficiency, though not severe, must be taken into consideration in certain situations.

Altitude effect. The atmosphere that surrounds the earth is a mixture of various gases. Although it is invisible and many people consider air as being a form of nothing, it does possess a definite weight that is applied uniformly all over the surface of the earth. A column of air 1 in. square and extending to the upper limit of the atmosphere weighs 14.7 lb at sea level, and, therefore, exerts a pressure of 14.7 psi. As elevations above sea level increase, the atmospheric pressure decreases at a rate of approximately ½ psi for every 1000 ft (Table 8-1).

The effect of variations in atmospheric pressure on the fire pump itself is not significant when the pump is being supplied by a hydrant or other source that provides a positive pressure at the pump inlet. When drafting, or when pumping fairly large volumes of water from a booster tank, where there is a vacuum created in the pump inlet, atmospheric pressure becomes increasingly important.

Weather conditions. Normally, rainy weather is accompanied by a reduction of atmospheric pressure. A difference in barometric pressure due to weather conditions will have the same result as a change in altitude. The difference in atmospheric pressure due to operation during a storm, instead of a cool clear day, could easily mean the difference in lift of 1 ft.

Effects of Altitude and Temperature on Engines

Low air temperatures make all internal combustion engines hard to start for several reasons. Lubricating oil becomes stiff and more viscous, so more power is required to crank the engine. A fully charged battery will lose 10% of its power at 0°F. Ice may form in gasoline lines, or diesel fuel may congeal on the walls of the tubing, reducing the flow of fuel to the engine. Condensation may form and freeze on the distributor points of gasoline engines, preventing ignition. Without a starting aid, the compression temperature in the cylinders of a diesel engine may drop to a point at which fuel will not ignite; cold weather starting aids include glow plugs, intake air preheaters, manifold heaters, and ether.

Engine performance depends on the amount of air that can be drawn into the cylinders, as well as on the fuel-air ratio, fuel quality, and several other factors. Energy is developed by burning a fuel-air mixture; the amount of fuel that can be burned during

TABLE 8-1. ATMOSPHERIC PRESSURE AT HIGHER ALTITUDES

ALTITUDE, ft	ATMOSPHERIC PRESSURE, psi	MAXIMUM THEORETICAL SUCTION LIFT, ft
Sea Level	14.7	33.9
1000	14.2	32.8
2000	13.7	31.6
3000	13.2	30.5
4000	12.7	29.3
5000	12.2	28.2
6000	11.7	27.1
7000	11.3	26.0

each power stroke of a piston depends on the quantity of oxygen available; anything that reduces the amount of oxygen available in the combustion chamber naturally also reduces the amount of power developed in the cylinders (Table 8-2).

TABLE 8-2. EFFICIENCY LOSS OF ENGINES AT HIGHER ALTITUDES

ALTITUDE, ft	VOLUMETRIC EFFICIENCY, %
Sea Level	100
1000	97
2000	93
3000	90
4000	87
5000	84
6000	81
7000	78

Increases in altitude and temperature reduce the horsepower developed by internal combustion engines because the air density decreases. As the air becomes thinner, a lesser weight of air enters the cylinders; the volume of air remains the same. At the higher altitudes and temperatures, this weight of air becomes insufficient to completely burn the fuel charge. Air density decreases at a rate of about 3% per 1000-ft increase in altitude and at about 1% per 10° F increase in air temperature. Very hot weather seriously affects engine performance because the air is expanded before it enters the combustion chamber; conversely, comparatively cold weather helps an engine to develop additional power because the air is more dense.

The power developed by gasoline engines decreases about 3½% for every 1000 ft above sea level; therefore, an engine that was just adequate at sea level would be almost 25% deficient at 7000 ft altitude. Normally aspirated diesel engines lose about 3% of their power for each 1000 ft altitude; they are more efficient than gasoline engines because there is usually some excess air available in the cylinders at rated conditions. Modern turbocharged diesel engines do not have a power loss until altitudes in excess of 4000 ft above sea level are reached, and then only about 2% per 1000 ft above that.

Effect of Relative Humidity on Engines

Relative humidity will seldom affect a pumper's performance, but it is well to be aware of the subject. The higher the humidity, the less oxygen will be made available to burn the fuel; this is caused by the water vapor replacing some of the air in the atmosphere. Normally, the variation in power due to this factor is less than 1%; it is significant only if the temperature and the relative humidity are both quite high.

Water Temperature

The temperature of the water to be pumped can affect performance, but normally it is not warm enough to be of any concern. If a drafting pit is being used long enough so the water becomes heated, or if water in a booster tank is very warm, the water temperature can have a noticeable effect. When water is subjected to a reduced pressure, the boiling point is lowered. If the water is too warm (over 85° F), sufficient water vapor may be given off to affect the pump's operation. High water temperature, low atmospheric pressure, or a combination of the two, will have the same effect on the maximum pump capacity as increasing the height of the lift when drafting.

Summary

To sum up, pumper performance will be reduced by the following environmental factors:

1. High altitudes decreasing the atmospheric pressure
2. Storms decreasing the barometric pressure

3. Higher than normal relative humidity
4. Hot air temperatures
5. Warm or hot water to be pumped

Pump Priming Failure or Priming Loss

When difficulties occur in priming the pump, check the suction gage to see whether it registers low or high vacuum. Following are the possible causes for each condition, and the remedies to employ.

Suction Gage Indicating Low Vacuum

Air leaks. Clean and tighten all the suction connections. Make sure the suction hoses and gaskets are in good condition. Use the following procedure to locate any air leaks:

1. Remove all the attached hose lines, cap all the suction and discharge openings (including wagon battery).

2. Close all the drain valves, radiator fill valves, and other valves connected to the pump.

3. Operate the priming device.

4. Study the suction gage to determine the maximum vacuum developed; this should be at least 20 in. of mercury. If the priming device fails to produce a sufficient vacuum, it may be defective, or the leaks may be too large for the primer to handle.

5. Close the priming valve, stop the priming device, and attempt to hold the vacuum in the pump. If the vacuum drops more than 10 in. in 10 min, there is a leak in the pump assembly. This leak may be in a valve, draincock, piping, casing, or pump packing.

6. A defective booster-tank pump valve may leak air and be the culprit. This would occur after prolonged drafting when the tank becomes empty, allowing air to enter the pump.

7. The leakage may be located by listening for air movement. If the leaks cannot be heard, apply engine oil to all suspected points; the leaks will be indicated by breaks in the film of oil as air is drawn into the pump.

8. Connect the suction hose to the hydrant, open the discharge valve, and flow water through the pump until all air has been evacuated. Close the discharge valves and hydrostatically test the pump; water will leak out where air entered. Do not exceed a pressure of 110 psi.

9. If the pump is carried dry and has not been operated for some time, the packing may be dried out. Tightening the packing gland may be indicated. The acceptable leakage of water varies from 10 drops per minute to 1 or 2 drops per second.

No oil in priming tank. With a positive displacement pump primer, oil is required to maintain a tight seal. Check the priming tank oil supply and replenish it if necessary.

Defective priming valve. A worn or damaged priming valve may leak and cause the pump to lose prime.

Improper clearance in rotary gear primer. After prolonged service, wear may increase clearance of primer parts and reduce efficiency.

Engine speed too low. Refer to instructions in manual for correct priming speeds. Speeds much higher than those recommended do not accelerate priming and may actually damage the priming pump.

Check valve in bypass line stuck open. If a bypass line is installed between the pump discharge and the water tank to prevent the pump from overheating with all its discharge valves closed, look for a check valve in the line. If the valve is stuck open, clean it and replace it, or temporarily block the line until a new valve can be obtained.

Excess carbon on exhaust primer valve seats.

Refer to the exhaust primer instructions for methods of carbon removal.

Priming pump not operated long enough. With the pump dry, apparatus equipped with a pump of up to 1500 gpm capacity is required to take the suction and discharge the water with a lift of 10 ft in not more than 30 sec through 20 ft of suction hose of appropriate size. Pumps of 1500 gpm or larger capacity are required to take the suction and discharge the water in not over 45 sec. If insufficient time is allowed, all the air in the pump and suction hoses is not discharged.

Suction Gage Indicating High Vacuum

Suction lifts too high. Do not attempt lifts exceeding 22 ft except at low altitudes and with the equipment in new condition.

End of suction hose not deep enough. Although the suction hose might be immersed enough for priming, pumping large volumes of water may produce whirlpools, which will allow air to be drawn into the suction hose and cause the pump to lose prime. Whenever possible, place the end of the suction hose at least 2 ft below the water surface. The water should be at least 4 ft deep where the strainer is located to provide clearance below the strainer and sufficient depth above it.

High point in suction line. If possible, avoid placing any part of the suction hose higher than the pump suction inlet. If a high point cannot be prevented, close the discharge valve as soon as the pressure drops and reprime. This procedure will usually eliminate air pockets in the suction line, but it may have to be repeated several times.

Dirt on suction strainer. Remove all leaves, dirt, and other foreign material from the suction strainer. When drafting from a shallow water source with a mud, sand, or gravel bottom, protect the suction strainer in one of the following ways:

1. Suspend the suction strainer from a log or other floating object to keep it off the bottom. Anchor the float to prevent it from drifting into shallow water.
2. Remove the top from a barrel. Sink the barrel so its open end is below the water surface. Place the suction strainer inside the barrel.
3. Make a suction box, using a fine mesh screen. Place the suction strainer inside box.
4. Place a ladder on the slope of a bank. Allow the ladder to support the suction hose, thus preventing the strainer from resting on the bottom.

Suction gage showing gradually decreasing vacuum. If vacuum decreases, and there is no loss of prime, it indicates that the height of the lift is decreasing or that a partial blockage is disappearing. When drafting from tidal waters, a rising tide will give this indication.

Suction gage showing gradually increasing vacuum. Increasing vacuum will indicate a gradual blockage of the strainer or an increasing height of the lift. When drafting from tidal waters, a falling tide will give this indication.

Insufficient Pump Capacity

When the pump is not performing at its rated capacity, look for the following probable causes. Procedures for locating the trouble are given, as well as remedies to correct the insufficiencies.

Engine and Pump Speed Too Low at Full Throttle

Low engine horsepower. Engine horsepower too low is the most likely cause. This is especially true when engine speeds tend to be lower than usual. The following are possible causes of low engine power:

1. The throttle linkage is not opening the carburetor throttle fully.
2. The ignition timing is incorrect.
3. The fuel flow is restricted due to a clogged filter bowl or some other restriction.
4. The engine is running too hot.
5. The governor is improperly set.

The above causes are relatively simple to correct and sometimes can be adjusted immediately so that pumping can proceed. Causes that are more serious and indicate that a thorough engine tune-up or other repairs are necessary include:

1. Slipping clutch
2. Restricted exhaust
3. Bad spark plugs, coil, condenser, or points
4. Poor compression in one or more cylinders
5. Poor carburetion
6. Leaking or sticking valves
7. Worn piston rings

Altitude effects. Power generated by internal combustion engines decreases about 3½% for each 1000 ft of elevation. Therefore, an engine that was just adequate at sea level would be about 25% deficient at 7000 ft altitude. Adjusting the carburetor or changing the carburetor jets may improve engine performance.

Governor set improperly. If the governor is set too low, it will decelerate the engine before the desired pressure is reached.

Relief valve set improperly. If the relief valve is set to relieve below the desired operating pressure, water will bypass and reduce capacity. Adjust the relief valve in accordance with instructions.

Transfer valve set improperly. Place the transfer valve in volume (parallel) position when pumping more than 70% of rated capacity. For high pressures, change the transfer valve to pressure (series) position. When shifting the transfer valve, make sure it travels all the way into the new position. Failure of the transfer valve to move completely into the new position will seriously impair pump efficiency.

Road transmission in gear too high. Consult instructions for the correct pumping gear. The pump usually works best with the transmission in direct drive. If the apparatus is equipped with an automatic transmission, it should be locked in pumping gear.

Engine and Pump Speed Higher than Specified for Desired Pressure and Volume

Air leaks. A frequent cause of low performance, recognizable by excess engine speed, hose pulsation, and unsteady pressure gages, is air leakage. To find the leaks, check the pump for holding vacuum. Sometimes these leaks are in the booster tank or the wagon battery plumbing. The leaks may be heard when the engine is stopped. Priming will be delayed by an excessively leaking packing gland.

Restricted suction. Restricted suction causes higher than normal engine speed and reduced capacity; it also causes fluctuation of the pressure gage and a high vacuum reading on the compound gage. Restricted suction may be the result of one or a combination of the following conditions:

1. Suction hose too small
2. High altitude
3. Suction lift too high
4. Improper type of strainer
5. Clogged suction strainer at pump or at end of suction hose

6. Aerated water

7. Water too warm (over 85° F)

8. Leaks on suction side of pump at intake or couplings

9. Collapsed or defective suction hose

10. Flow that has reached capacity of hydrant outlet

Suction hose too small. The full capacity of the pump may not be able to pass through the suction hose. This may be improved by using a larger suction hose or by adding a second suction hose connected to one of the other suction inlets.

Effect of altitude. When drafting water, the pump produces a vacuum at the top of the suction hose, and normal atmospheric pressure on the surface of the water forces the water into the suction hose and pump. However, as the elevation of the pumping site increases, atmospheric pressure decreases, creating a problem when drafting. Loss of lift at various elevations is shown in Table 8-3.

TABLE 8-3. ALTITUDE EFFECTS ON LIFT

ELEVATION ABOVE SEA LEVEL, ft	LOSS OF LIFT, ft OF WATER
1000	1.22
2000	2.38
3000	3.50
4000	4.75
5000	5.80
6000	6.80
7000	7.70

A difference in barometric pressure due to weather conditions will have the same result as a change in altitude. The difference in barometric pressure due to operation on a rainy day instead of a cool, clear day could easily mean a difference in lift of 1 ft.

Lift too high. Lift is the difference in elevation between the water level and the center of the pump when a pumper is drafting water. Lift too high will cause a high engine speed, high vacuum, pump roughness, and a pulsating pressure gage. The maximum lift is the greatest difference in elevation at which the pumper can draft the required quantity of water under established physical characteristics of operation; these include the design of the pump, the adequacy of the engine, the condition of the pump and engine, the size and condition of the suction hose and strainers, the elevation of the pumping site above sea level, and the temperature of the water. Theoretical values of lift and maximum lift must be reduced by entrance and friction losses in suction hose equipment to obtain the actual or measurable lift.

Internal strainer clogged. The strainer between the pump and the suction hose may be clogged or displaced.

Improper type of strainer. The wrong type of strainer will restrict the amount of water passing through.

Clogged suction strainer. Obstructions may be foreign matter such as grass or leaves on the suction hose strainer or in the pump suction tube strainer. To check the strainers, shut down and open a discharge valve very slightly, letting water run back down the suction hose slowly. This will often allow foreign matter to be flushed out so that it can be observed and the cause of the trouble determined.

Aerated water. The boiling point of water is lowered by aeration. A vacuum will further reduce the boiling point. Under these conditions, sufficient air and water vapor may be given off to affect the maximum capacity of the pump.

Water too warm. When water is subjected to a reduced pressure, the boiling point is lowered. If water is too warm (over 85° F), sufficient water vapor may be given off to affect the pump's operation.

Leaks on suction side of pump. Leaks on the suction side of the pump will allow air to enter the pump. This will aerate the water, and air, instead of water, will occupy space in the pump.

Collapsed or defective hose. The obstruction may be caused by a collapsed suction hose lining. Old or defective suction hose may have a loose lining that is pulled inward by a vacuum, substantially reducing the flow through the hose. It is difficult to see because the lining usually goes back into place when the hose is removed. Usually, however, there appears to be a low blister on the lining where it has pulled away from the carcass of the hose. A defective hose must be replaced. Sometimes the inner lining becomes so rough that it causes enough friction loss to prevent the pump from reaching capacity.

Water supply limitation. A high vacuum indicates a water supply limitation. The pump can deliver no more water than can be obtained from the source. Limitation is determined by the quantity of water that can be supplied to pump suction by water mains and hydrants. Quite often suction is drawn on just the hydrant outlet, and there is residual pressure left in hydrant barrel. When this is encountered with a double hydrant, additional water may be obtained from another outlet.

Foreign matter in the impellers. Foreign matter in the impellers causes higher than normal engine speed and reduced capacity. However, it does not cause abnormally high vacuum on the compound gage. Cleaning foreign matter out of the impellers usually requires the removal of half of the pump body and pushing obstructions out of the impellers with a rod. When the pump is open, it is well to check clearance, or sealing, rings for abnormal wear. Back-flushing a pump may dislodge debris; flow water backwards through the pump by connecting the discharge outlet to the water source.

Road transmission in gear too low. Consult the vehicle instructions for the correct pumping gear. The pump usually works best with the transmission in direct drive. Check both the engine and pump speed, if possible, to be sure the transmission is in direct drive.

Worn clearance rings and impeller hubs. Since a clearance ring replacement requires a pump disassembly, it is advisable to thoroughly check other possible causes of low performance before assuming that the clearance ring wear is the cause. Clearance, or sealing, rings hold to a negligible amount the internal bypass of water from the discharge side of the pump back to the suction. The radial clearance of the rings is only a few thousandths of an inch when new, which effectively prevents a large bypass of water. In clear water they continue to effectively seal for hundreds of hours of pumping. In dirty or sandy water the impeller hub and clearance ring will wear faster. The more they wear, the greater the bypass and the lower the performance. Also, the greater the pressure at which each stage is operated, the larger will be the bypass and the more the performance will be reduced.

When new, the radial clearance between the impeller hubs and clearance rings is from 0.005 to 0.007 in. An increase will allow more bypass and lower performance, but when the pump is adequately powered, it should not be necessary to replace the clearance rings and impellers until the average clearance reaches 0.015 to 0.020 in. or more, as measured by a feeler gage.

Since the clearance rings normally wear faster than impeller hubs, it is frequently necessary to replace only the clearance rings; thus, they are often referred to as wear rings. Replacement of the rings alone will largely reduce bypass (or slip as it is sometimes called) and restore the pump to near original performance. A complete restoration requires that the impellers also be replaced. When the pump is partly disassembled for checking, clearance rings, shaft, packing, and bearings should be checked for wear.

Pump Pressure Trouble Shooting

The characteristic indications of pump pressure problems are listed in the following, along with methods to pinpoint the sources of malfunction.

Insufficient Pump Pressure

In general, the remedies given for low pump capacity will also correct low pump pressures. In addition, check the following possibilities:

Pump capacity limits pump pressure. Do not attempt to pump a greater volume of water at the desired pressure than the pump is designed to handle. Exceeding the pump capacity will reduce pressure in an inverse ratio. Surpassing maximum recommended pump speed will produce cavitation and will seriously impair pump efficiency.

Pump speed too low. Check the pump speed with a tachometer. If the pump speed is too low, refer to the engine manufacturer's instructions for the method of adjusting the engine speed governor. Determine if the engine clutch is slipping.

Transfer valve. Pumping in volume position instead of required pressure position can be checked by partially closing off all discharge valves until only a small flow is observed, opening the throttle until the desired pressure is reached, then slowly opening all discharge gates and increasing the throttle as necessary to maintain pressure until the desired volume is obtained. An improperly adjusted or inoperative transfer valve may prevent a building up of adequate pressure.

Check valve stuck open. When the pump is in pressure (series) position, the discharge will bypass to the first stage suction. Operate the pump at 150 psi pressure (unless the pump has a maximum pressure at which the transfer valve is to be changed). Rapidly switch the transfer valve back and forth between positions. If this fails to release the valve, remove the suction hose and the opposite suction cap. Remove the suction tube strainers. Reach into the second-stage suction by hand, or with a rod, and push the check valve to determine freeness. When released, the check valve should strike the seat sharply. If not, there may be foreign matter on the seat; it must seal tightly. Remove the foreign matter. A leaking check valve allows bypassing of the first stage and reduces pressure.

Pressure Not Relieved with Discharge Valves Closed

Sticking pilot valve on relief valve. Disassemble and clean.

Incorrect assembly or plugged tube lines. Disconnect the lines and inspect. Reconnect according to the instructions supplied with the relief valve or governor.

Sticking mechanical linkage on governor. Check all joints and pulleys for freeness. Lubricate freely.

Pressure Not Returning to Original Setting After Discharge Valves Are Reopened

Sticking pilot valve on relief valve. Disassemble and clean. Lubricate with a light oil.

Sticking main valve on governor. Disassemble and clean. Lubricate with a light oil.

Incorrect installation. Check the instructions

and diagrams to be certain that the installation is correct.

Fluctuating Pressure (Hunting)

Sticking pilot valve. Disassemble and clean. Lubricate with a light oil.

Hypersensitive control cylinder on governor. Adjust the needle valve.

Water surges with relief valve. Hunting can result from a combination of suction and discharge conditions involving the pump, relief valve, and engine. When elasticity of the suction and discharge system and response rate (reaction time) of engine, pilot valve, and relief valve are such that the system never stabilizes, hunting results. With the proper combination of circumstances, fluctuation can occur, regardless of the make or type of equipment involved. Changing one of more of these factors enough to disrupt this timing should eliminate hunting.

Slow Response

Needle valve closed too much on governor. Adjust the needle valve.

Plugged filter or line. Clean the lines and filter.

9 Engine Company Procedures

An engine company is a basic fire fighting organizational unit headed by an officer and equipped with one or more pumpers, hose, nozzles, and other tools and appliances. The principal function of an engine company is to obtain, deliver, and apply water on a fire. Depending upon the hazards and the organizational structure of the fire department, engine company personnel may consist of a fully paid crew of one to six or more fully paid fire fighters, or a wholly volunteer team. An engine company will normally consist of an officer, an engineer or pump operator to drive and manage the apparatus, and one or more personnel to handle the hose, direct fire streams, and perform the myriad of other rescue and fire extinguishing duties on the fire ground.

A fire department may be composed of a single engine company, or there may be a complex organization of many engine companies, truck companies, fire boats, and special purpose companies. The smaller fire departments are often organized into mutual aid plans in which they assist each other when the fire or other incident is beyond the capabilities of a small fire fighting force.

The objectives of any fire fighting effort, listed in the order of importance, are: (1) rescue, (2) protecting exposures, (3) confining the fire, (4) extinguishing the fire, and (5) overhauling the debris.

While it is an old saying in the fire service that no two fires are alike, it is just as true that in many respects all fires are similar. A study of the behavior and characteristics of burning materials shows that even though there are variables in every situation, sufficient common properties are so closely identical that the sequence of events in a fire can be predicted with reasonable certainty. A complete understanding of fire chemistry and behavior, the extinguishing properties of water, and the effective use of apparatus and equipment are necessary if a department is going to efficiently and consistently prevent, control, and extinguish fires.

Strategy and Tactics

Regardless of the number of fire fighting books that have been written or the amount of discussion that has been devoted to the subject, many fire officers do not fully comprehend the importance of developing plans for fire fighting strategy, or even adequate tactics for carrying out operations involving the coordinated use of various companies on the fire ground (Figure 9-1). Too many fire departments have merely followed the plan of hitting the flames wherever they may appear, regardless of whether or not the blaze has been adequately surrounded and its spread intercepted. All too frequently flames are driven through a structure or allowed to spread in several directions through failure to coordinate the operations of various company units. Basic strategy and tactics must be understood, because at a major fire in the community, the officer in charge must effectively direct all companies reporting to assist. In even a one-company fire department the same basic considerations and tactics are required.

The basic purposes of any fire fighting effort are to: (1) locate the fire, (2) confine the fire, and (3) extinguish the fire. As a rule, all three actions are accomplished simultaneously. To effectively maneuver and use the apparatus, equipment, and personnel on the fire ground to control the fire as quickly as possible and hold the property loss, injuries, and deaths to a minimum, a system of strategic planning and efficient tactics must be adopted.

Essentially, strategy is concerned with deciding what should be done; it is a consideration of water requirements, apparatus and equipment, fire fighters, and how to utilize them in the most efficient and effective way. Tactics is primarily concerned with deciding how to accomplish the planned actions and then applying strategic theory. Officers should be encouraged to use their initiative and to take action that is consistent with the strategic and tactical objectives of a fire fighting operation.

Size-up

The size-up, or estimate of the situation, is made by the officer in charge of a fire or other emergency incident by which he decides what to do and how to accomplish it. This is a continuous process throughout the entire emergency until operations are completed. A size-up should consider the facts and probabilities of the fire; the personnel, apparatus, equipment, and water supply at the officer's disposal; and additional help which can be summoned if needed. After consideration of the apparent facts, as well as the likely probabilities and the possibilities that could occur, a decision should be made about the proper strategy and tactics to be employed; then, orders should be issued to place the plan of operations into action.

When sizing up the situation, primary consideration must be devoted to operations required to rescue and remove any endangered persons from an involved building or other hazardous position and convey them to a place of safety. Then in the following order, attention should be given to the protection of exposures, the confinement of the fire, the extinguishment of the blaze, and the overhaul of the property.

The officer must estimate what the fire conditions will be at the time decisions are put into operation. The plan may be to devote all of the first alarm companies to rescue operations and leave the fire control to later arriving units, or he may decide that the best methods of preventing casualties is to conduct an immediate, massive attack on the fire. Officers in charge should be constantly aware that a rapid and aggressive assault on the heat and flames will usually reduce, and often eliminate, all of the other problems.

Remember, a size-up is a continuous working procedure from the time of the original alarm, throughout the entire fire fighting operation, and until the last fire unit returns to its station. Constantly changing conditions on the fire ground could dictate alterations in strategy and tactics. Unless the officer in charge remains alert and flexible, a fire fighting maneuver that was strategically brilliant when it was planned could become a death trap for fire fighters before it can be executed.

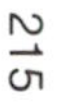

Figure 9-1. The coordinated use of a number of fire companies on the fire ground requires a systemized size-up before the correct strategy and tactics can be decided upon. (Courtesy of the Los Angeles City Fire Department)

Rescue Operations

When the fire presents a serious threat to human life, rescue takes precedence over fire extinguishment. However, it must be continuously kept in mind that whenever possible, positioning fire streams between the blaze and the potential victims to prevent stairways, halls, and other avenues of escape from becoming untenable provides the best protection to human life. When the fire is controlled, most hazards are eliminated.

Size-up and attack often become simultaneous activities on the fire ground when it is instantly obvious that a rescue must be made or an exposure should be protected. The rescue of trapped or endangered persons should be the first concern upon arrival at a fire; the officer in charge must quickly determine beyond doubt whether or not there is a possibility of anyone still being in the involved structure. If the need for rescue is readily apparent, the initial efforts of all companies may have to be concentrated on the life saving. This means that the main body of fire may continue to burn unchecked and the exposures remain jeopardized because the first hose lines are utilized to make the rescues possible. If a rescue operation is not too difficult and the initial fire fighter response is sufficient, the officer in charge may be able to divide the forces and accomplish both rescue work and the protection of exposures simultaneously. This, of course, is done whenever possible because it will lead towards confining the fire; this may be necessary to facilitate rescue operations.

Some years ago a rapidly spreading fire erupted on the first floor of a high-rise hotel in a large city. The first arriving fire companies exclusively directed their initial efforts to evacuating the residents on the upper floors, instead of simultaneously laying hose lines and attempting to confine the blaze. The fire's intensity quickly accelerated; heat, fire, and lethal gases spread up through stairways and elevator shafts; fire fighters narrowly escaped being trapped; and one of the greatest hotel catastrophes in history ensued.

Ventilation of heat, smoke, and toxic gases from an involved structure is the specified responsibility of truck companies, as is the rescue of trapped and endangered persons. At times it may be necessary for aerial ladder or elevating platform operators to ignore potential victims leaning out of windows and threatening to jump, because of the priority need to place truck fire fighters equipped with axes and power saws on the roof to cut large openings and remove skylights for the purpose of ventilating the structure. Efficient ventilation procedures will quickly relieve the building of lethal quantities of heat and toxic vapors. This removes a threat to the victims' lives and allows prompt and effective advancement of interior fire streams. Do not allow an immediate concern for a few vocal potential victims to overshadow an obligation to protect a larger number of unseen persons who may be trapped in the heat and smoke.

Fire fighters cannot rely on the word of hysterical bystanders, nor can they ignore them. If a distraught man in night clothes excitedly announces that his wife and children are still in the building, get as much information about their possible locations and launch an immediate rescue effort. If the same man says that his entire family escaped and are at a neighbor's house, the chances are that he is reliable. However, inquire if there was anyone else in the building; he may have forgotten about an overnight guest. If a bystander or neighbor claims that people are still in the house, question him closely; his information cannot be considered as absolutely dependable, but it cannot be disregarded.

Rescue operations also involve removing persons from hazardous locations in exposed structures; survey the entire area to ascertain if there are any endangered persons in adjacent buildings. When there is an explosion hazard from a vapor cloud or a flammable liquid spill, or a toxic gas atmosphere, a large section of the countryside may require total evacuation.

Even if the residents of an endangered or burning structure unanimously proclaim that all persons are out of the building and are accounted for, a thorough search will have to be conducted in case there was an unauthorized guest or an intruder. All structures and involved areas on the fire ground must be routinely searched to make certain that no victim is overlooked.

Exposure Protection

The protection of exposures is an essential element of confining the fire and preventing it from extending its boundaries. If the fire only involves one room when the first companies arrive, the primary objective is to hold it to that room; if one floor is involved, then the goal is to confine the heat and flames to prevent them from spreading from that floor; if an entire structure is ablaze, then the fire should be confined to that building.

When the fire is obviously confined, it is under control; then the extinguishment phase of the attack begins. Tension and pressure are relieved because the major problems are under control. As fire extinguishment progresses, hose lines and nozzles should be reduced in size or shut down as soon as they are not needed. The extinguishment activities of fire fighting gradually progress into salvage and overhaul operations.

Ventilation Provision

Fire streams are commonly thought of as a means of fire extinguishment, and this is their most important attribute. However, an improperly placed stream of water can actually make interior structural fire fighting impossible.

Every fire fighter has suffered the experience of being driven out of a building when the quantity of heat, smoke, and fire gases suddenly accelerated and became unbearable. This could have been the result of a normal buildup of the fire due to combustion, or it may have resulted from the improper use of a stream of water. One of the more common causes is conflict of streams when two crews are advancing hose lines in a structure simultaneously from opposite directions. When this happens, the heat and smoke are trapped between the two opposing forces and cause unnecessary discomfort to both crews. The best method is to attack the fire from the uninvolved portion of the structure and push the heat and smoke out of the building. Avoid attacking a fire from two sides.

Another common error is to direct a stream of water into a ventilation opening in a roof. Whether the fire vents itself by burning a hole in the roof or because a fireman has created an opening to allow the heated smoke and fire gases to escape from the structure, the ability of the combustion products to rise will depend on the difference in temperatures between them and the outside air. If this rising column of smoke and gas is cooled by a fire stream, it will become heavier than the surrounding atmosphere and settle down; the roof opening will be blocked and the interior of the structure will be rendered untenable. No fire stream, whether it is from a booster line or from a heavy stream appliance, should ever be directed into a roof opening.

Standard Operating Procedures

Modern, effective, intelligent fire fighting requires the adoption and use of standard operating procedures, often referred to in the fire service by the abbreviation SOP. These procedures will create a smooth running, efficient fire fighting force (Figure 9-2); without them, the chances of delay, indecision, and failure could occur without warning at a time of crisis. Standardization should not freeze or make mandatory any techniques or practices that would prohibit independent initiative and action under unusual circumstances. Standard operating procedures are not a substitute for experience, knowledge, and common sense. The use of standard methods of operation will give the officers and line fire fighters self-assurance during the stress of an emergency and reduce delays and indecision. Preplanned strategy and tactics enable the entire fire department to be most effective in protecting life and property.

Many departments develop their standard operating procedures so they can handle their usual emergencies in an efficient manner. When an uncommon incident occurs and the situation is out of the ordinary, they have problems. Plans should be devoted more to controlling the potential unusual incidents, instead of the common.

This book will not attempt to suggest which practices, procedures, and equipment are correct and which are deficient. This type of subjective thinking could not possibly consider all of the likely ramifications for every fire department; several different methods could be equally effective. Procedures work better if the personnel who will be guided by them have a voice in designing and adopting them.

The correct procedures will enable the responsible personnel with their assigned apparatus and equipment to place the nec-

Figure 9-2. Before the fire fighting and rescue operations of any engine company can be effective, standard operating procedures must be adopted and practiced until everyone is competent. Notice that every fire fighter on this German crew is busy with a preassigned job.

essary quantity of water at an effective nozzle pressure on a fire in the most advantageous location in a minimum length of time. Fire departments should study their fire fighting hazards, analyze their evolutions, experiment with various procedures, and drill under realistic conditions until this objective has been attained. The perfect solution to a problem will never be discovered so companies should constantly strive to improve their methods.

Rigid guidelines are neither practical nor workable. Prefire planning and standard operating procedures may delineate the duties and responsibilities of the first-in and later arriving companies of the first-alarm assignment for a particular occupancy or large industrial property, but these predetermined decisions may be rendered useless if the companies are not available or are delayed because they are out of quarters on fire prevention or are working on another fire when the alarm sounds. These options could affect which companies would respond and their arrival times on the scene.

Standard operating procedures should be formulated by each department because the personnel, fire fighting apparatus and equipment, types of hazards, and a multitude of other complicating factors may vary for different departments and must be considered. A practice in one city may be unworkable in an adjacent town. There are a few rules, however, which are generally adopted throughout the country and are recommended by all authorities.

Driving Speed

Apparatus responding to a fire or other emergency incident should be driven as fast as is consistent with safety, but high speeds are rarely justified. Numerous fire fighters, other motorists, and innocent bystanders have been killed and maimed because vehicles were being driven at a speed where they were not completely under control.

The fire officer who makes certain of an alarm location and requires the driver to proceed in a safe and sane manner will arrive with a company prepared for action. The excitement incidental to dangerous speed will be lacking if the vehicle is driven in a smooth and methodical fashion; the officer will arrive on the fire ground in a calm state of mind so that the size-up can be made in a cool and deliberate manner, and the crew will carry out orders similarly. Undue haste and excitement in all fire operations arouses and agitates the fire fighters; this breeds inefficiency. Excitement also tends to increase fatigue, making the fire fighters less attentive and more susceptible to injury.

Initial Attack

A widely accepted standard operating procedure is that in high value districts or where a serious life hazard may exist, the first engine company should always lay at least one large supply line, connect to a hydrant, and be prepared to pump anytime there is fire or smoke showing. If there is no indication of fire, either the first or the second engine company should lay a supply line, connect to a hydrant, and be prepared to pump if necessary. This is a good technique because many small fires have rapidly grown into holocausts when the initial attacks were made with small booster lines and no backup water supply was available to combat a larger-than-anticipated blaze.

In residential districts, supply lines should always be laid between the pumper and the water source. If fire or smoke is observed as the first engine company approaches the scene, and the last hydrant has not been passed, this company should lay the line. If there is no indication of fire, the first company may proceed to where the fire was reported and be prepared to use small streams from the booster tank; the second company would then be responsible for laying a supply line, connect to a hydrant, and be prepared to charge the line.

These procedures presuppose that the engine companies are equipped with a single pumper, that the second company will come from a station that is reasonably close and would arrive to supply water before the first pumper's tank becomes empty, and that a hydrant or other source of water is fairly near. If the first company is equipped with either two pumpers or a tank wagon and a pumper, then one apparatus can make an immediate attack on the fire while the pumper is laying a supply line and connecting to a hydrant. If a delay in the arrival of another company is anticipated, or if an adequate source of water is not immediately available, then prefire planning should anticipate these problems and workable procedures be adopted to resolve them.

Probably the most difficult procedures to prepare are those that attempt to state when hose of a certain size and streams of a specific caliber should be used on a fire, but there should be some guidelines. An analysis of large-loss fires has shown that a major factor responsible for the failure of the initial response of companies to control a fire was the faulty preliminary size-up and the choice of hose size. Inasmuch as most fires are minor and require only small hose for extinguishment, there is a strong tendency to use minor fire streams on large fires also. Some departments are known to use 1½-in. hose for just about any and all fires, even when the structure fire is so intense that the building cannot be entered. Heavy stream appliances are required and used so seldom that they are not always promptly employed when a large fire occurs. The value of making an accurate size-up of a fire should be constantly emphasized.

Fire Stream Placement

A standard operating procedure delineating the correct placement of fire streams is very important because more property is protected and a greater number of lives are saved at fires by the strategic positioning of fire streams than by all of the other fire fighting techniques combined. The procedures most generally practiced are simple and, if faithfully followed, will automatically place the streams in the correct positions.

The first fire stream should be placed between the fire and any endangered persons; this will prevent the fire from spreading to any means of escape of persons trapped by the fire. The best way to accomplish this is to take possession of the main path of egress, usually the main stairway, and hold it as an escape route for the trapped occupants on the upper floors. Possession of the stairway also serves as a base of attack to control the fire and conduct the necessary rescue search. When possible, place the first stream into operation on the fire floor; if two or more floors are burning, the fire should be knocked down as an advance is made up the stairway. Many fire fighters have been trapped and injured or killed when they advanced above the fire before the blaze was knocked down or another line was in position and working on the floor below.

For basement and cellar fires, the first fire streams should be used to prevent the fire from spreading vertically up shafts and stairways. Confine the fire to the basement or cellar before attempting to extinguish it.

If a structure is in flames upon arrival of the first company, place the first fire stream between the burning building and the most severe life-endangering exposure. If both a vacant factory and a crowded school were jeopardized, naturally the stream would be placed in a position to protect the people in the school, even if the factory was the most severely exposed of the two. Primary concern must also be given to rescue operations for potential victims in exposed buildings.

In an older building, people are often trapped on the exterior fire escapes above the fire; they cannot descend because flames are lapping out of the windows on the fire floor. The correct procedure is to stretch one line to the interior stairway because it is extremely important to protect the normal means of entrance and egress of a structure; most occupants will attempt to exit from a building in a time of emergency the way they entered. Another hose line should simultaneously be stretched in front of the fire escape to protect the trapped people until rescue can be made.

When no life is endangered, the first fire stream should be positioned between the fire and the most severe exposure. On interior exposures, place the stream between the fire and its direction of travel. If the fire is traveling in two directions at once, for example up a stairway and up a shaft, protect the stairs, because control of the main stairway greatly simplifies fire fighting operations.

On exterior exposures, position the first fire stream to protect the greatest amount of property. Assume that a fire was endangering a pile of waste lumber and at the same time threatened to involve, but not quite as severely, an oil storage yard. In this case, the stream should protect the oil storage, which is potentially more hazardous, even though it may be in less danger of becoming involved. When possible, take a position that will not only allow the fire stream to protect the serious exposures, but will also enable the stream of water to be used for extinguishing the main body of the fire. Often an exposure is best protected by devoting all initial efforts to knocking down the fire. In cases where the fire is beyond the extinguishing capability of the initial stream of water, alternate playing the stream on both the fire and the exposure. As long as the exposed structure and its glass windows are kept wet, there is no danger of fire communicating; deluging a structure with vast quantities of water will not protect it any better than keeping it moist. A major error in covering exposures is to put too much of the water on the exposed properties and not enough on the burning materials that are generating the heat and causing the fire to spread. Even if it is not immediately possible to cool the entire fire area, the parts of the fire that are threatening other property should be cooled sufficiently, if possible, to reduce radiated heat and allow the fire fighters to operate effectively in the area.

The second fire stream should be advanced to the same location as the first one to back it up; its purpose is to be instantly available in case the first hose line bursts or proves inadequate for the volume of heat and fire encountered. The second line should have enough slack hose so it may be extended above the fire floor; often, when it is apparent that the main body of fire is contained, the second line is stretched above the fire to prevent any vertical extension.

The third fire stream should be advanced to the secondary means of egress, such as a back stairway, rear door, fire escape, etc., and placed in operation if needed. The third line should not be used until the officer in charge of the company stretching the third line is certain that its use would aid, and not hinder, the overall fire fighting operation. Fire fighters should always remember that a fire stream can force heat, smoke, and fire ahead of it into uninvolved portions of the building. Fire streams advancing from opposite sides of a structure can actually make the building untenable for both crews. The improper use of a third line may stop the advance of the first two lines and, perhaps, cause serious injury. Always bear in mind the presence of fire fighters working opposite to the third line.

Avoid stretching two or three hose lines down the same aisle or up the same stairway simultaneously because confusion will be rampant, lines will become entangled, and mobility will be lost. When multiple lines are advanced through the same path of egress together, the advance will undoubtedly be slowed down, and it could be stopped entirely. The best method is for all crews to work together and stretch the first hose line; the first crew then handles that line and attacks the fire while the rest of the personnel stretch the second line; if there is a third line, the second crew handles the second hose and the third crew can then work undisturbed.

Try to avoid stretching hose lines in the aisles and stairways being used by escaping occupants, unless a fire stream is necessary to protect that means of escape. If a protective line for the occupants is not needed, it would be best to use other routes or methods of supplying fire streams to combat the blaze.

Aircraft Fires

A standard operating procedure for combating aircraft fires should be adopted. Even though the strategy and tactics would be similar to those utilized for combating tank truck, bus, and other large vehicle incidents, often the suddenness of a major catastrophe will momentarily overwhelm the first-in company. Because of the vast number of different types, models, and sizes of aircraft, it would not be practicable to attempt to cover this subject too thoroughly. However, any fire unit may become involved in aircraft crash fire fighting because the majority of airplane crashes occur away from airports. Following are the general principles to be utilized if a single engine company becomes concerned with an airplane fire. If the plane is large, or ground structures and casualties are involved, the problems will become increasingly complicated. When a greater number of types of fire fighting apparatus and personnel are available, the general operating procedure may be delegated out to the various units.

Regardless of complexities or whether ignition of the aircraft has occurred, the suggested placement of apparatus and use of hose lines would be to lay one or two supply lines immediately when approaching the airplane (Figure 9-3), if a water source is readily available; otherwise, utilize the booster tank. If flames have not erupted, operate on the premise that the flammable fuel spill could ignite at any moment. Position the apparatus and men with the wind at their backs, if this is possible; the worst position would be with the wind blowing the flames, heat, and vapors directly towards the crew. Depending upon the number of fire fighters on the company, lay out as many 1½-in. lines equipped with spray nozzles on both sides of the aircraft as can readily be operated. Approach along the fuselage while wetting it down to keep it cool. As a

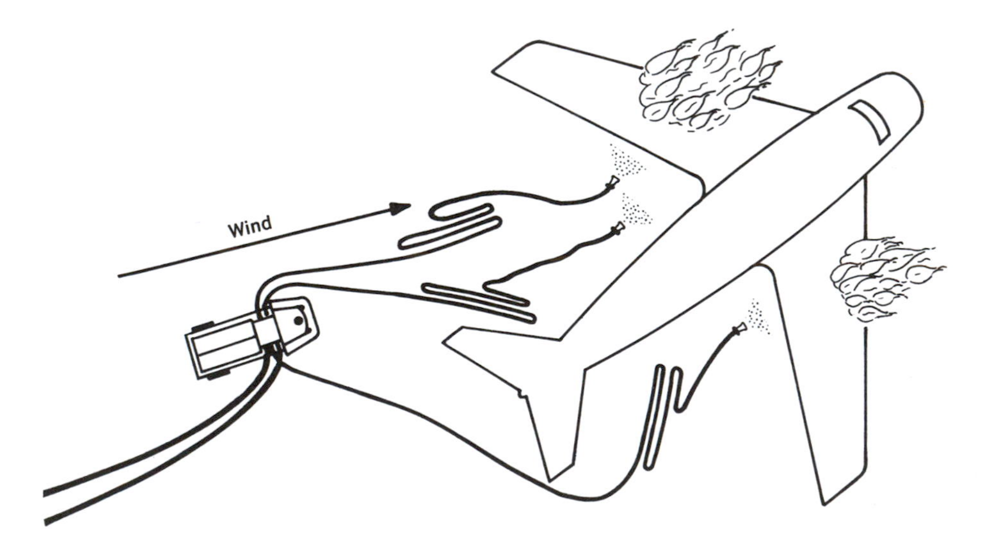

Figure 9-3. Combating an aircraft fire with conventional apparatus is similar to structural fire fighting. If practicable, attack from the windward side, approach along the fuselage, and drive the flames away from the passengers. Keep the fuselage wet and cool and protect the rescue exits of the occupants.

general rule, the main fire will initially involve the wings because the fuel tanks are integral with their structure. Drive the flames and heat outwards toward the wing tips. Passengers and crew will usually escape from the front and rear of the aircraft; protect these rescue paths. Later arriving engine companies should arrange backup water supplies and lay additional hose lines to help protect and rescue the passengers and airplane personnel.

Remember: experience has shown that there is a maximum of 5 min from the time of the crash to rescue the occupants in the majority of airplane wrecks and fires.

Ventilation Operations

The fire may be small, yet the heat and smoke it generates may be so dense as to prevent fire fighters from reaching it. The principle of ventilation is simply letting the heat and smoke out and bringing in cooler air so the fire fighters can work inside the building. Ventilation prevents further ignition of contents by removing the heat and gases of combustion.

Fire fighters should have a clear concept of what, where, and how to ventilate a building. This book will not attempt to reach the subject, or even discuss the various techniques of ventilation, but a standard operating procedure–taking into consideration the tools and equipment that a company is supplied with–should be formulated and adopted. Of greatest importance is emphasizing the basic principle that ventilation operations should not be started until hose lines are in place and fully charged, and hose crews are ready for an immediate attack on the fire. Fire fighters have been killed and structures lost because hose streams were not available to attack the increased ferocity of the fire when additional oxygen became available. Of equal importance, back draft explosions and flashovers have been caused by fire fighters making entry into a structure before the heat and flammable gases were removed from the building by intelligent ventilation methods.

Sprinkler Systems

Automatic sprinklers are undoubtedly the greatest fire protection devices that exist today. Very seldom does a sprinkler system fail to extinguish, or at least control, a fire within a structure which is so protected. Sprinklers will do most of the fire fighting in protected properties, thus relieving fire fighters and apparatus of undue stress and danger. However, they do require ample quantities of water in order to operate efficiently.

An automatic sprinkler system will detect the location of the fire and place water in the correct location to extinguish it, providing that the system does not run out of water. A proper plan should be adopted so that the best use is made of a building's built-in fire protection. Many fires have caused large property loss because the sprinkler control valve was closed or the water supply was insufficient and no pumper was connected to the supply inlets. The standard operating procedure should call for either the first or the second arriving engine company to connect two or more 2½-in. or larger hose lines to the sprinkler siamese connection and to start pumping from a hydrant at a pressure of 150 psi on orders, or when there is an indication of fire. When there is no indication of fire, some departments initially pump at 100 psi pressure until the hose lines and piping are charged. Concern for the proper utilization of the automatic sprinkler system should include assigning a fire fighter to the sprinkler control valve to make certain that this valve is open and to be ready to operate it when ordered to do so by the officer in charge.

Many large structures have two or more independent sprinkler systems. In this case it is possible to connect to and pump into an inlet that does not supply the heads applying water on the fire. If there is any doubt about the inlets, connect to and pressurize all of the systems. It could be possible that more than one system is involved; if not, this precaution will not cause any harm.

Since the automatic sprinklers will do the major job of fire fighting in the area they protect, when supplied with adequate quantities of water, sufficient sprinkler pressure must be maintained. If the fire department depletes the water supply by attempting to supply too many hose lines, the sprinkler system will be rendered useless. History is full of cases in which fire fighters took so much water from the mains for their hose streams that they depleted the supply to the sprinklers, which were effectively applying water directly on the fire. There have been unfortunate incidents in which the sprinklers were holding the fire under control and almost had it extinguished when the fire department arrived and pumped more water from the adjacent hydrants than the system could effectively supply. The result was that the sprinkler heads were robbed of water and ceased to function, the fire accelerated in volume, and the hose lines were merely utilized to protect the exposures and cool down the debris. In these cases, it is sad but true that these fires would have been controlled with a minimum of loss, and the structures saved, if the fire departments had never been called.

When responding to a fire in a sprinklered occupancy, a certain quantity of water must be reserved exclusively for sprinkler operation and never used for anything else until pumpers are connected to and are pressurizing the sprinkler system inlets. The best method of preserving a guaranteed sprinkler supply is to determine the minimum residual pressure below which the water main pressure should not be lowered. A good "rule of thumb" method is to figure that a 15-psi flow pressure is required for the highest sprinkler head in the system; to this pressure, add the back pressure between the pump gages and the highest head in the structure. To ensure that the highest sprinkler head in a 5-story structure would have an adequate flow, the water main supplying this system should not be pulled down below 40 psi. This calculation does not consider the friction loss in hose lines and piping, but it is sufficiently accurate to prevent a catastrophe.

Occasionally, piping will become ruptured; flow from the broken pipe will divert water from the operating sprinkler heads and render them useless. An alert pump operator will notice the increased flow of water when the pump discharge pressure drops. Unless the pressure drop is catastrophic, which indicates the rupture of a large size pipe, the engineer should increase the engine speed to regain the correct pressure and notify the officer in charge of this action. The fusing of additional heads will also increase the volume of water required from the pump and cause a low or moderate drop in pressure. If a building has more than one sprinkler system, the ruptured system should be isolated by the sprinkler control valves, if possible, while water is supplied to the remainder of the sprinklers.

Another standard procedure is to restore the sprinkler protection after a fire by replacing the fused heads and performing the other necessary operations to make the system serviceable.

While working out standard operating procedures for sprinklered occupancies, it is necessary to study the water distribution system and other nearby sources of water so that the companies will know which hydrants in the vicinity can be used for fire fighting without robbing the sprinkler system of essential water.

When supplying a sprinkler or standpipe system, the pump will overheat when there is no flow of water. A pump operator should periodically check to be sure that water from the pumper is being utilized; this can be done by closing one of the discharge gates without any pressure control in operation. If water has been flowing, the discharge pressure will rise as the valve is slowly closed; if the pressure remains constant, no water has been flowing through the pump. To avoid overheating the pump, either flow some of the water onto the ground from a bleeder valve or another discharge outlet. If the inactivity will continue for awhile, disengage the pump.

Standpipes

Like sprinkler systems, wet and dry standpipe systems should also be supplied by one of the first arriving pumpers to assure an adequate flow of water at the proper pressure for effective fire fighting. Modern strategy calls for fire crews to carry hose packs of 1½- and 2½-in. hose into tall buildings and use the standpipes so hose lines will not need to be stretched up stairways and fire escapes. For their maximum protection, fire fighters should always connect the first line to the standpipe outlet on the floor below the fire. They can then advance the fire stream into the fire area after the line has been charged and the water is flowing. The correct pump discharge pressure is determined by adding up the friction losses in both the supply lines to the standpipe inlets and the fire fighting hose, the nozzle pressure, the back pressure between the ground and the nozzle, and a standpipe allowance of 25 psi for friction loss in the piping. This is usually only an educated guess, because a pump operator at a working fire will seldom know what number and sizes of hoses and nozzles are being used to combat a fire in a large multistoried structure. The operator must, however, remain constantly aware of the various factors involved.

Effective Fire Streams

One of the primary requirements for prompt and effective fire extinguishment is the proper application of fire streams. All fire fighters should constantly adhere to the three basic rules of proper fire fighting:

1. Use the correct size fire stream.
2. Operate from the most effective position.
3. Apply water for the shortest necessary time.

If all three of these rules are rigidly adhered to, the fire will be extinguished in the most efficient way. If any of the three principles is ignored, deficiencies such as the following will emerge:

1. An incorrect size fire stream will result either in control of the fire with an excessive amount of water damage or noncontainment of the fire, with a consequent spread of the blaze.
2. Incorrect positioning could cause the proper size fire stream to be ineffective. A stream of water that enters a building but does not strike the fire is a complete waste. Fire containment is not accomplished, and the useless water could weaken the structural integrity of the building and increase damage to the contents of the building.
3. Applying water for an excessive period could result in structural weakening of the building and excessive water damage to the contents.

Fire fighters who have not adhered to the three basic rules of fire fighting are guilty of increasing fire losses and contributing to the number of deaths and injuries on the fire ground.

It is essential to select the correct size line and nozzle for the proper water application rate and shut down the line the moment it has knocked down the fire. Fires can be divided into three groups: (1) those that can be controlled by small hand lines; (2) those that require a large flow from 2½- or 3-in. hose lines; and (3) those that call for a massive attack with heavy streams. The decision of the first-in officer as to the size and number of lines to be stretched is very important because it is the first operation in confining and extinguishing a fire. If the officer's estimate is faulty, the battle may be lost, because the fire streams are not capable of absorbing heat faster than it is being generated by the fire.

One of the most serious errors that an officer can make on the fire ground is to overestimate the capability of a fire stream; regardless of size or quantity of discharge, nozzles are limited in extinguishing ability. Ignoring this fact could endanger the fire crew, as well as jeopardize rescue of endangered occupants and control of the fire with the least damage.

According to volume, pressure, and type, fire streams are generally assigned potential cooling ability, reach, mobility, and other attributes. However, these can be reduced or eliminated by circumstances on the fire ground. Officers concerned with the placement and use of a fire stream must be concerned with both the actual status of the fire and the potential developments. What will be the situation when the lines have been laid, the water supply attained, the nozzles placed, and effective fire streams flowing at efficient pressures and quantities? Will the water supply be capable of furnishing the correct volume of water required for fire control? Will the wind have sufficient effect on the reach, direction, and effectiveness of fire streams that an attack from the windward side should be launched? The officer must keep in mind that the fire status is not static; conditions are constantly changing.

Small fire streams must be backed up at the earliest possible moment by one or more lines of 2½-in. or larger diameter hose in case the initial attack is not capable of controlling the fire. The cooling capacity provided by a large flow from big hose lines is required to cope with fires in the advanced stages.

When charging hose lines, the discharge gates should be opened slowly; this precaution is so important for the safe handling and maneuvering of fire streams that it cannot be overemphasized. Water turned into hose lines too quickly will not only present a hazard to the nozzle operators, but it may burst the hose or make the line difficult to move at a time when the crew is trying to place the hose line in an effective position. Injuries from a sudden increase in nozzle reaction have resulted when the fire fighters were in a treacherous position, such as working on a ladder or with slippery footing.

The pump operator should be certain that the correct discharge gates are opened. An attempt should be made to find out where each line attached to the pump is working in case a sudden emergency requires remedial action. Some departments place tags on each hose line so they may be identified by number.

Hose kinks should be straightened out because they can impede the flow of water or cause the hose to burst. Hose operators are responsible for clearing any kinks out of the hose before water arrives, and the engineer should see that the line is sufficiently straightened out near the pump before opening a discharge gate. Minor twists and kinks will be removed by the flow of water. Every pump operator should observe the hose lines supplied by the apparatus and be alert for signals from the nozzle operators.

Stretching Hose by Hand

Hand stretching of hose lines on the fire ground is a laborious task even when adequate help is available. Under fire conditions, the crew is usually shorthanded and the situation becomes further complicated by excitement, confusion, and other unanticipated difficulties. The problems can easily lead to frustration unless some prior thought and practice has been devoted to designing an effective solution.

There is no one method that is universally the simplest and easiest way. Hose line procedures must be individually devised for each department after taking into consideration hose sizes, crew requirements, apparatus hose beds, types of fire ground terrain likely to be encountered, and many other factors. Obviously, a pumper equipped with hose reels will require procedures quite different from those for stretching lines from beds or trays. An apparatus from which all hose must be removed from the rear must operate contrary to procedures adopted for a pumper equipped with cross lay hose beds.

Preconnected hose lines with a nozzle already attached to the leading male coupling of a predetermined length of hose and the female coupling connected to a discharge outlet can result in unexpected problems if the fire is beyond the reach of the hose line. On the other hand, preconnected hose lines will greatly expedite fire fighting operations in the great majority of fire ground operations. It is important that some forethought should be devoted to preplanning procedures if some unexpected dilemma should arise. For one company that carries two preconnected 1½-in. lines in adjacent hose beds, a standard operating procedure to extend the preconnected line is regularly practiced: one crew member grabs the nozzle and a few loops of hose and leads in to the fire; the next crew member disconnects the first line from the pump outlet, grabs the second nozzle, connects the two lines together, and helps drag the hose towards the fire; a third fire fighter if there is one, can either aid in stretching the hose line or start to take in a backup line.

There are several methods of stretching hose directly from the hose bed into a fire; one is to drape several loops of hose over a shoulder, and another is to slide an arm through several loops of hose and drag them. The method and the number of loops handled by each crew member will necessarily depend upon the size and weight of the hose, as well as the crew number and the distance that the hose will be stretched.

The nozzle operator should lead off first with sufficient hose to give working room to attack after reaching the fire; there should be enough hose flaked back and forth at the door of the structure or nearest point to the fire to cover the anticipated use of the fire stream. The nozzle operator, who will be responsible for determining the best route to the fire, should not carry an overload.

After the nozzle operator proceeds far enough that the first section of hose will not be in the way, the leader stops and allows the following crew member to secure the next section of hose and follow; this is repeated for the succeeding member. Each crew member must keep back from the preceding member to avoid stepping on the hose ahead.

When the last crew member in line stretches the last section of hose tightly, it is dropped a loop at a time while dragging the remainder of the load forward. This maneuver is repeated by each fire fighter in turn until the fire is reached.

When the crew number is insufficient to stretch the amount of hose in one procession, they should proceed in the regular manner for a short distance to clear the rear of the apparatus, carefully place their loops on the ground, and then return to the apparatus and secure another load in the original manner. After they have stretched their second loads, they will then return to where they placed their first loads and stretch them similarly.

The most essential requirement for the effective hand stretching of a hose line is a preplanned method. Each fire fighter should walk when dragging the hose; undue haste will usually take longer as the crew will tire faster and the hose will have a tendency to catch on obstructions. The hose should be carefully laid so it will not pile up or become entangled.

Flaking Out Hose

One method of stretching the larger sizes of hose is to first flake the hose back and forth on the ground behind the apparatus before stretching in to the fire (Figures 9-4 and 9-5). In Figure 9-4, the supply hose is shown in the middle of the street. In actual practice, lay all hose lines as close to the curb as possible and connect hose lines to the pumper on the curb side to interfere with traffic as little as possible. Often a preliminary preparation such as this will greatly

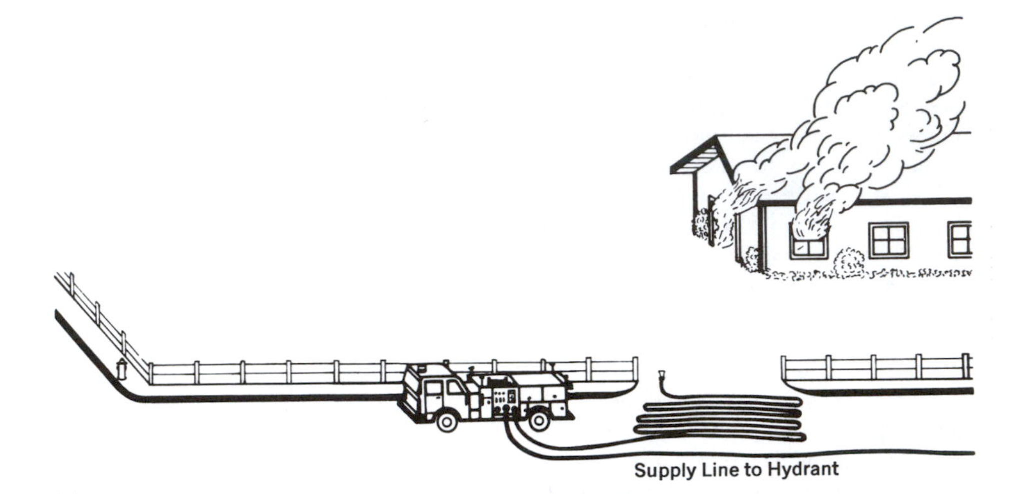

Figure 9-4. Supply-hose layout is shown in the middle of the street for illustrative purposes only. The hose is flaked out in 25-ft loops so it may be stretched with as few entanglements as possible.

lessen the work and speed up operations. There are several variations of this maneuver; one will be explained and different versions can easily be contrived.

A pumper lays one 2½-in. line from a hydrant and stops as close to the curb as possible and past the fire building; this line will be utilized to supply the pump through the gated suction inlet. When possible, connect all hose lines to the pumper on the curb side so traffic hazards will be reduced.

One 2½-in. hand line equipped with a 200-gpm spray tip will be stretched from the pump discharge outlet into the structure fire. The officer estimates that 200 ft of hose will allow for plenty of maneuvering room for the nozzle in the structure.

The first crew member takes hold of the first hose coupling in the bed, which will allow the ends to reach the pump inlets and outlets, breaks the connection, and hands the couplings to the engineer to connect up.

The next crew member grabs hold of a fold of hose and drags it back for about 30 ft, away from the side towards the fire; this should result in about 50 ft of hose flaked out and a 5-ft working area between the tailboard and the hose. Continue flaking out the hose until the predetermined quantity is laid out in an orderly manner; the desired 200 ft will result in 4 folds.

The hose is then broken at the next coupling; one end is placed in the hose bed to keep it out of the way, and the nozzle is attached to the other coupling.

The nozzle operator places the nozzle over one shoulder, grabs hold of a fold of hose, and leads in. The rest of the crew should take hold of the hose so the load will be equally divided, and follow the nozzle operator. All hose should be taken right up to the fire because it will be difficult to move after the line is loaded with water. The hose should be flaked back and forth in front of the structure so it may be stretched into the building with the least entanglements.

Reverse Lay, Double 2½-in. Lines

Occasionally a triple combination pumper will encounter a fire situation in which the officer will decide that two lines should be

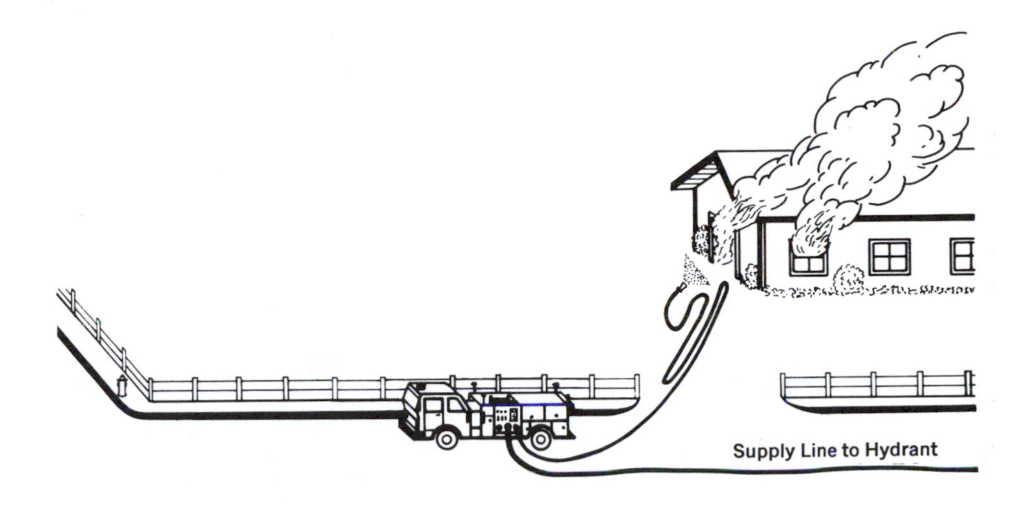

Figure 9-5. Stretch all of the hose as close to the fire as possible; flake the excess so it will not snarl when advancing into the structure.

immediately laid from fire to hydrant (Figures 9-6 and 9-7). The apparatus should stop past and on the same side of the street as the fire, staying as close to the curb as is convenient so traffic will not be unduly interfered with. Place a nozzle on the line towards the center of the street and flake out the desired quantity. If a 1½-in. wye assembly is wanted, attach it to the first line because there is more maneuvering room to work with.

If there are no parked automobiles or other hindrances, flake out the hose so it will lead directly towards the fire. If there are parked cars, flake the hose in the street straight behind the apparatus.

The second line then has a nozzle attached and the desired length of hose flaked out the same as was done for the first line. Some companies carry a nozzle already connected and a predetermined length of hose packed in a bundle and strapped together so it may be rapidly dumped in front of the fire.

The entire crew then assists in removing and placing on the sidewalk any additional equipment that may be needed for fire fighting. Crew members should stand on and anchor the hose lines while the engineer drives the pumper to the hydrant. The pump operator then removes sufficient hose to allow coupling up the hose lines to the pump outlets. The pump operator may utilize the booster tank to supply the fire streams until connection can be made to the hydrant.

Unless there is an extraordinarily large crew, it would be best to place a hose clamp on one hose line and then have the entire crew stretch the first line in and attack the fire. Nozzle operators are left to combat the blaze while the rest of the crew stretches and attends the second line.

Fire Hose Characteristics

A thorough knowledge of the theoretical aspects of fire fighting hydraulics is necessary for the effective development of fire streams and the proper utilization of all water sources. However, this theoretical knowledge must be coupled with a practical understanding of the physical forces that affect the movements of liquids through pipes and hoses.

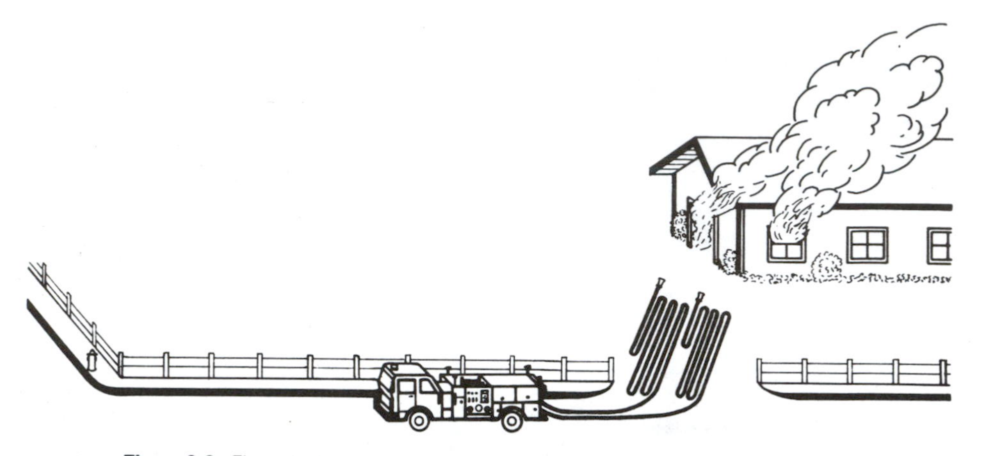

Figure 9-6. Fire to hydrant, double 2½-in. lines. If there is a driveway or no cars at the curb, flake the hose towards the fire. If there are parked cars, flake the hose straight behind the apparatus.

If strict reliance on the theoretical aspects of water movement is made, a student would believe that the bursting strength of fire hose is the only restriction on increasing the quantity of flow through a hose by increasing the engine pressure and that all hose of a certain size would have similar friction losses. From a practical point of view, this theory is not correct. Tests have shown considerable variation in friction loss in the various brands or makes of fire hose. The smoothness of the lining has a considerable effect on the friction loss; tests have determined that some new fire hose causes losses 50% greater than others. Also, age affects the smoothness of the lining; friction losses tend to increase as the hose becomes older. Some old fire hose has losses far less than other

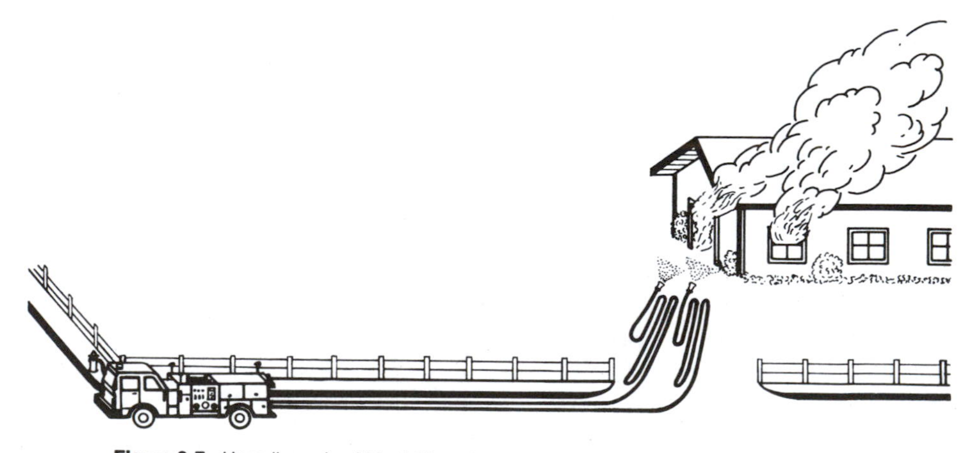

Figure 9-7. Hose lines should be laid as close to the curb as possible. If the hydrant has two or more outlets, position the pumper so that another apparatus can also connect up.

makes of new hose of the same size. A slight variation in diameter also produces a marked difference in friction loss; in the case of 2½-in. hose, a variation of 1/16 in. in diameter will result in 10% difference in friction loss. As some hose stretches and expands more than others when subjected to high pressures, the increased diameter will result in decreased friction loss. However, the friction loss tables and formulas in common use are sufficiently accurate to ensure good effective streams on the fire ground.

Critical Velocity

Regardless of what answers the friction loss formulas render, there is a maximum quantity of water that can be forced through a pipe or hose, regardless of the amount of pressure applied. Critical velocity is the point at which turbulence becomes a limiting factor in the quantity of fluid that can pass through a waterway. The maximum capacity of a hose line or pipe has been reached; added pressure will produce very little increase in flow.

When water flows through a hose or pipe, there is almost always a drop in pressure. Theoretically, the lost pressure between two points is caused by (1) friction between the moving water and the pipe or hose lining and (2) friction between water particles, including the turbulence created when the flow changes direction or when a rapid increase or decrease of velocity takes place, such as those caused by abrupt changes in hose diameter. A change in velocity results in some conversion of velocity head to pressure head or vice versa.

At low velocities, when the hose or pipe lining is smooth, all particles of the water move along in essentially straight lines and in concentric layers. The inner column of liquid is lubricated by the outer shell of liquid and it moves along more rapidly; friction loss occurs mainly in a thin layer at the hose lining. When very little turbulence is produced, the flow is called laminar (Figure 9-8).

As the velocity is increased, friction loss increases and the outer layer of the liquid becomes more disturbed; a point will be reached where friction becomes so great that the flow of the outer layers is broken up and the fluid becomes turbulent (Figure 9-9). The outer currents eddy and twist, overcoming the smooth flow of the inner currents; soon the entire stream is disturbed and the resulting flow and discharge pressure are decreased. This turbulent action creates a loss in pressure; it is the result of the relative roughness of the hose and the viscosity of the water. Viscosity is the internal resistance to flowing that is characteristic of liquids.

Tests have established that frictional resistance is:

1. Independent of the pressure in the fire hose or pipe.

2. Proportional to the amount and character of the frictional surface.

3. Variable with the velocity of flow. If the velocity is below critical, the resistance varies with the first power. When a critical velocity has been reached, the resistance is nearly proportional to the second power of the velocity.

Slippery Water

Small hose lines are easier to maneuver and control during interior fire fighting, but frequently the discharge rate is not adequate for effective fire extinguishment because friction losses become higher as hose diameters are reduced.

Friction losses may be reduced by adding a chemical substance; this slippery water requires less energy than plain water when it is being pumped under identical conditions. However, the additive is not equally effective under all conditions. When the velocity in the fire hose is such that there is a laminar flow and the water is proceeding in a nonturbulent, quiescent manner, slippery water is nearly indistinguishable from ordinary water. When the flow approaches the point of critical velocity, and the water becomes

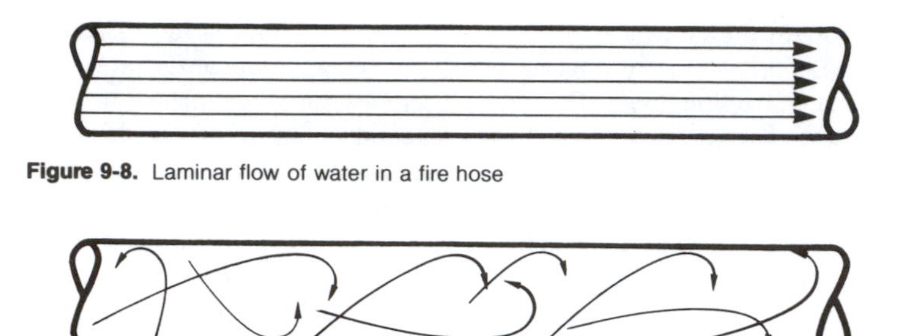

Figure 9-8. Laminar flow of water in a fire hose

Figure 9-9. Turbulent flow pattern of water in a fire hose

more turbulent, slippery water requires less energy, and the friction loss is reduced. It is believed that the long polymer molecules of the chemical uncoil and align themselves in the direction of the main flow of the water. They reduce the friction loss by acting like guide rails to change the turbulent action of the water into an efficient laminar flow.

One large fire department benefits from the advantages of using a slippery water additive because it permits utilizing 1¾-in. hose to achieve the same fire streams that they formerly obtained with 2½-in. hose lines.

Simulating Hose Layouts

Some companies counterfeit the friction losses created when large volumes of water flow through long hose layouts. The two most common methods are to use an orifice plate in the female hose coupling and to use fewer sections of small hose to create the anticipated friction loss. This simulation may be necessary at times when fire crews are shorthanded and taking the pumper out of service is impracticable because all of the hose has been stretched out. However, believing that this simulation will create realistic conditions, and that the satisfactory results of a counterfeit operation would be recreated when the actual hose layouts were made, could lead to a fire ground debacle. It is important to develop procedures by practice under the simulated conditions, but the procedures should be proved under actual fire conditions. That can be done by laying out the anticipated lengths of fire hose at the fire ground, utilizing the correct appliances and nozzles, and objectively pumping the desired quantities of water at the required discharge pressures to ascertain if the simulated hose layouts are really attainable and practicable.

Remember: there are physical forces other than friction loss and back pressure that affect the flow of liquids; some of these cannot be exactly predicted. A hydraulic situation could be mathematically possible but physically impossible.

To simulate friction loss, an orifice plate that resembles a large washer is placed in the female swivel coupling and the hose is then connected to the pump discharge outlet. As an example, an orifice of 1.348 in. will simulate 400 gpm flowing through 1000 ft of 3-in. hose; 1.362 in. will create friction loss equivalent to a flow of 500 gpm; and 1.369 in. will imitate 600 gpm. Remember to remove the orifice plate after the drill.

There is an old rule that friction loss in 1½-in. hose is 13 times the friction loss in 2½-in. hose for the same gpm flow. Therefore, the use of three lengths of 1½-in. hose is equivalent to using 39 lengths, or 1950 ft, of 2½-in. hose. Using 100 ft of 1½-in. hose is the same as using 1300 ft of 2½-in. hose. Since 200 gpm creates 10 psi friction loss in every 100 ft of 2½-in. hose, pumping the 200 gpm through 100 ft of 1½-in. hose would theoretically create a friction loss of 10 × 13, or 130 psi. Always use the correct reducers and increasers to connect the hose and fittings together.

Pressure Control Devices

Pumpers are equipped with relief valves and governors to protect fire crews from excessive nozzle pressures and the resulting reaction forces if another hose line is shut down. Nozzle operators must realize that these pressure control devices will offer no protection when a single supply line equipped with a wye is supplying two or more hand lines. A pressure control device is adjusted to maintain an engine pressure that takes into account the friction loss in the supply line and the hand lines, back pressure, and nozzle pressure. If one of the wyed lines is shut down, flow through the supply line will be reduced, which will decrease the friction loss.

The engine pressure will remain the same. With the reduction of friction loss in the supply line, the pump energy that was forcing water to both tips is now concentrated in flowing water to a single tip. The nozzle operators on the tip remaining in service could now have more pressure, with the consequent reaction forces, than they can handle.

Example: In Figure 9-10, a 500-ft, 3½-in. supply line is wyed into two 600-ft lengths of 2½-in. hose. There is a 1⅛-in. tip at 50 psi pressure on each hand line. Consequently, there are 500 gpm flowing through the 3½-in. hose and 250 gpm flowing through each 2½-in. line. When 500 gpm are flowing through 3½-in. hose, there is 10 psi friction loss created in each 100 ft of the supply line; therefore, friction loss in the 3½-in. supply line totals 50 psi. When 250 gpm are flowing through the 2½-in. hose, 15 psi friction loss is created in every 100 ft; therefore, friction loss in 600 ft of 2½-in. hose totals 90 psi. Friction loss of 140 psi plus 50 psi nozzle pressure will require an engine pressure of 190 psi.

If one nozzle is shut down, the flow through the 3½-in. hose and the consequent friction loss will be reduced. In the preceding example, the friction loss will decrease about 10 psi, so the nozzle pressure will rise approximately 10 psi. Nozzle reaction is calculated with the formula $NR = 1.5\ D^2 \times NP$. A 1⅛-in. tip at 50 psi will create a nozzle reaction of 95 lb; if the pressure is increased to 60 psi, the reaction force will rise to 114 lb. While this is not a massive increase in nozzle reaction, other hose layouts and nozzle combinations could cause safety problems. In fact, a 19-lb increase in nozzle reaction when the footing is treacherous or an operator is balanced on a ladder could be hazardous.

Heavy Stream Applications

There are times when small fire streams are not advisable for fire fighting because they will not supply sufficient water to confine the fire, they cannot provide the necessary reach to strike the burning materials, or

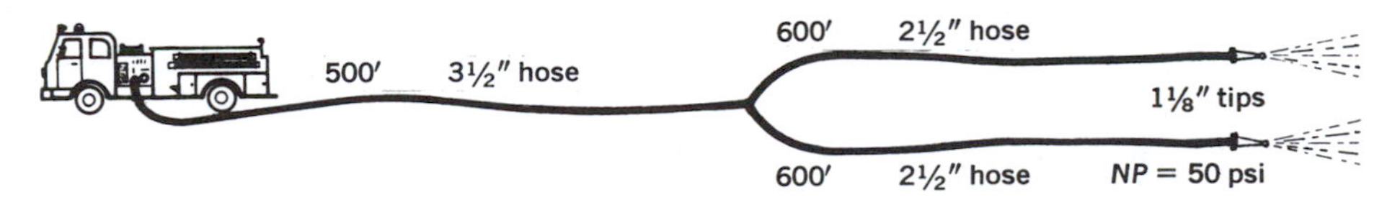

Figure 9-10. When two or more tips are provided with water by one supply, shutting down one nozzle will cause a pressure increase in the other.

there may be imminent peril of building collapse. If upon arrival of the first-in company the fire is obviously beyond control by hand lines or is threatening the stability of the structure, large fire streams may be necessary to knock down the fire without endangering the lives of fire fighters.

The largest fire stream that can be safely maneuvered and manually controlled by hose crews is from a 1¼-in. nozzle; this tip will discharge about 325 gpm when the nozzle pressure is 50 psi. Greater flows than this will create sufficient nozzle reaction that the hose line would be difficult to maneuver and hazardous to control under the slippery footing conditions present at every fire. From the formula for calculating nozzle reaction, $NR = 1.5\ D^2 \times NP$, a 1¼-in. tip at 50 psi pressure would create 117 lb of reaction. Water flowing from a 1¼-in. tip at 80 psi nozzle pressure is commonly considered to be a master stream because it delivers a greater quantity of water than can safely be maneuvered by hand; it will create a nozzle reaction force of 188 lb. The nozzle reaction of fog and spray nozzles depends upon quantity of flow and construction.

Large fire streams from ladder pipes, elevating platforms, water towers, wagon batteries, portable monitors, and other heavy stream appliances may be utilized for a fast, aggressive attack on a large blaze that requires massive volumes of water or a long range. Heavy streams are also referred to as master streams and large caliber streams. These large flows are effective for defensive purposes, such as wetting down exposed structures to prevent them from becoming ignited, setting up a water curtain between a burning structure and adjacent exposures, and cooling down a large mass of embers and debris. Often an immediate, massive, rapid, aggressive attack on the main body of a fire is the best method of protecting exposures; a 100% purely defensive fire fighting strategy is seldom justified.

A fire involving a large area of a building, a total structure, or a vast quantity of flammable materials, such as in a lumber yard or an oil refinery, generates a gigantic quantity of heat; only a substantial volume of water will effectively control the blaze. While combating a fire with fire streams, sufficient water must be applied to the burning materials so that the cooling effects of the water will be greater than the heating effects of the blaze; if this is not accomplished, the fire will continue to accelerate in intensity as long as there is sufficient fuel and oxygen.

Many officers of first-alarm companies are under the misconception that heavy streams are the responsibility of later arriving units. On the contrary, the sooner large fire streams can be placed in operation, the more effective they will be. Frequently, the need for a large stream is strategically passed by the time it has been set up and placed in operation because of the delays and complicated procedures which are necessary. Persistent and thorough practice drills will reduce the setting-up time to a tolerable amount.

Unless a building is so engulfed in flames that hand lines could not be maneuvered in close proximity to the blaze, or the structure has been weakened so there is imminent danger of collapse, application of heavy streams is seldom an effective tactic of interior fire fighting. Pouring on large volumes of water jeopardizes the structural integrity of a building and increases damage to the contents. Also, as compared with the maneuvering of hose lines and the close scrutiny of fire stream effects by hose crews on a hand line, the direction and control of a master stream is much less efficient.

When master streams are placed in operation on a structural fire, it should be realized that they are not expected to extinguish the blaze. They should be operated only until the fire has been knocked down and hand lines can move back inside the fire area to aggressively extinguish the fire. Three important rules apply to combating a structure fire with a heavy stream: (1) Before starting a large caliber stream, all personnel working inside the building must be withdrawn from the area where the stream is to be directed. If the fire is on an upper floor, everyone beneath the fire floor must be evacuated. If hand lines are in operation, they must be shut down and the crews ordered to a safe area. Only after these precautions have been taken should the heavy stream operation commence. Failure to observe this safeguard has resulted in the deaths of many fire fighters. (2) A fire fighter should be stationed in an advantageous position to effectively guide the application of the stream of water so that maximum benefits can be achieved. (3) The application of master streams should be of relatively short duration. Unless the officer in charge is concerned with the structural stability of the building, the large stream should be discontinued as soon as the fire has been knocked down sufficiently to allow the hose crew with hand lines to enter the fire area.

Hazardous Weight of Water

The primary concept of fire fighting is to apply sufficient water on a fire so that burning materials are cooled at a faster rate than combustion is generating heat. One of the primary rules of fire fighting is that fire streams should be shut down or reduced in size when the blaze is knocked down to avoid unnecessary property damage. These fire fighting and salvage principles are very important, but there is another effect of water that has caused a great number of deaths and injuries among fire fighters. The excess weight of water has been responsible for the collapse of many buildings and the sinking or overturn of boats and ships.

The extinguishing capability of water is an important attribute, but fire crews should also be constantly concerned and aware of its hazards. One cubic foot of water weighs 62.5 lb; one gallon weighs 8.35 lb. Water 1 in. deep would weigh 5.2 lb for every square foot of area (62.5 ÷ 12 = 5.208). Multiply the number of square feet of floor area times the depth of water times 5.2 lb and it can be seen that a structure-collapsing weight could easily accumulate. A 1⅛-in. nozzle at 50 psi pressure will discharge about 250 gpm, or over a ton of water every minute. Even a 100-gpm spray tip would discharge 835 lb of water every minute, or 1 ton every 2.4 minutes. A 1000-gpm tip on a water tower would direct over 4 tons of excess weight into a structure every minute.

Some years ago a city proudly accepted delivery of a new 85-ft water tower. Shortly after they received the new apparatus, a large fire erupted in a multistoried structure; the upper floors, occupied by a printing firm, carried the weight of heavy presses and paper stock. The officer in charge assigned the fire fighting duties on the upper floors to the new water tower, while fire crews with hand lines were simultaneously attacking the blaze on the lower levels. Such a huge quantity of water was poured into the top windows of the structure that observers in the street clearly witnessed water running over the 3-ft-high window sills. Suddenly the upper floors of the building collapsed, trapping and crushing many fire fighters underneath. It was later calculated that the 3-ft depth of water contributed an extra weight of 187 lb per square foot on the already heavily loaded floors.

Floating roof tanks are utilized in petroleum tank farms to prevent an explosive accumulation of flammable vapors. In one instance, several tanks were burning in a waterfront tank farm. A fireboat directed such a huge quantity of water from its monitors over an exposed uninvolved tank full of gasoline that the floating roof sank because the scuppers could not carry the water away as fast as it was being applied. For the duration of the conflagration, fire fighters had to maintain a thick blanket of foam over the gasoline to prevent ignition.

The floors and roofs of structures are engineered and designed for a certain load-carrying capacity; these requirements vary according to the type of occupancy. There is a margin of safety in building design, but this can rapidly be exceeded when the weight of water is added to the building contents. Simultaneously, the fire is probably also weakening the structure. Even if the building is so constructed that the water will rapidly flow out of doorways and down stairways and shafts so none accumulates on the floors, there is always the problem of absorbent materials becoming wet and increasing in weight. Bales of paper, fibers, and other absorbent materials rapidly accumulate tonnage sufficient to overburden a structure.

All officers and fire crews should constantly keep this hazard in mind as many fire fighters have been crushed in collapsing structures. No one should be working in a building while heavy streams are being applied; this would be a definite life hazard. Officers should watch for an accumulation of water on the floors; if this becomes precarious, reduce the use of water until some of the weight can be removed or the stability of the structure is assured.

From the aspect of fire fighting, salvage, and safety, the best use of a fire stream is to direct it to the correct location, in the right quantity, in the most efficient form, so the greatest quantity of water is completely vaporized. Cold water flowing from a burning building clearly indicates that the fire streams are not effectively being directed on the fire.

Water Supply

Water is so fundamental to fire fighting that a good water supply is the most important single factor in fire protection. When engineers from the insurance industry grade the fire defenses of a municipality, water supply and fire department receive equal recognition; each has a 39% relative effect on the final class of the city. This rating system emphasizes the relative importance of an adequate water supply to successful fire fighting; the accessibility and strength of this supply is a significant factor in the control of most large fires. Without sufficient quantities of water, the best equipped fire department with the most highly trained fire fighters can do very little toward bringing a fire under control and preserving life and property.

Because water is basic to fire extinguishment, it is the duty of every fire chief to definitely determine whether the water system can supply adequate amounts of water to every part of the city or locality. A fire chief who does not know the community's water system deficiencies cannot possibly utilize the available apparatus, equipment, and manpower with maximum efficiency. As a result, serious mistakes could be made in fire fighting strategy and tactics, resulting in lost time, ineffective fire streams, and eventual increased fire loss to the property. Likewise, it is the duty of every fire company officer to ascertain if there are any water supply problems in the district. If any problems are identified, practical solutions should be preplanned and prepared for.

Water Systems

Historic water systems employed wooden mains, and the first fire hydrant consisted of a wooden plug driven into a hole in a wooden water main that had been formed from a hollowed-out log. When the plug was withdrawn, the water would flow into a street depression. This primitive device produced the term *fireplug*. Sanitary hazards, the entrance of dirt into the mains, and the maintenance of street surfacing provided the incentives to develop a better fireplug.

The first significant improvement over the wooden plug followed the introduction of cast iron pipe for distribution systems. Early cast iron fireplugs consisted of a branch pipe riser with a wooden plug inserted. Canvas cisterns were later utilized to reduce water loss and to provide a better sump for the fire pumps to draft from.

A metal standpipe with a hose connection above ground and connected to the wooden water main proved more popular and efficient than the cisterns. As cast iron distribution systems became more common, higher water pressures were introduced. These higher pressures soon rendered all plug-type hydrants obsolete.

A well-designed water system for municipal fire protection makes use of large mains in grids whenever possible. The water mains are cross-connected at frequent intervals so that when pumpers are operating from hydrants the water is fed from several directions (Figure 9-11). This reduces friction loss in mains carrying a large fire flow and helps to maintain a satisfactory residual pressure at hydrants supplying pumpers and hose streams. If a water system is well gridded and the mains are of adequate size, friction loss may be reduced to a relatively few pounds, except in the hydrant riser and barrel.

When no water is flowing, the static pressure at various hydrants may vary from 30 to 80 psi or more; this is not the pressure that will be available when water is flowing. The more water that flows, the greater the friction loss in the water mains, and the pressure drops accordingly until no more water can be taken from a hydrant or group of hydrants.

If a hydrant can flow 600 gpm at 20 psi residual pressure and the fire department attempts to supply three 2½-in. lines requiring 750 gpm, it will experience low pressure on the suction side of the pump because it is attempting to take out more water than the system can supply at that point.

At large flows there may also be considerable friction loss between the hydrant and the pump unless a large-diameter suction hose and big pump suction inlet are utilized. Unless a large hydrant outlet is used, the flow may be seriously restricted when there is low residual pressure because of the demand for water.

At a large fire, the problem may be further compounded by the fact that more pumpers may be trying to take water from the hydrants near the fire than the mains in the area can possibly deliver. If the water system

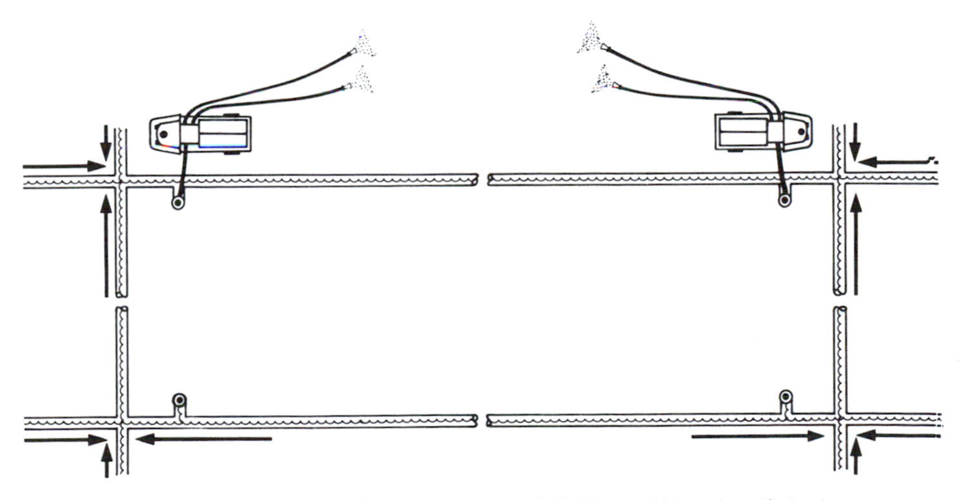

Figure 9-11. When large mains are cross-connected at frequent intervals so that water is fed from several directions when pumpers are operating at hydrants, a satisfactory residual pressure is better maintained.

can only supply a definite fire flow in a given part of town, the water consumption by pumpers must be limited to that amount. The available water must be used wisely and placed where it will do the most good.

At every major fire, a competent water control officer must be designated to take charge of the water supply to see that the available supply is used properly and that the system is not overtaxed. The gages on the pumpers already working will show whether the residual pressures at the hydrants are holding up and whether additional lines may be supplied.

However, most cities have locations where the mains serving the hydrants are not very well gridded because of terrain conditions or the system has not been adequately developed. In areas where the supply is severely limited, and a large fire flow would create a water deficiency, relaying water from hydrants on another main could be the best remedy (Figure 9-12). Normally the most effective supplementary pumping procedure would be to spot a pump at a hydrant where additional water is available, then pump this extra supply through hose lines to a pumper located at a hydrant where the residual pressure has fallen. It has been calculated that under fire conditions, and at the higher pressures, two 2½-in. hose lines are approximately equivalent to adding a 6-in. main to the water system grid.

Another method of improving the supply of water in a system is to pump from a hydrant on a different main into another hydrant near the fire. Never contaminate a city's water supply by pumping water from a static source into a hydrant. At a large waterfront conflagration a fireboat attempted to augment a low main pressure by pumping salt water into a hydrant. The water supply was improved, but it was not until many days after the fire was extinguished that any waterfront restaurant could brew a decent cup of coffee.

At many fires, water supplies would be much improved if, instead of placing all pumpers on hydrants close to the fire, one or more pumpers were utilized to supply additional water for pumpers near the fire so they could discharge their full rated capacity. Figure 9-13 illustrates the wrong way and a better way to go about this: in A, engine 1 is taking most of the water in the dead-end main, leaving little for engine 2; in B, engine 3 also lays lines directly to the fire, thus robbing engines 1 and 2 of water. A better procedure is shown in C, in which engine 3 lays supplemental supply lines to engines 1 and 2, which increases their water supply and enables them to lay additional hose lines into the fire. Laying additional hose lines between pumpers obtaining water from a main of insufficient size or carrying capacity will have the same effect as increasing the diameter of the water main; this would effectively decrease the friction loss in the pipe.

Water supply problem areas should be prominently identified on the station maps so the company will be warned and alerted while they are checking a street address; this would include districts of the water distribution system where an insufficient supply exists or where extraordinarily long hose lines would be necessary to obtain an adequate fire flow. "Forewarned is forearmed" is an excellent motto; if fire fighters know beforehand that water supplies are weak or nonexistent in a certain locale, they can preplan and take steps to make them adequate.

Hydrant distribution and installation recommendations are fairly well standardized, but the actual hydrant placement and performance may vary widely. Company officers should conscientiously study the hydrant maps, which indicate the water main sizes and hydrant distribution. When they receive a dispatch to a fire, they should know which hydrants have above average and below average water supply and pressure. They should be aware of which hydrants would probably be deficient because they are on a dead-end main.

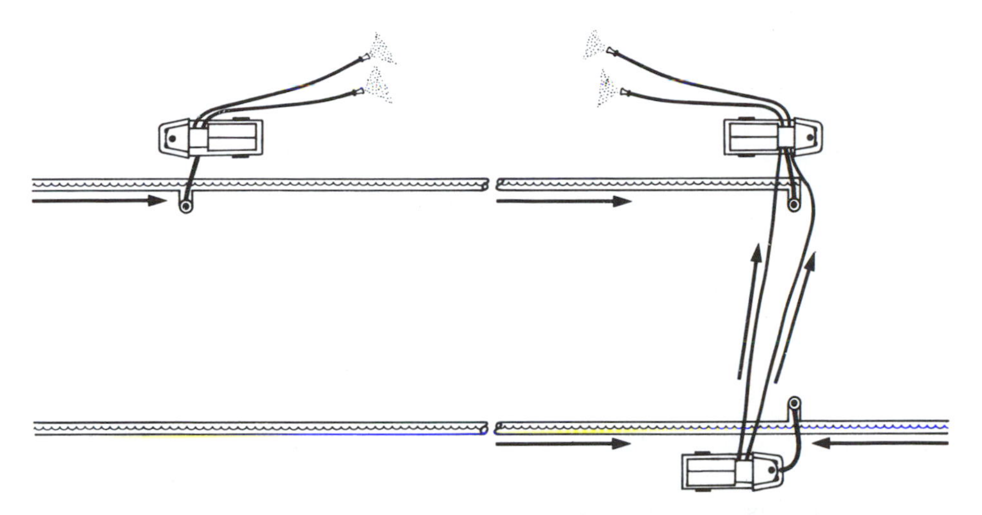

Figure 9-12. Dead-end mains provide a poor supply; the second pumper may receive little water. A supplemental pumper supplying two 2½-in. lines from an adjacent main provides the equivalent of a 6-in. grid.

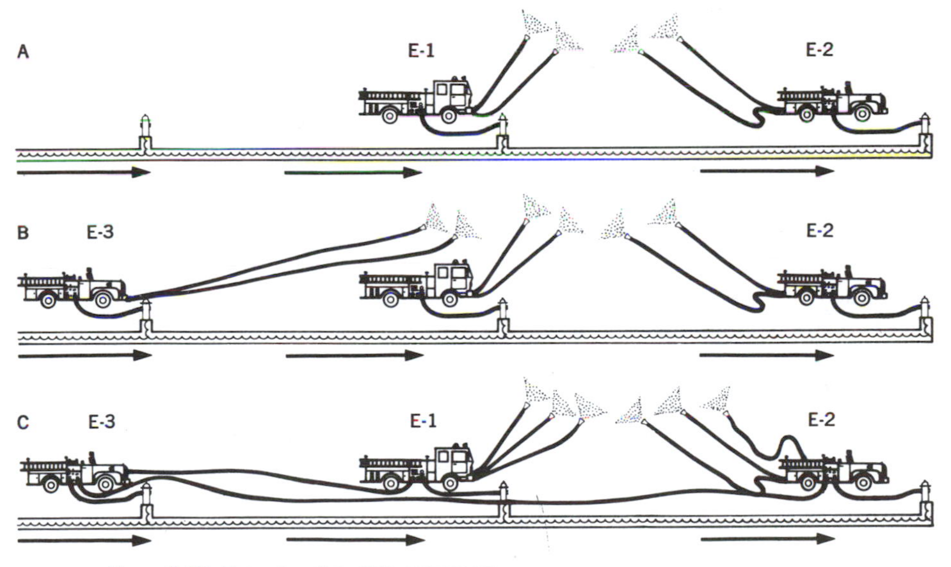

Figure 9-13. Examples of multiple pumper use

Dry Barrel Hydrants

In the colder climates of this country, the portions of fire hydrants that are above ground must be drained in the wintertime to prevent freezing. In these frost-proof hydrants, one valve controls the water supply to all outlets. An important practice is to place gate valves on the hydrant outlets that are not going to be immediately used before the water is turned on. This permits connecting an additional pumper or a second supply line without shutting down the hydrant.

Fire hydrants should be opened fully on every use. If a dry barrel hydrant is only partly opened, the drain valve in the bottom of the barrel will not close. This permits water under pressure to escape from this valve; the escaping jet of water can wash away all the earth from around the hydrant riser and barrel, possibly causing it to break away from the main. In addition, there is the problem of freezing in the winter as the hydrant would be full of water. Some hydrants are so constructed that they will open very little during the first few turns, as compared with the last few turns. The fact that a hydrant is not fully opened may not be readily apparent until it is necessary to use the full capacity of the pump.

Wet Barrel Hydrants

In the warmer climates, the hydrants may be continuously filled with water because there is no concern about freezing in winter. These hydrants are equipped with a separate valve to control water flow from each outlet.

Hydrant Performance

The greatest percentage of all fire ground pumping is done from a hydrant supply. Two fundamentals are essential for obtaining an adequate amount of water from a hydrant: (1) Attempt to pump from a hydrant capable of delivering the desired quantity of water, and (2) connect to the largest hydrant outlet with a suction hose big enough to provide for the full capacity of the pump. The static pressure of a hydrant when no water is flowing is not of any great significance; more important is the ability of the hydrant to deliver the desired flow at an effective pressure.

When a fire department wants to obtain large hydrant flows, it should take advantage of the carrying capacity of the ample-sized mains under the streets and avoid long hose lines. This can be accomplished by making maximum use of the hydrants nearest to the fire that have the best water supply. Pumpers should connect to multiple-outlet hydrants so that another apparatus can obtain water from an unused outlet (Figure 9-14A). This helps reduce friction loss by keeping the hose lines as short as possible. If the first pumper blocks the hydrant (Figure 9-14B), or the second pumper neglects to utilize it, the necessarily long hose lines may cause fire stream problems. Always make the maximum use of multiple-outlet hydrants when hooking up. Generally, it is considered to be poor practice to have several pumpers working at different hydrants, each delivering a relatively small gallonage through one or two hose lines; this is not taking full advantage of either the delivery capacity of the hydrant or the pumping capacity of the apparatus, unless the hydrants nearest to the fire are incapable of supplying the required flow.

A large multistoried hotel in the metropolitan area of a city was swept with fire shortly after the first-alarm companies arrived on the scene; the initial hand lines making an interior attack on the blaze were driven out of the structure. A great number of occupant deaths and injuries resulted. Fire fighting was then reduced to protecting the exposures and cooling down the debris with master streams from ladder pipes, elevating platforms, and portable monitors because the fire was too intense for a hand line approach. Huge quantities of water were demanded from the water supply system, which consisted of hydrants with two big streamer outlets closely spaced on a well-designed gridded network of extremely large mains. At the height of the fire fighting efforts, the commander in chief received numerous radio messages that the pumpers were having difficulties in supplying an adequate amount of water, so he dispatched additional engine companies to lay supply lines from adjacent mains and relay water into the deficient pumpers. A water supply officer simultaneously inspected the fire ground and found that the water system was capable of supplying the required quantity

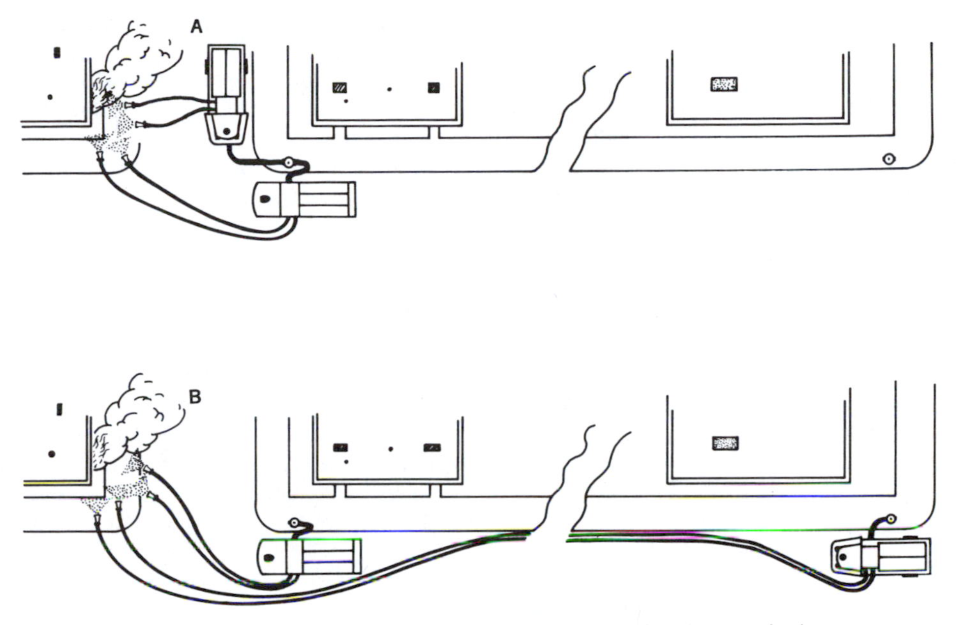

Figure 9-14. (A) Proper use of multiple-outlet hydrant. (B) Inefficient pumper hookup.

of water without the relay pumpers; there was ample water in the mains close to the fire that was not being utilized.

The water control officer determined that the water deficiencies were caused by five major mistakes in hydrant hookup and hose layout. Standard operating procedures that are thoroughly drilled into all personnel will help avoid fire ground errors that may accompany any major conflagration with the unusual distraction of victims jumping from windows. Following is a summary of the errors made in the example just cited.

1. Pumpers hooking up to the nearest hydrants spotted their apparatus so another pumper could not utilize the second hydrant outlet; there was an adequate quantity of water still in the mains to supply more than twice the flow being delivered to the fire.

2. The 1250- and 1500-gpm pumpers in this city carried a soft suction connected to the gated front suction inlet. The smaller size piping and fittings in these inlets were too restricted to allow the pump to obtain its full rated capacity. All engineers utilized these restricted front suctions, instead of hooking up to the large side suction inlets.

3. Insufficient supply lines were laid between apparatus; a great amount of the pump discharge pressure was used up in overcoming friction loss.

4. The 2½- and 3½-in. supply lines were connected into the 2½-in. gated suction inlets when available; others went into an increaser attached to the front soft suction. In no case did any engineer place a large siamese fitting on the side pump suction inlet to reduce restrictions.

5. All supply lines were equipped with a nozzle shutoff butt where it was connected into another apparatus so the water flow could be controlled. These shutoffs seriously restrict the quantity of water that can flow through a hose layout.

Officers should order additional pumpers to withdraw water from a hydrant only when they are certain that the system is capable of supplying an adequate quantity of water at the desired location without creating problems for the pumpers already connected up. An officer should be able to calculate how much additional water can be expected by the size of the pressure drop caused by the quantity of water already being pumped. A sudden drop in the pressure may indicate that other pumpers have connected to nearby hydrants, thus competing for the available water in the mains, or there may be a ruptured pipe or other failure of the water system. Placing an additional pumper in service may further reduce the residual pressure of the hydrant sufficiently to make it impossible to satisfactorily supply the hose layout.

It must be kept in mind that pump discharge quantities are reduced at the higher pressures. If a pump is rated at 100% of its capacity at 150 psi net pump pressure, it will discharge 70% of its capacity at 200 psi and 50% at 250 psi. With a 1250 gpm pumper, limitations will be reached at about 200 psi when discharging 875 gpm and at approximately 170 psi when supplying 1000 gpm. Since this is related to the net pump pressure, the discharge quantities would be increased when the pumper is aided by a strong hydrant supply; the discharge would likewise be decreased when drafting.

When pumps are discharging at the higher pressures, it may require two pumpers to supply a large-volume tip. Reducing the friction loss by laying an additional hose line between the engines may mean the difference between a practical hose lay and an impossible hydraulics dilemma.

The flowing pressure of a hydrant is equivalent to the benefits produced by adding another stage to the pump. If a hydrant supplies water in sufficient quantity and pressure, a total quantity may be produced that exceeds the rated capacity and pressure of the pump. As an example, a centrifugal pump rated to deliver 1250 gpm at a net pressure of 150 psi at draft could possibly deliver 1500 gpm at 180 psi from a hydrant without exceeding the allowable engine speed.

When operating from a hydrant, the net pump pressure of a centrifugal pump is the indicated (gage) discharge pressure minus the suction gage (hydrant flow) pressure. To put it another way, any centrifugal pump should theoretically be able to deliver at least its rated capacity at its rated pressure plus the flow pressure of the hydrant. To illustrate, picture a 1250-gpm pumper rated at 150 psi net pump pressure connected to a hydrant that will produce a flow of 1250 gpm with a residual pressure of 30 psi. The pump will be able to deliver its rated capacity at 180 psi.

Drafting from a Hydrant

Some departments connect their pumpers to hydrants with the hard suction hoses used for drafting; these hoses are usually heavy, stiff, and unwieldy. Other departments use soft suctions, which are lighter and more flexible. Advocates of the hard suctions claim that pulling a vacuum on a hydrant will allow a greater flow from a weak water system and that soft suctions will collapse when subjected to a suction. Actually, there is very little advantage gained by pulling a high vacuum on a hydrant; tests have shown that a pumper operating with vacuum of 30 in. of mercury showing on the suction gage will discharge less than 5% more water than it would with zero vacuum. In fact, it is recommended that pump operators maintain at least 10 psi on their suction gages for the most efficient functioning.

Drafting from a hydrant will subject the hydrant, mains, and plumbing in adjacent buildings to stresses for which they were not designed. There have been occasions where this practice has caused broken hydrants and ruptured mains.

Estimating Hydrant Performance

The pump gage to which operators have customarily paid the greatest attention when pumping from a hydrant has been the gage indicating the pump pressure being discharged through the hose lines. However, the compound gage on the suction or intake side of the pump is nonetheless important. A good operator who can interpret the compound gage reading correctly can generally get the maximum performance from the pump and can often keep out of trouble by not attempting to supply lines for which adequate water is not available at the hydrant.

To make proper use of the compound gage, note the pressure available on the suction side of the pump after the hydrant is opened but *before* a line is charged. This will indicate static pressure in the water system with no water flowing from the hydrant. After the line is charged and the proper pump pressure has been established, again note the reading on the compound gage when water is flowing. The difference between the static pressure and flow pressure at which the hydrant supplies the first line will give an indication of the total quantity of water that can be obtained from that hydrant outlet.

The most convenient way of estimating available flow is on the basis of the number of hose streams with a given size tip that a hydrant outlet can supply. For example, when supplying a 1⅛-in. tip on the first line, if the pressure drop is not over 10%, the hydrant is good for three additional 1⅛-in. streams. With a pressure drop of not over 15%, the hydrant should supply two additional 1⅛-in. streams. If the pressure drop is not over 25%, one additional 1⅛-in. stream could be supplied. Where the pressure drop exceeds 25%, the hydrant could not be expected to supply an additional 1⅛-in. line. These percentage figures may be applied to any size nozzle tip to indicate the number of additional streams of the same size available from the hydrant outlet.

To illustrate further, suppose the pump is connected to a hydrant that shows a static pressure on the compound gage of 100 psi, and suppose also that the first line to be supplied has a 1⅛-in. tip with 50 psi nozzle pressure desired. With water flowing and a proper pump pressure established, the compound gage indicates a hydrant flow pressure of 90 psi. The pressure difference, between the static and flow pressure of the hydrant, is 10%. In this case, the pump operator could expect to supply three additional 1⅛-in. streams or a total of 1000 gpm. (The gpm figure includes the original line.)

Now let us consider a second hydrant. Again we have a static pressure of 100 psi and a 1⅛-in. tip with 50 psi nozzle pressure on the first line. However, in this case, the compound gage indicates a hydrant flow pressure of 78 psi. The pressure difference between the static and flow pressure is about 22%. In this instance, the pump operator is warned that in all probability the hydrant can supply only one additional 1⅛-in. stream or a total of approximately 500 gpm.

It should be kept in mind that the estimated quantity of water available from a hydrant outlet does not necessarily indicate that the pump is capable of delivering a like quantity at working pressures. The capacity of the pump is subject to limitations of pounds-gallons pumped at net pump pressure. When pumping at a net pump pressure in excess of the rated pressure for capacity operation, pump limitations also become a determining factor in the number and size of lines that can be supplied.

Other pumps withdrawing water in the immediate area, particularly at large fires, may lower hydrant flow pressures after the pump operators have estimated the performance to be expected from the hydrant to which they are connected. For example, suppose a pump is operating from a hydrant indicated to be capable of supplying four lines equipped with 1⅛-in. tips. The pump is supplying three lines, and the pump operator notices the compound gage reading has fallen considerably since charging the third line. This reduction in hydrant flow pressure will, in all probability, prohibit handling a fourth line. Pump operators should check their gages periodically for any indications of changes that affect their operations.

A sudden drop in the pressure reading on the suction gage may indicate that other pumpers have connected to the water system at hydrants located in the neighborhood, or that there is a break in the main or some other failure in the water system.

Estimating the Static Pressure

If the static pressure was not noted when the hydrant was opened, it can be determined as follows:

1. Note the flowing pressure on the suction gage with the first line in operation.
2. Place another nozzle delivering the same gpm into operation and note the drop in flow pressure.
3. Divide the drop in pressure by 2 and add this amount to the flowing pressure that was noted when the first line was in operation. This is the estimated static pressure.

Example: With the first line in operation, the flowing pressure on the suction gage is 68 psi. Another nozzle delivering the same gpm is placed in operation, and the flow pressure on the suction gage is now 44 psi. Estimate the remaining available flow.

The first reading on the suction gage was 68 psi and the second showed 44 psi, which is a decrease of 24 psi. If 24 is divided by 2, it equals 12 psi. This 12 psi is now added to the flow pressure that was noted when the first line was in operation. The estimated static pressure is 80 psi (68 psi + 12 psi).

The decrease from the static pressure to a flowing pressure of 12 psi gives a drop of 15%. Therefore, two more like volumes is the estimated remaining available flow from that hydrant outlet.

Static Water Sources

The water systems of most cities are so excellent that many fire departments tend to ignore any water supplies other than hydrants. This is a mistake, because there are occasional situations where static water facilities such as reservoirs, lakes, rivers, oceans, ponds, and other natural bodies of water could readily be made available for emergency use.

Because of a lack of sufficient water for fire streams, many large, expensive homes in Southern California were destroyed during a brush fire conflagration in spite of an abundance of fire fighters and apparatus. Most of these homes had large swimming pools with sufficient water to preserve the structure, but there were no provisions for access of a pumper to the pool.

Occasionally a large recreational hotel located on the shore of a lake, river, or ocean will burn to the ground; the usual alibi of the fire department is that there were no close hydrants.

The fire department should have a knowledge of all the static sources of water that could be drawn on in the event of an interruption or curtailment of the normal supply. These emergency sources would include private well supplies, swimming pools, private reservoirs at industrial plants, the ocean, lakes and ponds, rivers and streams, and other natural bodies of water. All of these emergency sources should be identified on maps, inspected, and routinely supervised, because they could be as important during a fire as hydrants.

Every fire department should develop fire attack plans that include provisions for supplementing the available water supplies by relay operations, drafting from emergency static sources, or utilizing vehicles with large tanks to transport water. Access to these emergency water sources should be preplanned and arranged for. Gates or trap doors cut through a fence or wall in a strategic location could render a previously unobtainable supply of water easily accessible. At reservoirs and other bodies of water, suitable approaches and ramps should be provided so that pumpers may get close enough to the water for drafting without becoming mired down; in this case, a deep basin should be provided for the suction strainer if the water is shallow. Some pumpers carry three or four hard suction hoses so they may readily draft from water supplies too far off the road for normal operations (Figure 9-15). A dam can be constructed in a shallow stream to provide a sufficient depth for drafting. Where a pumper cannot approach close enough to the water for drafting, a permanent suction pipe and strainer can be installed. Access holes with hatch covers can be cut in wide wharves and piers when it is not advisable to drive the heavy apparatus out on the structures where they could drop their suction hoses over the edge.

For static water sources in areas subject to freezing, tapered wooden plugs constructed to be larger than the suction hoses can keep the water accessible in this way: As the cold season approaches, float the plugs with the large end down in the water. When the plugs become frozen in the ice, they can be driven down with a sledge hammer to provide access for drafting.

Figure 9-15. Pumper equipped with four hard suctions can draft water from emergency sources that could otherwise not be used. (Courtesy of Universal Fire Apparatus Corporation)

(Courtesy of the Houston, Texas, Fire Department)

10 Fire Ground Operations

The test, of course, of the principles and procedures discussed in chapter 9 is to apply them on the fire ground. Fire crews must be so well trained and drilled, and so familiar with the recommended practices and procedures that there will be an automatic response to the fire scene situation and safe and practical tactics quickly applied. The following sections discuss the more common problems met on the fire ground and the recommended solutions.

Maneuvering Fire Apparatus

The importance of maneuvering fire apparatus into the most favorable position for fighting the fire cannot be emphasized too strongly. It requires good judgment, the ability to make quick decisions, and skillful handling of the apparatus.

Approaching the Fire Ground

With the advent and universal use of two-way radio communications, there is no longer any reason for other than the bare minimum of companies to proceed directly to the address given, unless there is a definite fire showing and the placement of companies is either very apparent or has been previously determined at prefire planning sessions. Deviations are to be expected when the standard operating procedures of the department require one of the first-arriving pumpers to connect to the sprinkler or standpipe inlets, the second-arriving companies to proceed directly to the rear of the building, or some other definite prearranged assignment.

First-arriving companies should proceed directly to the address and park just past the front of the building. Before leaving the apparatus, the officer should notify the dispatcher of the conditions at the time of arrival; this will also inform the rest of the first-alarm assignment that a unit is on the scene. The message should include:

1. The address. This will allow the dispatcher to confirm that the correct location was responded to.

2. The approximate size, height, and occupancy of the structure.

3. The status of the situation.

4. If there is any doubt that the responding companies can handle the situation, request additional help. Fire records confirm that one of the most prevalent causes of large-loss fires is a long delay occurring before adequate apparatus and personnel were requested.

5. Instructions to the rest of the first-alarm companies.

Typical initial messages might be, "Engine 1 to Fire Dispatch. At 146 South Main Street there is nothing showing in a one-block square, two-story warehouse. We will investigate further. Engine 2, check the east side of this structure," or "Engine 2 to Fire Dispatch. At 1401 York Boulevard we have a large four-story hotel with the top floor well involved. Send two more engines and two trucks. Truck 1 open up the roof. Engine 1, pump into the standpipes." Standard operating procedure should be adopted that states what information is needed and the correct sequence for transmitting the known facts so that complete messages will be conveyed during the stress and excitement of major fires.

Later arriving companies should hold back and not proceed past the last intersection until the first-in officer has sized up the situation; the officer will then have an opportunity to place the companies in the most advantageous positions. If all companies responding on the first alarm cram into the street where the fire was reported, they could readily cause such a traffic jam that their full potential could not be utilized. It is no rarity for the officer in charge to discover that dispatch of additional companies to a more strategic location must be requested because the initial assignment became entangled in a cul-de-sac or blocked street and the reassignment of companies already on the fire ground would be delayed or impossible. The same predicament can be caused when companies pile into a narrow street from both directions; this practice could make it impossible for engine companies to stretch hose lines from the desired hydrant or for aerial ladders, elevating platforms, or water towers to be correctly positioned so the full potential of the apparatus can be realized.

Often a later arriving company approaching the fire scene from a different direction can detect an indication of fire or smoke that is not apparent to the officer in charge. Personnel should immediately inform the commanding officer of any observation or information that could affect the size-up.

There is no reason for officers to report to the command post for verbal orders. Companies responding on a greater alarm should radio for an assignment when approaching the fire ground. The message should state the approximate time of arrival and from which direction the unit is traveling. This will expedite assigning the company to the most advantageous position and will avoid the necessity of driving into the fire ground and then having to turn around to fulfill an assignment.

Positioning Apparatus

When rolling in on a fire, extreme care when positioning the apparatus is important. Attention must be devoted to the wind direction, ground slope, terrain, and other possible hazards or complicating factors. Placing the apparatus on the windward side of the fire will relieve many of the heat and smoke problems that would be encountered in a leeward advance where the breeze would be blowing the fire directly towards the fire fighters, and permit a closer approach to the fire with less punishment to the crew. If a windward approach is not practical, then launch an attack from the side of the fire that would offer the most protection. Ground slope is an important consideration when approaching a flammable liquid fire; be careful that crew and vehicles will not become engulfed if a tank or pipeline should suddenly rupture.

Remember that each type of apparatus has an essential purpose. Engine companies are staffed and maintained to lay hose lines and pump water. Aerial ladder trucks are constructed and equipped to ventilate heat, smoke, and toxic gases from the burning structure, furnish access to buildings, and provide aerial fire streams from ladder pipes. Elevating platforms and water towers must be spotted in the correct positions so they may be best utilized. This means that every driver must consider the fire fighting effort as a whole. Each apparatus should be positioned so it can be operated correctly, yet leave room for other vehicles to maneuver.

The first-in engine should stop just beyond the fire building; this gives the officer a chance to view three sides of the structure before getting out of the cab and away from the radio. This also gives a 75% chance of spotting the fire if it is visible at all from outside of the structure as viewed from the front and both sides. Before leaving the cab, the officer should radio a preliminary size-up. Spotting the pumper past the fire allows the hose lines to be pulled from the hose bed without binding on the sides; it is difficult to smoothly pull hose from the rear of an apparatus and lead into a building that is forward of the vehicle. Some departments get around this problem for small lines by using a traverse bed just forward of the large hose bed. Placing a pumper immediately in front of the fire building could take space that an aerial ladder or an elevating platform could use to better advantage.

Even hooking up to a hydrant may create a traffic problem. On narrow streets, if the pumper is angled into the hydrant too loosely, the rear end may project into the roadway enough to prevent the passage of other apparatus. This is something to be watched, especially when a front suction connection is used, because the tendency is to leave the rear of the pumper in the middle of the street.

Running Over Hose with Apparatus

When absolutely necessary at fires, apparatus may drive over fire hose when it is charged, and power is not applied to the driving wheels while they are in contact with the hose. However, severe damage to the hose is probable if it is lying empty or under low pressure (Figure 10-1). This damage may not be apparent, but examination by opening up the hose may show that its lining has separated from the jacket, or may even be torn apart. A vehicle is propelled forward by means of friction between the driving wheels and the street pavement. When the wheels reach the flattened line of hose, they are exerting a powerful force forward; immediately upon riding on the hose, all of the power required to push the apparatus forward is applied through the hose to the ground. Thus, the wheel braces itself on the edge of the hose as it begins to cross over it. This forces the lining into one edge of the hose, while the jacket is acted upon in such a manner as to peel it free from the lining. Avoid driving over or placing any pressure on the couplings.

Under any circumstance, driving over fire hose by any vehicle should be avoided if at all possible. Hose bridges should be carried on every piece of fire apparatus and should be utilized whenever any vehicles must cross fire hose. If available, build up an approach on both sides of the hose with lumber. On unpaved roads, it may be practicable to dig a ditch and bury the hose line.

Rural Areas

Unless they share a regional emergency radio network, companies from different departments responding to a mutual-aid alarm should receive their fire fighting assignments at the time they are dispatched. This forethought will eliminate the needless maneuvering of equipment and general traffic confusion on the fire ground that would become necessary if the responding companies had to obtain verbal instructions at the command post.

In many volunteer departments that use home alerting radio or telephone systems,

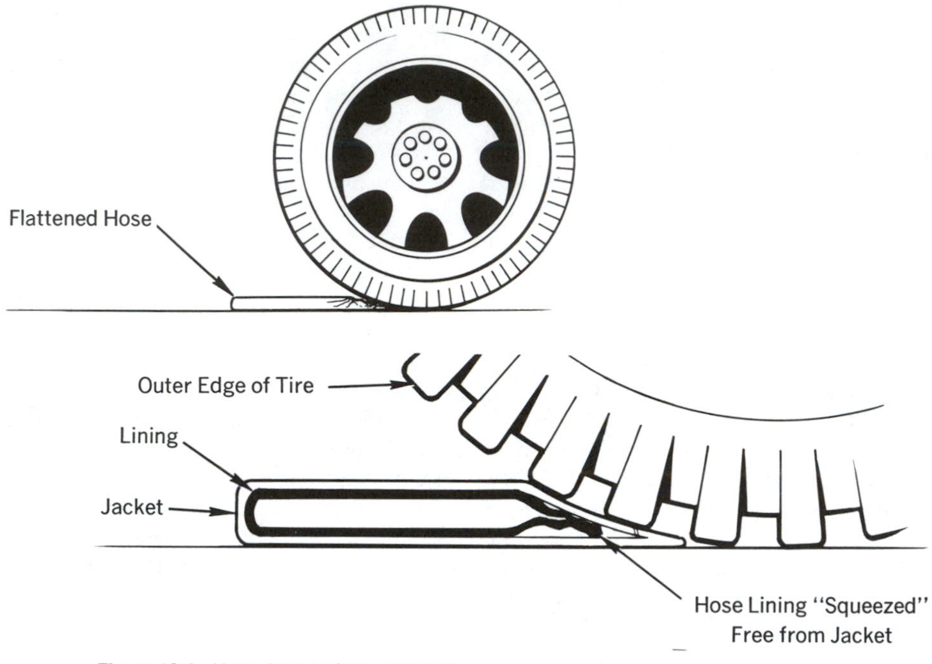

Figure 10-1. Hose damage from apparatus

personnel living closer to the alarm location than the fire station will drive directly to the fire; they may arrive before the engine company. It is mandatory that these fire crews be constantly aware that access and maneuvering room must be left for all responding apparatus. In areas without sidewalks and curbs, automobiles should be parked off the pavement. If terrain, mud, or snow prevents parking cars off the road, they should be left in yards or driveways away from the fire area.

When the first-in engine company sees no flame or smoke at the reported address, and therefore does not lay a line en route, the second engine may back down a short dead-end road or a long driveway so it can lay a hose line back to the nearest water source. On a long dead-end road with the fire beyond the last hydrant, the second engine should stop at the hydrant or other water source and be prepared to lay a line to the fire if it should be required. If the second pumper proceeded to the fire before the need for water became apparent, it should turn around and lay a line from fire to hydrant. On long country roads, the maneuverability of later arriving apparatus can be maintained by backing into a driveway and awaiting orders.

Keep apparatus movement to a practical minimum; avoid hose lays that require maneuvering vehicles on narrow rural roads where there may be drainage ditches or soft shoulders. Keep in mind that on rural lanes and long driveways it is often impossible to turn around and difficult to back up. In such a case, it may be desirable to have the pumper attacking the fire lay its line from the street to the fire and the second engine connecting to the end of the hose and laying to the water source. This same hose lay is often utilized by metropolitan departments; some refer to it as the "alley lay" or the "split lay," because urban departments also commonly encounter dead-end roads, narrow streets, and blocked alleys.

Apparatus Hazards

It is natural for an officer to select the hydrant closest and most convenient to a fire. In one way, this would be good thinking, because the pump will be able to discharge a greater volume of water when the hose layout is kept short. In addition, the tools and equipment carried on the apparatus would be more accessible and could result in the radio being more available for communications. However, fire fighters should consider the possibility of the apparatus becoming threatened or damaged if the fire should increase in intensity (Figures 10-2 and 10-3). Another consideration is the danger of the vehicles being struck or buried by a collapsing structure.

Some authorities assert that the first arriving pumper should never connect up to a hydrant that is directly in front of or closely adjacent to a fire structure. From the standpoint of the obvious hazards involved, and the fact that a pumper may be positioned in the most strategic location necessary for the effective use of an aerial ladder, elevating platform, or water tower, this is undoubtedly correct. On the other hand, perhaps a rapid and massive blitz attack on the fire by the first arriving engine company would succeed in extinguishing the blaze with a minimum of damage; whereas taking the time to lay long hose lines from a more remote hydrant could allow time for the fire to accelerate in intensity sufficient to be beyond the ability of the first-alarm assignment to control. The decision about whether or not to connect the first arriving pumper to the most convenient hydrant will necessarily be made by the officer as a result of the initial size-up.

Hazards to pumpers connected to a hydrant are not restricted to the first arriving apparatus. There have been many occasions

Figure 10-2. Apparatus spotted near the fire may become threatened if the blaze should increase in intensity. (Courtesy of the San Francisco Fire Department)

Figure 10-3. A fire fighter was trapped by a fast-moving brush fire while attempting to protect the tanker with a hose and was cremated.

when apparatus working on a fire—often quite a while after the initial alarm—were suddenly overwhelmed, engulfed, or crushed by a fire or collapsing structure. Many injuries and deaths of fire fighters have resulted from these accidents. If a pumper is connected to a hydrant, and the fire starts to increase in intensity, or the building shows evidence of structural collapse, the officer in charge should be notified. If the hose lines are suddenly shut down when the apparatus is moved or abandoned, the fire crew on the hose lines may be seriously endangered. Keep alert and reposition the hose lines and apparatus at the first evidence of danger.

The same precautions should be observed when operating aerial ladders, elevating platforms, and water towers adjacent to a burning building. As with the pumpers, all repositioning of any type of apparatus and equipment should be coordinated with the officer in charge because it would affect the overall fire ground strategy and tactics.

Remember: Before laying the initial hose line, take time to properly conduct an initial size-up of the fire. The most convenient hydrant may actually be the most hazardous choice, or the closest hydrant could truly be the greatest strategic option.

Hydrant Hookup

Some members of fire departments feel that hard and fast rules for spotting and hooking up can be made and applied unbendingly in any situation. This is a misconception. Just as there is no universal hose lay to combat every fire, there is no such thing as a universal spot and hookup. Since conditions of every pumping job vary, it is imperative that pump operators remain flexible in their thinking and operational procedures. The engineer must size up the situation and take whatever action is necessary.

Connecting Pumper to Hydrant

It is of the utmost importance that the roadway be blocked as little as possible. With a small amount of extra effort the pump operator can spot the wheels of the apparatus close to the curb and still not kink the suction hose. The supply hose between the hydrant and the pump should be kept reasonably straight or in a gentle S curve, with no sharp bends or kinks to restrict the flow. Experienced engineers have discovered that when an S curve is necessary under restricted conditions, twisting a soft suction hose 180° before tightening the connection will normally prevent a kink after pressure is applied.

A pumper should be spotted sufficiently close to the hydrant that the suction hose does not form a bridge; if all of the strain is on the two couplings, an undue strain may cause the hose to pull loose from a coupling or the hose could burst. When the suction hose does not touch the ground, place some type of support under the center part of the hose; if the chock blocks are not being used to stabilize the apparatus, they may be utilized.

When large flows may be needed, and the hydrant is so equipped, the large (steamer) outlet should always be used; normally, a single 2½-in. outlet will not provide full capacity except to small pumps. If the hydrant does not have a large steamer connection, which is usually 4 to 6 in. in diameter, it may be necessary to connect to two or more 2½-in. outlets of a multiple hydrant to obtain an adequate supply. When laying a single 2½-in. supply line for a long soft suction lay or utilizing a four-way valve in the initial hose layout, connect to the 2½-in. hydrant outlet and save the large steamer outlet for a pumper hookup.

Pump operators, in making their size-up for the correct spot and hookup, should consider hydrant variables, capacity variables over which they have no control, undue blocking of streets, connection requirements of other pumpers, and a multitude of other factors.

Variables Affecting Spotting and Hookup

Hydrant variables. The mounting of hydrants and the angle of the hydrant outlet or outlets to the roadway will vary considerably with each installation. The location of the hydrant itself will also vary. Most hydrants are located close to the curb, while others are found some distance away; there may even be a fence or wall between the hydrant and the pumper. Hydrants installed at street corners will be found at various distances from the intersection. Each hydrant variable–type, mounting, and location–will affect the spot and hookup procedure to be employed. Identical hydrants in different locations may require different spotting and hookup procedures. For example, a front connection to a hydrant located near a corner may place the pump in a position that will hamper fire fighting operations by unduly blocking the intersection; at a different location, a front connection may be desired.

Capacity variables. Many of the capacity variables that determine the quantity of water available for fire fighting are beyond the control of the pump operator. These include the choice of hydrant, the capacity of the hydrant, the capacity of the pump, and the type and length of the hose lay between the pump and the fire. Pump operators, however, generally have control over the manner in which pump hookups are made. The importance of proper hookups should not be taken lightly. Bear in mind that the manner in which the pump is connected will often determine the quantity of water available at the pump when operating within the limitations of the hydrant and the pump.

From a position at the hydrant, the pump operator can rarely determine the extent of a fire or the quantity of water that may be required to combat it. Therefore, pump hookups should normally be made in a manner that will enable pumps to supply the maximum amount of water whenever possible. The following factors relative to a pump hookup should be considered in obtaining the maximum quantity of water.

The direct suction connection should be made to the largest hydrant outlet so as to reduce the possibility of the pump cavitating. Where a choice of hydrants is available, avoid those on small mains.

The 4-in. water main is of very little value for fire fighting. Except for short branch or looped lines which may be 6-in. pipe, an 8-in. main is the smallest that gives a satisfactory supply for a pumper. One standard 250-gpm hose stream will cause a pressure loss of approximately 28 psi per 1000 ft of 4-in. main.

There are two fundamentals for successful pumping from hydrants:

1. Try to connect the pump to a hydrant capable of giving the desired supply.
2. Connect to the hydrant with a hose large enough to handle the full capacity of the pump. If a small suction hose is used, additional supply lines may be needed.

If a large, soft suction hose is used, the operation is facilitated if one end is carried preconnected to a pump inlet. The same advantage may be obtained by using a "squirrel tail" hard suction, which is carried attached to the pump by a swivel connection. In either practice, there should be a valve in the suction connection so the pump can use water from the booster tank without filling the suction hose.

Because of their construction, with small waterways, and numerous bends, front and rear pump inlets will reduce the output of the pump. This is particularly true at low intake pressures. Kinks in soft suctions restrict the flow of water to the pump.

Size-up by pump operator. Every pump operator should become familiar with hydrant and capacity variables peculiar to the local district. Where unusual or exceptional conditions are found, there should be preplanning to determine the best methods of spotting and hookup.

Pump operators, to be efficient, should size up the situation at each emergency to determine the proper spot and hookup. A good size-up will enable pump operators to evaluate conditions at the hydrant and avoid confusion and delay in obtaining water.

Even in cities that have excellent water systems with closely spaced hydrants and large mains, there are occasions when water supply becomes a problem; this is seldom a fault of the water system. A common error is for the first arriving pumper to be spotted in such a manner that a later arriving pumper cannot utilize the unused connections of a multiple outlet hydrant. Engineers should spot their apparatus close to the curb and in such a manner that other companies can also connect to double and triple hydrants. If it is a dry barrel hydrant, place valves on the unused outlets before turning the hydrant on; this will allow later arriving companies to connect up without shutting the hydrant down.

Pump operators should not assume that the capacity of the water system has been reached merely because the pump suction gage reads zero. While it is true that such a reading indicates that the capacity of that particular hydrant outlet has been reached, a second outlet, in the case of double hydrants, will often supply sufficient water for a second pump to operate at capacity. For example, if a 1500-gpm pumper were to connect to the 2½-in. outlet of a double hydrant on a large main, there would be no way that a volume of water sufficient to allow this pumper to attain its maximum rated capacity could flow through the small 2½-in. opening. However, the full capacity of the hydrant may be utilized by connecting hose lines between an unused hydrant outlet and a gated pump suction inlet.

Remember: A vacuum indication on the pump suction gage only shows that the capacity of that particular hydrant outlet has been reached; the hydrant and mains may still be capable of supplying more hose lines.

Connection Procedures

Pump operators must develop the ability to properly spot their pumps so that a connection can be made to the pump inlet of their choice without kinking the soft suction. The procedure for hooking up must be both rapid and smooth to assure that adequate amounts of water will be quickly available for fire fighting. It is seldom an excusable situation when the nozzle operators must wait for water after they have stretched their hose lines into position; at the same time, engineers must avoid loading the hose lines too soon or they would become unmaneuverable.

When developing a procedure, consideration should be given to making suction hose connections first to the hydrant, unless the hose is carried preconnected to the pump suction inlet. This method makes it easier to reposition the pumper if it becomes necessary to correct the spot. Each pump operator must develop a procedure that is best suited to the requirements of the particular apparatus in use.

Four-Way Valve

A four-way valve, which also has various trade titles, is designed to provide a means of changing from hydrant pressure to pump pressure and vice versa without shutting off the flow of water in the hose line (Figures 10-4 and 10-5); thus, the flow of fire streams is not interrupted. With this device, a hydrant stream can be used on the fire until the pumper returns to the hydrant and the pressure can be increased. In other words, the four-way valve may be utilized whenever a line is laid to or from a hydrant and a pumper is not immediately available to pump into the hose line. Figure 10-4 shows one model of this device, which automatically changes over when pump pressure is applied. The products of other manufacturers may appear radically different, but the purposes and water flow will be similar. Four-way valves can also be used as relay valves. In this case, they would be inserted in a long single-line hose lay so that a pumper can later hook up to the valve and help to relay the water.

Front Gated Suction Inlet

Most modern pumpers are equipped with a front suction inlet to simplify hydrant hookup and drafting spots. Hydrant operations are additionally facilitated if a soft suction is carried preconnected. These inlets are valuable for the majority of fires. However, at that occasional blaze where the full capacity of the pumper is needed, piping and fittings of these inlets will cause an unavoidable amount of restriction and friction loss. The photograph in Figure 10-6 clearly illustrates the smaller size piping, elbows, and other fittings that will severely limit the maximum quantity of water that can be conveyed through this waterway.

Supplemental Suction Connections

When operating from a double hydrant and when the capacity of the outlet to which the pump is connected has been reached, it is generally possible to obtain additional water from the second outlet. Figure 10-7 illustrates the use of two 2½-in. lines for providing a supplemental suction that enables a pumper to use both outlets of a double hydrant. Supplemental suctions can also be made up of hard suctions (it will generally

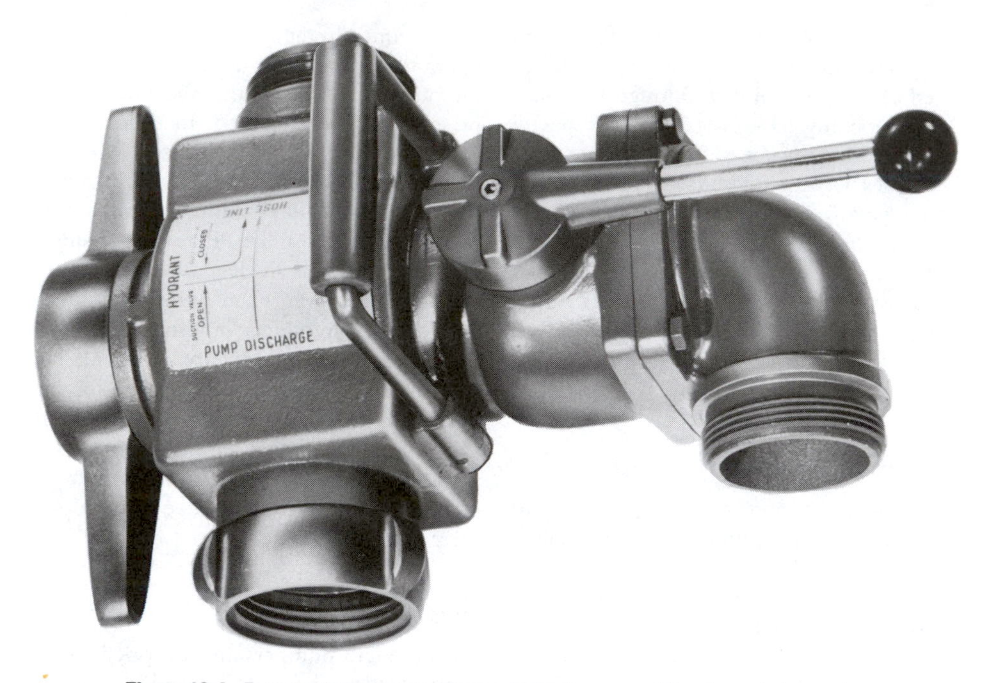

Figure 10-4. Four-way hydrant valve allows a hydrant-pressured fire stream to be used on the fire until the pumper can return and increase the pressure. (Courtesy of Akron Brass Company)

be necessary to obtain one or possibly two lengths from another pump in the area) or of single lines of 3½-in. or 2½-in. hose, provided the proper fittings are available. The additional water obtained by the use of a supplemental suction may make the difference between effective or ineffective hose streams.

The need for a supplemental suction is usually determined after hose lines are in operation and it is found that one hydrant outlet will not supply the required flow. Under these conditions, it is most desirable that supplemental suction connections be made without shutting down the pump or interrupting the flow from operating hose lines. This can generally be accomplished by first laying out the hose lines or hard suctions to be used and making the connection to the unused outlet of the double hydrant. After the supplemental suction is laid out and the connection to the hydrant completed, advance the throttle until the pump is drawing a vacuum, as indicated by the soft suction becoming soft or flabby and by the reading on the compound gage. A pump inlet cap can then be removed. Little or no water will issue from the uncapped pump inlet as long as the pump is operating at a vacuum. The supplemental suction can then be connected to the pump. After all connections are complete, open the second hydrant outlet and readjust the pump.

Indications of Maximum Volume From Hydrant Outlet

When the pump is connected to a hydrant, it is desirable to maintain at least 10 psi of pressure on the suction (compound) gage. However, this is not always possible under conditions in which a maximum hydrant flow is required. Within close limits on either side of a zero reading, compound gages are not very accurate.

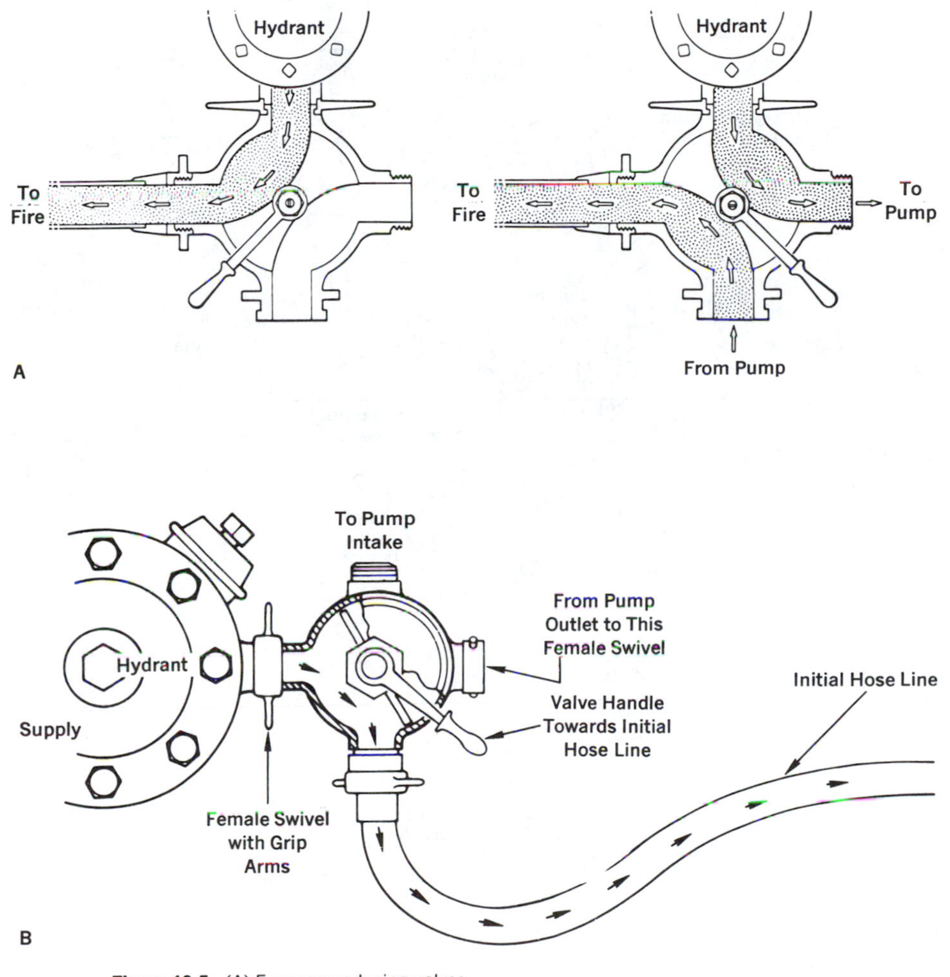

Figure 10-5. (A) Four-way reducing valves
(B) Phase I of a four-way valve

When using a soft suction, the point at which the maximum volume of water is being received from the hydrant outlet can be determined by feeling the suction hose where it enters the pump inlet. If the hose is soft and flabby when squeezed, it indicates that the maximum flow from the hydrant outlet has been reached and the pump is beginning to run away from the water. These same indications are true when engaged in a relay operation. Some experienced operators will maintain contact by keeping one leg

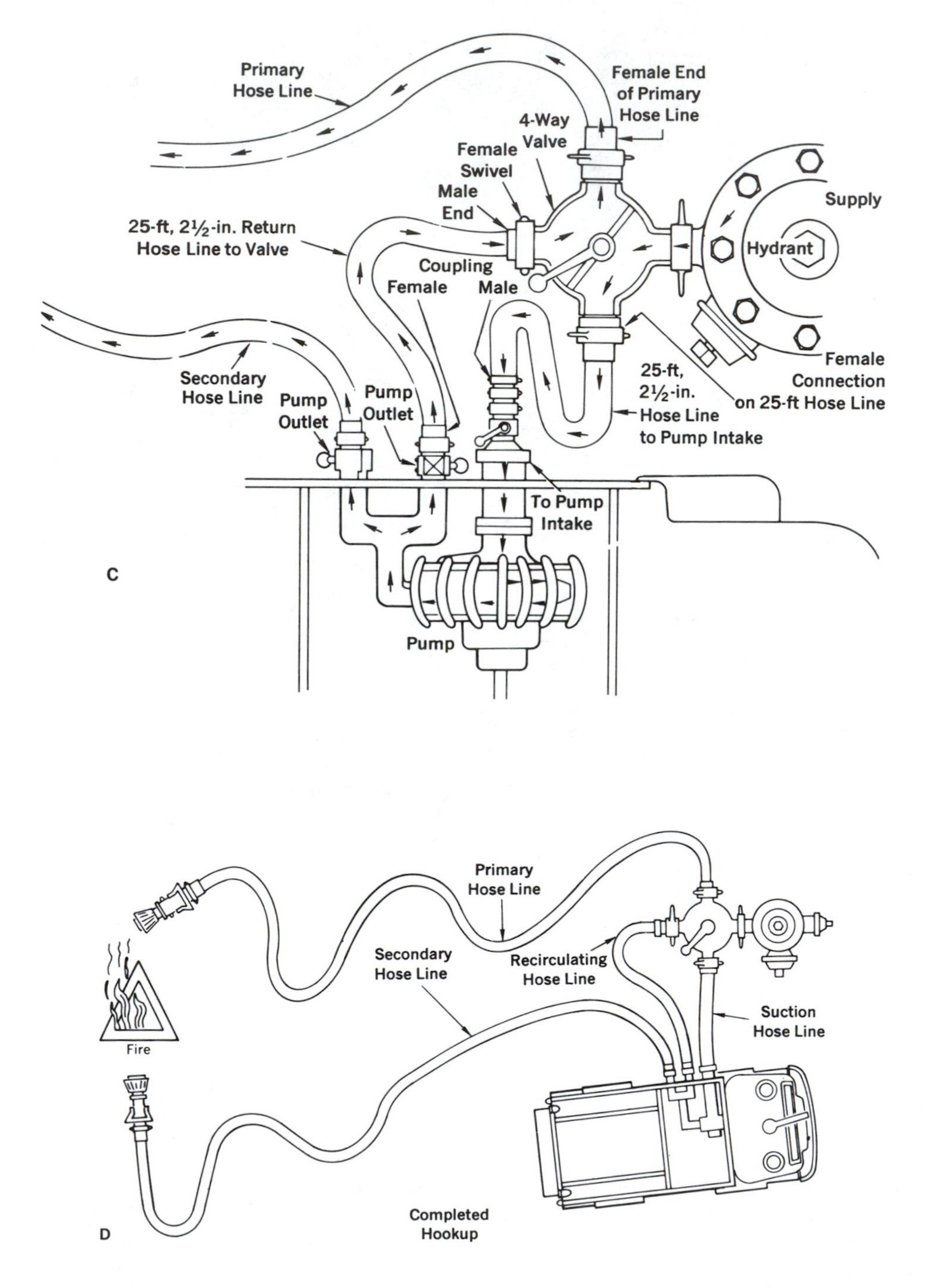

Figure 10-5. (C) Phase II of a four-way valve
(D) Phase III of a four-way valve—completed hookup

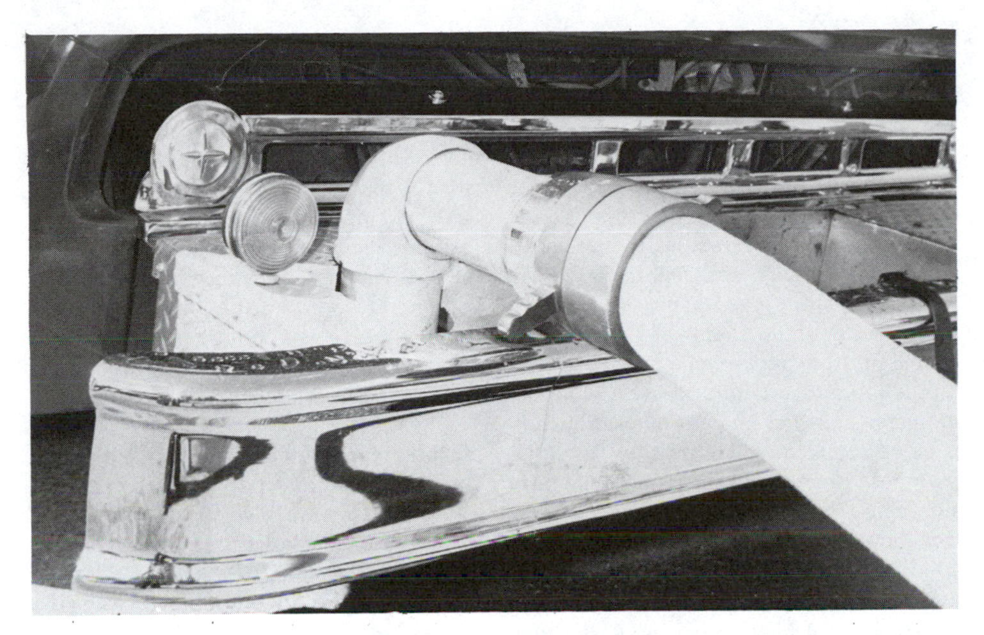

Figure 10-6. Front-gaged suction inlet

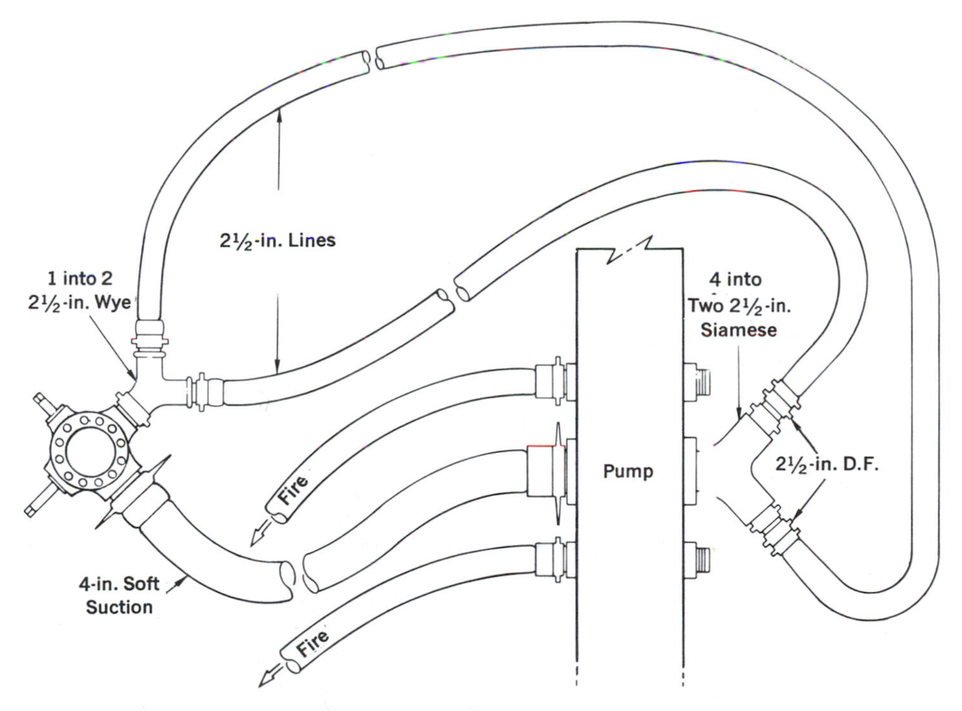

Figure 10-7. Supplemental suction connection

pressed against the suction hose; they will be instantly warned by a quivering of the hose when it is starting to collapse. As long as the line is hard or firm, the pump is getting enough water to meet the demands of the hose lines; when the hose starts to get soft, the volume available from the supply lines in use has been reached. If the engineer slows down the engine speed, the hose will regain its firmness; the situation can usually be corrected with a slight reduction in discharge pressure and volume.

It is futile to attempt to force a pump to discharge a greater amount of water than it is being supplied; it will probably result in poor hose streams and cavitation could possibly cause irreparable damage to the pump. Cavitation, as explained in Chapter 3, is a condition in which vapor cavities are formed in the pump and can cause serious damage. Cavitation can also occur during drafting operations if the engineer runs the pump faster than the suction hose can deliver water to the pump. A reliable way to determine the cavitation point of a pump under any suction condition is to open the throttle gradually, watching the engine tachometer and discharge gage. An increase in engine speed without a corresponding increase in pressure indicates cavitation (assuming the clutch is not slipping). Nothing is gained by running at a higher rpm.

Blocked Hydrants

Any hydrant so obstructed that the pump cannot be spotted close enough to make a hookup with a standard suction hose is considered a blocked hydrant (Figures 10-8 through 10-12). Blocked hydrants, while not encountered frequently, can cause considerable delay in completing the pump hookup.

Pump operators should determine if the obstruction can be quickly removed and a normal hookup made. In the event the obstruction cannot be quickly removed, some means of connecting the pump to the hydrant must be improvised. Most likely, every pumper has the right fittings and hose to complete the extended hookup; engineers should give some prior thought to these types of problems. Pump connections to blocked hydrants can be completed in most cases by utilizing hard suctions or hose lines.

The examples illustrated utilize wet barrel hydrants and four-way valves, but these suggestions may be applied to any type of equipment with variations. Engineers should preplan what actions they will take to overcome any hookup difficulties. Figures 10-8 and 10-9 illustrate possible uses of hard suctions in conjunction with a soft suction in making pump hookups to blocked single and double hydrants respectively. In both cases, a four-way valve is used to supply water while the connections are being completed. Figures 10-10 through 10-12 illustrate the possible uses of hose lines when the required reach is greater than the combined length of all the suctions carried. A single 3½-in. line or two 2½-in. lines will, in most cases, supply enough water for one and possibly two good working fire streams. Again, in all cases, a four-way valve has been utilized to supply water until the pump hookup has been completed.

Whenever connections are made to blocked hydrants, side suction inlets should be used, if at all possible, because of the increased friction losses developed in the extended suction employed. The exact manner in which pump connections are made to a blocked hydrant will depend on conditions at the hydrant, the type and amount of equipment carried on the pumper, and most important, the ingenuity of the pump operator.

Two Four-Way Valves on Double Hydrant

In a situation when two companies with triple combination apparatus lay from the same double hydrant and neither pump is immediately available, two four-way valves may be used. A wet barrel hydrant is illustrated in Figure 10-13, but the operations at

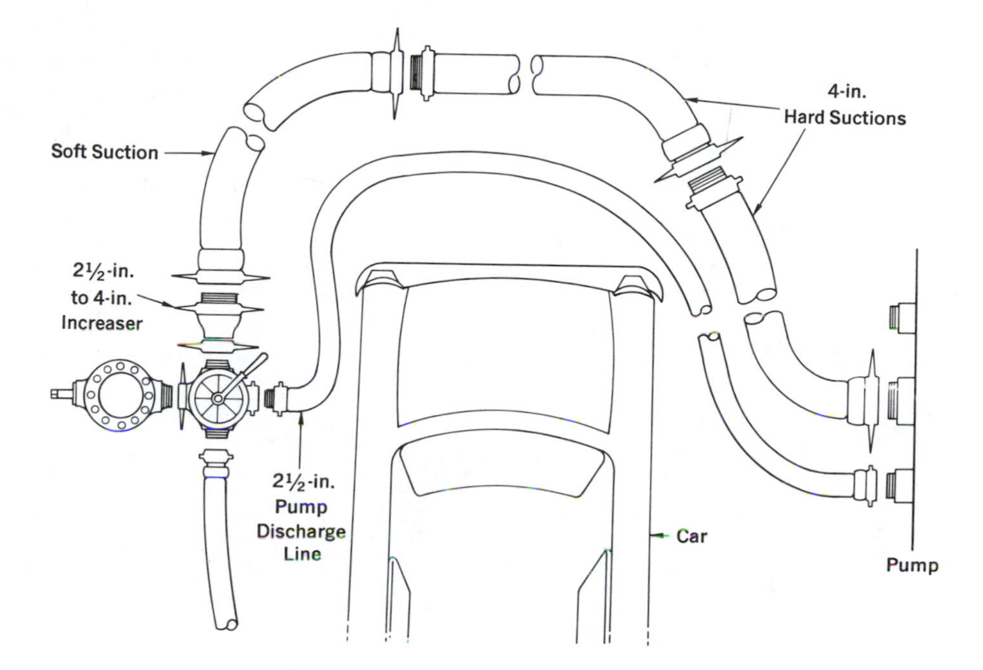

Figure 10-8. Single 2½-in. hydrant; block hydrant; pump connection using 4-in. suctions

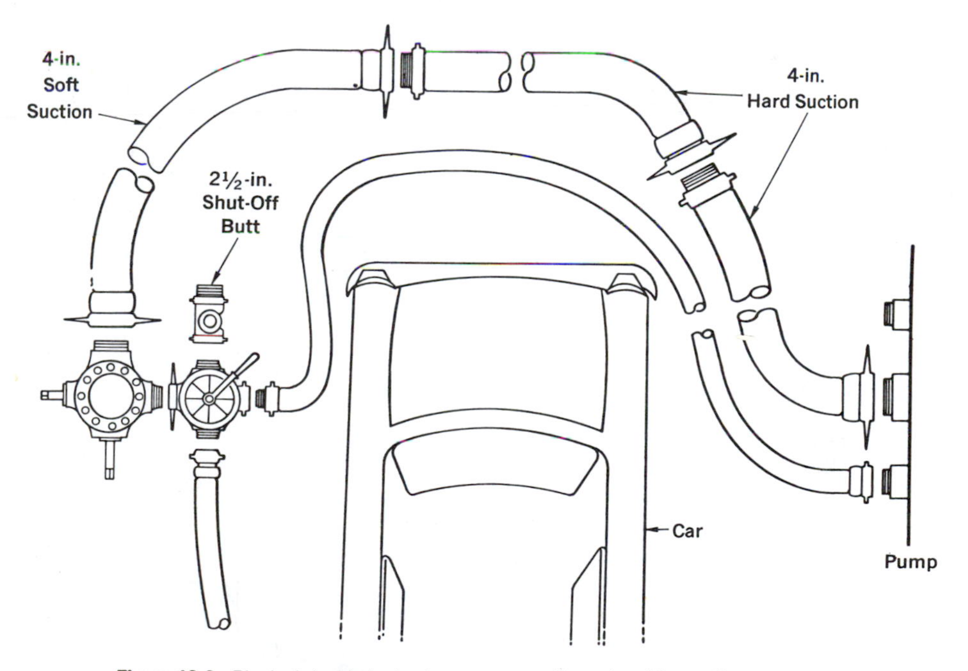

Figure 10-9. Blocked double hydrant; pump connection using 4-in. suctions

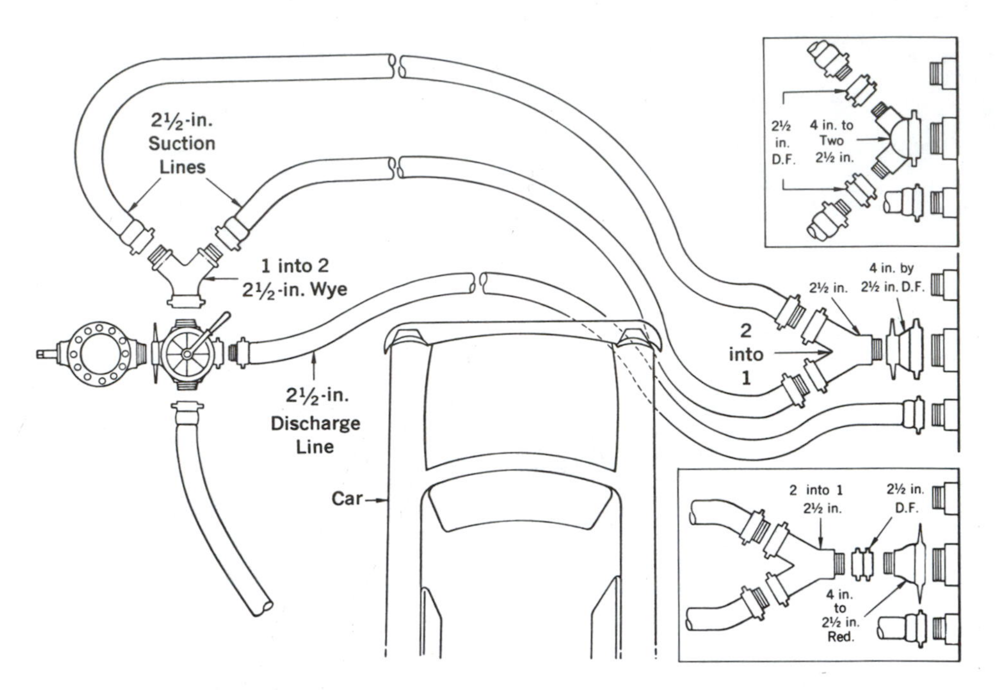

Figure 10-10. Blocked single hydrant; pump connection using two 2½-in. hose lines

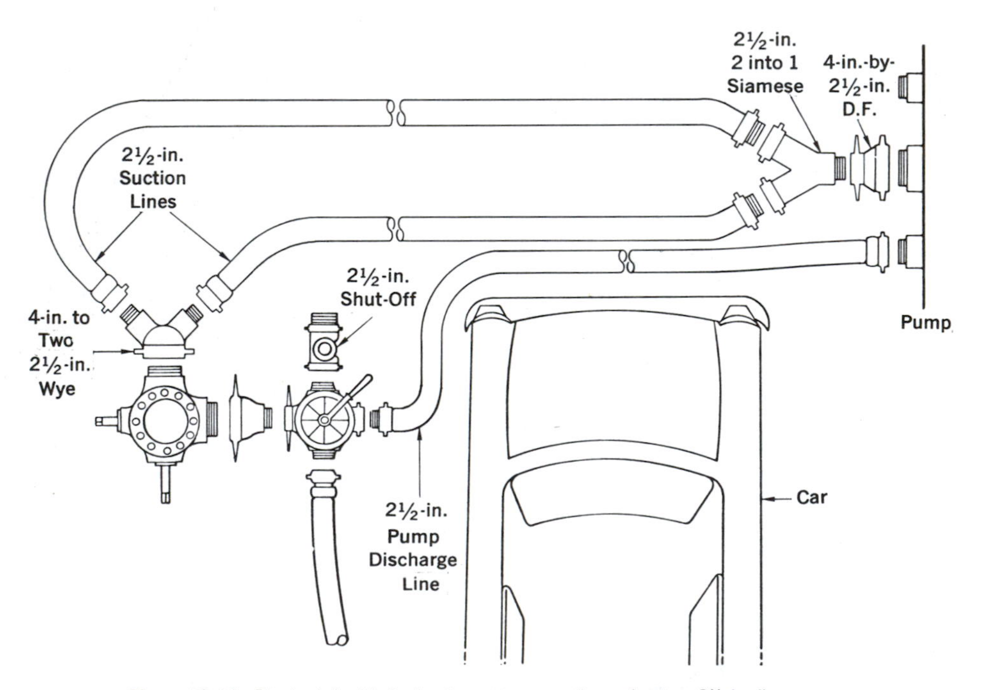

Figure 10-11. Blocked double hydrant; pump connection using two 2½-in. lines

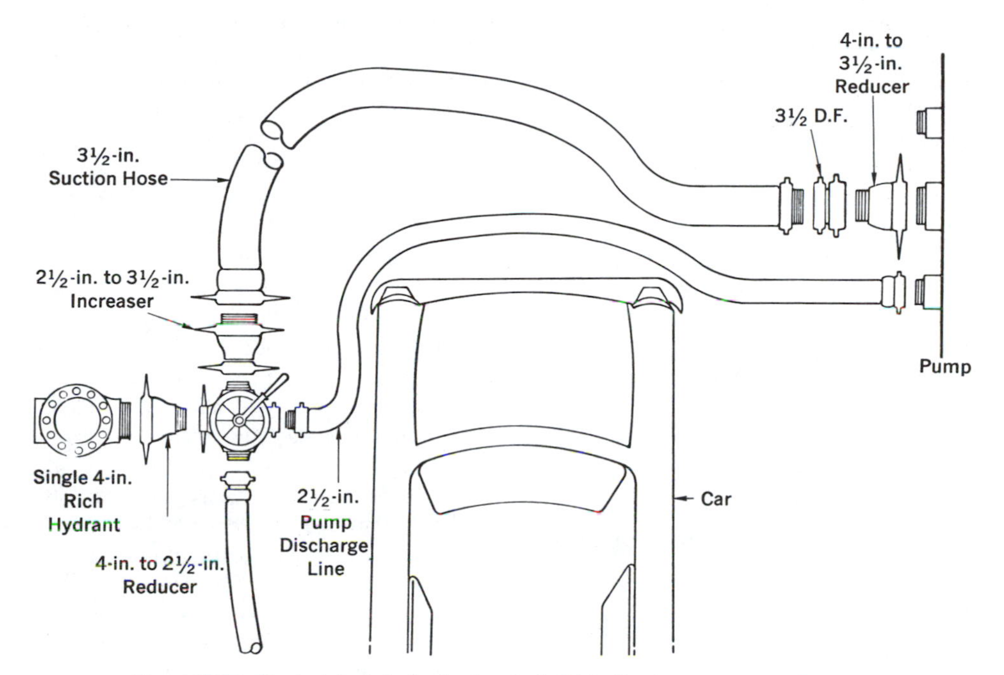

Figure 10-12. Blocked single hydrant using single 3½-in. line for pump connection

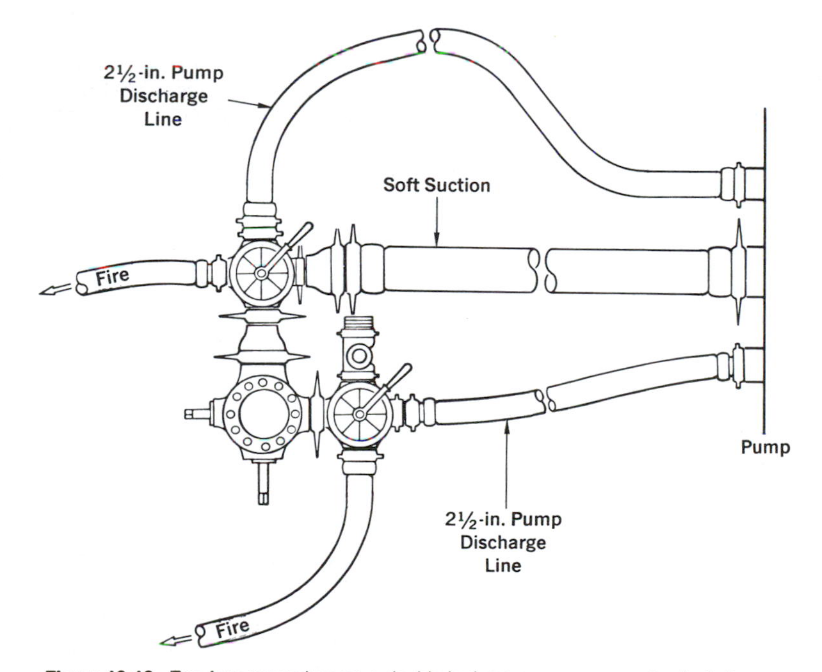

Figure 10-13. Two four-way valves on a double hydrant; one pump supplies both lines

a dry barrel hydrant would be similar. Each company should connect a four-way valve to the hydrant in the proper manner and supply water when its hose line is laid in position. The fact that two four-way valves are on the same hydrant does not change the responsibility for supplying water at the proper time. When the first pump returns to the hydrant, members from both companies should assist the pump operator in making the necessary connections. The illustration shows the required connections for one pump to provide pump pressure to both lines.

Soft Suction Hose Failure

Failure of a soft suction may be partial or complete. Where partial failure occurs, such as a small hole or leak developing at a coupling, continue normal operation as long as possible. In the meantime, make all possible preparations to reconnect the pump to the hydrant should complete failure occur. Sometimes complete failure can be prevented by reducing the pressure of the hose as much as possible; the pump intake pressure can be reduced to near the zero mark by increasing the pump discharge pressure or by partly closing the hydrant.

When complete failure of a soft suction suddenly occurs, the first consideration of the engineer is to get water on the fire again as soon as possible. If a four-way valve is on the hydrant, water can be immediately supplied by switching the valve back to hydrant operation until a reconnection to the hydrant is completed. Otherwise, shut down the hydrant and replace the soft suction hose as quickly as possible. Any disruptions of the water supply must be immediately reported to the officer in charge so the nozzle operators can be informed.

Tandem Pumping

Occasionally there is more water available in a hydrant to which additional pumpers cannot spot and connect. In these cases, tandem (dual) pumping will make more water available without requiring the laying of longer hose lines to remote hydrants.

This procedure requires the second pumper to connect to the suction inlet of the first pump at the hydrant without shutting down operation of the first pumper. If there is a gated suction inlet available that would permit connecting without affecting the supply of water to the pump, it should be used. However, if such a gate is not available, the following procedure will allow removal of the cap on the intake without shutting down operation of the pump.

Since the success of this procedure is based on the principle that no change will be made in the discharge quantity from the pump at the hydrant, connections must be completed as rapidly as possible. Shutting down or changing a nozzle setting will affect the success of this operation. The procedure is as follows:

1. The first pump operator should partially close down the hydrant or intake gate. At the same time advance the throttle, if necessary, to draw the suction pressure down to zero.

2. Check the pressure by trying the pump inlet cap to determine whether it turns easily before removing it. Should the cap become hard to turn while unscrewing, stop and check for the reason and correct the condition before continuing. Shutting down the lines or lowering the discharge pressure will cause an increase of intake pressure and cause the cap to tighten.

3. Remove the cap and quickly make the necessary connections.

4. Open the hydrant or suction valve; the operator will make the necessary adjustments of pressure.

5. The second pump will operate exactly as if it were connected to a hydrant.

Tanker Water Supply

Multicompany relay operations can effectively deliver water over a considerable distance for a long period of time, but they involve committing apparatus and laying long hose lines that may not be available in the early stages of a fire. Unless large quantities of water are required, water is best moved in a tank truck when the distance between the fire and the water source exceeds 2000 to 2500 ft. A trained crew with a properly designed tanker should be capable of sustaining a flow of approximately 100 gpm when the water source is 1 mile from the fire. Where additional tankers or larger capacity nurse tankers are available, more distance may be covered or the amount of water transported may be substantially increased. Upon arrival of the first unit at a rural fire, additional tank trucks should be requested if there is a possibility that they will be required. Every effort should be made to have an excess of water available, rather than to wait until the need is apparent and, as a result, run out of water before additional apparatus arrives.

There are three general methods of utilizing tanker water: (1) pumping water directly onto the fire, (2) pumping water into the suction inlet of a relay pumper or the engine supplying the hand lines, and (3) dumping the water into a portable tank with a pumper drafting the water as needed (Figure 10-14). For long, sustained operations, a portable tank offers the most constant supply for rural fire fighting activities; the tanker can dump its load and return to the source with the least delay. The efficiency of this operation is enhanced considerably if the tanker has a large rear dump valve to expedite unloading the water.

Figure 10-14. At rural fires, a pumper can draft from a portable tank supplied by tankers or hose lines.

Obtaining Water With an Eductor

An eductor, also known as a syphon or a syphon ejector, will permit water to be picked up from streams, lakes, swimming pools, and other static sources where the distance from, or the height above, the water makes it impracticable or impossible to draft with hard suction hoses (Figures 10-15 and 10-16). In the example in Figure 10-15, a single 2½-in. line is supplying the eductor from a discharge outlet and two 2½-in. hoses carry the water supply to the pump suction inlet. An additional benefit is that an eductor pushes the water up, instead of lifting it by vacuum, so the normal limits of drafting heights do not apply.

When using an eductor, limit the lengths of supply and discharge hose lines as much as possible. Both back pressure and friction losses have a great effect on the efficiency of these devices. Where the hose layout will be long, it will be necessary to increase the hose diameters or use siamesed lines. Plan ahead; in the initial layout, provide sufficient hose to allow later movement of the eductor to a deeper location if that should become necessary.

The eductor may be utilized to refill a booster tank, or to supply a fire stream of the same hose diameter as the supply line. Following are the procedures to operate a 1½-in. eductor; other sizes and makes will operate identically or with minor modifications.

1. Spot the apparatus as close to the water source as possible.

2. Connect a 1½-in. line between the pump discharge outlet and the eductor inlet; open the discharge valve.

3. Connect a 2½-in. line between the eductor discharge and the pump suction

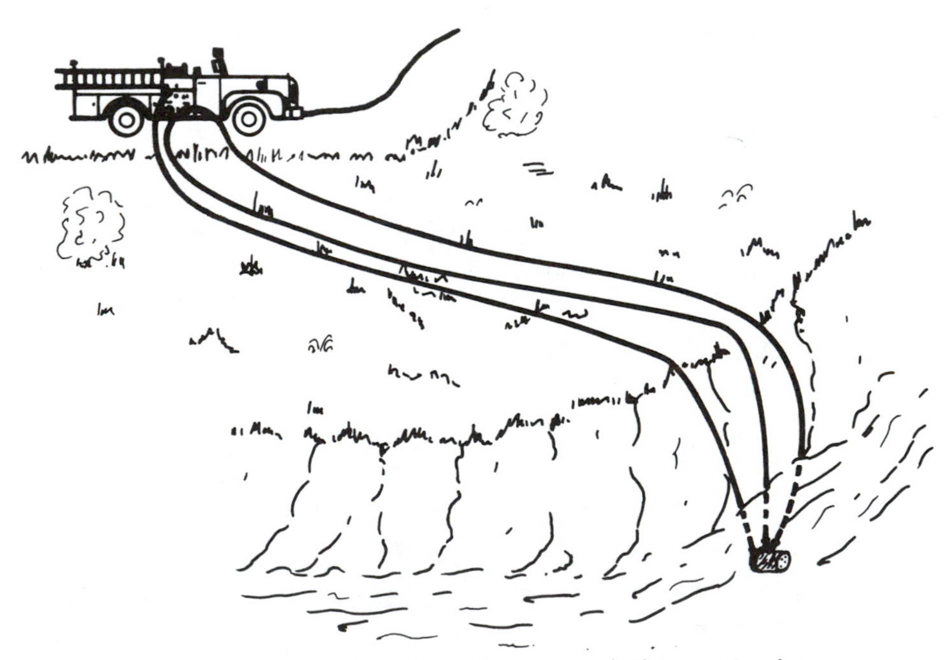

Figure 10-15. An eductor can obtain water from a source that is too remote or from which the lift is excessively high for drafting.

inlet. If more than two sections of hose are required, it will probably be necessary to use a 3½-in. or two 2½-in. hose lines to obtain an adequate flow. Friction loss has an overpowering adverse effect on an eductor's efficiency.

4. Place the eductor in the water so the strainer is below the surface. Prevent the strainer from resting on the bottom if mud or sand may be picked up.

5. Water from the booster tank will be needed to initially prime the lines and eductor. Open the tank dump valve so the pump, supply line, eductor, and return line will fill; if the eductor is so equipped, the foot valve will prevent the escape of water. The dump valve may be closed after the eductor is in operation, or it may be left open to supply some additional water for fire fighting. Until the tank is empty, this would prevent the pump from running away from the water; however, it would be a gamble because no water would remain in the tank to reprime the eductor and lines if the prime were lost.

If the eductor does not have a foot valve, start pumping as soon as the dump valve and the discharge gate valve are opened. Otherwise, water from the tank will escape from the strainer. Close the dump valve as soon as pressure is indicated on the pump discharge gage. Refill the booster tank at once in case the prime is lost.

6. If the water is to be used for refilling the tank, open the fill valve or place a line from a discharge gate into the tank. If the tank is full and it is desired to lay a line to the fire or another apparatus, operate a line directly from a discharge gate. A 1½-in. eductor should adequately supply a 1½-in. line working on a fire. Theoretically, about one-half of the water from the eductor is required to supply the needs of the eductor and one-half can be used to combat a fire.

7. Place the pump in gear and increase the engine speed. The eductor is more efficient at the higher engine pressures.

8. To keep the pump from running away from the water, watch the suction gage. Do not allow the gage pressure to drop below zero or the eductor will cease to operate. If there is no suction gage, feel the 2½-in. line into the pump inlet; do not allow it to collapse. To prevent cavitation, control the amount of water being discharged into the

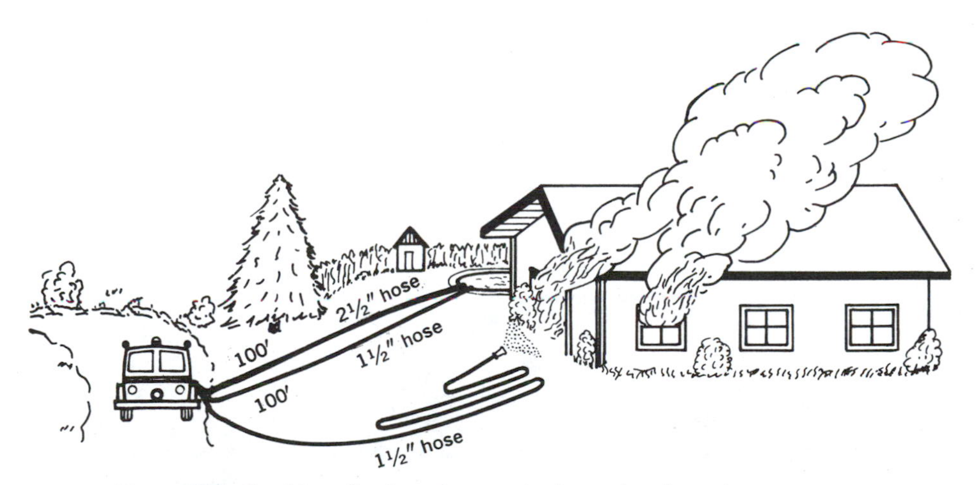

Figure 10-16. Supplying a fire stream from a swimming pool or other water source when a hydrant is not available is made possible with a 1½-in. eductor.

tank or out of the nozzle by partly closing the tank fill valve or the pump discharge gate. The friction loss in the return line will have the same effect on reducing the flow of the eductor as discharge head loss (back pressure).

To obtain the maximum flow out of an eductor, an experienced engineer will constantly monitor the suction gage and flaccidity of the line into the pump inlet while adjusting the flow to the nozzle or tank to maintain a slight positive pressure on the suction gage. The suction gage reading will reflect a change every time the nozzle flow is increased or decreased.

Dewatering with an Eductor

An eductor is very effective for removing water from a basement, boat, ship, or other flooded area (Figure 10-17). Even if it were possible to draft the water with hard suction hoses, debris and other contaminations could clog or otherwise damage the internal parts of the fire pump. Utilize the applicable recommendations given in preceding information on obtaining water with an eductor. Following are procedures to operate a 2½-in. eductor; other makes and sizes will function similarly.

1. Spot the pumper at a hydrant or in a position to draft. Place the apparatus so the hose lines into the flooded area are as short as possible. The length of the discharge hose lines are especially critical to the operation.

2. Connect a 2½-in. line between the pump discharge outlet and the eductor inlet; open the discharge valve.

3. Connect two 2½-in. or 3-in. lines, or one 3½-in. hose line, between the eductor discharge and the closest place the waste water can be dumped. Place the open hose butt in the gutter, over the ship's side, or other location where the waste water will not create additional problems. It may be possible to discharge the waste water into a drain opening inside the structure. If more than two sections of hose are required, it will probably be necessary to increase the number of lines or the hose diameters to obtain an adequate flow.

4. Use a barrel or other rounded object where the hose passes over the ship's railing, fences, or other places where the discharge hose may be kinked or compressed.

5. If the elevation between the eductor and the discharge point is over 30 ft., head loss can be reduced by lowering the discharge end of the hose to the lowest possible point so a syphon effect aids the flow of waste water. This can be accomplished by lowering the discharge end of the hose over the side of the ship until the coupling is just above the water, by lowering the discharge coupling down a storm drain, or any other way that the overall difference of elevation

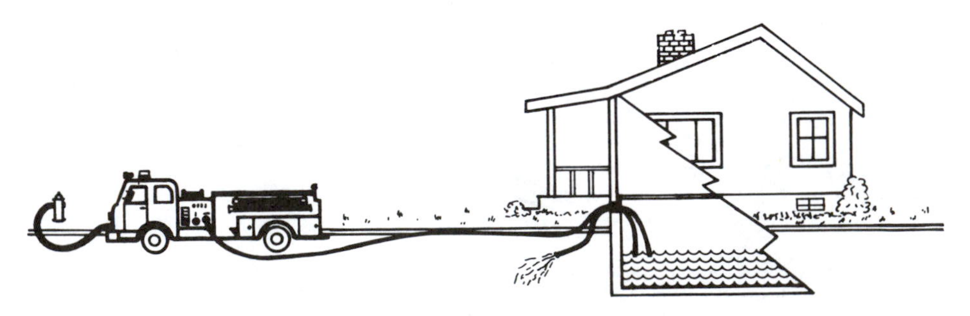

Figure 10-17. Dewatering a basement with a 2½-in. educator, the discharge hose should be a minimum of two 2½-in. or one 3½-in. line.

between the eductor and the waste water discharge point can be reduced.

6. Place the eductor in the water; keep the intake end well below the surface. This will prevent drawing air and floating debris into the line. Avoid placing the eductor on the bottom until near the finish of the operations to avoid clogging and to maintain a clear intake port. If the strainer has a tendency to clog frequently, and the water is being discharged from an open hose butt, remove the eductor strainer. With the strainer removed, small debris will be picked up readily and carried out with the waste water and will cause no trouble. A large wire basket, perforated garbage can, or other screen can be employed to maintain the eductor free of the larger pieces of foreign matter.

7. Connect the pump suction to a hydrant or draft from a static water supply. Place the pump in gear and increase the engine pressure; eductors operate most efficiently at the higher pressures.

8. A flow of water in the direction of the eductor is an excellent indication of efficient operation. If the eductor becomes clogged, and it is not equipped with a foot valve, it is not necessary to remove the device from the water to clear the blockage; merely shut down the supply line and permit the head of water in the supply and discharge lines to backflow and clear the strainer or eductor. If the eductor is equipped with a foot valve, it will be necessary to clear the blockages manually.

Initial Attack

The objective of every engine company responding to a reported fire should be to operate in such a manner that they are prepared to mount a full scale attack on a fire in progress or, if necessary, to establish an effective defensive operation. When making the initial size-up, it is a good idea to slow down the apparatus or momentarily come to a full stop so the first-in officer has an opportunity to completely gather the facts. It is indefensible for the company to enter the vicinity of the fire ground at such a high speed that the apparatus will have to back up or engage in a lengthy maneuver to lay a line from the closest hydrant or to determine the most strategic location for the initial attack. If there is the slightest doubt that the first-alarm assignment of companies can control the fire, additional units should be requested immediately. Whether it is stretching hose lines or calling for additional companies, the strategic approach is to do immediately what will have to be done eventually.

Modern pumpers are designed and arranged for quick action so they can deliver their full capacity at once. The only limiting factors should be the duration of the booster tank supply and the speed of obtaining sufficient water from a hydrant or other source. It is the duty of the officer and the crew to put this extinguishing force to work quickly and effectively; fire fighting efficiency and performance depend upon how the company, and the entire fire department, have developed procedures and trained for these operations. The possibility of developing new and more efficient techniques offers a continual challenge to a progressive fire department.

Because most fires are minor, every company must be constantly alert for the big fire that requires maximum effort by every fire fighter and the optimum use of apparatus and equipment. The first-in officer sets the operational tempo, and must be quick to recognize a working fire and order initial operations that will be the foundation for the eventual size of the attack. Some departments place the odds more in their favor by adopting the standard operating procedure that at least one supply line should be stretched from the nearest source of water to every structure that has a reported fire. If the fire is extinguished with a few gallons of water from a booster tank, or there is no fire upon arrival, the line is never charged. In areas where hydrants are closely spaced, only a few lengths of 2½-in. or larger hose will have to be picked up. However, if a fire is discovered after the building is entered, effective fire streams can be applied on the blaze more rapidly.

Some authorities assert that engine companies should initially lay parallel hose lines when responding to any reported structure fire, whether or not there is any visible indication that a fire is actually in progress. This is very good advice up to a point, but these experts must never have responded to an incorrect address, because the resident reporting the fire gave his own address to the dispatcher and then directed the fire fighters to a neighboring location, or because an excited citizen was confused between north and south or east and west, or any others of a myriad of occasions where an actual fire was in progress, but at a different address. Laying an unnecessary hose line under these circumstances will delay the company and actually make them unavailable for fire fighting until the hose can be reloaded.

Standard operating procedures should make strong suggestions concerning the number and sizes of hose lines to be laid under various circumstances, but they should avoid definite orders, because too many variable factors must be considered when deciding about the hose layout. Experienced officers believe that it would be a serious faux pas for an engine company loaded with 2000 feet of 2½-in. hose to attempt stretching parallel lines between a hydrant and a burning structure that is 1200 ft away, regardless of what orders are contained in a standard operating procedure. Also, one 3½-in. or larger hose line may be sufficient. Therefore, whether the first engine company lays one or two lines will depend on the anticipated quantity of water necessary for fire fighting and the amount of hose carried on the apparatus.

Hose Layouts

Any number of different hose layouts can be devised by an engine company, depending to some extent on the type of apparatus and how it is equipped. In Figure 10-18, a triple combination pumper is equipped with hose reels and cross lay beds. The booster reel carries 200 ft of 1-in. hose; each of the large reels carries 600 ft of 4-in. hose. The 2½-in. hose in the traverse hose beds is preconnected with the nozzle attached. This hose layout arrangement allows a single engine company to make a massive attack on a fire with a minimum of personnel.

The hose reel arrangement on the Wheeling, Illinois, Fire Department pumper shown in Figure 10-19 allows the fire crew to lead in with a multiple-outlet manifold and 4-in. supply line from the pumper. This tactic places the equivalent of a multiple-outlet fire hydrant directly adjacent to the fire, which reduces the necessary quantity of supply lines and decreases the friction losses.

On the fire ground, many factors influence the ultimate decision of which layout would be best for a specific incident. Of the many options, select the one that is the most practical and possible to attain. There could easily be two or more equally desirable methods of accomplishing a task; it is the duty of the company officer to size up the fire, analyze the factors, and decide on the appropriate strategy and tactics. To facilitate these fire ground operations, hose layout standard operating procedures should be formulated; these would be based on the apparatus, hose sizes and amounts, types of fires usually encountered, water supply system, and a myriad of other facts and probabilities.

When discussing hose layouts for supply lines, in the interest of brevity this book uses the terms *one line* and *two lines.* However, it is to be understood that one 3½-in. or larger hose will supply more water at less friction loss than two 2½-in. lines; a single 3½-in. hose will supply the same volume of water at essentially the same friction loss as a parallel layout of one 2½-in. and one 3-in. line; and a single 4-in. or larger hose will supply

Figure 10-18. Triple combination pumper equipped with hose reels and cross-lay hose beds (Courtesy of Hannay & Son)

Figure 10-19. Fire crew of the Wheeling, Illinois, Fire Department leading in with a multiple-outlet manifold and 4-in. supply line from the pumper. (Courtesy of Hannay & Son)

substantially greater quantities of water than parallel lines of 2½- and 3-in. hose.

Also, remember that a second hose line of equal size and length will reduce the friction loss to approximately one-fourth of the original single line; a third hose line of equal size and length will reduce the friction loss to one-half of the friction loss in two lines.

Where large flows are required, some fire fighters have the mistaken belief that increasing the pump's discharge pressure can overcome deficiencies in the hose layout. It should be quite clear from the above comparisons that the important factor in moving large quantities of water is the size of the waterway, rather than the pump capacity or discharge pressure. Many fire departments are adopting larger hose for laying supply lines because it has long been recognized that the low frictional losses of large-diameter hose would allow the movements of large volumes of water over long distances. Of course, this is nothing new. Large-diameter hose has always had this capability, but conventional hose construction was a deterrent to its adoption because of excessive bulk and weight; 3½-in. hose was about the largest that could be handled.

The advent of truly lightweight, large-diameter, flexible hose allows a new fire fighting concept. Placing a light, portable, multioutlet manifold at the fire and connecting it to the nearest hydrant with a large-diameter hose moves the hydrant right up to the fire ground for all practical purposes. These low friction losses result because a relatively small increase in hose diameter produces a very large increase in cross-sectional area. Practically speaking, this means that a 5-in. hose has almost four times the area of one 2½-in. hose; when cross-sectional area is related to friction loss, for a 15.2 psi pressure drop, almost six times as much water can be moved through 100 ft of 5-in. hose as through a similar length of 2½-in. hose (Table 10-1).

Apparatus

Fire apparatus assigned to engine companies may radically differ between various cities, and even between stations within a city. These variations will mainly be dictated by the hazards to be protected, the fire department budget, personnel, and the water supply system. Fundamental engine

TABLE 10-1. FRICTION LOSS, psi PER 100 ft

FLOW, gpm	HOSE INSIDE DIAMETER, in.					
	2½	3	3½	4	5	6
200	10.1	3.9	1.8	0.82	0.28	0.12
250	15.2	5.9	2.7	1.4	0.47	0.18
300	21.2	8.2	3.7	1.9	0.66	0.24
400	36.2	14.1	6.3	3.3	1.1	0.42
500	55.0	21.2	9.5	5.0	1.7	0.62
700	—	39.5	17.7	9.6	3.3	1.2
800	—	50.5	22.7	12.4	4.3	1.7
1000	—	76.5	34.3	19.2	6.6	2.6
1100	—	91.5	41.0	23.0	7.9	3.0
1200	—	—	—	28.5	9.5	3.5
1500	—	—	—	41.5	14.5	5.7
2000	—	—	—	75.0	26.0	12.0

Courtesy of Snap-tite, Inc.

company procedures may be developed around a triple combination pumper, two triple combination pumpers, a hose wagon and a pumper, a tanker and a pumper, a high-pressure fog pumper, a quad, a quint, or any other type or combination of vehicles that carry hose and are equipped with a pump. Standard operating procedures should be developed, practiced, experimented with, improved upon, and adopted so the assigned apparatus can be utilized to its best advantage.

Hose Beds

Modern fire pumpers utilize divided hose beds which allow two separate hose lines to be laid simultaneously, or they may be connected together to make one continuous line. Also, a variety of hose diameters can be accommodated; this allows a great flexibility of engine company operations on the fire ground. Examples of hose beds are shown in Figures 10-20 and 10-21. The pumper in Figure 10-20 carries both 2½- and 3½-in. hose; the left-hand bed carries a wye assembly consisting of two 100-ft lines of 1½-in. hose. Note how the four-way valve, nozzles, and other fittings are readily accessible. Engine companies should load their hose in the beds in such a manner that their routine and most common lays are performed with the least confusion, fewest operations, and a minimum of fittings necessary. The principle objective of any hose lay is to place the required quantity of water on a fire at an effective nozzle pressure in a minimum length of time. Any operation or procedure that interferes with these objectives should be eliminated.

Figure 10-20. Split hose bed allows two hose lines to be laid simultaneously, or they may be connected together for a longer single-hose layout.

Figure 10-21. The 800 ft of 4-in. hose being carried in the right bed of this pumper will allow an adequate-capacity supply line to be laid when required.

The nozzles and fittings that may be required should be readily accessible in the necessary sizes and quantities immediately adjacent to the rear of the apparatus so that no time will be wasted in searching for and obtaining the necessary equipment. A compartment reserved just for fittings is a good idea because it will protect the equipment from abuse and damage.

Whether a hose lay is considered to be forward or reverse depends mainly on how the hose is loaded on the apparatus. When the last coupling in the hose bed is a female, the logical layout would be from hydrant to fire. This would be a forward lay because no fittings would be required. From fire to hydrant would be considered a reverse lay because a double male would be needed for the nozzle and a double female for the hydrant. This is the most common method of loading hose because it is not good practice to pass up the last hydrant without laying a line when smoke or fire is showing, or there is a good indication that a fire stream will be needed.

When the last coupling in the hose bed is a male, the logical way to lay a line would be from fire to hydrant because no fittings would be required. From hydrant to fire would be considered a reverse lay because a double male would be needed for the hydrant and a double male for the nozzle.

Companies that load one hose bed in one direction and the other in the opposite way, thus requiring fittings to connect the two lines together when it is necessary to lay one long continuous line, should think in terms of *hydrant to fire* and *fire to hydrant* instead of *forward* and *reverse.*

Some departments have adopted sexless, quick-connect hose couplings that are identical on both ends. The advantages of these couplings are many. There is no need to match male and female ends, there is no requirement for double male and double female adaptors, and they couple up quickly and simply. Some states have passed a law that requires all fire departments to standardize on the same hose threads. With such standardization, mutual aid between cities will allow concentrated attacks on large fires that are beyond the capability of any one of the individual cities or communities. A great number of conflagrations could have been controlled much sooner if standardized hose threads had made possible a coordinated effort by the surrounding fire departments.

Some companies carry the fittings already attached to the end couplings in the beds in case a reverse lay is required; often with a cap to protect the male hose thread. In fact, a few even use both a double male and a double female fitting to connect the two hose beds together. If the hose is loaded in the beds in a direction to accommodate the most common hose layouts, these extra fittings will complicate the operations and lengthen the time required to get water on the fires in the majority of cases. Utilizing both a double male and a double female to connect the two hose beds together is not very logical because it presupposes that each lay will be exactly the length of one bed of hose; this could lead to pulling off a few extra sections of hose to reach the necessary fitting, or they both may be wasted in the middle of a long hose lay. Some companies routinely run with a double male attached to one nozzle and a double female on one of the pumper discharge gates; thus, they are immediately ready for a reverse lay.

Large Siamese Fitting

A large siamese fitting on the main pump suction inlets (Figure 10-22) will allow the greatest quantity of water to enter the pump with the least amount of restriction or friction loss. The initial hose lines at any fire where the pumper may eventually be required to discharge large quantities of water should be connected into this inlet. However, these inlets are usually between 4 and 6 in. in diameter, so a direct hose connection is not possible. All pumpers should carry at least one large siamese fitting; one end will connect to the pump suction and the other end should be equipped with two or three female couplings so the hose lines can be attached. It is most convenient if the siamese inlets are equipped with gates (valves) or clapper check valves so additional hose lines may later be attached without shutting down the pumper; otherwise, place caps on the unused openings.

Gated Suction Inlets

Modern pumpers are equipped with at least one 2½-in. or larger size gated inlet (Figure 10-23), which is connected by piping and elbows to the suction side of the pump. This pipe may be as small as 2-in. internal diameter, and the fittings will further reduce this waterway; these restrictions often cause sufficient friction loss that the flow is reduced to less than one-half of that attainable by connecting the hose lines directly into the main pump inlet.

Because of the restrictions caused by the smaller size pipe and fittings, these inlets should not be utilized for the initial hose layout when large volumes of water may be required. These gated inlets are valuable for connecting additional supply lines after the apparatus has started pumping from the booster tank or is being furnished water through hose lines connected to the main pump suction inlet. If all the gated inlets are utilized for the initial hose lines, the pump must be shut down before any additional supply lines can be connected.

Some departments regularly run with a gated or clappered siamese fitting permanently attached to the 2½-in. gated inlet so two or more lines may be connected; this is a good idea if the piping between the inlet and

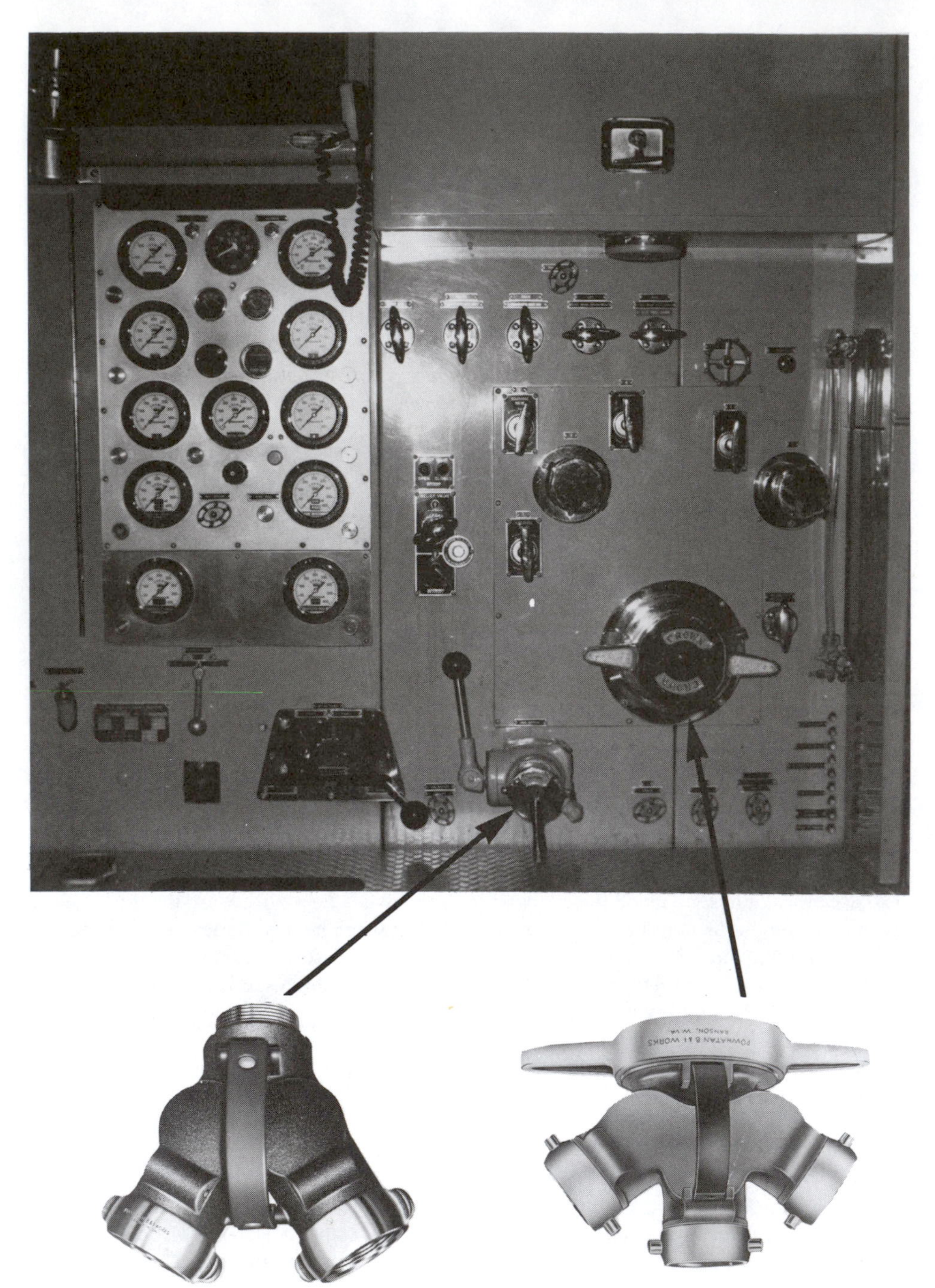

Figure 10-22. Large siamese fitting on the pump suction inlet. If the piping between the 2½-in. gated suction inlet and the pump is sufficiently large, a siamese fitting would allow the connection of two lines. (Courtesy of Badger-Powhaten; panel photo by author)

Figure 10-23. Modern pumpers are equipped with large piping and full-flow valves between the gated suction inlet and the main pump inlet to reduce the supply restrictions. (Courtesy of Hale Fire Pump Company)

the pump is sufficiently large that the friction loss created by substantial flows is held to a practicable minimum. However, in many pumpers the plumbing between the inlet and the pump offers a considerable amount of restriction to the flow from one 2½-in. hose line; it is difficult to visualize that in these apparatus the 2½-in. gated suction inlet could effectively convey the 500 to 600 gpm that the two hose lines would probably want to discharge.

Shutoff Butts

Many departments attach nozzle shutoff butts to all hose lines before they connect to sprinkler inlets or pumper suctions. Their contention is that this allows them to control the flow of water when the hose needs to be removed, and it prevents squirting water if the line is loaded before it has been attached. Most departments have two-way radio communications between the officer and the pump operator, so the need for this procedure has passed into history; the engineer can now be quickly informed of the need to shut down or load any hose line.

Some shutoff butts have a waterway as small as 1 in. in diameter; this restriction, as a rule, cannot be compensated for. When hose lines are being used to supply a heavy stream application or a sprinkler system, do not use shutoff butts or other valves. Tests have shown that these restrictions can reduce the volume greatly. At a recent relay drill, several pumpers with a discharge pressure of 200 psi were supplying a 1½-in. tip with 80 psi nozzle pressure. With the hose layout remaining identical, but with the shutoff butts removed, 200 psi engine pressure supplied 80 psi nozzle pressure to a 1¾-in. tip. Thus, with no change in pump pressure or hose layout, removing the shutoff butts increased the volume by 200 gpm. Other departments should conduct this and similar drills and the results will probably be astounding; the effects will undoubtedly differ because of varying butts, hose diameters, and other variances, but they will be noticeable.

Laying Hose Lines

When laying hose lines, the speed of the vehicle and the location of the lay are important. Excessive speed when laying lines causes the hose to whip, jerk, and flop. This action may cause damage to the hose and apparatus; crew members riding on the tailboard could be injured. If one of the crew is to be dropped off to take a hydrant or to lay a line from another pumper, stop the apparatus to give ample opportunity to snub the line around the hydrant, under a wheel, or over a bumper; this would prevent an accident if the hose should become snarled in the hose bed.

Much unnecessary crossing of hose lines may be avoided by following a uniform method of laying hose lines at fires. Where possible, when approaching a fire and stretching the first line, a hydrant on the same side of the street as the fire should be selected. Lay the line as close to the curb as feasible; if two lines are required, they can be laid parallel along the curb. Spot the apparatus as close to the curb as possible to prevent interfering with the traffic as much as possible. It is best to connect the hose lines into the suction intakes on the curb side. If a deliberate effort is made to position all hose lines and apparatus as close to the curb as possible, apparatus maneuvering in front of the fire scene will not have to drive over the hose lines.

If no hydrant is available on the same side of the street as the fire, then a hydrant on the opposite side of the street must be used. The line should be stretched alongside the curb on the same side of the street as the hydrant until a point opposite to the fire is reached; the hose line should cross the street in front of the fire. By crossing the street with hose lines only in the immediate vicinity of the fire, it is unnecessary for apparatus to drive over hose already stretched in the street unless they have to pass the fire scene to reach the far side.

Laying the hose lines on the hydrant side of the street often requires driving opposite to the normal traffic flow. Operate the apparatus cautiously and motion for the opposing drivers to pass on the clear side of the vehicle. When the traffic is dense, it may appear that this method of laying hose lines is exceptionally slow, but the overall fire ground operations of the engine company will be simplified.

Attack Pumpers

Many departments use an attack pumper tactic (Figure 10-24) of employing the booster tank of the first-in pumper to get water on the fire as rapidly as possible to achieve a quick knockdown. It is a valuable technique that enables a single pumper with a minimum of personnel to make a fast aggressive attack on a small fire that might accelerate to major proportions if there were the time delay to lay large multiple lines for the initial attack. If the size-up is faulty, it is easy to become embarrassed when the fire is larger than anticipated and a supply of water is not available before the booster tank is emptied; this could allow an unchecked fire to increase in intensity.

The primary precept of an attack pumper tactic is to direct one or more streams of water on the fire as rapidly as possible. For this maneuver to be both practical and effective, standard operating procedures

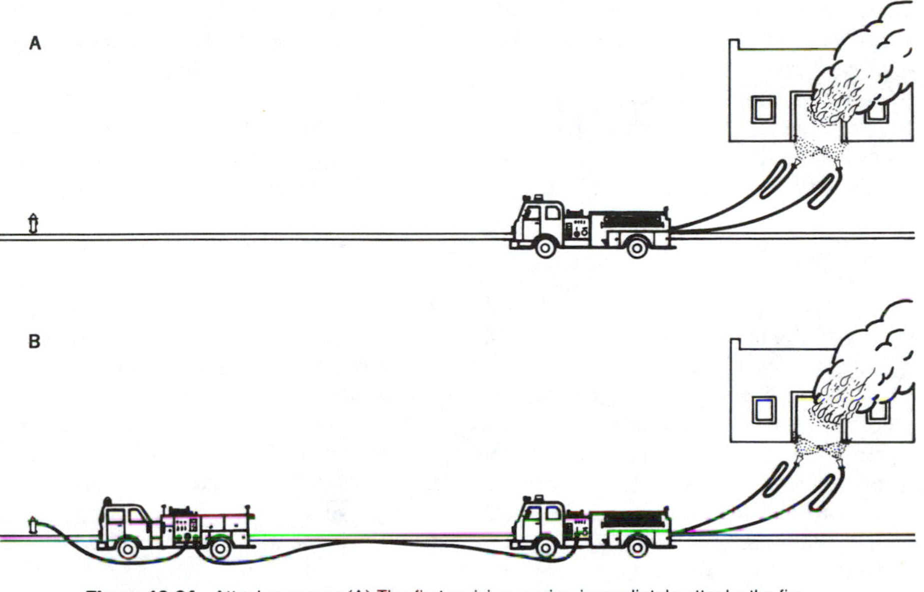

Figure 10-24. Attack pumper: (A) The first-arriving engine immediately attacks the fire with water from the booster tank. (B) The second engine lays supply line close to the curb, connects to a hydrant or other source, and provides water before the tank becomes empty.

must be formulated, adopted by all companies that may be involved, and practiced sufficiently often that everyone concerned will automatically know what must be done and how to accomplish it.

The attack pumper responds directly to the fire and supplies one or more lines from its booster tank. The success of this technique rests upon either quickly extinguishing the fire or obtaining additional water before the booster tank becomes empty. Nozzle operators must be extremely prudent and thrifty with the water, making sure that none is wasted. Some departments always lay at least one hose line if there is a clear indication of fire or smoke before they pass the last hydrant; others never become involved in any action that would delay their initial fire attack; they rely on the next engine company to lay the supply lines. There is considerable merit in both theories, but a line should always be laid by the attack pumper to a definite fire when there is a clear indication that the second pumper could be delayed.

In any event, hose lines should be stretched immediately by the second-in pumper to ensure an adequate supply of water for any pumper operating on a fire, even if the blaze appears trivial or under control. Unless the closest source of water is quite remote, it is never recommended that the second pumper responding to the fire also use its booster tank; if the blaze is beyond the control of the booster tank water carried on the first-in pumper, then hose lines should be laid. The hose does not necessarily need to be loaded; one or more lines can be kept dry and only loaded if additional water is required. The advantage of keeping one or both lines dry and only loaded if additional water is required by the attack pumper is to avoid unnecessary work. Dry hose is much easier to pick up; therefore, companies can be made available for response to another fire more rapidly. Of course, if there is the slightest indication that more water is needed, there should never be the slightest delay in loading the lines.

Inline Pumping

Inline pumping, which is also known as the long soft suction lay, is a procedure employed by many engine companies that are equipped with a single triple combination pumper. This technique allows the pumper to remain at the fire to supply fire streams from the booster tank for an immediate attack on the fire, while simultaneously, a hose layout is being prepared to assure a water supply from the nearest hydrant. This procedure is further enhanced if a four-way valve is used to connect the supply line to the hydrant; this would better enable a later arriving pumper to increase the pressure in the supply line. Unless the hose layout is short, and the water system is exceptionally strong, it is best not to depend entirely upon hydrant pressure.

Figures 10-25 and 10-26 illustrate examples of inline pumping procedures. Typical procedure where two-piece engine companies are in use is shown in Figure 10-25: In (A), the first-in engine company lays a line from hydrant and attacks the fire with booster tank water while the engineer is connecting supply line to the suction inlet. Hydrant pressure should supply additional water before the tank is empty. In (B), the second engine company connects to hydrant and boosts supply line pressure. This operation is simplified if a four-way valve is used on the hydrant by the initial pumper so the second engine will not have to interrupt the flow of water in the supply line.

In the example of the long soft suction lay in Figure 10-26, the hydrant is capable of supplying 1000 gpm; the static pressure is 100 psi. In (A), a supply line of 400 ft. of 2½-

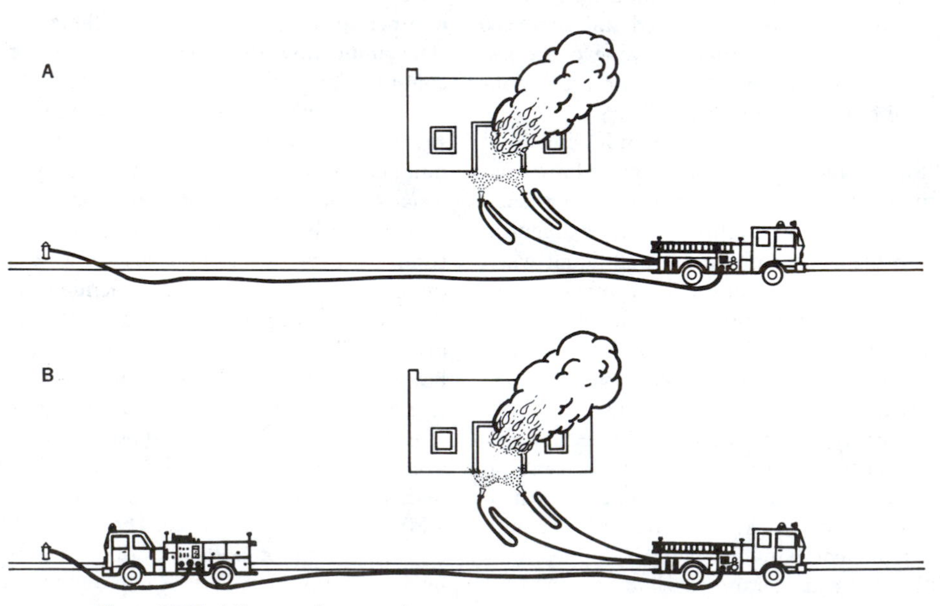

Figure 10-25. Inline pumping procedures

in. hose is laid from the hydrant and connected into the pump inlet. Two 1½-in. lines with 100-gpm fog tips at 100 psi are used on the fire. When the 200 gpm flow is attained, the residual hydrant pressure is 50 psi. This is the limit that this hydrant can supply because the friction loss in the 2½-in. hose is 40 psi; this leaves 10 psi pressure on the pump.

In part (B) of the illustration, the same hydrant and 2½-in. hose layout is attempting to supply a 300-gpm fog tip with a nozzle pressure of 100 psi. When 300 gpm are flowing through a single 2½-in. hose line 400 ft. long, 84 psi friction loss is created. Even if the hydrant could still maintain a 50-psi residual pressure after the 300 gpm were flowing, it would not be capable of overcoming the friction loss in the supply line. A parallel 2½-in. line or a larger supply hose would overcome this difficulty. In inline pumping procedures, it is important not to exceed the capability of the hydrant.

In some departments, it is a standard operating procedure to start a preconnected 1½-in. line into the fire as soon as the apparatus stops. The engineer charges this line the moment it clears the hose bed, and then two more 1½-in. lines are placed in service. In structure fires, the first line attacks the main fire or is used for rescue purposes; one of the later fire streams goes above the fire to prevent vertical extension and the other nozzle backs up the initial preconnect line. An attack using inline pumping generally is restricted to the use of 1½-in. or booster lines, although it is possible to use a 2½-in. line which is flowing 200 gpm or more under ideal conditions.

The value of this tactic lies in the ability of the hydrant attender connecting the supply line and the pump operator coupling the other end of the line to a gated pump inlet quickly enough so that hydrant pressure will provide a continuous supply of water before the booster tank is emptied. It is possible to provide full capacity for a pump spotted near the fire and far from a hydrant if

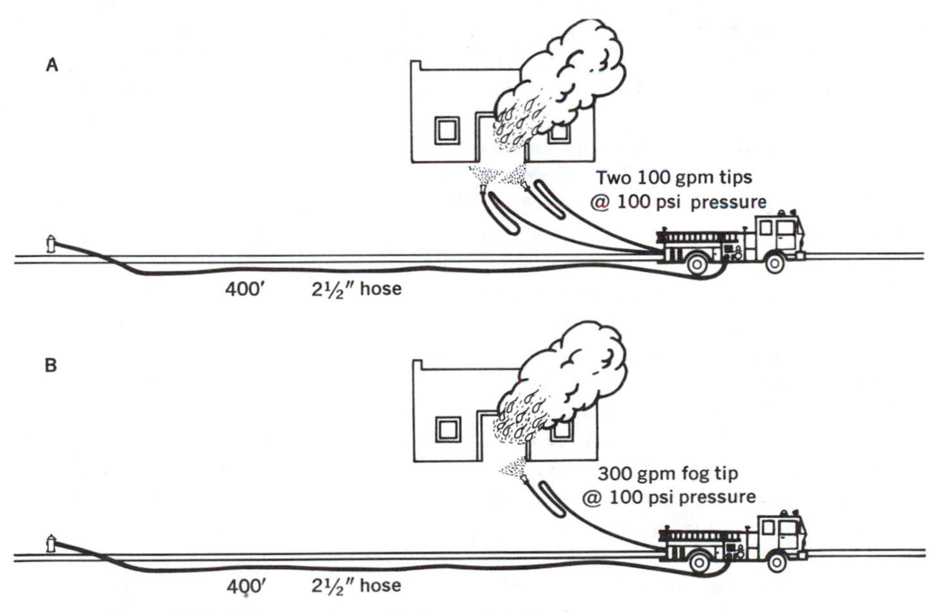

Figure 10-26. Inline pumping, or the long soft suction lay

enough hose lines of adequate diameter are connected to the inlet side of the pump and are supplied under sufficient pressure. When 2½- or 3-in. hose is used for the supply line, the friction losses will correspond to losses encountered when the same size hose supplies fire streams. Friction loss in 2½-in. hose supplying 250 gpm will be about 15 psi per 100 ft, so a hydrant having a residual pressure of 50 psi could supply this volume to a pumper through no more than 300 ft of single 2½-in. hose. Laying two or more lines initially, or utilizing larger hose than 2½-in., will further improve this tactic.

One officer ordered a single 2½-in. line laid to a structure where fire and smoke were clearly apparent. When the apparatus arrived in front of the fire, the officer noticed that the interior of the building was fully involved; the crew was ordered to place the wagon battery equipped with a 500-gpm spray tip into action to knock down the fire. The 400-gal booster tank was rapidly emptied by the 500-gpm fire stream and there was no way that a 50 or 60 psi hydrant pressure could overcome 55 psi friction loss in each 100 ft of a long hose layout. The structure completely burned to the ground before later arriving pumpers could boost the pressure in the initial hose layout and lay additional lines.

One textbook that suggests solutions to fire ground problems shows the advantages of inline pumping by illustrating a hose layout of a single 2½-in. line supplying 450 gpm to one 2½-in. and two 1½-in. fire streams. This would be an impossible answer to a difficult problem unless the hose line was very short and the hydrant could maintain an extremely high residual pressure after the 450 gpm were flowing. That quantity of water would create a friction loss of about 45 psi in every 100 ft of single 2½-in. hose, and it is questionable whether the critical velocity of this quantity of water attempting to flow through a single 2½-in. hose would not further make this an impossible solution to the dilemma. All companies should conscientiously drill under authentic fire ground conditions to make sure that standard operating procedures they adopt and the fire fighting tactics they employ are safe, realistic, and practical.

Dead-End Streets and Driveways

When responding to an address on a dead-end street, alley, long driveway, industrial complex or housing development roadway, or other location remote from a hydrant and where vehicle maneuvering room and parking space is limited or nonexistent, only the first engine should proceed in and make the size-up when there is no visible indication of fire or smoke. If the engine company is equipped with two pumpers or a tanker and a pumper, one pumper should hold back and remain capable of laying any necessary hose lines.

Examples of alternatives in engine positioning in dead-end locations are illustrated in Figures 10-27 through 10-29. In Figure 10-27, with no fire showing, engine 1 drives directly in; engine 2 either drives in and turns around, or backs in; engine 3 remains on the main road. In case fire is discovered, engine 1 attacks the fire with water from the booster tank; engine 2 lays one or two supply lines out to the nearest water source; and engine 3 assists where needed. Truck company either drives in first or remains on the main road.

In Figure 10-28, with fire showing, engine 1 lays one or two supply lines from the entrance to the main road and attacks the fire with water from the booster tank; engine 2 connects to the hose coupling and lays a line to the nearest water; engine 3 is available to help out where needed.

In Figure 10-29, with fire showing, as an alternative, engine 1 drives directly in, laying one or two lines from the entrance to the main road and attacks the fire with water from the booster tank; engine 2 immediately connects to the hose line of engine 1 and supplies its booster tank water; engine 3 lays a line from engine 2 to the nearest water

Figure 10-27. Positioning for dead-end street or driveway; no fire showing

source. Additional pumpers could be utilized to parallel these hose lines or to extend the relay to a remote water supply.

When a truck company is responding along with the first-in engine company, the truck may be sent in first to make the preliminary size-up; in this case, the pumper should stop at the nearest hydrant ready to lay hose lines. Once apparatus have been positioned at the fire location down one of these dead-ends, they will not be able to be repositioned for the duration of the fire. When a fire is discovered, the first-in pumper should make an immediate attack with water from the booster tank; if additional water is required, it will be the responsibility of later arriving pumpers to lay the necessary lines.

Figure 10-28. Positioning for dead-end street or driveway; fire showing

Figure 10-29. Alternative positioning for dead-end street or driveway, with fire showing

When there is the slightest indication that a fire is in progress, the first-in pumper should always lay one or two lines from the closest hydrant when approaching the fire ground. If there is no convenient source of water, the engine should drop one or two lines at the entrance of the dead-end and lay on in to the fire. The second pumper will then connect to the dropped hose lines and lay to the nearest water source. There should be at least two sets of double male and double female fittings, as well as adaptors to join various hose sizes, on every pumper so there will be no difficulty in connecting the two hose layouts together. Whether the first-in pumper lays one or two lines will depend on the anticipated quantity of water necessary for fire fighting and the sizes and amount of hose carried on the apparatus.

In the hose layouts illustrated in Figures 10-28 and 10-29, there are three vital areas of action: the attack pumper, the midway point where the hose layouts of the two engines are coupled together, and the supply engine. Immediately upon arriving at the fire location, the attack pumper should break the hose lines and connect them into the gated suction inlets; if this is not done, utilize a hose clamp or shut-off to prevent prematurely arriving water from filling the hose still on the apparatus (loading the hose bed). The crew members left at the road entrance must have the necessary adaptors and fittings to join the two hose layouts together. The supply engine should continue laying the supply lines to the closest water source and load the lines as rapidly as possible. If this procedure is correctly planned and practiced, water will be supplied to the attack pumper well before its booster tank is emptied.

In a city that has a multitude of expensive estates where serious fires may not be readily apparent because the residences are screened from the road by extensive landscaping, high fences and hedges, and curving scenic driveways that effectively shield and protect the owner's privacy, the following method was adopted to lay supply lines when required. When no signs of fire or smoke are apparent, the first engine company proceeds up the driveway; the second pumper backs up the driveway and takes up a position tailboard to tailboard. If a fire is discovered, the first engine immediately launches a booster line attack. The second pumper leaves some of its crew at the fire to help stretch hose lines and combat the blaze while the apparatus is laying one or two hose lines back to the nearest water source. The truck company will either enter first and proceed past the structure, or it will stay out on the street until the second pumper has laid out the supply lines.

Blitz Attack

The blitz attack in fire fighting is theoretically identical to the World War II military maneuver, the blitzkrieg; it is a fast, massive, all-out assault on a fire. When used in an attack, a large fog stream will penetrate where a solid stream cannot. Upon contact with intense heat or fire, water particles will rapidly absorb heat energy and be converted into steam; the more finely divided the fire stream is, the faster the heat absorption. The steam then is carried upward by the heat convection currents to help ventilate the structure. This rapid steam generation absorbs tremendous amounts of heat energy, which renders maximum efficiency from the water. With this technique, it is possible to obtain maximum effectiveness with limited supplies of water. In a limited number of fires this tactic would result in a quick knockdown with a minimum of crew.

If the conditions are favorable, the blitz attack is a strategic option. However, there are some limiting factors and several disadvantages that could ensue.

Fundamentally, a blitz attack is made when the first arriving engine company makes an immediate assault on a fire with a large-volume fire stream. Speed is essential, so the booster tank is utilized as a water supply for a deck gun or a water tower equipped with an adjustable pattern spray nozzle. Hand lines should be simultaneously stretched and ready for immediate advancement into the fire area when the large stream is turned off. A blitz attack could also be made by utilizing the booster tank to supply one or two 2½-in. hand lines equipped with spray nozzles; the lines should be kept short, because the effectiveness of the blitz depends on how quickly the assault can be launched.

Most blitz attacks depend on a 30- to 60-second assault with a 300- to 600-gpm fire stream. Because the limiting factors are the capacities of the booster tank, the booster tank, and the piping between the tank, pump, and wagon battery, this tactic should be preplanned and practiced thoroughly to make it workable. Obviously, a 400-gal tank with piping that will only flow a maximum of 250 gpm could not be expected to adequately supply a 600-gpm spray nozzle. The officer has to decide whether the best operation is a fast, massive attack with a large-volume fog nozzle, a more conservative attack with small lines and nozzles, or to protect the exposures and depend on later arriving companies to control and extinguish the main fire.

The blitz attack is a gamble because it depends on the ability of a large fire stream to knock the fire down before the tank is emptied. Much of the risk is removed if the pumper has laid one or two supply lines from the nearest hydrant while approaching the fire. Then, while one crew member is directing the deck gun on the fire, the rest of the crew would be connecting up the hose lines; unless there were complications, this would assure a constant water supply. It is also a valuable tactic when a later arriving pumper lays supply and backup lines.

On structures in which the fire was on the ground floor, this maneuver has been successful when the apparatus was driven right up on the sidewalk and a full spray stream directed through a window or doorway; a straight stream is used when more reach is needed.

Disastrous brush fires have been prevented on steep mountainous slopes when the first arriving pumper used its booster tank water and wagon battery to knock down an incipient brush fire on the side of a steep hill. If time were taken for the fire crew to drag hose lines up that difficult terrain and apply water with small nozzles, the fire would have traveled up and over the top before they ever got near. Hand lines, however, should be simultaneously stretched to complete the extinguishment.

Water tower and elevating platform nozzles can be utilized for a blitz attack to achieve a quick knockdown on fires in the upper floors of a building, especially if the elevating platform or water tower is mounted on a pumper equipped with a water tank.

The advantage of a blitz attack is that it allows a rapid knockdown of a large fire with a minimum of crew and water. Used correctly, it is a valuable addition to any engine company's procedures. However, the conditions must be ideal. For structure fires, the building must still be closed up tightly as this is a variation of the indirect method of fire fighting. If the structure has vented itself, if there are open piles of combustibles burning, or if the volume of fire and heat are greater than can be handled by the available amount of water, then the more conventional methods of fire fighting should be adhered to.

Among the disadvantages of the blitz attack is that the force of the fire stream may drive the fire into uninvolved portions of the structure. Normally, a well-planned assault on a fire will be launched from the uninvolved portion of the building; this will prevent further extension of the blaze. A misdirected fire stream could easily drive the fire into uninvolved portions of the structure.

If there are still occupants in the building, they could be endangered and, possibly, killed or injured. The forcing of heat and fire gases could make the stairs and passageways untenable or lethal.

If the size-up were accurate, and all conditions were agreeable, the blitz attack could achieve spectacular success; an error in judgment, or a mistake in execution, and the results could be disastrous.

11 Relay Operations

Occasionally, an officer is confronted with the task of supplying a heavy stream application at a large fire, delivering sufficient water to nozzles working on a remote brush or rural fire, or meeting some other special demands that would require relaying water. Because of the quantities of water involved, the terrain, and the distance between the fire and the water source, coordinated operations with other companies or fire departments are usually necessary. This procedure is of paramount importance to mutual aid groups in rural areas.

Situations Calling for Relays

In relaying of water by a fire department, two or more pumping engines operate in series, the discharge of one pump passing through another. A relay is employed when supply requirements for an effective fire stream exceed the performance ability of a single pumper. This situation could arise because of excessive friction loss in a long hose layout, or because of back pressure when the point of consumption is at an elevation greater than the source of supply. In the example shown in Figure 11-1, additional hose would require a higher engine pressure, a lower nozzle pressure and discharge, or both. Back pressure would further decrease the length of the hose layout.

As demonstrated in Figure 11-2, adding a second pumper to relay the water allows larger quantities of water to be pumped for longer distances. At an engine pressure of 200 psi at the source pumper, 200 gpm can be delivered through a single 2½-in. hose line for 1900 ft and still have a residual pressure of 10 psi at the second pumper. The second pumper can now supply 100 psi nozzle pressure to a 200-gpm fog tip through 1000 ft of single 2½-in. hose. With the addition of one relay pumper, the 200 gpm of water can be delivered 1900 ft further. Of course, back pressure will affect somewhat these effective distances. The engine pressure of the second pumper will be correspondingly lower if the hose line to the nozzle is shorter.

The need for a relay operation may occur anywhere. Under normal conditions, the need for relaying water would not be expected to arise in built-up areas because hydrants should generally be available. However, in rural and undeveloped areas,

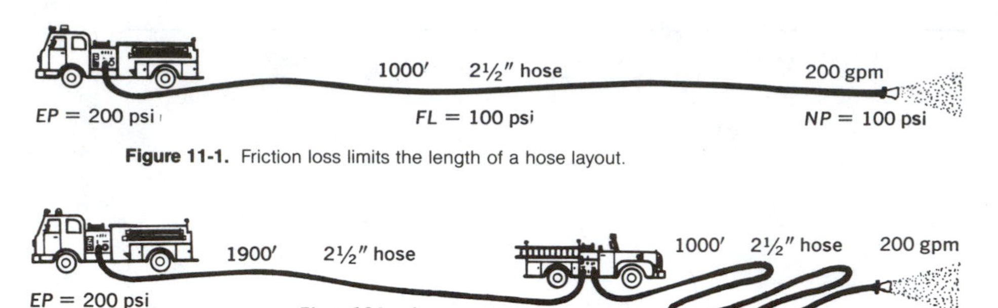

Figure 11-1. Friction loss limits the length of a hose layout.

Figure 11-2. Adding a second pumper to relay the water will allow larger quantities of water to be pumped for longer distances.

hydrants are often widely spaced, or the available supply of water from a hydrant is insufficient for a large fire. Under emergency conditions brought on by an earthquake, enemy action, or any other disaster, mains could well be broken and relaying might be necessary even in congested value districts. Usually, however, conditions requiring relay hookups will be found at brush and forest fires and in the rural areas of the country.

It is doubtful that any two relay operations necessary to combat actual fires will be the same. Therefore, it is of little value to set up rigid conditions governing the placement of pumpers in a relay, the laying of the lines, or the setting of the pump discharge pressures. All that is attempted in this book is to present some basic information, general requirements, and limitations that can be tailored to fit situations as they occur.

Factors Affecting Relay Setup

When it becomes necessary to resort to a relay operation in order to get water on a fire, many things must be taken into consideration. Principal relevant factors are:

1. Topography
2. The time necessary to set up a given relay problem
3. The total amount (in gpm) of water estimated to be necessary at the point of ultimate consumption
4. The size and length of hose necessary to carry the amount of water needed or, in the event the proper size is unavailable, the number of siamesed smaller lines necessary to carry the water
5. The capacity of the apparatus available to do the pumping
6. The adequacy of the water source

Topography

If the ground is level, it may be possible to lay several thousand feet of hose without relaying. In the hills and mountains, differences of elevation might call for a pumper every 300 ft. Terrain can easily call for a different spacing of pumps in the same relay. The Underwriter's formula calls for the placing of a second pump in the line two-thirds the distance to the nozzle; it will work well only if two pumps are to be used and the lay is on level ground. Where more than two pumps are to be utilized, or where head is a major consideration, it cannot be used.

Time Element

Since it will always take longer to lay a relayed line than a straight lay, consideration must be given to how long it will take, the route to be followed, and the size of the fire when the lay has been completed.

Amount of Water Necessary

The amount of water in gpm to be delivered in a line of relay pumps is an important factor in determining a relay setup. The volume of water flowing, together with the total length of the hose layout, determines the total friction loss in the layout.

Size and Length of Hose

Once the quantity of water is decided upon, the problem is to select the hose to carry the water. The use of hose lines less than 2½-in. in diameter is impractical between pumpers of a relay operation because of excessive friction loss.

The length of the hose lay beyond the last pumper is immaterial as far as the relay operation is concerned, except as it affects the total amount of water to be delivered to the nozzles. The operator at the last pump in line calculates incoming line pressure and volume just the same as at a hydrant.

New hose is usually tested to 600 psi, but as it grows older it may be expected to stand less pressure; hose in service is tested annually or semiannually to 250 to 300 psi. The weakest link in the hose layout should stand this annual test pressure. Because the closing of nozzles increases the pressure in the hose, every effort should be made to keep individual pumps in a line to 200 psi maximum.

It is good practice to assume a required inlet pressure of 10 psi at the pumps; there will remain 200 − 10, or 190 psi, for overcoming friction loss and back pressure that may be encountered.

Certain general requirements can be set up for lines to be used. Up to a maximum flow of 325 gpm (necessary to supply one 1¼-in. tip at 50 psi nozzle pressure), one line of 2½-in. hose laid on the level will use up 200 psi in 800 ft to overcome friction loss. For quantities in excess of 325 gpm, additional or larger lines should be used. For flows up to 650 gpm (two 1¼-in. tips), two lines of 2½-in. hose will use up 200 psi in 800 ft. If 3½-in. hose is used, it would take approximately 1200 ft for friction loss to use up 200 psi of pressure. Keeping the maximum pressure at 200 psi, 650 gpm is approximately the capacity of a modern 1000-gpm pump (70% at 200 psi).

Apparatus Available for Pumping

An up-to-date inventory of all pumpers, hose, equipment, and crew assigned to companies and departments in mutual aid groups should be immediately available to the dispatchers and units who will be expected to work together in a cooperative fire ground operation. This information is especially valuable for prefire planning in rural areas. Advance knowledge will allow the officer in charge to request a certain unit which has large diameter hose, an extra large water tank, a plentifully long load of hose, a substantial main pump, or any other desirable feature that would make a certain fire company more desirable than the others. On the other hand, adverse features could cause one fire company to be less desirable than the others.

The sizes of the pumps will have to be taken into consideration when figuring the number of pumpers required for a large-quantity relay. It must be remembered that when pumpers are placed in line so that 200 psi net pressure is required from each pump, 70% of each pump's capacity is all that can be expected. If possible, pumps of not less than 1000-gpm capacity should be used in any relay hookup requiring flows greater than 500 gpm. If large flows of water are necessary, pumps must necessarily be spaced closer together to decrease their pressure output and correspondingly increase their quantity output.

Adequacy of Water Source

It would be useless to set up an elaborate

relay only to find out that the hydrant or other source cannot supply the desired quantity of water. Conversely, a large flowing pressure from a hydrant can make it possible for a single pumper to deliver sufficient water at a satisfactory pressure.

Necessary Exchange of Knowledge

Unless a pump operator possesses clairvoyant powers, it is difficult to know the size of the tip, the elevation of the nozzle, and the number of hundred feet of hose between the pumpers being used by another fire company several thousand feet away. Therefore, there is frequently too little pressure and a poor fire stream because the engineer at the pumper closest to the fire runs away from the water before building up a sufficient discharge pressure for nozzles at the fire. It is mandatory that the officer in charge of the fire inform all personnel of every condition that may affect the delivery of water.

Water Control Officer

In view of the many variables encountered in relay operations, it should be apparent that no simple universally applicable rules can be given. Since the urgency of a fire situation and other factors such as communications problems, unfamiliar locality, an absence of a good water supply system, a shortage of apparatus and fire crew, and a myriad of other complications could lead to errors and delays that could seriously affect the overall fire fighting action, a water control officer should be designated at any complex operation to supervise the placement of apparatus and the best utilization of hydrants and static water supplies. This officer is absolutely essential to effective rural mutual aid relay implementation. The water supply officer would be responsible for supplying the nozzles or heavy stream appliances effectively. The officer should estimate the quantity of water needed, the necessary number of pumpers, the number of hose lines between the pumpers, and the spacing of the pumpers. This makes a thorough knowledge of hydraulics mandatory for all company commanders and chief officers.

The officer in charge of setting up the relay is justified in delaying initiating the operation until the best available source of water can be determined. A closer water source may cause a relay to be unnecessary, or the required volume of water could make the closest hydrant impracticable.

Prefire Planning

The water control officer is the key to planning and implementing an effective relay operation.

Prefire planning inspections will allow the officers and crew an opportunity to prepare their fire fighting strategy and tactics before the fire occurs. Officers should anticipate the fire ground problems that they could encounter and conduct complex multicompany drills under realistic conditions. Many factors that will affect the adequacy of fire streams are not indicated by a study of fire ground hydraulics; here is where drilling, experiments, tests, and experience pay off.

Prefire planning should identify the occupancies and locations where an adequate fire flow is not immediately available within a reasonable distance. Fire flow is the anticipated quantity of water required for fire protection, measured both in gpm and in specific duration. Preplanning would then establish the closest and most accessible source of water and the most effective method of delivering adequate volumes of water to the fire ground.

The anticipated necessary fire flow, the companies that will be dispatched on the first alarm, the backup companies that can be requested if needed, the types of apparatus and equipment, available crew, the distance between the fire ground and the water source, the roadways, geographical terrain, and the time delay that the first-in unit may anticipate before additional help arrives must all be considered in varying degrees when deciding what is the best method of combating a fire. If the structure is a considerable distance from the water, and large volumes are not required, trucking the water in tankers may be the most practicable method. If a water source is reasonably close and accessible, or if large volumes are needed, then a relay requiring several pumpers may be wanted. In other cases, a combination relay utilizing both tankers and pumpers may be the fastest and most efficient. So many factors and variables affect the delivery of adequate volumes of water to combat a fire that no one method could be considered the best for every fire and department. Every situation must be analyzed and the factors and variables considered before a tentative plan is adopted; then the plan must be tried and modified.

Drills

Companies should practice and drill under realistic conditions. The fire should be simulated, the anticipated hose lay and nozzles should be used, and the pumpers operated by the engineers who will be required to combat the fire. Some departments simulate relay and large-volume operations at a drill by using 1½-in. hose to replace the 2½-in. or larger hose lines. They defend this practice by asserting that the shorter lengths of smaller hose will duplicate the friction loss of longer lines of large hose and will alleviate the necessity of laying out large quantities of hose. There is no substitute for making the actual hose layout and pumping the desired quantity of water through the correct nozzles. Often the theoretical answer is not the practical solution to a problem.

Typical Drill Procedure

The following drill procedure produced fairly consistent results, even when a variety of companies, apparatus, and crews participated. Four pumpers and an aerial ladder with a 1½-in. tip on a ladder pipe were used; a pitot gage was employed to measure the nozzle pressure. The first engine spotted to a hydrant. The next pumper laid two 2½-in. lines from the supply engine's discharge gates until its entire load was out; they were then connected into the 2½-in. gated suction inlets. The third engine laid two lines of 2½-in. hose from the second pumper's discharge gates until its entire load was out, and then connected them into its 2½-in. gated suction inlets. The fourth pumper laid two lines from the third engine, connected them into its gated inlets, and supplied the ladder pipe.

Nozzle butts were placed on all male couplings where they were connected into the pumpers' 2½-in. gated suction inlets. A 200-psi discharge pressure on the relay pumpers would not create an 80-psi nozzle pressure on the 1½-in. tip, so other companies were ordered to lay a third hose line between the relay pumpers. The first difficulty encountered was that the pumpers had utilized all of their gated suction inlets, so the relay had to be shut down while the operators connected the third line into the large pump suction inlet. The three hose lines produced the 80 psi pressure at a standard relay discharge pressure of 200 psi.

When the nozzle butts were removed and the original two hose lines were connected into a siamese fitting attached to the large pump suction inlet, these two lines easily supplied the ladder pipe with 80-psi nozzle pressure at a standard relay pressure of 200 psi. Therefore, two properly attached hose lines were as effective as three improperly connected lines.

Nozzle Selection

Nozzles for relay operations should be selected that will produce good fire streams within the capacity of the relay. Use of nozzle tips that are too large may cause the pump at the fire to run away from its water supply and result in low nozzle pressure and poor reach. When making heavy stream applications, one of the most prevalent mistakes is to attempt to supply too large a nozzle in the early stages of a fire. Usually a 500-gpm spray nozzle or a 1½-in. tip is the most practical in the beginning, unless the water will be delivered through multiple 3½-in. or larger hose lines in the initial attack. *Remember:* A second hose line of equal size and length will reduce the friction loss to approximately one-fourth of the original and a third hose line of equal size and length will reduce the friction loss to one-half of that in two lines.

All nozzle operators on fire streams supplied by a relay must be instructed to close nozzles very slowly to avoid building up pressure surges in the hose faster than the pressure relief devices on the pumpers can handle.

Relay Pumpers

It is important that officers and pump operators know the rated capacities of their pumpers at 150, 200, and 250 psi net pressure. Table 11-1 gives these ratings. Then, assuming that the pump is operating in the correct transfer valve position, it will be possible for the machine to deliver the required volume at the appropriate engine speed, taking advantage of whatever additional pressure the hydrant may be able to provide.

The size of the pumps will have to be taken into consideration when figuring the number of pumpers required for a large quantity relay. It must be remembered that when pumpers are placed in line so that 200 psi net pressure is required from each pump, 70% of each pump's capacity is all that can be expected. If possible, pumps of not less than 1000 gpm capacity should be used in any relay hookup requiring flows of greater than 500 gpm. When a large volume of water is the prime consideration, it may be necessary to limit the discharge pressure to 150 psi because most engines would be pumping at close to their rated capacities. When this is necessary, either or both of the following remedies will be needed: (1) Space the pumpers closer together to decrease their pressure output and, correspondingly, increase their quantity output, or (2) lay additional or larger hose lines to reduce friction loss.

When supplying a single-line relay at 200 psi discharge pressure, the pump is probably discharging less than 70% of its rated capacity. The engine speed will probably be lower if the transfer valve is in the series (pressure) position.

Pump at Water Source

While minimum relay requirements could be supplied by any capacity pumper at the water source, the capacity of the source pumper is the limiting factor for the entire relay operation; it is obvious that the max-

TABLE 11-1. LIMITATIONS OF PUMPER CAPACITY

RATED CAPACITY	gpm AT 150 psi	gpm AT 200 psi	gpm AT 250 psi
500	500	350	250
750	750	525	375
1000	1000	700	500
1250	1250	875	625
1500	1500	1050	750
1750	1750	1225	875
2000	2000	1400	1000

NOTE—Ratings refer to net pump pressure, which is the difference between the inlet and the discharge pressures.

imum relayed flow depends on the capacity of the first engine. A 500-gpm pumper could not deliver enough water to allow larger engines to operate at capacity. Therefore, whenever possible, the largest pumper should be at the water source. It can assist the smaller centrifugal pumps by providing good residual pressure on the suction inlet of the second pump.

Pumps in Center of Relay

The role of the pumpers in the center of a relay operation is to accept the incoming water flow and move it along, adding pressure as needed.

Pump Nearest Fire

The pump nearest to the fire should adjust its discharge pressure to that needed to supply the fire streams. The work of the engine at the fire is very similar to the job of pumping while being supplied by a soft suction at a hydrant, except that the supply lines are longer.

Standard Engine Pressure

It is efficient and practical to have each engine, except the one at the fire, start a relay with the same discharge pressure. This eliminates mental calculation by the pump operators, who have only to provide the previously chosen pressure for all relay work. The standard engine pressure for relays should be selected after considering the diameter and amount of hose carried on the apparatus, the condition of the hose, and the amount of water the relay is expected to supply. Most relays, except those supplying heavy stream applications in a metropolitan area, are originally set up with a single line because the amount of hose is limited and the distance between the fire and the water source is often several thousand feet.

A modern technique of relay pumping is to have the supply pumpers maintain a good pressure on all lines between the pumpers, instead of attempting to calculate the friction loss and back pressure. The only precaution to be taken is not to exceed the maximum inlet pressure to which the receiving pump has been tested. In general, it is recommended that the pressure be increased to 200 psi as soon as the line is safely charged. If this pressure should prove to be 25 to 50 psi too much, no harm is done and the work of the engineer nearer to the fire is made easier. After all, pump operators do not have to use all of the pressure available on their intake suction gages any more than they have to pull the hydrant pressure down to zero at every fire.

When figuring pressures to be pumped to a relay pumper or supply reservoir, an intake pressure of 10 psi should be used in place of the 50 to 100 psi normally allowed for nozzle pressures. This 10 psi allowed for intake pressure, subtracted from the net working pressure of 200 psi, leaves 190 psi pressure available to overcome friction loss and back pressure.

There are some who think that pumping at higher than necessary pressures may rupture hose lines. It is not the recommended 200 psi discharge pressure that is likely to burst the hose, but surges caused by opening and closing nozzles and valves too quickly.

When maximum flow is wanted, the relay engines can increase their pressure, and consequent volume, as long as the hose remains firm where it enters the suction inlet of the pump. If more water is wanted, the operator of the supply pumper at the water source can be directed by radio or hand signals to increase pump pressure, or another supply line can be laid. When the 200 psi standard pressure is insufficient, the officer in charge may order the relay pumpers to increase their discharge pressures to a maximum of 225 psi. If nozzles are opened and closed very slowly, the pressure control devices are properly set, and the hose is in good condition, this higher pressure should not impose any danger to a relay operation that is already working at 200 psi.

Required Number of Pumpers

The officer in command of the relay operation will determine the quantity of water to be delivered to the last pump, decide what hose is to be used, and estimate the total back pressure to be overcome. When this information is available and the total length of hose is known, the number of pumps, in addition to the last pumper, can be determined by dividing the maximum pressure of 200 psi into the total pressure requirement. At fires where pumps cannot all be spaced equally, allowance can be made. It is always better to place one more pumper in the line than is absolutely necessary, particularly in long relay lines, than to have too few.

In relay operations, the total required engine pressure determines the number of pumpers needed. This may be stated as:

$$\text{Number of pumpers} = \frac{FL + BP + NP}{200}$$

Example (Refer to Figure 11-3): Suppose a 1⅛-in. nozzle is being used 4200 ft from the nearest source of supply. Using a single 2½-in. hose for relaying, how many pumpers are required and what is the spacing?

A 1⅛-in. tip delivers 250 gpm, which has a 15 psi friction loss in 100 ft of 2½-in. hose; this is a total of 630 psi pressure (15 × 42). To this, add 50 psi nozzle pressure for a total of 680 psi. The number of pumpers is determined by dividing 680 psi by the maximum working pressure of 200 psi, which equals 3.4, or a total of four pumpers needed to deliver the required quantity. If elevation is involved, head pressure will require an increase or decrease accordingly.

Furnishing 10 psi intake pressure to each pumper in the relay requires 30 psi; added to the 680 psi, this is a total 710 psi required engine pressure. To determine the distance between relay pumpers, and the engine pressure distribution among them, divide the maximum net pump pressure (190 psi) by the friction loss in each 100 feet of hose; 190 psi divided by 15 psi equals 12.66. Since hose comes in 50 foot sections, round off 12.66 to 12.5. Therefore, relay pumpers delivering 250 gpm through a single 2½-in. hose should be spaced 1250 feet apart. The first three pumpers have an engine discharge pressure of approximately 200 psi each. Three times the 1250 feet spacing places the fourth pumper 3750 feet from the source; it will be 450 feet from the nozzle; it will receive 250 gpm at a 10 psi intake pressure and discharge at a pressure of about 120 psi (4.5 × 15 psi + 50 psi).

Suction Pressure

When a pumper is connected to a hydrant or other source of water supply under pressure, it is desirable to maintain 10 psi or more pressure on the suction gage. However, with large flows, the friction loss in the hydrant and suction hose may be 5 to 10 psi or more. When using a soft suction, the point at which the maximum volume is being received can be detected by feeling the suction hose at the point where it enters the pump intake. If a soft suction or incoming hose is soft and flabby when squeezed, it indicates that the pump is beginning to run away from the water, since there is no pressure left to keep the suction hose solid. Retard the throttle until the supply lines are reasonably firm where they enter the pump. This will prevent damage to the pump from cavitation, which may occur when the pump cannot obtain as much water as it requires (runs away from the water). Cavitation can be detected by a chattering sound from the pump.

To attempt to get more water without additional supply lines is futile and may result in poor hose streams as well as possible damage to the pump due to cavitation. Cavitation is a condition in which air cavities are formed in the pump; when they implode, pump damage occurs. Cavitation can also occur during drafting operations if the engineer runs the pump at a higher speed than that at which the suction hose can deliver a adequate volume of water. The suction gage readings at pressures near the zero mark are not very accurate; the accuracy may be determined approximately by observing the gage when the pump is taking all available water.

Relay Relief Valve

In a relay, as during any pumping operation, the relief valve or governor should be set to protect the crew, the pump, and the hose. An alert engineer will watch the gages and be prepared to quickly open a spare discharge gate if any pressure surge is detected. The ordinary pressure regulating device operates on the discharge side of the pump; it does not protect the suction side of a relay pump. If one of the pumpers shuts down, varies its discharge pressure quickly, or a line is shut off, the pressure immediately rises; it may create a fluctuating pressure surge that could damage pumps and hose, because pressure surges tend to build up and become more serious in long lines of hose. The surges are not confined to the hose lines between two pumpers; they can travel the whole length of the relay line because centrifugal pumps have an open waterway that would allow a pressure fluctuation to pass through from suction inlet to discharge with no hindrance. The safest protection is a relay relief valve installed at the suction inlet and set to operate at a predetermined pressure; it prevents pressure surges from entering the pump. With any pressure rise, the relay relief valve dumps the excess volume of water on the ground and prevents a high-pressure buildup.

With some relay relief valves it is necessary to keep the suction pressure below 10 psi to prevent the valve from opening and dumping water; others are set to operate at 40 to 50 psi. When the valve is closed, it indicates that the pump is utilizing all of the water being delivered by the relay.

If the pump at the fire is not supplying water to fire streams when the relay hose line is being charged, it is a good practice to open drains or an unused discharge gate to bleed air from the hose. When the pumpers are fitted with a relay relief valve, some models will evacuate the air automatically.

A relay relief valve may not be necessary if the pumper's relief valve or governor is properly set and if the nozzle operators use reasonable care in opening and closing the nozzles slowly.

Four-Way Valve

In a single-line relay, it is a good practice to place four-way valves in the line in the

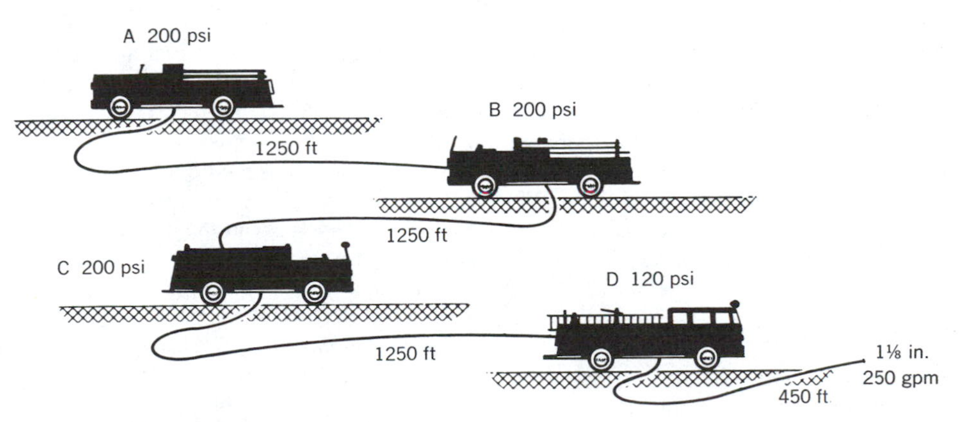

Figure 11-3. Pumpers relaying water

locations where pumpers are to be spotted. This will be beneficial in several ways:

1. If a pumper breaks down, it may be replaced without shutting down the relay.

2. A pumper may be removed when it is not needed.

3. Placement of a four-way valve may be made at strategic points in the line when it is being laid; thus pumpers may be assigned later as they arrive.

Siamese Inlet Fittings

Most pumpers have an inadequate number of gated suction inlets. If the first lines laid in a relay utilize all of these gated suction inlets, or if the apparatus has none, subsequent lines laid to reduce friction loss may require shutting down the flow of water to add more supply lines.

It is a good practice to connect the first-laid lines directly into the main pump suction inlet for the following reasons:

1. This eliminates the large amount of friction loss present in gated suction inlets.

2. Gated suction inlets can be used to connect additional hose lines without shutting down the relay.

Siamese fittings (Figure 11-4) equipped with clapper valves or gates are the best, otherwise the unused inlets must be capped. If the fitting is heavy, it is best to use a suction hose between the pump inlet and the siamese to reduce the amount of suspended weight. Necessary fittings will vary according to the size and type of suction hose, the fittings carried on the apparatus, and the size of the fitting on the pump suction inlet.

Pumping Procedures

As soon as a hose line is coupled to a discharge valve or a gated suction inlet, the valve should be fully opened. This will accomplish several objectives: it prevents forgetting to operate the valve later, it reduces the possibility of the valve jamming closed when the lines are pressurized, it allows air in the lines to escape, and it speeds up the entire operation. If there is any possibility of delay in attaching an incoming line to the suction inlet, place a hose clamp or shutoff butt on it at once.

The source pumper should start pumping as soon as water has been obtained and the discharge lines connected. The other relay pumpers should have the inlet and outlet valves open and be ready to place the pump in gear as soon as water is received. Open an unused discharge gate or bleeder valve to exhaust the air from the hose line; close it when water starts to be discharged.

Relay lines should be charged slowly to get the air and kinks out of the hose; these could result in burst sections due to excessive initial pressure. As soon as the line has been charged, the pump discharge pressure should be slowly increased to the desired figure. The relief valve or governor is then set. After the operator nearest to the fire receives water from the relay, the throttle can be advanced as required to supply the fire streams or until the pressure on the suction gate falls below 10 psi, whichever occurs first.

After a relay is in operation, pump operators must remain continually on the alert. In addition to procedures normally attended to on any pumping job, they must always remain aware that they are a part of a team operation requiring continual coordination with other pumping units. Basically, however, relay pumping is no different from any other pumping job. Water comes in the suction side and goes out the discharge side with a boost in pressure.

Any change in the operation of any pump in a relay operation, such as raising or lowering the pressure, or closing the discharge gates, will affect the operation of other pumps. This can be detected by observing the suction and discharge gages. The

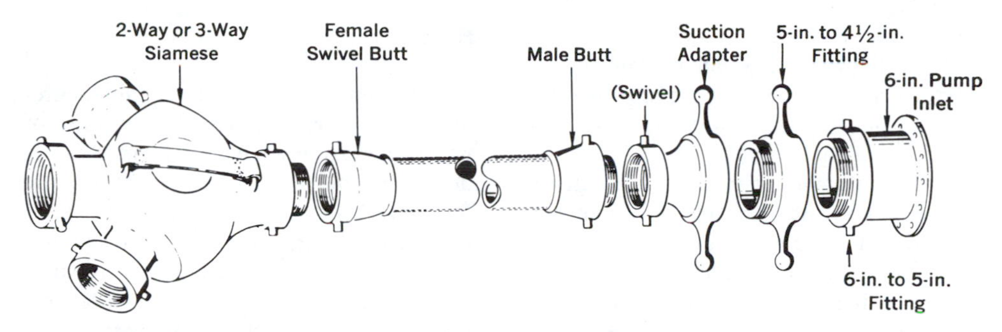

Figure 11-4. A large siamese fitting will allow two or three hose lines to be attached directly to the main pump suction inlet.

sudden closing of the nozzles or the discharge gates on an intermediate pump can cause a severe water hammer that may rupture hose lines or damage pumps between the point of shutdown and the source of supply; therefore, any shutdown, whether from nozzles or from a pump in the line, should be made very slowly to give pump operators a chance to prevent excessive pressure buildups.

Ruptured hose lines outside the vision of pump operators may be detected at the pump behind the rupture by a rapid drop in its discharge pressure. If operating through two lines, slowly close the discharge gate supplying the burst line, make the throttle adjustment and continue pumping through the single line at the same pressure while the damaged section is being replaced. To identify the line having the ruptured section, note the discharge pressure immediately upon closing the discharge gate. If the proper valve has been closed, there will be a substantial increase over the original pressure. The other pump operators will have to make throttle adjustments to avoid excessive pressure or vacuum. The pump closest to the nozzle from the burst section will probably have to operate at zero suction pressure in order to obtain as much water as possible while the damaged section is being replaced.

Unaccountable increases in pressure may indicate debris on the hose restricting the flow; other causes may be a closing of the discharge gates on a pump nearer the nozzle, or a closing of a nozzle.

The pumper gage customarily given the greatest attention is the one showing the pressure being discharged through the hose lines. But, the gage on the suction, or intake, side of the pump is also important. An operator who can interpret this gage correctly can get maximum performance from the machine and keep out of a lot of trouble by not attempting to supply hose layouts for which there is not adequate water.

To make proper use of the suction, or intake, gage, notice the pressure on the suction side after the hydrant has been turned on but before a line is charged. This will indicate static pressure in the water system with no water flowing from the hydrant. The difference between this static pressure and the pressure with which the hydrant supplies various flows will indicate the amount of water that can be obtained from the hydrant and the residual hydrant pressures at which various flows may be obtained.

A convenient way of estimating available flow is on the basis of the number of fire streams a hydrant can supply. When a 300-gpm spray tip is utilized, consider the flow in multiples of 300 gpm; if two 100-gpm fog nozzles are being used, think in multiples of 200 gpm. When the pressure drop is not over 10%, the hydrant should be good for three more streams; if the pressure drop is not over

15%, two more streams will be possible; a drop of not over 25% will indicate one more stream; while a drop of over 25% will show that some additional water in lesser quantities is possible.

If the pumper at the fire cannot develop adequate fire streams on the hose lines being used on the fire without running away from the water, he should notify the officer in charge. If the supply relay cannot increase its pressure or lay additional hose lines to reduce the friction loss, one or more of the discharge nozzles will have to be shut down. Additional water may be secured by opening the tank dump valve. Water tanks permit application of water at higher rates for brief periods than where a direct relay is the sole supply.

Tank Operation

If a hose in the relay should burst, the operator of the pump at the fire would revert to tank operation until the ruptured section is replaced.

When the pump at the fire is supplying attack lines from its booster tank, the water received in the relay can be used to supplement the tank supply. This permits much greater flexibility in the use of water than where the relay is used to supply a stream directly on the fire.

The operator of the pump at the fire, upon receiving water from the relay, should close his tank dump valve and operate from the relay supply. If the discharge lines are not using all the water being received, the pump operator can open the tank fill valve and refill the tank. This is a good precaution in case hose in the relay should burst and require replacement.

When no more water is needed from the relay, notify all pumpers before shutting down so they may refill their water tanks. Pressure surges are best eliminated if the shutdown is originated at the source pumper and proceeds in order of pumpers to the last unit in the line.

Hose Layouts

In relay operations, the essential requirement is to keep the necessary volume of water flowing between the pumpers so the desired nozzles will have sufficient pressure for effective fire extinguishment. The key to good fire fighting is to make the maximum use of the capacity of the pumping apparatus; this can be accomplished by using large hose, or several smaller lines, to reduce the friction loss to the smallest possible figure (Table 11-2).

Ordinarily, hose will be laid from apparatus working on a fire to the water source. However, in many circumstances, it will be advantageous to stretch lines from the water source to the fire. In some cases, one engine will lay its hose in one direction and the other pumper will extend the line by laying in the opposite direction. The methods of hose laying employed and fire ground situations may be determining factors. All pumpers should carry adaptors, double male, and double female fittings to allow reversing normal hose lays and to permit joining of lines laid by different apparatus with a variety of hose sizes. Allow at least 50 ft of reserve hose at each pumper for emergency use in replacing a burst section or to respot the vehicle.

Different hose layouts will produce different results at a given pump discharge pressure, if the pump and water source are capable of providing the desired flow. When a good residual pressure from a strong hydrant or a large-capacity pump is available at the source, the pressure may be increased to help the downstream relay pumpers.

The hydraulic formulas are reasonably accurate only when the most efficient sizes of hose are being utilized for the desired volume of fire flow. Regardless of what answer a mathematical problem would indicate, water will always flow from the end of a hose line unless elevation is an insurmountable obstacle.

TABLE 11-2. FRICTION LOSS (FL) IN VARIOUS HOSE LAYOUTS PER HOSE LENGTH IN FEET

				SINGLE 2½-in.					
Flow, gpm	100	150	200	250	300	350	400	450	500
FL	2.5	5.8	10.1	15.3	21.2	28.4	36.2	45.2	55.0
				TWO 2½-in.					
Flow, gpm	300	350	400	500	600	700	800	900	1000
FL	5.9	7.9	10.1	15.2	21.2	28.3	36.2	45.2	55.0
				THREE 2½-in.					
Flow, gpm	400	500	600	700	800	900	1000		
FL	4.7	7.1	10.1	13.4	17.2	21.2	25.8		
				ONE 2½-in. AND ONE 3-in.					
Flow, gpm	400	500	600	700	800	900	1000		
FL	5.9	9.0	12.7	16.8	21.5	26.7	32.4		
				ONE 3-in.					
Flow, gpm	200	250	300	350	400	500	600	700	800
FL	3.8	5.9	8.2	11.0	14.1	21.2	29.9	39.5	50.5
				TWO 3-in.					
Flow, gpm	400	500	600	700	800	900	1000		
FL	3.9	5.9	8.2	11.0	14.1	17.5	21.2		
				ONE 3½-in.					
Flow, gpm	400	500	600	700	800	900	1000	1100	
FL	6.3	9.5	13.4	17.7	22.7	28.2	34.3	41.0	
				TWO 3½-in.					
Flow, gpm	600	700	800	900	1000	1100	1200	1300	
FL	3.5	4.7	6.0	7.6	9.3	11.2	13.3	15.6	
				ONE 4-in.					
Flow, gpm	500	700	800	1000	1100	1200	1500	2000	
FL	5.0	9.6	12.4	19.2	23.0	28.5	41.5	75.0	
				ONE 5-in.					
Flow, gpm	500	700	800	1000	1100	1200	1500	2000	
FL	1.7	3.3	4.3	6.6	7.9	9.5	14.5	26.0	
				ONE 6-in.					
Flow, gpm	500	700	800	1000	1100	1200	1500	2000	
FL	0.62	1.2	1.7	2.6	3.0	3.5	5.7	12.0	

On the other hand, it is impossible to overcome high friction losses when flowing large volumes of water through relatively small hose by increasing the discharge pressure. The formulas indicate that a pumper could deliver 500 gpm through a single 2½-in. hose line with a friction loss of 55 psi per 100 ft; do not depend upon it. Many factors influence the flow of water through hose lines: the actual diameter of the hose, the internal roughness, the sinuous winding position of the hose, and the amount the hose increases in diameter at high pressures. In addition, there is a factor which severely limits the quantity of water that can flow through any hose line: the critical velocity of the water. When a fluid reaches its critical velocity, regardless of how much additional pressure is applied, the velocity will not increase; if the velocity will not increase, the quantity cannot increase.

The laying of hose will be facilitated if the pumpers have divided hose beds, which allow laying two or more parallel lines at the same time. In a practical operation, distance between pumpers should be limited to the hose carried in the divided hose bodies. Thus, engines that carry 1500 ft of hose would be spaced at 1500-ft intervals for a single-line relay and at 750-ft spacing when parallel lines are utilized.

When sufficient pressure cannot be maintained, or a larger tip is desired, lay another hose line. This is easiest accomplished by using a pumper to lay the additional hose line from the discharge gate of one engine to the next gated suction inlet. If necessary, spare hose on other apparatus on the fire ground may be hand laid.

As soon as the hose line or lines have been laid and connected, they should be loaded. Water should be waiting at the last pumper by the time it has been connected up.

Relay operations with a single line of 2½-in. hose permit long distances between pumpers, but because of the limited capacity of this hose size, the amount of water moved is relatively small considering the total capacity of the pumpers employed. Also, with a single line the entire relay is cut off with the failure of any hose section in the layout.

When 2½-in. hose is used, the flow is generally in the 200 to 250 gpm range. The flow can range up to 400 gpm with 3-in. hose, but with 3½-in. hose, 500 gpm can be supplied at about the same engine pressures and distances between pumpers as 200 gpm with 2½-in. hose. Or to look at it another way, an engine can deliver a certain quantity of water nearly 6 times as far through 3½-in. hose without any increase in pump pressure than it could through 2½-in. hose. The virtues of hose for relay operations multiply as the diameter increases. Four-inch hose is about 11 times more efficient than 2½-in. in terms of waterway capacities, and 5- and 6-in. hose is approximately 32 and 80 times more effective, respectively.

Communications

It is essential for the success of a relay operation that communication be maintained among the various relay elements. Difficulties encountered in attempting through communications to coordinate the pump operators' activites and individual discharge pressures was the principal reason that a standardized pump discharge pressure was adopted.

Radio

The best method of communication between pumpers during relay work is by radio. If radio communication is not available or is too jammed with messages, and pump operators are too far apart for ready communications, it is advisable to use standardized signals when any change in working conditions is desired. If the engineers are out of sight of each other, crew members can be stationed at strategic points to receive and relay the signals between pumps.

Hand Signals

Stand facing the person being signalled. The person receiving a signal should answer with the same signal received, except when unable to comply with the change indicated by the signal. The following signals are commonly used:

Hold the water. Place the hands overhead and wide spread, with palms toward the person being signalled in the universal STOP signal (Figure 11-5).

Charge lines or increase pressure. Wave the outstretched arm up and down (Figure 11-6).

Lower pressure. Wave the outstretched arm from side to side. This signal is also used by the operator at the source of supply to indicate inability to increase pressure (Figure 11-7).

Shut down but not break up. Place the arms overhead with hands touching, making an inverted V (Figure 11-8).

Figure 11-5. Hold the water.

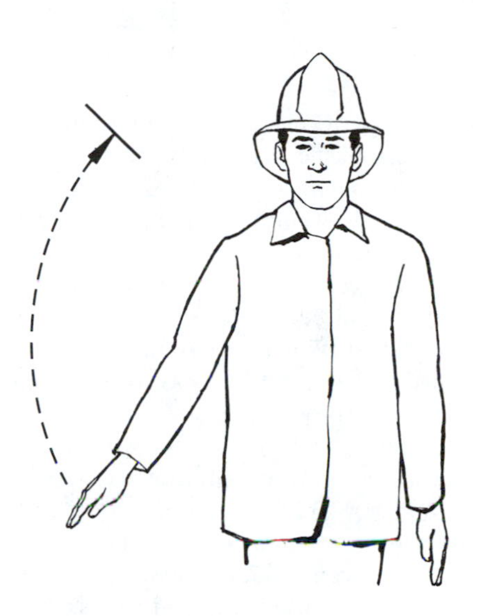

Figure 11-6. Charge lines or increase pressure.

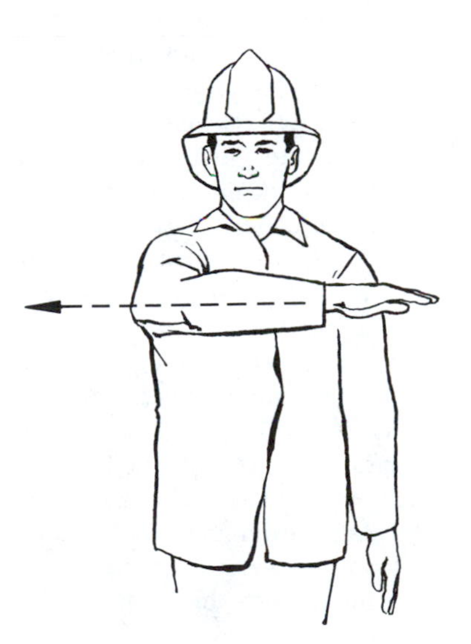

Figure 11-7. Lower pressure.

Figure 11-8. Shut down or break up.

Break up. Same signal as above while already shut down.

Rural Relays

In many rural areas where water supplies are limited, specific operations are needed for fire control. Officers in charge must decide quickly how to employ apparatus: whether a single pumper will pump directly from the water source to nozzles at the fire; whether a relay should be employed, and if so, what type; or whether better results will be obtained by using tank trucks to move water from distant sources.

Generally, relays are employed for two reasons: to provide a continuous supply for pumpers or tankers working at a fire and to move water through hose layouts that would require excessive pump pressures if a single pumper were used. A good rule is that a relay should probably be employed whenever over 200 psi engine pressure is required to supply effective hose streams. Without a relay, some water will be delivered through long lines, but possibly not in adequate volume or pressure. For example, to supply 100 psi pressure to two 100-gpm spray nozzles through 2000 ft of 2½-in. hose, a single pumper would have to maintain a 300-psi engine pressure. By using a two-pump relay, the average discharge pressure per pumper would be cut in half. As the length of the line between pumps in a relay is reduced, additional water can be relayed at a given pressure. With pumpers 1000 ft apart, 200 psi at the source pump will relay about 300 gpm through a single line of 2½-in. hose.

Open Relays

At rural fires, open relays with portable tanks (Figure 11-9) have proven very useful. When an open tank is used, the main problem is to balance the nozzle output with the supply. The larger the tank, the better the results, whether the water is delivered by a hose line or by a tanker shuttle.

Reservoirs have the advantage of permitting incoming tankers to discharge their water immediately upon arrival, without waiting, and to proceed immediately for more water. This permits a maximum use of the pumpers and tankers in supplying hose lines; by using large capacity reservoirs, lines may continue to operate for some time when the tankers do not return immediately with water. Open relays also reduce the effects of fluctuating and surging pump discharge pressures. In a closed relay, pressure fluctuations can result in burst hose lines, damage to the pump, or ineffectual fire streams. In open relays, such pressure fluctuations merely result in a small change in the water levels of intermediate storage facilities. Reservoirs may be improvised by using salvage covers, ladders, or other materials on the fire ground.

Tanker Shuttle

Tank trucks shuttling water to a fire may utilize any one of several methods: they could pump directly into the hose lines supplying the fire streams, they could replenish the booster tanks of pumpers operating on the fire, they could dump their loads of water into a reservoir that a pumper is drafting from, or they could pump their water into a hose line which is supplying a pumper working on the fire. This last operation would be a closed relay; it is very satisfactory where narrow roads would render tanker traffic directly into the fire ground impracticable.

In the example illustrated in Figure 11-10, the first-in engine company places a gated or clappered siamese fitting on the end of the hose and lays a single line up the long driveway to the fire; an attack is made on the blaze with water from the booster tank. Tankers can shuttle water from the nearest source and pump into the siamese, thus avoiding travel on the narrow dead-end road or driveway. When one tanker becomes empty, it will disconnect the hose from its discharge outlet and another tanker can

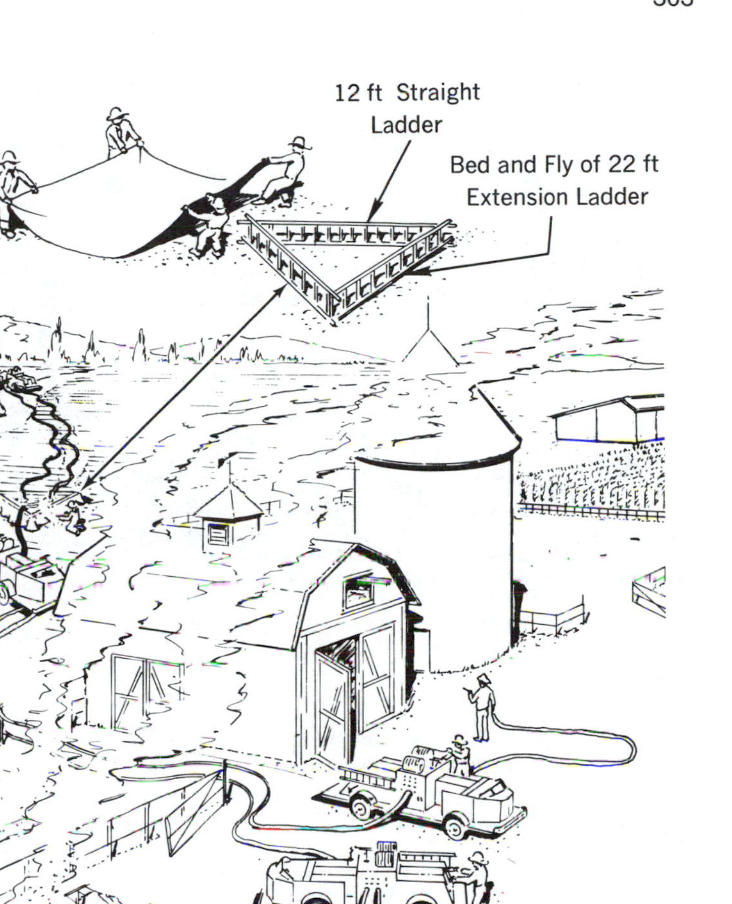

Figure 11-9. Open relay utilizing portable tanks and improvised reservoir

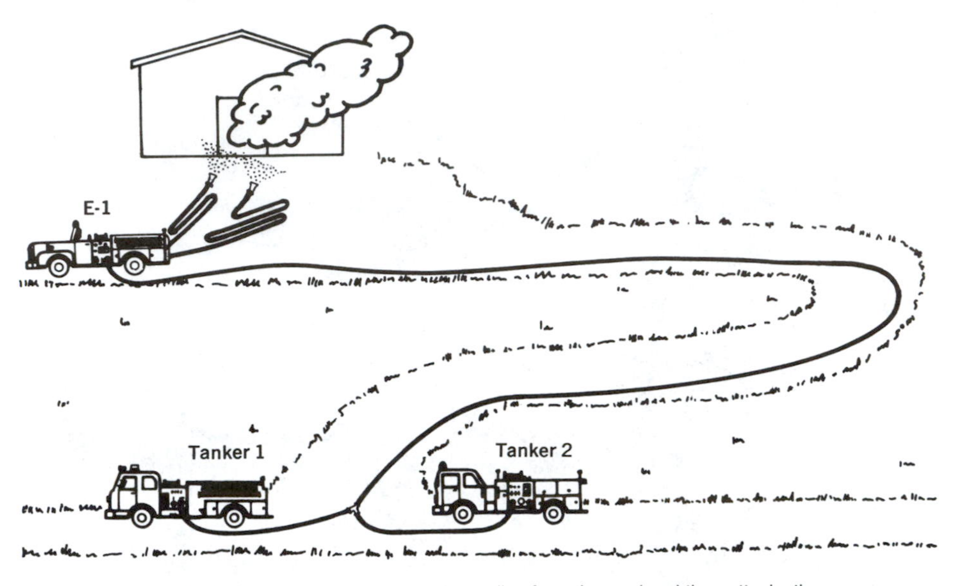

Figure 11-10. First-in engine company lays a line from the road and then attacks the fire with booster tank water.

replace it. It is best if only one tanker at a time pumps into the hose line; this will allow it to empty and go back for another load more quickly.

Small Quantity Relays

When a relay operation is set up for an ultimate consumption of water less than 200 gpm, friction loss is a much smaller problem than if large flows are expected. It is more likely that pump relays will be used at mountain and rural fires than anywhere else. At this type of fire, the ultimate consumption of water will be relatively small.

For purposes of illustration, assume that the ultimate delivery is through a ¼-in. tip at 50 psi (13 gpm). Friction loss for this flow in a 1-in. hose is about 5 psi per 100 ft. If the nozzle pressure on a ¼-in. tip is dropped to 35 psi, a good stream will still be produced, but the flow will be cut down to about 11 gpm with a corresponding drop in friction loss. Without giving any consideration to back pressure for the moment, it may be shown that an almost unlimited length of 2½-in. hose may be used without resorting to relay because the loss per 1000 feet is only 1.15 psi.

This is probably the smallest amount of water consumption that might be used. The same demonstration may be made with another illustration, using more water. Assume six or seven 1-in. hoses leading from one 2½-in. supply line; this gives a probable consumption of about 100 gpm, with a loss in the 2½-in. line of about 3 psi per 100 ft, or 30 psi per 1000 ft. Still not counting back pressure, it should be possible to go at least 5000 ft without resorting to relaying.

Carry this still further and assume that water is finally delivered at a nozzle pressure of 75 psi to one 100-gpm fog tip on a 1½-in. line; the loss in the 2½-in. supply line will still only be about 3 psi per 100.ft, so it would still be practical to go at least 5000 ft before setting in relay pumps.

Maximum delivery would probably be through two 1½-in. fog tips at 75 psi nozzle pressure on 1½-in. lines, and loss in a 2½-in. hose now begins to be a considerable factor. Loss in the 2½-in. hose for this flow (200 gpm) is 10 psi per 100 ft, or 100 psi per 1000 ft. At the point where the 2½-in. line is wyed down to the smaller lines, there still must be sufficient pressure to carry water through the small lines and expel it with sufficient pressure at the nozzle. With this flow then, friction loss in the smaller lines will have to be considered before deciding how far it is possible to go with the 2½-in. hose without relaying. Without giving consideration to back pressure and with a water consumption of 200 gpm or less, it will be quite safe to assume that an engine company with 1500 ft of 2½-in. hose could lay out all this hose in one line without having to resort to relay operations.

Where long lines of hose are used at rural fires, back pressure must be considered. In fact, where the ultimate use of water is less than 200 gpm, consideration given to back pressure will often be more important than that given to friction loss in determining the positioning of relay pumps. Terrain plays a large part in the placement of pumps for relay operations, and it may not always be feasible to place all pumps in such a manner that they can all pump within 200 psi. This applies particularly where relay pumps are being placed along narrow roads. It is necessary to keep the roads open for other fire department traffic, so relay pumps must be placed at passing points.

The determination of head in mountain areas is difficult. There is, however, one sure way of finding what the head is at any or all relay pumps in the line. To do so requires a complete shutdown for a few seconds, but fire conditions may permit this action.

To determine the back pressure (head):

1. Load the line from the hydrant to the nozzles.

2. Close the hydrant–have all pumps disengaged–allow a few seconds for the gages to steady.

3. Read the pressure gages at each pump in the line; they will register the head in psi at that point.

With the pump disengaged, the pathway through a centrifugal pump is continuous and uninterrupted; therefore, the same pressure readings will be observed on both the suction and discharge gages. The pressure readings on the gages of each pump in the layout are caused by the head between that pump and the highest point in the relay. In Figure 11-11 the elevation between the highest point in relay D and pump C is 30 ft. This will cause a pressure reading of approximately 13 psi at pump C. The elevation between point D and pump B is 65 ft and will cause a pressure reading of about 29 psi at pump B. The elevation between point D and pump A is 100 ft, which will cause a pressure reading of approximately 43 psi at pump A.

To determine the head by this method between any pump and the next pump in the relay, it is necessary to close the discharge gates of the next pump. The pressure equivalent of the head can thus be read on the gages. As an example, in Figure 11-12 pump C has closed its discharge gates and the pressure reading on its gages is zero. The elevation between pump B and pump C is 35 ft, which will be shown by a pressure reading of approximately 15 psi at pump B. Pump A will have a pressure reading on its gages created by the head from pump C to pump A.

At the beginning of a relay operation, it probably will be inadvisable to shut down long enough to determine back pressure by the above method.

Another method, which is not so accurate, but which will give a fairly close approximation, is to assume a certain maximum percentage of grade and allow accordingly. Relay lines will probably be laid along roads of one kind or another, either on streets and highways or along rural roads. The steepest grades will be encountered on mountain roads, though a definite effort has been made to hold their grade within 12½%. In some locations, and generally only for very short distances, this percentage may be exceeded. Based on this information, back

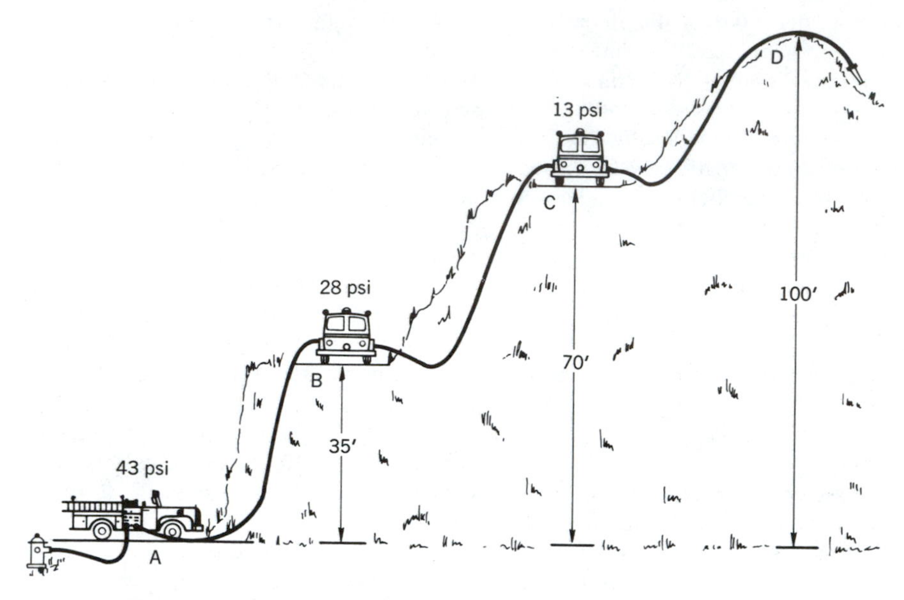

Figure 11-11. Figuring back pressure

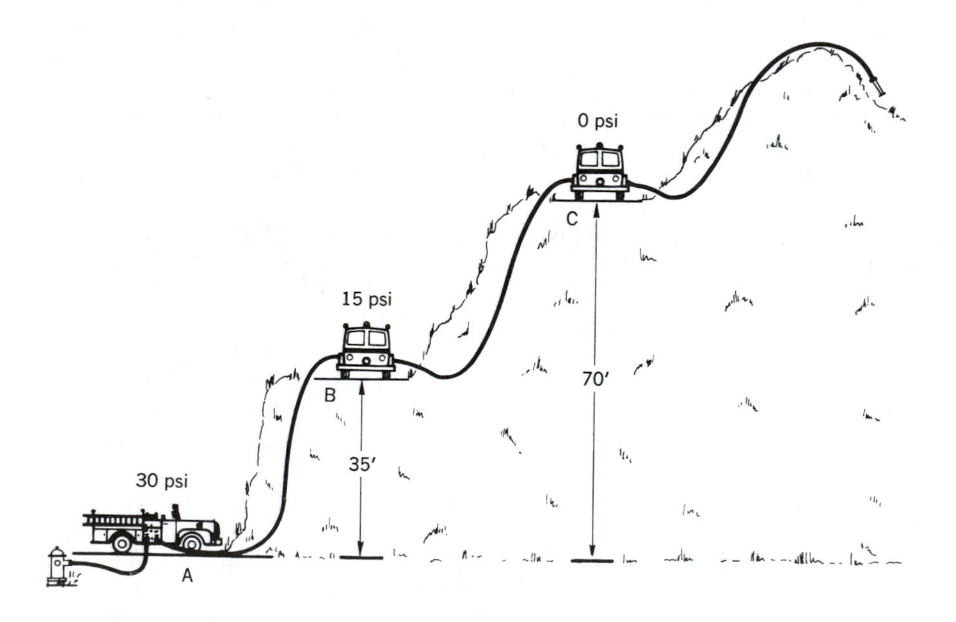

Figure 11-12. Figuring back pressure—alternate method

pressure may be determined with a fair degree of accuracy: When setting up a relay operation uphill along a mountain road, assume a 10% grade—10 ft of rise to each 100 ft of hose laid out. To put it another way, for each 1000 ft of hose laid uphill along a road, assume 45 psi back pressure.

Once a relay pumping layout has been set up, the best method of determining the necessary pump pressures at each pump in line is by radio communication.

Large Quantity Relays

When it is necessary to pump in relay and deliver quantities of water in excess of 200 gpm, the problem of overcoming friction loss becomes more and more critical as quantity increases. Everything that has been said previously concerning overcoming back pressure still holds true because increased quantity does not cause any change in the head. There is one important fact to consider, however, in relation to overcoming head and at the same time delivering increased quantity—to overcome a head of 100 ft, a pressure of 43.4 psi is necessary regardless of flow; but, as quantity increases, the power necessary to maintain the same pressure (43.4 psi) also increases at the same rate. For ease of calculation, allow 4½ psi per 10 ft.

Since pressures at each pump in the line should be held to a 200 psi maximum, the proper spacing of pumpers will be determined by the balancing of three factors: quantity, friction loss, and head. Quantity is governed by the need for water at the fire, the capacity of the pumps, and the available water supply. Friction loss is governed by the quantity of water and the size of the hose. Head is determined by terrain.

For purposes of illustration, in applying the above, assume a hypothetical suitation. The last pump in a relay line requires approximately 250 gpm to supply hose lines combating a fire (Figure 11-13). This pump is 2000 ft distant and 200 ft above the water supply. How many pumps and what hose is necessary to complete this relay? Assume that intermediate pumps can be spaced anywhere along the 2000 ft of roadway (which many times will not be possible under fire conditions).

The total head to overcome is 200 ft; converted into pounds pressure, this amounts to approximately 90 psi. If a single line of 2½-in. hose is used, total friction loss for 2000 ft will equal 300 pounds when 250 gpm is flowing. At least 10 psi pressure should be maintained on the inlet sides of the pumps. The total pressure requirement for this lay amounts to 400 psi.

$$\begin{aligned} FL &= 300 \text{ psi} \\ BP &= 90 \text{ psi} \\ &+ 10 \text{ psi} \\ \hline EP &= 400 \text{ psi} \end{aligned}$$

One pump could not deliver this amount of water to the pumper at the fire without exceeding its pressure limitations on hose and equipment.

By using a pumper in the center of the hose line (Figure 11-14), each pump would exceed the 200 psi limitation by a small amount; this is permissible when necessary.

$$\begin{aligned} FL &= 150 \text{ psi} \\ BP &= 45 \text{ psi} \\ &+ 10 \text{ psi} \\ \hline \text{EP} &= 205 \text{ psi} \end{aligned}$$

When dividing the total pressure requirements by 200 psi to determine the number of pumpers to place in the line, do not count the last pumper at the fire as one of the relay pumpers. In a single line relay, it would be better to place two intermediate pumpers in line about 700 ft apart. This would lower the discharge pressure of the relay pumps.

If two intermediate pumpers were used in the hose line, three pumps would be doing the work; therefore, the friction loss and back pressure would be divided by three:

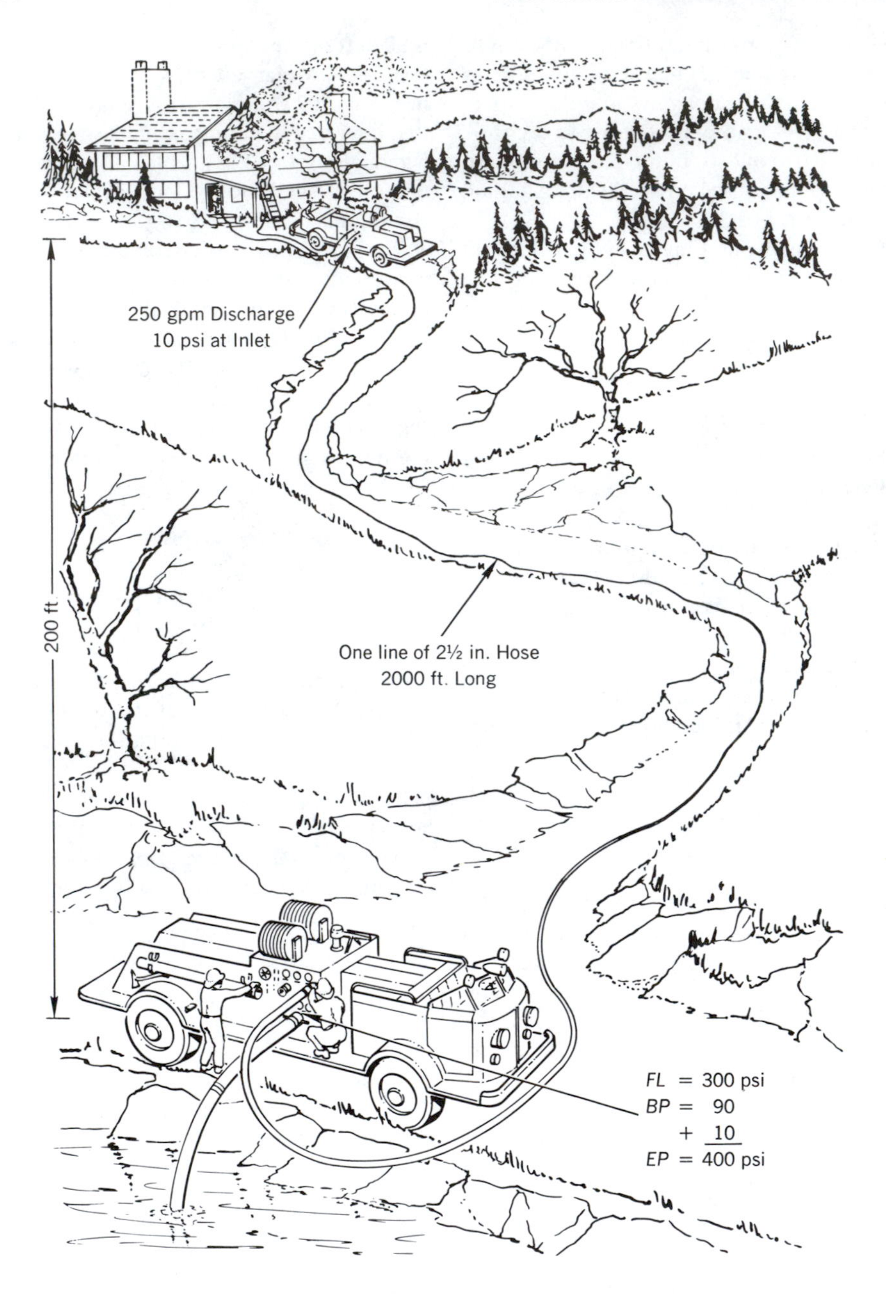

Figure 11-13. Closed relay; single line of 2½-in. hose, two pumps

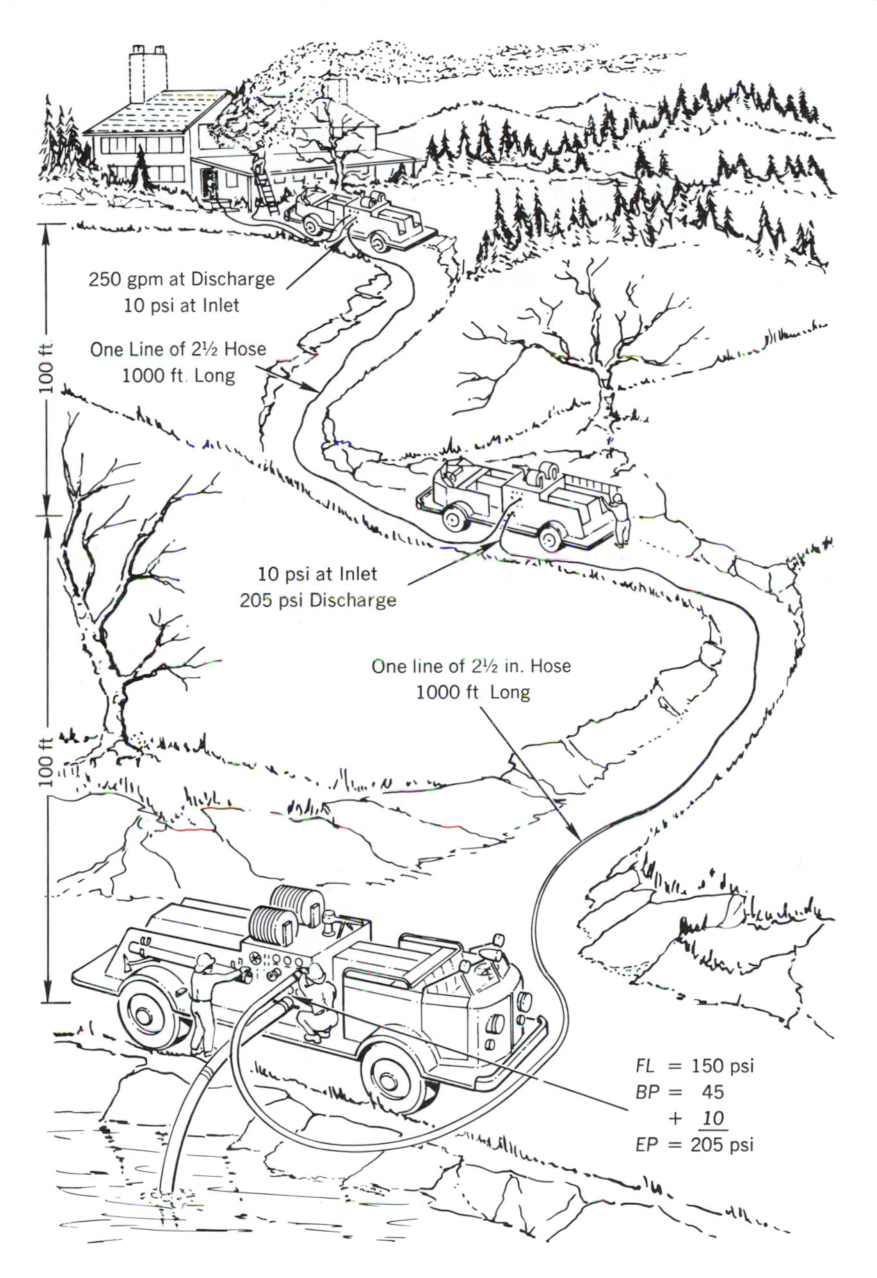

Figure 11-14. Closed relay; single line of 2½-in. hose, three pumps

$$\begin{aligned} FL &= 100 \text{ psi} \\ BP &= 30 \text{ psi} \\ &+ 10 \text{ psi} \\ \hline EP &= 140 \text{ psi} \end{aligned}$$

By using two parallel lines of 2½-in. hose, the friction loss is reduced to approximately 80 psi, as only 125 gpm are flowing through each line (Figure 11-15). Back pressure remains the same, and 10 psi is still desired at the inlet of each pump:

$$\begin{aligned} FL &= 80 \text{ psi} \\ BP &= 90 \text{ psi} \\ &+ 10 \text{ psi} \\ \hline EP &= 180 \text{ psi} \end{aligned}$$

By placing an intermediate pumper in the relay using two parallel lines of 2½-in. hose, the discharge pressures will be further lowered (Figure 11-16):

$$\begin{aligned} FL &= 40 \text{ psi} \\ BP &= 45 \text{ psi} \\ &+ 10 \text{ psi} \\ \hline EP &= 95 \text{ psi} \end{aligned}$$

Supplying Heavy Stream Appliances

Heavy stream appliances and multiple large fire streams require such a huge quantity of water that relay operations are usually needed to supply the required volumes. Unless sufficient hydrants supplied by a well-designed water system are immediately adjacent to the fire ground, multiple company operations will be the most practical method of furnishing the water.

Ladder Pipe

In the example in Figure 11-17, a ladder pipe equipped with a 700-gpm fog nozzle is being operated on a fire; the nozzle is elevated to a height of 40 ft. The heavy stream is being supplied by 700 ft of 3½-in. hose. A good practical figure of 80 psi is generally utilized to compensate for the back pressure and friction loss in such an elevated set-up: the siamese inlet, the fittings, 100 ft of 3-in. hose running up the ladder, and the ladder pipe. The figure of 80 psi would apply regardless of the quantity of water flowing or the height to which it is raised.

The standard nozzle pressure of a fog nozzle is 100 psi. A pumper of at least 1000 gpm capacity must be placed adjacent to the base of the ladder in order to deliver this pressure to the nozzle. And there must be another pump at the water source in order to move 700 gpm through 700 ft of 3½-in. hose; with this quantity of flow, the friction loss in every 100 ft of the 3½-in. hose will be about 18 psi. Thus, the total friction loss to be overcome by the pumper will be 126 psi. To this is added the intake pressure of 10 psi for the second pump, for a required engine pressure of 136 psi.

Large-Volume Hose Layouts

When setting up a relay to move large volumes of water through a hose layout, much thought will have to be devoted to designing and adopting good practical standard operating procedures that will work under a variety of fire ground conditions. Following are several effective hose layouts that have proved to be efficient.

Figure 11-18 illustrates two ways to position six engines in a relay setup. In part A, two separate and independent relays waste pumpers and crews; six engines are required to supply two fire streams. Part B shows a better utilization of the companies. The fourth and fifth engines should lay an additional hose line between the first relay pumpers. The sixth engine could be used to protect exposures, patrol downwind for spot fires, or held in reserve to lay additional hose lines.

Figure 11-19 shows a three-engine relay hose layout. Engine 1 stops at the nearest

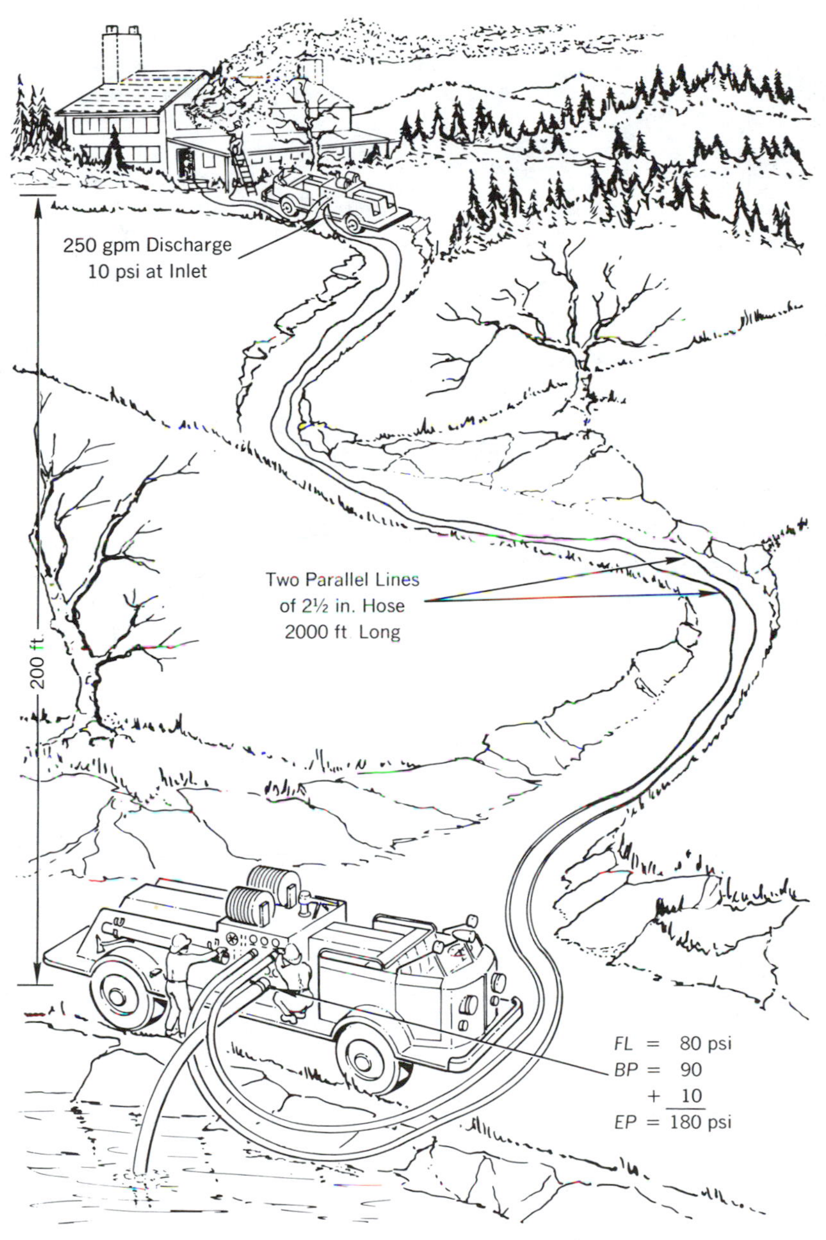

Figure 11-15. Closed relay; parallel lines of 2½-in. hose, two pumps

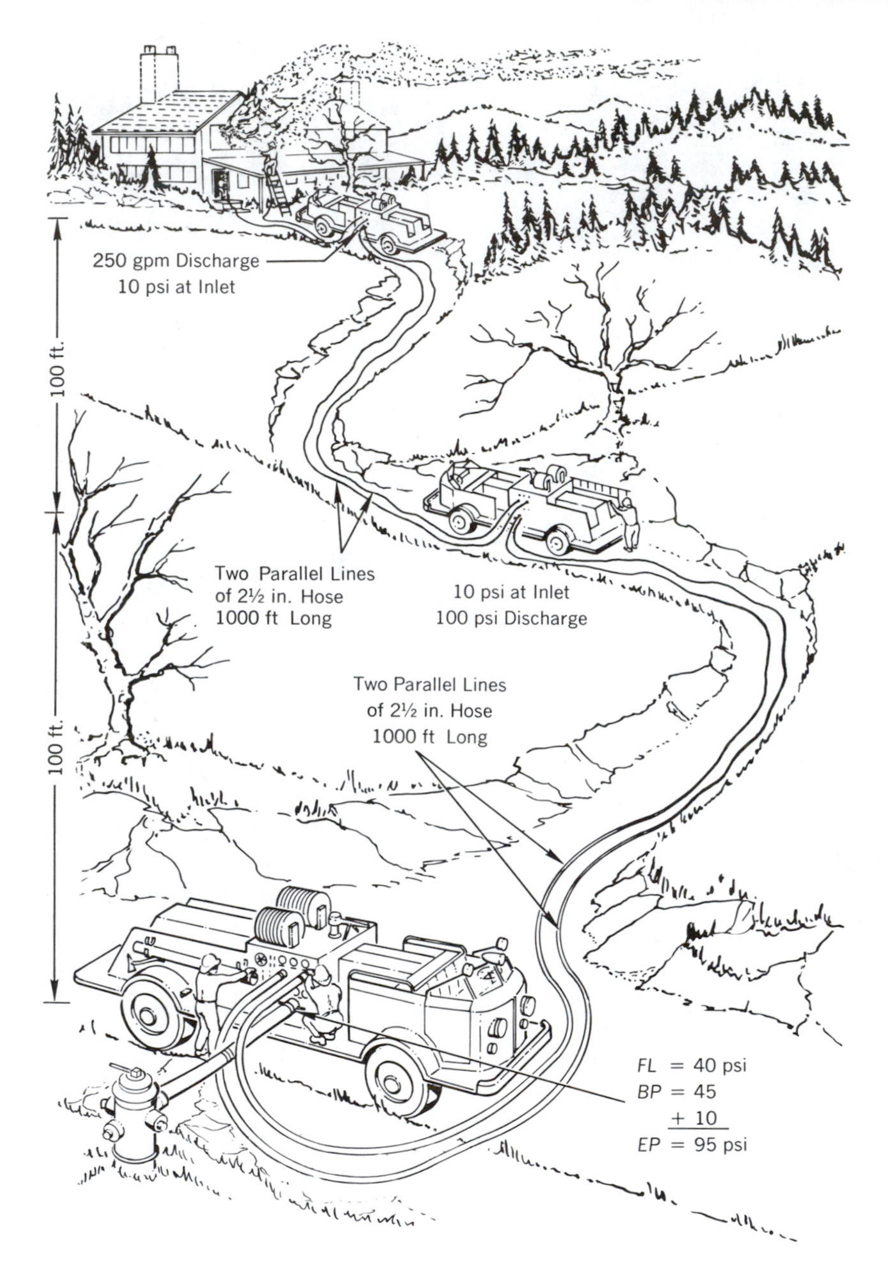

Figure 11-16. Closed relay; two parallel lines of 2½-in. hose, three pumps

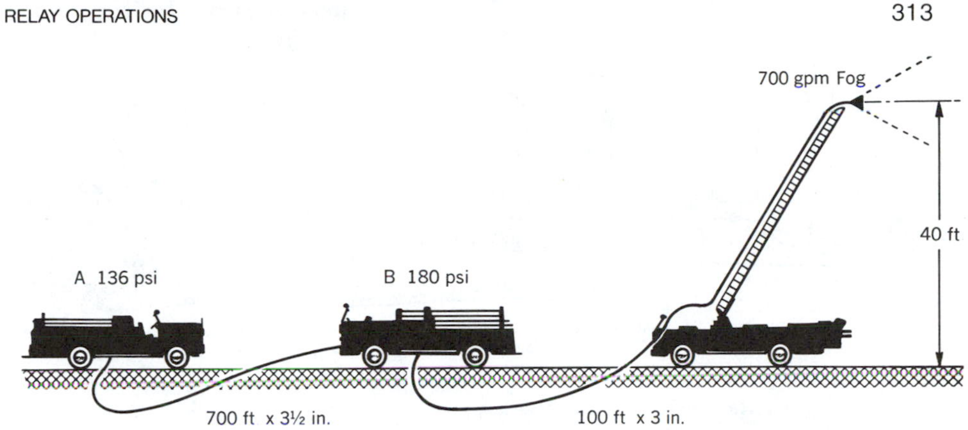

Figure 11-17. Pumpers relaying to ladder pipe

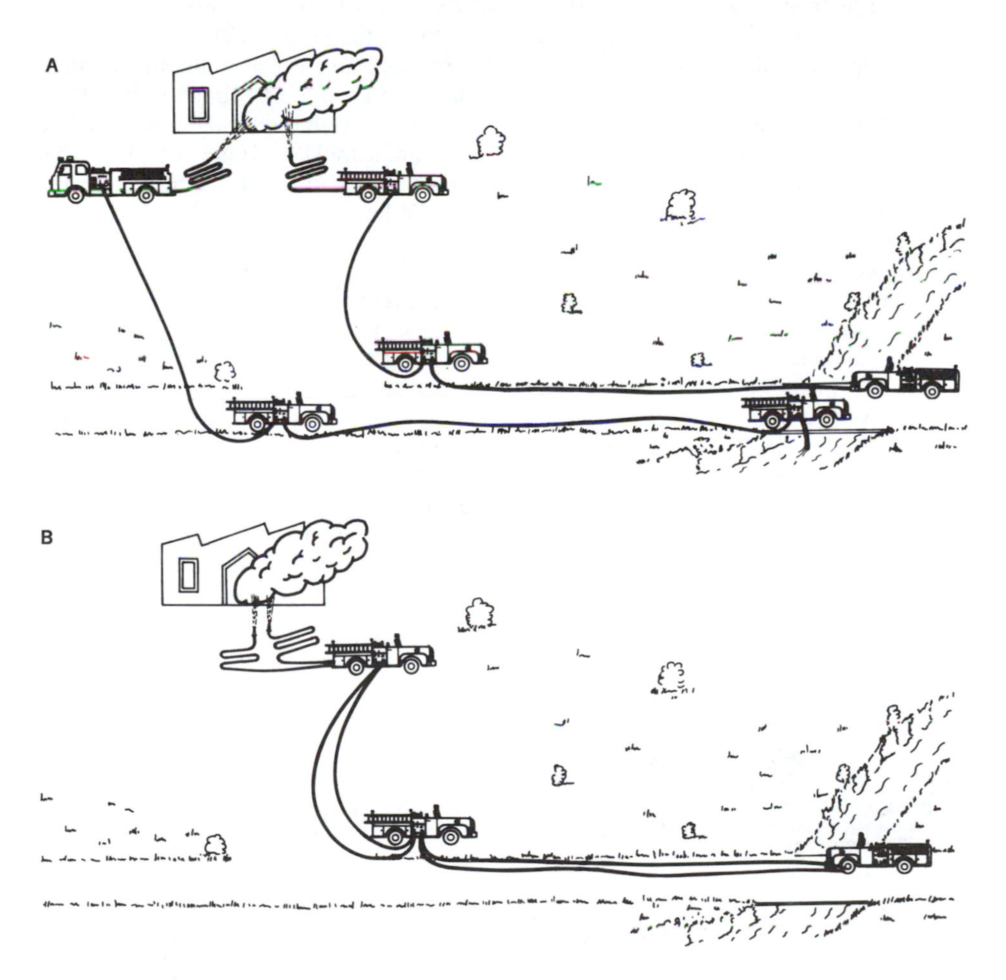

Figure 11-18. Examples of hose layouts with a six-engine setup

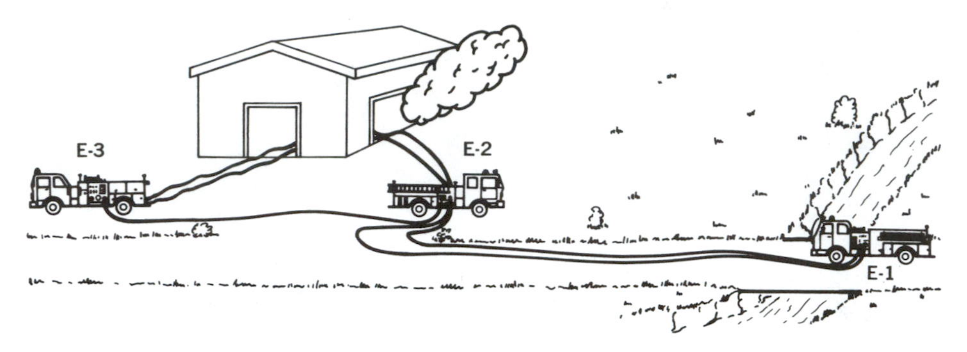

Figure 11-19. Three-engine relay hose layout

water source and prepares to draft or connects to a hydrant. Engine 2 lays one large hose line or two 2½- or 3-in. hose lines to the fire. Engine 3 lays a line from engine 2. Both engines 2 and 3 then attack the fire with booster tank water until engine 1 can furnish them an additional supply. The effectiveness of this operation will depend entirely on having sufficient hose layout water-carrying capacity to adequately supply the fire streams required. If engines 2 and 3 do not receive an adequate supply of water, engine 4 can lay one more hose line between the relaying pumpers.

A large-volume relay is illustrated in Figure 11-20. The first-in engine starts the operation by spotting at either water source or the fire. The second pumper lays two 2½- or 3-in. hose lines, or one larger line, from the first engine until all hose is laid out. The third and any additional pumpers repeat this operation until the fire or water source is reached. All pumpers should initially pump at 200 psi discharge pressure.

If the initial hose layout will not supply the required quantity of water, use another pumper to lay one of two additional hose lines between the relaying pumpers, as shown in Figure 11-21. Hand lines should be stretched and tended, ready to attack the fire after the heavy stream has knocked down the blaze.

The best method of reducing friction loss is to increase the hose size or lay a parallel hose line, as shown in Figure 11-22. Two parallel lines of hose will have one-fourth the friction loss of a single line of the same diameter and length carrying the same quantity of water. Three lines will have one-ninth the friction loss of a single line, and four lines will have one-sixteenth the friction loss. Adding the third hose line of equal size and length will reduce the friction loss to one-half of that created in two lines.

Figure 11-20. Large-volume relay

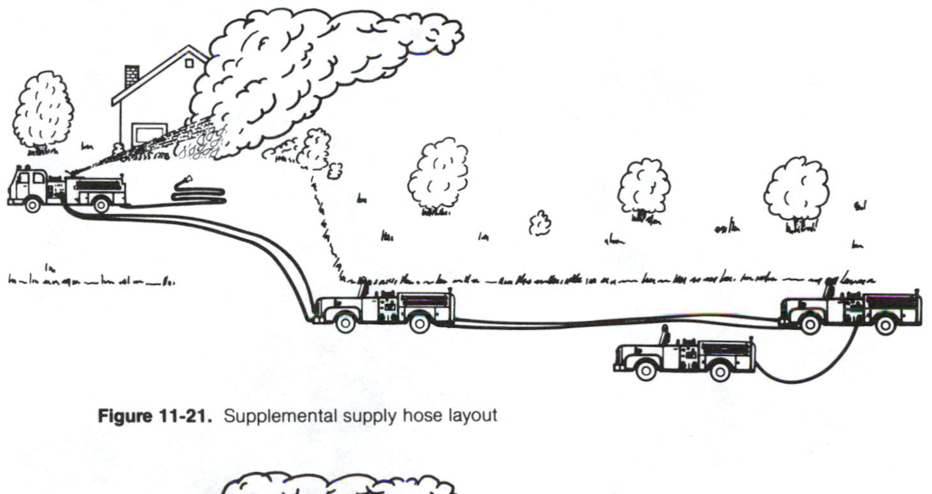

Figure 11-21. Supplemental supply hose layout

Figure 11-22. Parallel hose lines reduce friction loss.

(Courtesy of Crown Fire Coach Company)

12 Drafting Operations

Normally, most fire departments have a dependable water system with good hydrant distribution and accessibility to provide an efficient source from which pumping operations can be carried out. However, many rural organizations are not this fortunate; they have to utilize static sources of water such as rivers and streams, lakes and ponds, oceans, private reservoirs, stock ponds, irrigation ditches, and other natural and artificial bodies of water. Also, metropolitan areas that have excellent water systems may occasionally need to draft water because of an earthquake, tornado, or other disaster that could disrupt the normal use of fire hydrants. In fact, fire fighters in one large city had to draft flood water from the streets to combat a multistructure conflagration because the water was over the tops of the hydrants thus hiding them.

Fire departments should routinely inspect and utilize all auxiliary water sources for prefire planning and wet drills. They should also assure access to them by building approach roads, grading pads for the pumpers, dredging the water deeper to provide an adequate depth for the suction strainer, constructing permanent suction piping so the apparatus will not have to drive on treacherous ground to get close enough to the source, or make any other alterations that will ensure a dependable water supply in case of a nearby fire. In some areas, irrigation wells and other private water systems are fitted with adapters so that a pumper may obtain a supply. A list of these auxiliary sources of water should be carried on all pumpers that might have occasion to need them. It is mandatory that all fire fighters be thoroughly familiar and well drilled in obtaining water by drafting (Figure 12-1).

Success in drafting is assured by practice and a good preventative maintenance program to keep fire fighting apparatus in perfect operating condition. One piece of advice that is good for all pumping operations, but is especially true when drafting, is "do not try to hurry too much." Better results will be obtained by methodical, deliberate movements. When charging lines from a pump, the discharge valves should be opened slowly as pressure is increased; otherwise, it is possible to lose the prime. Also, there is less likelihood of damaging hose or injuring hose personnel when the lines are charged slowly.

Figure 12-1. Extensive waterfront fires require fire streams from fire boats as well as pumpers drafting water. (Courtesy of the U.S. Coast Guard)

Principles of Lift

It is commonly thought that a pumper, when drafting, lifts water from an open body of water to the pump inlet. Water has no tensile strength and therefore cannot be pulled upward. It requires an additional factor—atmospheric pressure—to move the water.

Definition of Lift

Lift is the difference in elevation between the water level and the center of the pump when a pumper is drafting water. Maximum lift is the greatest difference in elevation at which the pumper can draft the required quantity of water under the established physical characteristics of the operation.

Theory of Lift

In reality, nature plays an important role in drafting operations. Since water cannot be pulled upward, some force is required to push water through the suction hose into the pump. Nature, in the form of atmospheric pressure, provides this push.

The diagram in Figure 12-2 shows how this works: priming pump B has removed air from centrifugal pump C until a pressure of 10 psi remains within the centrifugal pump casing. As the pressure at point A is 14.7 psi (normal atmospheric pressure at sea level), the 4.7 psi differential will cause the water to move up suction D toward main pump C. If this difference of pressure is great enough to push the water into main pump C, it will travel through it to priming pump B and out of the discharge outlet of pump B. The operator, noticing that the discharge of priming pump B is clear of air, can close the valve between priming pump B and main pump C; the main pump is now primed.

Lift Limitations

From the foregoing demonstration it is evident that the maximum height to which a

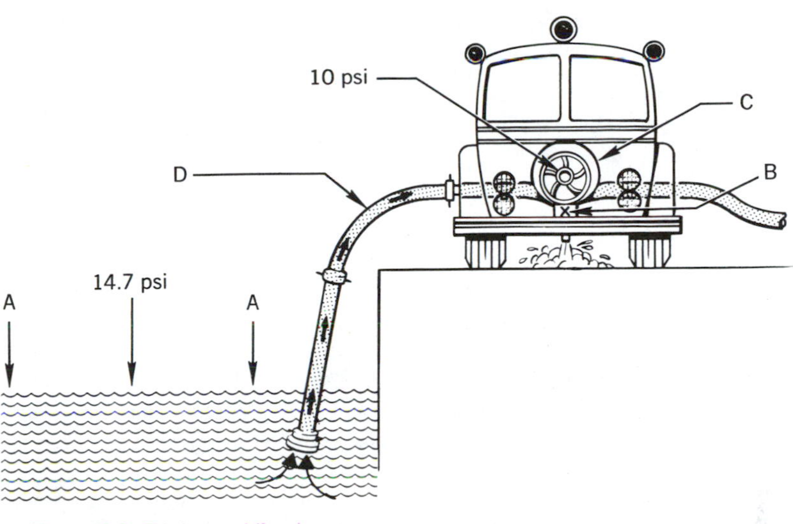

Figure 12-2. Priming centrifugal pump

pumper can lift water is determined ultimately by the atmospheric pressure. It follows that the tightness of the pump and the ability of the priming pump to create a vacuum will also affect maximum lift. At sea level, atmospheric pressure is approximately 14.7 psi; this pressure is capable of sustaining a column of water 33.9 ft in height. This figure is arrived at by multiplying 14.7 psi by 2.304, the height of water that will exert 1 psi pressure. Thus, if a pump could produce a perfect vacuum, the maximum height it could lift water at sea level would be 33.9 ft. But no pump can produce an absolute vacuum. Taking into consideration this inability to create a perfect vacuum because of minute air leakage and frictional resistance through the suction hose, a common practice is to use two-thirds of the theoretical distance, or 22⅔ ft at sea level, for the practical limit of a draft.

Water will rise 2.304 ft for each pound of pressure removed from the pump. The height in feet that water will rise can be calculated by multiplying the reduction of pressure within the pump by 2.3. As the suction gage on a pumper is calibrated in inches of mercury, any formula used to determine how far water will rise in the suction hose must be expressed in inches of mercury.

Height water will rise in suction hose = 1.13 Hg (Hg = inches of mercury).

Example: When the compound gage indicates 15 in. of vacuum, to what height will water rise in suction hose?

$$\begin{aligned} \text{Height} &= 1.13 \text{ Hg} \\ \text{Hg} &= 15 \text{ in.} \\ \text{Height} &= 15 \times 1.13 \\ &= 16.95 \text{ ft} \end{aligned}$$

A simplified rule is, "Water will rise one foot in the suction for each inch of indicated vacuum." The lower amount is based on the inaccuracy of the gage at near zero readings, friction in the suction hose, and other variables.

Although it has no practical use, the following formula is used to determine how many inches of vacuum is required to lift water to a desired height:

$$\begin{aligned} \text{Hg needed} &= 0.885 \text{ h} \\ \text{h} &= \text{height} \end{aligned}$$

Example: How many inches of vacuum will it take to raise water 16.95 ft?

$$\begin{aligned} Hg &= 0.885\ h \\ h &= 16.95\ \text{ft} \\ Hg &= 0.885 \times 16.95 \\ &= 15\ \text{in. of vacuum} \end{aligned}$$

Effect of Altitude

As lift depends upon the pressure difference between the vacuum within the pump and the atmospheric pressure, a problem is created as the elevation above sea level of the pumping site increases. Atmospheric pressure is reduced approximately ½ psi for each 1000 ft of elevation above sea level; this results in a loss of lift at various elevations.

Power generated by internal combustion engines decreases about 3½% for each 1000 ft of elevation. Therefore, an engine which was just adequate at sea level would be about 25% deficient at 7000 ft altitude.

Effect of Weather Conditions

Normally, rainy weather is accompanied by a reduction of atmospheric pressure. A difference in barometric pressure due to weather conditions will have the same result as a change in altitude. The difference in atmospheric pressure due to operation on a rainy day instead of a cool, clear day could easily mean a difference in lift of 1 ft.

Net Pump Pressure

When water is drafted, it requires the same amount of horsepower for the engine to lift the water from the source of supply to the pump intake as it does to force water an equal distance upward from the pump. For instance, it requires just as much power on the part of an engine to raise 500 gal per minute 15 ft by suction as it does to force 500 gal per minute to a height of 15 ft by pressure. An engine operating at a 15 ft lift and forcing the water that was lifted up an additional 35 ft is doing the same work as if it were forcing water up 50 ft. This will be reflected in the net pump pressure.

The net pump pressure is the pressure actually produced by the pump. When operating from a hydrant, the net pump pressure is the difference between the pump discharge pressure and the incoming pressure from the hydrant. For instance, if the pump discharge pressure is 150 psi and the pressure shown on the suction gage is 50 psi, the net pump pressure is 100 psi. As 1 psi of pressure will force water to a height of 2.304 ft, 100 psi will force water to a height of 230.4 ft. This means that the pump in this case is doing enough work to force water to a height of 230.4 ft.

When drafting, the net pump pressure is the sum of the pump discharge pressure, the suction pressure (inches of vacuum converted to pressure), and the friction loss in the suction hose. For example, if the pump is discharging water at 120 psi, the friction loss in the suction hose is 8 psi, and the suction gage is reading 20 in. Hg, then the net pump pressure will be:

120 psi pump discharge pressure
8 psi friction loss in suction hose
10 psi approximate pressure equivalent to 20 in. Hg
138 psi net pump pressure

This means that the pump is doing enough work to force water to a height of 138 psi × 2.304 ft, or 317.95 ft.

Hookup Procedures

When selecting a spot for drafting on other than wharves or piers, choose a spot that will enable the pumper to be placed on level and solid ground that will bear the weight of the pumper under conditions of considerable vibration. The pumper should be placed as close to the water's edge as possible. It is usually more convenient to have the pump panel on the side of the pumper away from the water. If the pumper cannot be placed close to the water's edge, it may be possible to draft water through a long horizontal suction, but with decreased efficiency due to increased friction loss.

In most cases, one 10-ft length of suction hose will not be sufficient to reach the water supply. Two 10-ft lengths are normally considered minimum, and in some cases three lengths are necessary. Suctions should be completely assembled with strainer and tie rope before being attached to the pump and lowered into the water. When selecting the number of lengths of suction hose to use in the hookup, the effect of the following conditions should be considered.

Depth of Water

The suction strainer should be submerged at least 2 ft below the surface. Water should be at least 4 ft deep where the strainer is located to provide clearance below and sufficient depth above it. The 2-ft distance below the surface helps prevent eddies and whirlpools from forming; these whirlpools could cause the pump to lose its prime or produce poor fire streams due to entrapped air. Where eddies cannot be prevented by submerging the strainer deep enough, floating a piece of plywood, a door, or any other large flat surface on the water over the strainer will be helpful. The clearance below the strainer helps prevent foreign materials from being forced into the pump or clogging the strainer.

Ocean Tides

On long pumping operations in tidewater, the rising and falling of the water due to tidal movement must be considered. If the tide is going out, the original number of sections may in time be insufficient.

Flowing Streams

When drafting from flowing streams, brooks, and irrigation ditches, the depth can be increased by making a dam to impound the water. Ladders and salvage covers are often helpful for constructing dams.

Hard Suction Layout

Making up a hard suction layout for drafting normally requires the cooperation of the entire engine company. All suction connections must be absolutely air tight. To insure this, before placing the suction hoses in water, a rubber mallet should be used to tighten the coupling lugs. A new pliant rubber gasket in a suction hose is one of the least expensive items used by a fire department, yet one of the most important.

After the suctions have been coupled, there is some choice about the manner in which they are to be placed in the water. If possible, the suctions should be connected to the pump inlet before lowering over the side. Sometimes, however, because of the pump's location, it is necessary to get the suctions started over the side before connecting them to the pump.

To secure the suction hoses, to take weight off the coupling, and to help raise the hose out of the water, a tie rope is used. The following tying procedure offers good control in both the lowering and the raising of suctions, and the weight will be well distributed when the suction hose is in position:

1. A clove hitch should be used at the bottom, next to the strainer; the hitch must be separated so as to be placed on both sides of the male coupling lugs. On long strainers, leave a sufficient end to permit connecting into the eye at the base.

2. With the other end of the rope, tie a running bowline on the first coupling above the strainer end. To prevent the rope from slipping down, the bowline may be separated to straddle the coupling lugs. When removing the suctions from the water, pull on the rope attached to the first coupling above the strainer until the strainer clears

the water. This will allow the suction hose to drain.

3. Secure the loop of the rope to the pumper or to any other convenient place.

Larger pumps are rated at capacity using a 6-in. suction hose and a lift of 10 ft. Due to the friction loss in suction hoses and other factors, any time that a suction hose less than 6 in. in size or a lift exceeding 10 ft in height is encountered, the pump cannot deliver rated capacity. The restriction imposed on capacity by small suction hoses can be partially offset by using two sets of hard suctions. Two sets of 4-in. suctions are equivalent to a single suction of 5.6 in. and may be connected to both the side and the front inlet connections.

Air Locks

When pumping from a draft or a booster tank through a small line, the flow through the pump is so small compared to its capacity that occasionally it is difficult to scavenge all air from the pump. This results in unexpectedly low pressure. It is good practice when this occurs to momentarily open a 2½-in. discharge gate. The larger flow for a few seconds while building up pump pressure completes removal of the air.

Figure 12-3 shows a typical air trap—an inverted air trap in the suction hose. Water rises in the suction hose, flows over the hump, and runs into the pump, leaving air trapped at the top of the hump. The pump, full of water, shows pressure. The discharge valve is opened and flow through the suction hose increases, sweeping air down into the pump and resulting in a temporary loss of prime.

When the pump is primed stationary (impellers not rotating) from the top of the pump discharge, each time the primer is shut off and the pump is engaged, some of the air is swept into the eye of the impeller, breaking the prime. If this should happen, stop the pump; start priming again until water runs from the priming pump; shut off the primer, start the pump, etc., until a complete prime is accomplished. This procedure may take more than one try to achieve priming, depending on the size of the air trap. It is for this reason that air traps are more easily handled by priming from the eye of the pump with the impellers rotating than by priming from the top of the pump with the impellers stationary. This is an inherent feature designed into the pump and is beyond the control of the operator, but in operation it should be understood that pump design features must be compensated for.

Drafting Drills

Before engaging in drafting drills, pump operators and company commanders should determine whether or not the water contains sand or other abrasive material, and whether the suction strainer is free from debris. At fires where this cannot be determined, reasonable precautions should be taken, if at all possible, to draw only clean water into the pumps.

Priming Procedures

Rotary gear and rotary vane pumps are vacuum-producing units; in other words, they are positive displacement pumps capable of displacing air or water. These pumps are capable of creating their own vacuum and of lifting water. Centrifugal pumps, on the other hand, are not capable of creating a vacuum because there is an open waterway through the pump. In order for centrifugal pumps to be used at a draft, priming pumps or priming devices must be utilized. These devices evacuate air from the centrifugal pump and allow atmospheric pressure to push water up into the pump casing. Once filled with water, the centrifugal pump can then draft by its own power.

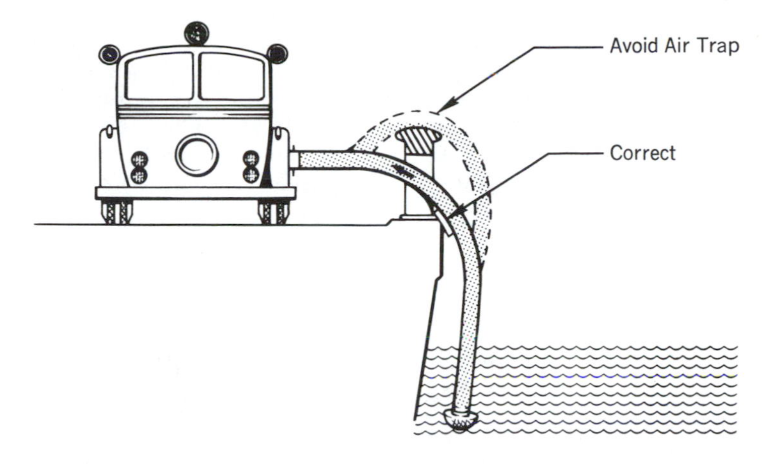

Figure 12-3. Inverted air trap in suction hose

Rotary gear primers are designed to operate best at comparatively low engine speeds of about 700 to 1200 rpm.

Rotary vane pumps operate best at moderate engine speeds of about 1200 to 1500 rpm.

Exhaust primers depend upon the velocity of exhaust gases discharging through a venturi tube for efficiency. Engine speeds of 1500 to 2800 rpm will be needed.

Priming mechanisms utilizing vacuum developed by the intake manifold of the engine are most efficient at a fast idle.

Positive displacement rotary gear pumps driven by an electric motor operate at 1500 to 1700 rpm. This type of primer is independent of the main fire pump and its engine for operation.

Priming procedures on apparatus vary slightly, but the following may be considered as good general procedure:

1. Check the gaskets (inside the female coupling) while assembling the hard suctions.

2. All suction hose connections must be air tight. Use a rubber mallet to tighten the couplings.

3. All openings to the pump such as drains, discharge gates, valves and bleeder cocks must be closed to prevent the entrance of outside air. This applies equally to both the suction and discharge sides because air can readily flow through a centrifugal pump.

4. The transfer valve should be in volume position because air will usually be evacuated from the pump quicker and more completely. This is not always convenient because many times it is desired to pump in pressure positions; this requires changing the transfer valve after water is obtained. Some pumps prime equally well in either position.

5. When ready to prime, engage the priming device and advance the throttle to the engine speed recommended for the type of primer provided. Some pumpers require that the main pump be engaged during priming operations; others cannot possibly engage the main pump while priming.

6. Continue to operate the priming device until water is discharged free of air.

7. As soon as the pump is primed, disconnect the priming device and engage the main pump, if this has not already been done.

Caution: If the priming device does not discharge water within 30 sec, do not continue to operate it; check for the difficulty. Pumps of 1500 gpm or larger capacity are required to take suction in not over 45 sec; smaller pumps are not to exceed 30 sec.

Priming pumps run in a bath of oil. Failure to secure a prime within a reasonable time may indicate that the priming pump is dry and needs lubrication. Failure to secure a prime may also be caused by an air leak; if the leak is in the suction hose, it is usually at the first coupling below the pump inlet.

A high reading on the vacuum gage, with no water entering the pump within the proper time, indicates that the suction lift is too high or that the strainers may be obstructed.

8. Open a discharge gate *slowly,* at the same time advancing the throttle until a steady flow of water is established.

9. Note the readings on the pump gages. The suction gage should indicate a vacuum and the discharge gage should show a pressure reading. Should the engine speed increase without a corresponding increase in pressure, either the most efficient point has been passed or the pump has not been completely primed. If the most efficient point has been passed, close the throttle slowly until the pressure gage indicates a pressure drop, then operate at this pressure. When the pump is not completely primed, first try retarding the throttle until a low pressure is obtained and shift the transfer valve from volume to pressure and back to volume. This will often clear air out of the pump. If it does not, then it will be necessary to reprime the pump until a solid stream of water is discharged from the priming pump. If the pump still fails to prime properly, look for air leaks.

Auxiliary Methods of Priming Pump

At fires, it occasionally happens that one or more of the pumpers at the draft position cannot draft water because of a failure of the priming device, leaks in the pump or some of its connections, or any other of the multitude of factors that affect priming centrifugal pumps. Unless the defect is serious, there is seldom an occasion where ingenuity will not triumph. Following are some of the methods that may be utilized:

1. Open the tank dump valve to prime the pump. This is successful when the pump and suction hoses are small. However, because of the restricted size of the dump valve and the piping, water will not flow in a sufficient quantity to prime the larger pumps. A foot valve on the end of the suction hose will facilitate priming by this method because it prevents water from running out of the strainer.

2. Bend the suction hose up out of the water into a U shape so that the strainer is higher than the suction hose or the pump. Open the tank valve and allow water to fill the pump and the suction hose. Use cloth or another material to seal the strainer end of the hose and thus maintain the prime. Place the pump in gear and allow a small amount of water to be discharged, leaving the tank dump valve open. Quickly place the suction strainer in the water, remove the sealing material, and increase the pump pressure. Then close the booster tank dump valve.

3. Bow the suction hose out of the water into a U shape so that the strainer is higher than the suction hose or the pump. Use a bucket to fill the end of the suction hose; slightly open a discharge gate to remove air from the pump. When the pump and hose are filled with water, seal the strainer end of the suction with cloth or another material, place the pump in gear, and quickly place the hose strainer in the water. Increase the pump pressure.

4. If there are any other pumping apparatus in the vicinity, a discharge line from another pump will prime the defective one. This apparatus can use either pump pressure

from its booster tank or a working line from a hydrant or draft. This line will enter the defective pump either through a gated suction inlet, or through a suction with an *increaser* and valves, so that the line may be disconnected later.

5. Another pumper or a portable pump that can develop sufficient suction can be connected to a discharge gate or a gated suction inlet of the disabled apparatus by a hard suction and the proper fittings to make the coupling. The priming device of the second pump then removes the air from the first pumper until the pump and suction hose have been filled. Continue pulling a vacuum, with the pump on the disabled pumper also in gear and operating, until water flows from the second pumper's primer outlet. Then close the valves, remove the suction hose between the two pumpers, and the second pumper is free.

Pumping Procedures

Even though most manufacturers recommend that their pumps be primed with the transfer valve in the parallel (volume) position, many modern centrifugal pumps will prime equally well in either series or parallel. When priming in series, keep the primer engaged until both stages are filled with water. This can be determined by sound as well as by observing the discharge from the priming device under the apparatus. There will be a pause after the first stage is primed, and then water will be discharged from the second stage of the pump.

After the engine has been brought up to the correct speed for priming, engagement of a mechanically driven priming pump tends to slow down the engine and a further decrease in engine speed occurs when water reaches the pump. If the operator keeps one hand on the throttle, the engine speed can be maintained at the optimum rpm. With electric priming pumps, the initial problem of engine speed decrease does not occur until water fills the pump.

When the pump is fully primed, some discharge pressure will be showing on the discharge gage; the exact amount will depend on the pump speed. If the pump is primed with the engine at idle speed, the throttle should be advanced as full prime is attained so that a discharge pressure of at least 20 psi is registered. The priming should not be discontinued the moment that engine pressure begins to show on the gage or when water is sprayed from the priming pump. To stop priming before a full prime is obtained produces a false prime. With less than a full prime, the pump has an air lock and will not discharge its rated capacity. When this situation occurs, the pump must be completely primed again.

Slowly open each discharge gate with an attached line; slowly advance the engine throttle to maintain or slightly increase the pump discharge pressure. This should be a two-handed operation; one hand is slowly opening the discharge gate while the other hand is simultaneously increasing the engine speed. If the discharge pressure is allowed to drop excessively, the prime may be lost. The operator should also note that the suction gage will indicate a gradually increasing vacuum as the discharge increases. This is due to the friction loss in the strainer and suction hose as the water flows; as the pump discharge increases, so does the velocity of the water in the suction hose.

Most pumpers have slight air leaks in their valves, packing, and plumbing; this is not unusual, and seldom is it sufficient to affect pump priming. This air leak can be enough to cause a loss of prime when no water is moving, but it is often too slight to cause difficulty if water is discharged immediately when drafting at moderate or high lifts.

Sometimes the pressure drops as a discharge gate is opened despite an increase in engine speed. Stop opening the gate immediately and watch the pressure gage closely. If the pressure starts to pick up slightly, hold the gate where it is until pressure is regained.

This gives time for the reduced amount of water flowing from the pump to carry along, or wash out, a small amount of air left in the pump as a result of failing to obtain a full prime. If the pressure gage does not show an immediate increase when the gate is left stationary, move the valve slightly toward the closed position and leave it there for a few seconds to give the pump an opportunity to regain pressure.

If the pump still fails to recover the pressure lost when the hose gate was first opened part way, leave the valve partly open, slow down the engine to the correct rpm for priming, and operate the priming device for about 5 sec. Often this is all that is required to progress into a successful pumping operation from draft as it rids the pump of air that the flowing water could not wash out. The fact that a discharge gate is partly open will usually speed, rather than prevent, a full prime under these conditions. If a complete prime still cannot be obtained, then close all valves and start over.

Back Pressure and Friction Loss

The loss in pressure due to both elevation and friction loss can be observed by watching the suction gage at the pump inlet. At approximately 10 ft of suction lift with the pump primed but with no hose stream flowing, it will be noted that the gage shows a suction reading of about 9 in. of mercury. The vacuum side of the gage is graduated in inches of mercury; each inch is equivalent to 0.491 psi, or 1.133 ft of pressure head. This shows the loss of atmospheric pressure due to elevation difference between the surface of the water and the pump chamber.

Next, start the water discharge through the hose lines, opening the throttle slowly. As the flow increases, the pressure loss on the suction side will increase. This is due to the increase in friction loss in the hose and fittings on the suction side of the pump. If there is sufficient flow, all the available atmospheric pressure will be used or the point of maximum capability of the pump to maintain a vacuum will be reached. For a pump in reasonably good condition, it may be that a vacuum of 22 in. of mercury can be attained. If a pump is drafting with a lift of 20 ft, it will use about 18 in. of vacuum just in overcoming the elevation and it will have 4 in. left on the suction gage to overcome friction loss.

When drafting, the vacuum gage must be watched closely. An increase in the reading could indicate that the suction strainers on the end of the suction hose or in the pump intake may be clogged, or it could be a warning that the water level in the drafting source may be falling.

Pump Cavitation

When drafting at a high lift, the pump operator must be always alert for signs of pump cavitation. If the pump sounds as if it has a lot of small stones rattling around inside, cavitation, which is the implosion of air bubbles, is slowly destroying the impellers. When a corresponding pressure increase is not shown on the discharge gage as the engine is speeded up, it is another indication of cavitation. This condition should be immediately corrected by slowing down the engine speed to reduce the volume of discharge.

Water Temperature

When the water supply is from a small pond, lake, or swimming pool and summer heat has increased the water temperature to above 85°F, trouble can be anticipated both in priming and in obtaining capacity discharge. At a water temperature of 95°F, the lift capability is greatly reduced and only a small portion of the rated capacity can be delivered; the water flow will be erratic. This problem is due to insufficient difference between the vapor pressure of the water in the pump and atmospheric pressure; as water is heated, its tendency to give off vapors is increased.

Shutdown Procedures

For temporary shutdown, the procedures are as follows:

1. SLOWLY retard the throttle (by turning it clockwise). Use the panic button in the center of the throttle only for emergencies.
2. SLOWLY close the discharge gates.
3. AT DRAFT. Maintain approximately 35 psi discharge pressure to avoid losing prime.
4. If shut down for an extended period of time, take the pump out of gear or open a discharge gate or bleeder valve to prevent the water from heating due to the churning of the pump.

Permanent shutdown procedures are as follows:

1. SLOWLY retard the throttle (by turning it clockwise). Use the panic button only for emergencies.
2. SLOWLY close the discharge gates.
3. Take the pump out of gear.
4. Open a discharge gate bleeder valve to drain the hose lines.
5. Open the pump drain.
6. Reposition the transfer valve, if necessary.
7. Return the pressure regulating device and all other valves to their normal positions.
8. Apparatus with pump transmission behind road transmission:

   a. Place the Road-to-Pump Shift Lever in road position.

   b. Place the road transmission in neutral.
9. Stop the engine.
10. Disconnect the hose lines.
11. Disconnect the suction hoses.
12. Pick up the chock blocks before moving apparatus.

Maintenance

The maintenance of the apparatus following any drafting operation is particularly important. The number of items that should be checked and the proper sequence for checking will vary from apparatus to apparatus. Following are a few things that should be considered:

1. Check the amount and condition of the oil in the crankcase and in the priming oil reservoir.
2. Check and add water in the radiator as necessary. Flush if salty or dirty water was used in cooling.
3. Check the lubrication chart carefully to see what items require lubrication.
4. If salty or dirty water was used, completely flush the pump and all equipment used.
5. Seal and lubricate the priming pump. This is usually accomplished by engaging the priming pump with the priming valve closed. The engine should be at a fast idle, and the operation should be continued until oil is discharged from the priming pump.

Unusual Drafting Situations

Occasionally it may be necessary for a pump operator to draft when hydrants are readily accessible but no body of water from which to draft is available. Such a situation can occur when it is necessary to pump into piping systems that might contaminate the water supply.

An example of such a situation occurs when a material such as butadiene or casing head gasoline under high pressure is escaping from a break in a line. By pumping into the system it may be possible to push the material back beyond the break and replace with water the dangerous flow of material from the break. The danger of connecting to such a system is that the pressure of the material may be greater than the pressure of the water, with the result that the material may force itself into the hose line, into the pump, back through the suction hose, and into the hydrant and main system.

This is not theory alone; it actually has happened. At one large metropolitan waterfront fire, sea water was drafted into the mains to increase the pressure; the next day a nearby restaurant found it was using sea water to make its morning coffee.

When this kind of situation arises, some sort of arrangement is needed that will deliver water from the water main to the hose line, but that will break the connection in case of a backup so that the contaminant will not flow into the fresh water supply. These arrangements are known as broken or remote connections. They are usually combinations of hydrant supply and drafting, and the forms they take in each particular case are left to the initiative of all concerned. The following are offered as suggestions:

Use of Storm Drains

Figure 12-4 illustrates the use of a storm drain for making a broken connection. The use of a storm drain requires that:

1. The outlets from the storm drain be dammed.
2. The water from the nearest hydrant(s) be directed into the storm drain.
3. The suction be taken from the storm drain.

Use of Barrels

Figure 12-5 illustrates the use of a barrel for making a broken connection. The use of barrels requires that:

1. A barrel of sufficient size be obtained.
2. A line be connected to a hydrant or pumper and water discharged into the barrel.
3. The suction be taken from the barrel.

Use of Improvised Reservoir

Figure 12-6 finally illustrates the use of salvage covers for improvising a reservoir. This method requires that:

1. A hole of sufficient size be provided. Normally the hole should be a minimum of 3 ft deep. The larger the reservoir, the easier it is to maintain satisfactory operations.
2. A line be connected to the hydrant or pumper and water discharged into the reservoir.
3. The suction be taken from the reservoir. This method is used in many areas to insure smooth operations among mutual aid departments when a variety of threads is encountered.
4. Ladders carried on the pumper can be used as supports for salvage covers when improvising a reservoir. Separate the extension ladder, stand the sections on their beams, and use rope or ladder straps to form the ladder sections into a triangle. This will support a salvage cover to form an adequate open tank.

Use of these systems requires cooperation between the crew members filling the reservoirs since they must avoid aiming the discharge from the hydrant line at the strainer end of the suction hose; this would cause aeration of the water and a consequent loss of prime. Coordination is also required between a member filling the reservoir and the operators of the hose line; a system of signalling must be adopted to make the necessary adjustments of flow and to allow for variations in the pump discharge. If the supply in the reservoir becomes too low, the pump will lose its prime due to eddies being formed.

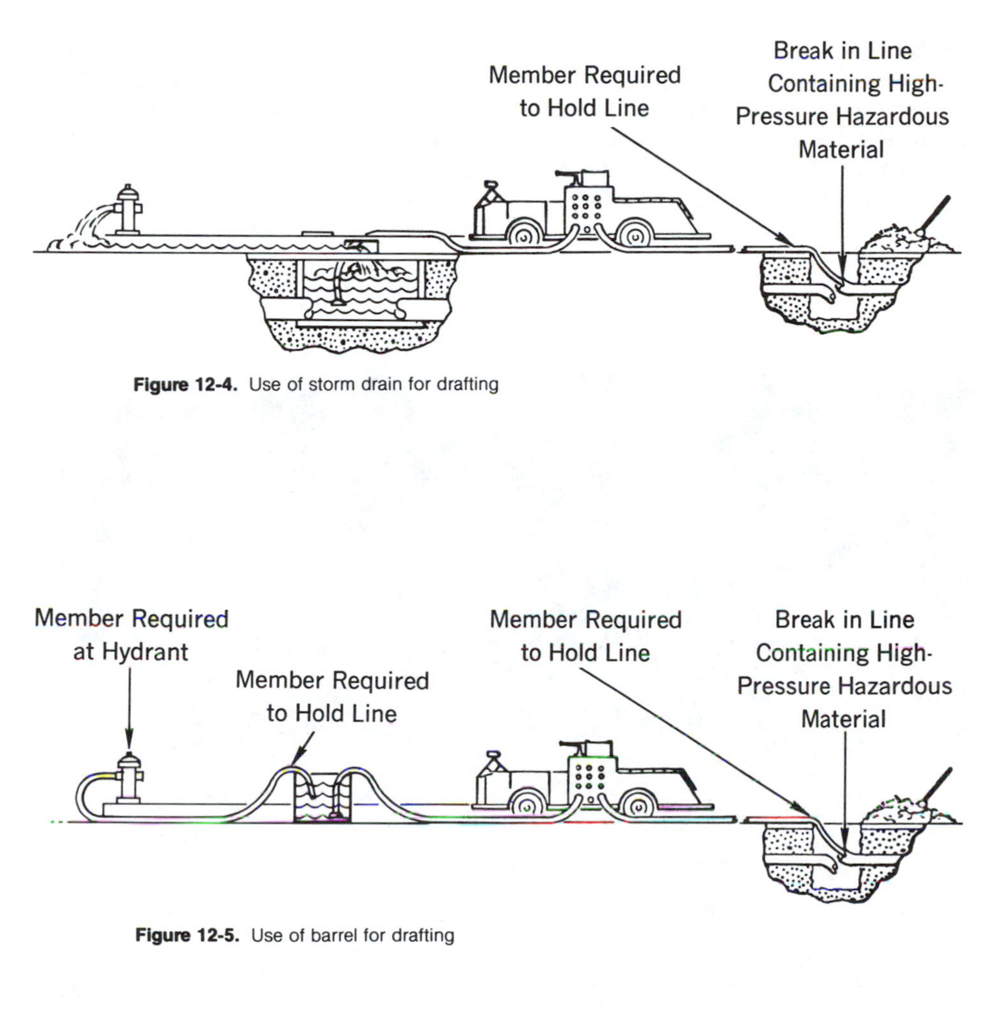

Figure 12-4. Use of storm drain for drafting

Figure 12-5. Use of barrel for drafting

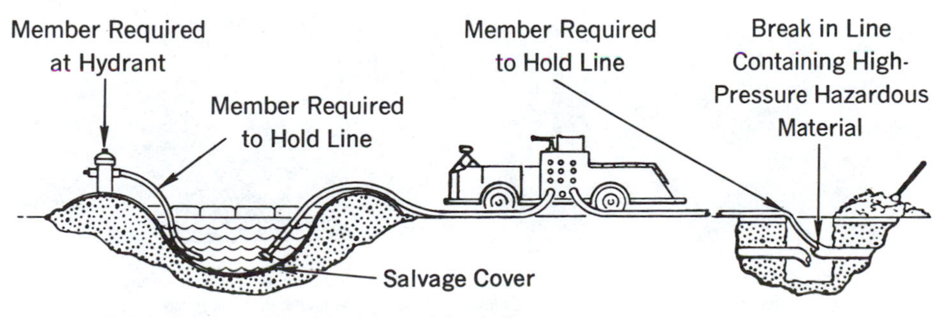

Figure 12-6. Use of improvised reservoir for drafting

Portable Pumps

Portable pumps have become standard equipment in rural fire fighting. These pumps may be used to supply water from rivers and lakes that are impossible for vehicles to reach (Figure 12-7). They are valuable for relaying water from a static source to the fire, freeing apparatus for other duties. Helicopters have been used to airlift portable pumps, hose, and crew members into areas inaccessible to fire apparatus. The helicopters are invaluable as airborne command posts, from which command officers can survey conditions and deploy fire apparatus.

Figure 12-7. Portable pump used for drafting and relay operation when pumper cannot be spotted close enough to the water source for drafting (Courtesy of Hale Fire Pump Company)

13 Aerial Ladders

The development of the modern power-operated aerial ladder has greatly influenced fire fighting, ventilation, and rescue performance; aerial ladders are invaluable to rescue trapped occupants from a burning building, to allow access for ventilating heat and smoke from a structure, and to allow fire fighters a direct route for extending hose lines to the upper floor of a building. In addition, aerial ladders provide a mobile base for elevated fire streams. In many cities the modern aerial ladder truck has become a multipurpose rescue and fire fighting unit (Figure 13-1)

Aerial Ladder Performance

The expansion of aerial ladder capabilities, which has occurred over a period of years, has necessitated a revision of old theories and practices in aerial ladder operation. The modern aerial ladder is no longer merely an instrument for ascending and descending from one level to another, but has many varied and useful purposes. The wide range of uses makes it impossible to set exact standards of operation. Each manufacturer builds his aerials under rigid specifications with wide margins of safety. The type of emergency involved, the make of apparatus and its limitations, and countless other factors will dictate the operational procedures for any given situation. However, regardless of such factors, for any emergency situation requiring aerial ladder operation, certain practices have general application.

The apparatus may consist (counting duals as one wheel) of a four-wheel, two-axle chassis; a six-wheel, three-axle chassis; or a six-wheel, three-axle tractor-drawn truck. To receive full credit as a truck company, the apparatus must carry a full complement of manually operated ground ladders in addition to an aerial ladder.

Each type has its advantages. The four-wheel chassis is shorter and takes less storage space in quarters. By using a four-section ladder it can carry an aerial ladder up to 100 ft, and no tiller operator is required. The tractor-trailer type has several significant advantages: it can be maneuvered with greater ease in traffic and on the fire ground and can reach locations which the two-axle vehicle cannot; it provides a more stable base when the ladder is extended, which reduces the danger of overturning; it provides more space for necessary tools and equipment; and the

Figure 13-1. Ladder pipe and handlines are being used to knock down a fire. (Courtesy of the San Francisco Fire Department)

driver and tiller operator can position the truck faster than a driver alone.

Quads, or quadruple pumpers, are combination pumper and ladder apparatus that consist of ground ladders, fire pump, water tank, and fire hose; all of these must comply with National Fire Protection Association Publication 1901, *Automotive Fire Apparatus.* When an apparatus has an aerial ladder in addition to all of the above-named, it is referred to as a quint, or quintuple pumper. These are excellent for furnishing ladder pipe service on the fire ground because the water pressure is supplied and controlled right at the base of the ladder. One trouble with many of the quints is that they cannot effectively furnish both pump discharge pressure for the water supply and hydraulic pressure for ladder operation simultaneously; some manufacturers have solved this problem by adding a small auxiliary engine to power the hydraulic pump.

Aerial Ladder Construction

Aerial ladders have been in use since the days of horse-drawn fire apparatus (Figure 13-2). Modern vehicles are now equipped with a tractor and pneumatic tires, and many improvements in aerial ladders have been made.

Manually Operated Aerials

Manually operated aerial ladders are usually constructed of wood and consist of two

Figure 13-2. Manually operated wooden aerials usually consisted of two ladder sections. With the advent of the gasoline engine, many of these horse-drawn vehicles were modernized with a tractor and pneumatic tires. (Courtesy of the Los Angeles City Fire Department)

sections with suitable metal truss rods and braces to gain adequate strength. Truss rods are long, flexible metal rods, one over each beam, attached to either end of the ladder. Tension on truss rods and braces gives the extended ladder added strength and permits it to carry its maximum load.

It is good practice to place the top of a wooden aerial 6 to 12 in. from the part of the building on which it is to rest. (This distance depends on the flexibility of the ladder.) The weight of fire crew or hose lines causes the ladder to bend and to rest on the window sill or roof cornice. This position places the trussing in tension for maximum support and strength.

When a wooden aerial ladder is placed against a building and then loaded, the weight places the main beams in tension and the trussing in compression. The trussing then has very little or no load-carrying ability, and the safe load capacity of the ladder is sharply reduced.

The operating mechanism for manually operated aerial ladders is shown in Figure 13-3. The initial lifting power is supplied by two large compressed coil springs housed within the two steel tubes at the bottom of the photo. The force of these coiled springs raises the ladder from the bed to a semivertical position; this motion is controlled by a friction brake. Additional elevation is accomplished by turning the large threaded shaft. The ladder is extended by cranking on the two large wheels at the top of the photo; this usually requires hard work by three or four crew members. The turntable is rotated by turning the wheel at the lower right. The turntable is lined up in the bedding position by dropping the pin, which is projecting up just below the rotation wheel, into a hole in the lower turntable assembly.

A clutch, consisting of a split nut operated by a handle, is mounted on a steel cross member of the main ladder. When the clutch is engaged, the split nut meshes with a large worm screw geared to a cross shaft. Detachable cranks on the cross shaft provide positive control of further vertical movement.

Crank wheels operate a cable drum mounted on the main ladder to extend and retract the fly. A turntable control wheel rotates the turntable to any desired position.

Power-Operated Aerials

Power-operated aerial ladders fall into one of three general classifications: the hydraulic ladder, which is entirely powered by fluid from a hydraulic pump; the hydromechanical, which uses hydraulic power to operate the elevating cylinders and mechanical linkage from the engine to drive the turntable and ladder extension; and the electric motor ladder, which uses electrical power to elevate the ladder, extend the fly sections, and rotate the turntable. Hydraulically powered and controlled aerial ladders are most generally used.

Turntable Construction

A turntable is constructed of steel plate welded into a section of great strength and rigidity. It is the base on which is mounted the aerial ladder and all of the mechanism for operating and controlling its various movements.

On the bottom are machined ball races that register with similar races in a large-diameter, heavy steel ring bolted to the chassis frame. The races contain a large number of hardened steel balls that carry

Figure 13-3. Operating mechanism for manually operated aerial ladder (Courtesy of the Los Angeles City Fire Department)

the weight of the ladder and the mechanism and permit free rotation of the turntable. Suitable gear teeth are cut on the periphery of the ring and are engaged by gears driven by the rotation motor, which furnishes the power for rotating the turntable. Another steel ring, bolted to the under side of the turntable, has a ball race that engages inversely with another in the gear ring. Steel balls in this race take the reaction or tipping load of the aerial ladder.

The aerial ladder operates on a shaft that permits vertical movement. Horizontal movement or rotation is provided by a turntable at the base of the ladder. The larger the diameter of this turntable, the greater the ladder stability and the greater the reduction in ladder load on the turntable bearing surfaces. When a ladder is resting against a building there is little torque stress load on the turntable. The load is static weight and includes the weight of the ladder and the operating mechanism. In a cantilever position, that is with the upper end unsupported, the loading of the turntable is maximum, being subjected to both static and torque stress.

Metal Aerial Ladder Construction

The metal aerial ladder consists of three or more sections constructed entirely of rust-resistant steel or aluminum alloy (Figure 13-4). The cantilever bridge design gives it almost equal strength in either the tension or compression condition. However, it will support its capacity load only while the main beam and the trussing are in line with the pull of gravity, that is, with the rungs in a horizontal position. Any deviation from this position causes its strength to fall off rapidly; hence side stress should be avoided.

Ordinary care should be used when handling the ladder because a sharp blow on the bottom of the beams, as would occur if the ladder were dropped abruptly over a building edge or window, may damage the ladder and affect its smooth operation over the rollers.

Figure 13-4. Modern tractor-trailer aerial ladder truck equipped with a four-section aluminum alloy ladder of cantilever construction (Courtesy of Peter Pirsch & Sons Company)

Axle Locks

Tractors of the semitrailer type aerials are equipped with lever operated clamps or jacks for immobilizing the rear springs, thus locking the tractor frame rigidly to the tractor rear axle. This locking renders the chassis springs inoperative and helps prevent excessive sag to one side when the aerial ladder is in use.

Stabilizing Jacks

According to NFPA 1901, *Automotive Fire Apparatus,* apparatus equipped with an aerial ladder should be provided with at least two ground jacks, one on each side of the vehicle at the turntable location, with the widest practicable spread between their feet when in the operating position. The ground contact area of each jack should not be less than 140 sq. in. and, in addition, foot plates of ample strength shall be provided with each jack that is capable of increasing the ground contact area of each jack to not less than 320 sq. in. However, many of the older trucks have far less ground contact area than the above recommendations; this should be recognized and compensated for when operating the aerial on thin pavement or soft terrain.

When an aerial ladder is rotated in an arc to a position at right angles to the truck, a load is added to the chassis, springs, and tires, causing additional tire compression on the side of the ladder's movement and a lightening of the load on the opposite side. This roll or torque action in the ladder may be dangerous. To overcome this condition and to provide the required stability, ground jacks are provided. Their function is to control the shift in weight and to limit the transfer of the load to either the springs or tires on the operating side.

These stabilizing jacks may be mechanical, hydraulic, or a combination of the two. Never set the ground jacks other than straight up and down. If they are set at an angle and the ladder is extended to one side of the truck, the jack on the opposite side will crawl toward the vertical; then, when the aerial ladder is swung in line with the truck, the trailer will be forced into a tilted position. When correctly set vertically, the off-side jack may leave the ground an inch or more but will reset itself correctly when the ladder is brought in line with the truck.

Any movement of the ladder caused by the sudden sinking of the ground jacks when placed on ground that is not solid may cause a great enough force at the end of the ladder to tip the truck over (Figure 13-5). When jack pads are small in area and must be placed on soft ground, large support plates or blocks should be placed under the stabilizer feet.

Ladder Cables

Hoisting and retracting cables are usually made of galvanized plow steel, and are many times stronger than necessary to carry the load imposed. Power may be applied to the cables by a roller chain driven by a hydraulic motor (Figure 13-6), an extension drum driven by a hydraulic motor, or a hydraulic cylinder.

Ladder Locks

A small lever, mounted at the base of the ladder, is connected by a cable to dogs or pawls near the top of the main section. It is also connected to pawls near the top of intermediate sections by means of cables and pulleys.

When the lever is pulled forward, these pawls are held clear of the rungs of the adjacent ladder sections, but when the lever is moved up and back, the pawls engage under rungs of the other sections. In this manner, all load is taken off the extension cables when the ladder is in use. The pawls are so arranged that when they are engaged with the rungs, all rungs line up, making the ladder easy to climb.

Figure 13-5. When it is necessary to operate an aerial ladder on soft ground, careful consideration must be devoted to stabilizing the truck.

Figure 13-6. A hydraulic aerial ladder hoisting and operating mechanism, control stand, and turntable. The extension drum is in the foreground and the two hoisting cylinders are in the rear. (Courtesy of Peter Pirsch & Sons Company)

Operating Procedures

Because hydraulically powered and controlled aerial ladders are most widely used, operating procedures for this type are given in the following sections. In most respects, truck company operational procedures are similar, regardless of the type of apparatus (Figure 13-7).

Operator's Position

All the controls and instruments necessary for normal operation of the aerial hoisting mechanism and ladder are located on or near the control pedestal (Figure 13-8). On the side of the ladder is the angle limit indicator to show the safe operating extensions. Adjacent to the inclinometer is the lever that controls the ladder locks.

There are usually three main control levers; one controls the hoisting and lowering of the ladder, one extends and retracts the ladder, and one controls the rotation of the turntable. The position of the levers and the direction of movement varies according to the manufacturer.

There are usually engine and throttle controls. On a truck equipped with an automatic transmission, the supplementary engine starting control will function only when the transmission is in the neutral position, unless this is compensated for. This will cause it to be inoperative in the case of a quintuple apparatus when the main pump is in operation and is being driven through the output shaft of the automatic transmission. Locking hydraulic pressure in the hoisting cylinders is accomplished with a conveniently placed lock valve. Turntable rotation may be locked with a friction brake or lock valve; when a hydraulic motor drives the turntable gear by means of a worm gear, no locking device is necessary.

Figure 13-7. Aerial ladder pipe helps combat a flammable liquids blaze. (Courtesy of the Los Angeles City Fire Department)

Figure 13-8. The operator's position has clearly marked switches, levers, and valves for easy operation. (Courtesy of Peter Pirsch & Sons Company)

Gages. Gages vary on different models; all have one showing the height reached by the ladder when it is at approximately an 80° angle from the horizontal, or the equivalent extension at any angle. Hydraulic pressure is shown on a pressure gage; it indicates the operating pressure of the hydraulic system. Some manufacturers include a gage that indicates the pressure at the bottom of the hoisting cylinders when the control lever is in the neutral position. When the ladder is unsupported at the tip end, this gage provides an indication of the load on the ladder.

Control pedestal gages. Understanding and using the control pedestal gages is necessary before an operator will know what can be accomplished under a particular set of circumstances. The use of the ladder extension gage allows the operator to estimate the number of feet needed before the ladder can be lowered into the building. For public displays, a trial run before the event will allow a spot to be selected, the correct inclination to be noted on the inclinometer as the ladder is raised, the correct extension of the ladder to reach the objective, and how best to sight along the ladder beams to make the demonstration fast and smooth.

Pump pressure gage. The pump pressure gage indicates the discharge pressure that the hydraulic pump is providing for the system; this pressure is governed by the relief valve and the engine speed. Under full load and with the control lever at its maximum range, the system pressure should be up to the manufacturer's recommended relief valve setting. When an aerial truck is new, the ladder can be hoisted to its maximum angle, extended to its full length, and rotated 360° in a definite time period. The hydraulic system performance should be checked often to be certain that the ladder is fully effective. If it is not, the hydraulic pump may not be developing the correct pressure or the engine may not be accelerating to the right speed to develop maximum pump pressure and volume.

Cylinder back pressure gage. Some manufacturers place another gage on the panel to indicate the pressure being exerted down by the ladder on the oil in the hydraulic hoisting cylinders. The cylinder back pressure gage indicates to the operator the load placed on the ladder while it is unsupported. As long as the cylinder back pressure is less than the known pressure that can be expected in the system from the hydraulic pump, the ladder can still be raised. For example, if the cylinder back pressure gage shows 500 psi and the system pressure is set to operate at 1000 psi, the ladder can be raised when the control lever is activated. If, in another instance, the ladder is exerting 1000 psi downward on the cylinder and the system pressure is 1000 psi, no movement could be expected as both pressures would be equalized. Hoisting could not be accomplished until the ladder was retracted or a load taken off so that it would exert a lower downward pressure than could be expected from the hydraulic system. At least 50 psi greater pressure in the system is necessary before movement can be expected; for example, if the maximum hydraulic pressure that can be expected is 1000 psi, then movement can be expected anytime the back pressure gage indicates 950 psi or less.

The hoist cylinder back pressure gage indicates to the operator what amount of control can be exerted over the ladder; this gage is also valuable during ladder pipe operation because it shows what effect the nozzle reaction is exerting. Because the gage pressure will depend on the hoist cylinder diameters, hydraulic pump pressure, ladder extension, size of nozzle, nozzle pressure, and a myriad of other factors, operating parameters will have to be developed by each operator through experience. For example, one operator noticed that when all control levers were in neutral, the back pressure gage would indicate about 500 psi when the ladder was raised between 70 and 75° with no water flowing. As the stream was turned on, the nozzle reaction started to reduce the gage pressure. It is good practice to attempt to retain 100 psi on this gage so that control over the ladder can still be exerted by the operator; if the downward pressure drops too low, there is danger of the ladder being forced back past the vertical position. If the gage reading drops below 100 psi due to nozzle reaction, the ladder's inclination can be lowered slightly to cause the ladder weight to help combat the nozzle reaction; the nozzle pressure can also be lowered to reduce the reaction forces. Keeping 100 psi on the hoist cylinder back pressure gage is a safety measure that has been found to work very well. There are several factors, however, that can wipe out this safety margin in a hurry, such as a pressure surge or a rapid change in the vertical direction of the ladder pipe stream.

Safety Limits

Aerial ladder manufacturers have devised various warning and safety devices to alert operators when the hoisting, lowering, extending, and retracting limits have been reached. If the operator does not voluntarily stop the travel before the safety limits are reached, a warning horn will usually sound and the control lever will be returned to neutral. The operator should not depend on these signals of approaching limits; as the ladder nears its safe restrictions, the operator should ease up on the control lever and slowly bring the ladder to a halt without any shock. There are also hydraulic cushions and controls to prevent excessive pressure buildup or stress.

Emergency Operation

Provisions for the emergency operation of aerial ladders vary with the make and model, but some system of lowering the ladder into the bed is necessary in case the engine or hydraulic power should fail while the ladder is raised. All ladder operations can be carried out manually, though at a reduced speed, but a great dependence is placed on gravity. Most models are dependent on enough hydraulic pressure being developed manually to hoist a supported ladder from a building, to extend the ladder a sufficient distance to release the locks, and to rotate the turntable to line up the ladder for bedding. Moving the hoist and extension levers to the down position will permit them to descend by their own weight. If the elevation angle is sufficiently low that the fly sections will not descend by themselves, manual operation of the hydraulic pump will help.

The emergency manual procedures are only intended to allow a raised and extended ladder to be lowered; it would not be very practicable to expect to obtain normal operation of a disabled truck by manual operation of the hydraulic pump.

Hydraulic Systems

Hydraulic systems vary between manufacturers, yet they are similar in theory and operation. One model (Figure 13-9) will be explained so the various aspects may be discussed, but it must be understood that even different models built by this same manufacturer will vary.

Power application. Power for operating the aerial ladder hoist, and also the ladder extension and rotation, is derived from the truck engine driving a conventional power takeoff attached to the side of the transmission. This in turn drives a hydraulic pump through a universal joint and shaft. From the hydraulic pump, the transmission of power is entirely hydraulic.

Hydraulic circuit, idling. From the hydraulic pump, oil passes through tubing to the control valves. If all of the control levers are in the neutral position, the bypass valve will be

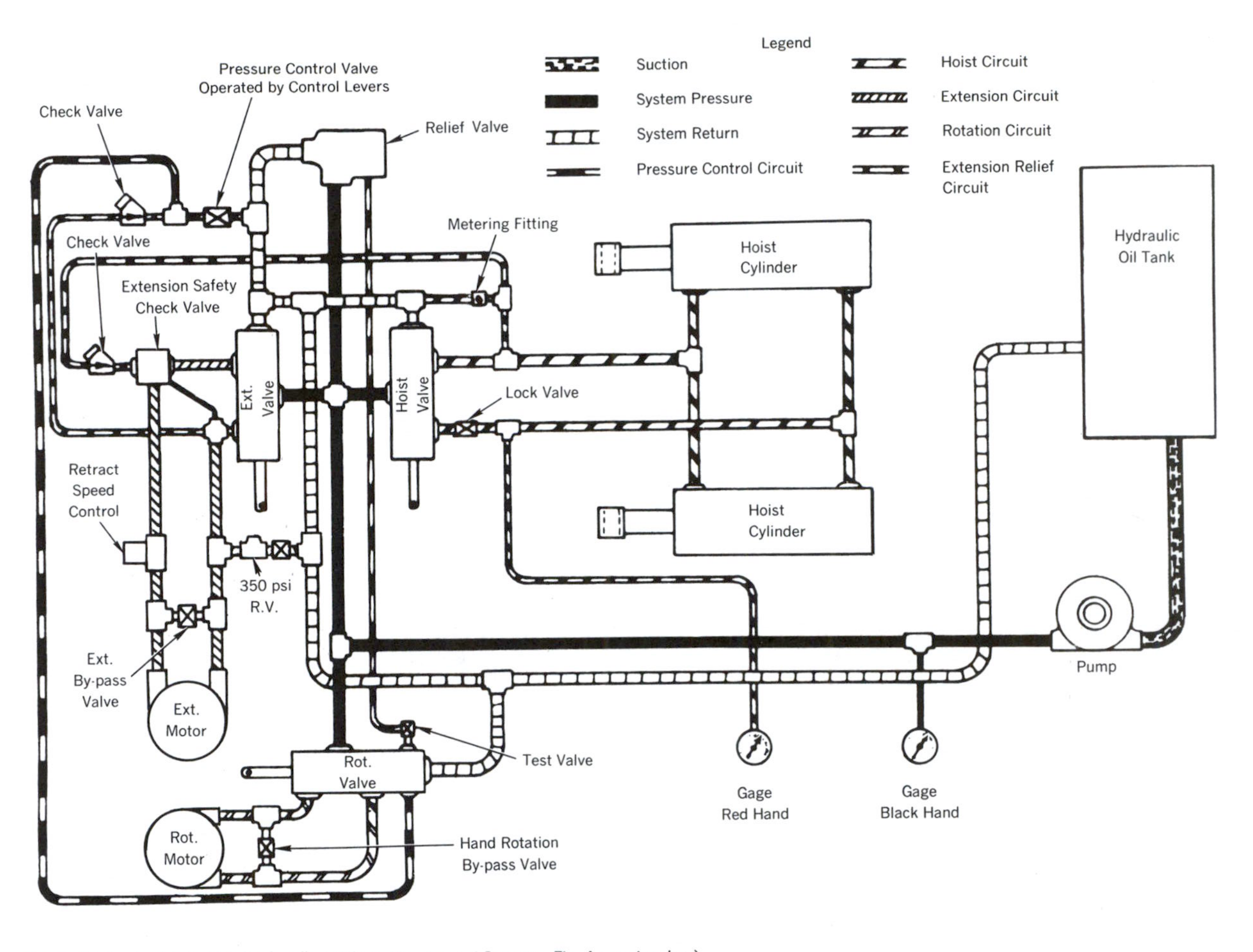

Figure 13-9. Aerial ladder hydraulic system (Courtesy of Seagrave Fire Apparatus, Inc.)

open. This permits oil from the relief valve, under a pressure of 60 to 125 psi, to bypass back to the hydraulic reservoir, and from there to the pump, completing the cycle.

Automatic throttle control. To utilize full pump speed while operating the ladder, and to slow down the pump to prevent excessive heating of the oil when idling at low pressure, an automatic throttle control is provided. This consists of an electric solenoid, mounted on the engine, arranged to act upon the throttle. The first slight movement of the operating controls makes electrical ground contact through adjustable commutator quadrants. This connects, through a commutator ring on the bottom of the turntable, with the electric solenoid; when the solenoid is energized, it pulls the throttle open and increases the engine speed.

The linkage connecting the solenoid to the throttle shaft is provided with an adjustment for regulating the increase of the engine speed. The engine speed should be adjusted to the maximum attained without a chattering noise being produced by the hydraulic pump; the exact engine speed cannot be stated because of variations in different apparatus. If, for any reason, the automatic throttle control should fail to function, use the hand throttle on the cab dash panel to set an engine speed just under that producing a chattering noise.

Hydraulic motors. Hydraulic motors which provide power for extending and rotating are of the eccentric rotary vane type; they are designed to operate in either direction by simply reversing the oil flow (Figure 13-10). Vanes are held in contact with the housing by means of coil springs and retaining plungers. Extension and rotation motors vary only in their lengths and, consequently, in the power developed.

Control valve. A bypass valve is provided in the hydraulic system. When all the control levers are in the neutral position, the control valve is open and permits oil from the relief valve, under pressures of 60 to 125 psi, to bypass back to the hydraulic oil reservoir.

When any one of the control levers is moved to a position requiring high pressure, the bypass is closed, causing pressure in the system to increase to the point at which the relief valve is set. The bypass valve is not closed at any point in the lower travel of the hoist control lever. Therefore, high pressure is not normally available for lowering the

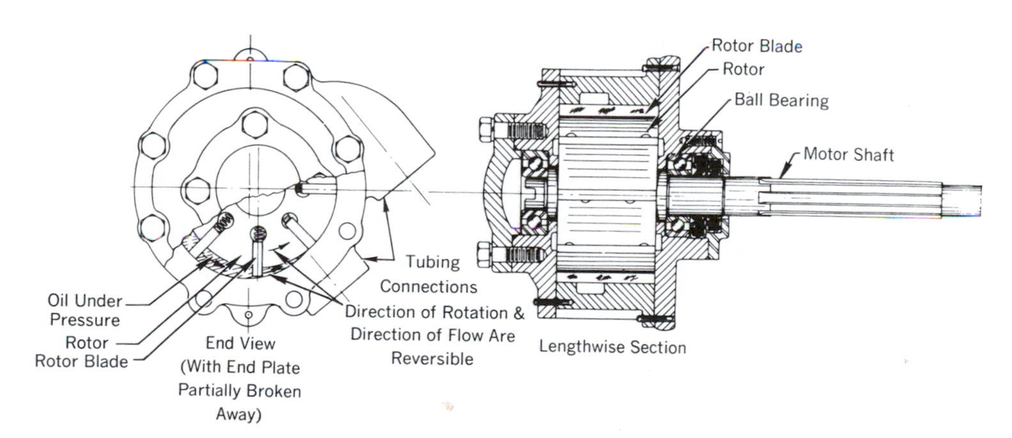

Figure 13-10. Hydraulic motor (Courtesy of Seagrave Fire Apparatus, Inc.)

ladder, the initial pressure being ample for all ordinary purposes. The object of this arrangement is to prevent high pressure that would force the ladder down against a building or any other object, possibly damaging the ladder.

Hoisting and Lowering

Following are the operating procedures for hoisting and lowering aerial ladders.

Cylinders. Hoisting cylinders are constructed of heavy seamless steel tubing, accurately honed to a smooth finish on the inside. Force produced by hydraulic fluid under pressure acts on closely fitted pistons that travel in the cylinders. The total force developed is determined by the hydraulic pressure and the area of the pistons. Hydraulic fluid forced into a cylinder at a pressure of 1000 psi would create a total force of 100,000 pounds when acting on a piston of 100 sq. in. To allow necessary movement of the hoisting cylinders, hydraulic oil lines are attached to the cylinder trunnions by means of metal swivel joints.

Ladder-raising hydraulic circuit. Figure 13-11 diagrams the hydraulic circuit that raises and lowers the ladder of an American LaFrance model. Other makes may vary in details. When the control lever is pulled back toward the operator, the ladder is raised. As the ladder rises the control valve directs hydraulic fluid through the hollow piston rods in the raising cylinders to the space above the piston. The fluid in the space below the piston returns to the reservoir through the piston rod outer passage and the control valve.

When the lever is pushed forward, away from the operator, the ladder is lowered as the control valve directs low-pressure fluid through the piston rod outer passage to the space below the piston. While the ladder is being lowered, fluid in the space above the piston returns to the reservoir through the piston rod inner passage and the control valve. Orifices within the cylinders control the lowering speed of the ladder and prevent rapid fall in the event of hydraulic line failure. The ladder descent is not powered but returns to its bed under the force of gravity.

Hoisting. When the hoist control lever is moved slightly from the neutral position toward the operator, a simple cam and lever mechanism is actuated which closes the bypass valve, building up oil pressure in the hydraulic system to the level at which the relief valve is set. Further movement of the hoist control lever moves the hoist control valve to a position that permits oil under pressure to flow through the tubing and lock valve to the bottom of the hoisting cylinders where it acts on the pistons; this forces the piston rods out and raises the ladder. Oil that is in the cylinders above the pistons flows out of the passages in the cylinder heads through the tubing to the control valve chamber, and then to the hydraulic oil reservoir and back to the pump. If the hydraulic pump supplies oil in excess of that required for hoisting the ladder, the surplus oil passes through the relief valve and back to the reservoir. When the hoist control lever is returned to neutral, the oil flow is stopped and the ladder is held at whatever angle of inclination it may have attained.

Lowering. When the hoist control lever is moved away from the operator beyond the neutral position, the control valve takes a position that permits oil in the bottom of the cylinder to return through the tubing and the lock valve to the control valve chamber and back to the reservoir and pump. At the same time the initial, or low-pressure, oil flow is diverted through the tubing to the top of the hoisting cylinder, thereby gently assisting gravity to lower the ladder.

Safety check valve. A safety-check valve is installed in the inlet at the bottom of each hoisting cylinder. This valve has a calibrated

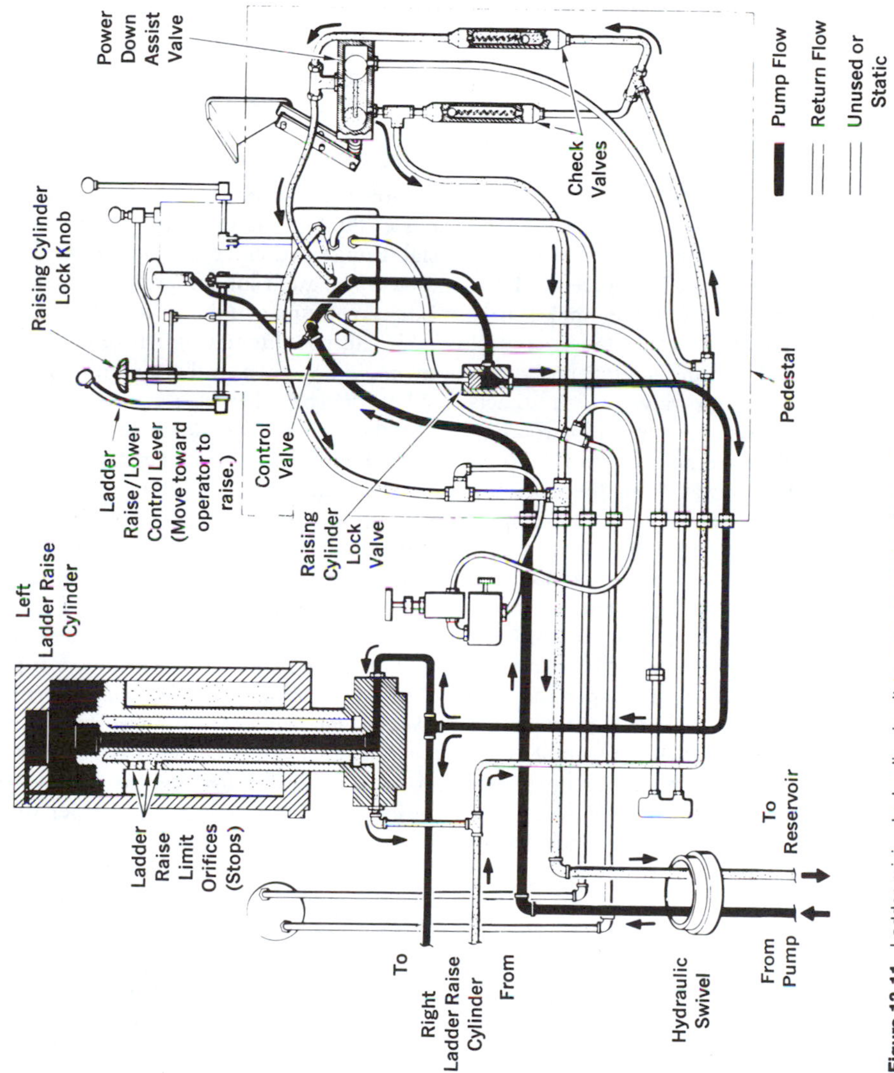

Figure 13-11. Ladder-raising hydraulic circuit (Courtesy of American LaFrance)

orifice that would prevent rapid falling of the ladder even if the tubing connected to the cylinders burst or was completely severed.

Lock valve. A lock valve is used to lock the pressure in the hoisting cylinders; this prevents the ladder from lowering and should be closed if the ladder is to be left unsupported in an elevated position for an extended period of time.

Rotation

A diagram of the ladder rotation hydraulic circuit of an American LaFrance model is shown in Figure 13-12. Other makes may vary in design. When the rotation control lever is moved from the neutral position it acts upon a control valve which directs hydraulic oil under pressure to the rotation motor. As the turntable control lever is moved clockwise away from the operator, hydraulic fluid is directed by the control valve to the turntable rotation fluid motor, which rotates the turntable clockwise. The hydraulic fluid returns to the reservoir from the turntable rotation fluid motor through the control valve. When the control handle is moved back, toward the operator, the hydraulic circuit is reversed, rotating the turntable counterclockwise. Turntable rotation is in the same direction as the movement of the turntable control lever. The turntable may be locked in any position by pushing the turntable lock lever away from the operator.

Warning. Rotation control, as with all other ladder movements, must be accomplished smoothly and very slowly. Abrupt motions can damage the ladder and endanger personnel in the area.

Extension and Retraction

Procedures for extension and retraction of aerial ladders must be followed with extreme care to avoid damage to the ladders. The schematics and descriptions of hydraulic circuits in the following discussion refer to the specific American LaFrance model in the illustration. Other makes will have similarities and differences to some degree.

Hydraulic circuit. A diagram of the ladder extension and retraction hydraulic circuit is shown in Figure 13-13. When the extension control lever is moved from the neutral position toward the operator, the same cycle of operation is accomplished as in the case of hoisting, except that the hydraulic oil passes through the extension control valve, stabilizing valve, and tubing to the extension motor. The extension motor then rotates the extension drum, and the ladder is extended; oil is returned through similar tubing to the reservoir, and then to the hydraulic pump. When the control lever is returned to neutral, the hydraulic oil circuit is stopped and the ladder remains at the height attained.

When the control lever is moved from neutral away from the operator, the extension motor is reversed and the ladder is retracted. High pressure is not available for retracting ladder extensions as long as the extension control lever is only moved to the first, or safety, notch of the retract position. *This is important since the ladder sections should under no circumstances be lowered against the ladder locks with high pressure.* As long as the control lever is not pushed beyond the safety notch, the ladder drifts down gently. The only times that the lever should be pushed beyond the safety notch to the high-pressure position is when it is necessary to retract the ladder very rapidly or when it is in a near-horizontal position.

The extension motor not only furnishes power to extend the ladder but also acts as a means of holding the ladder in an extended position against gravity and of controlling the speed of the ladder when retracting. In the latter two cases the extension motor ceases to act as a motor, but performs the function of a brake.

When the extension motor is used as a brake for retracting the ladder under certain

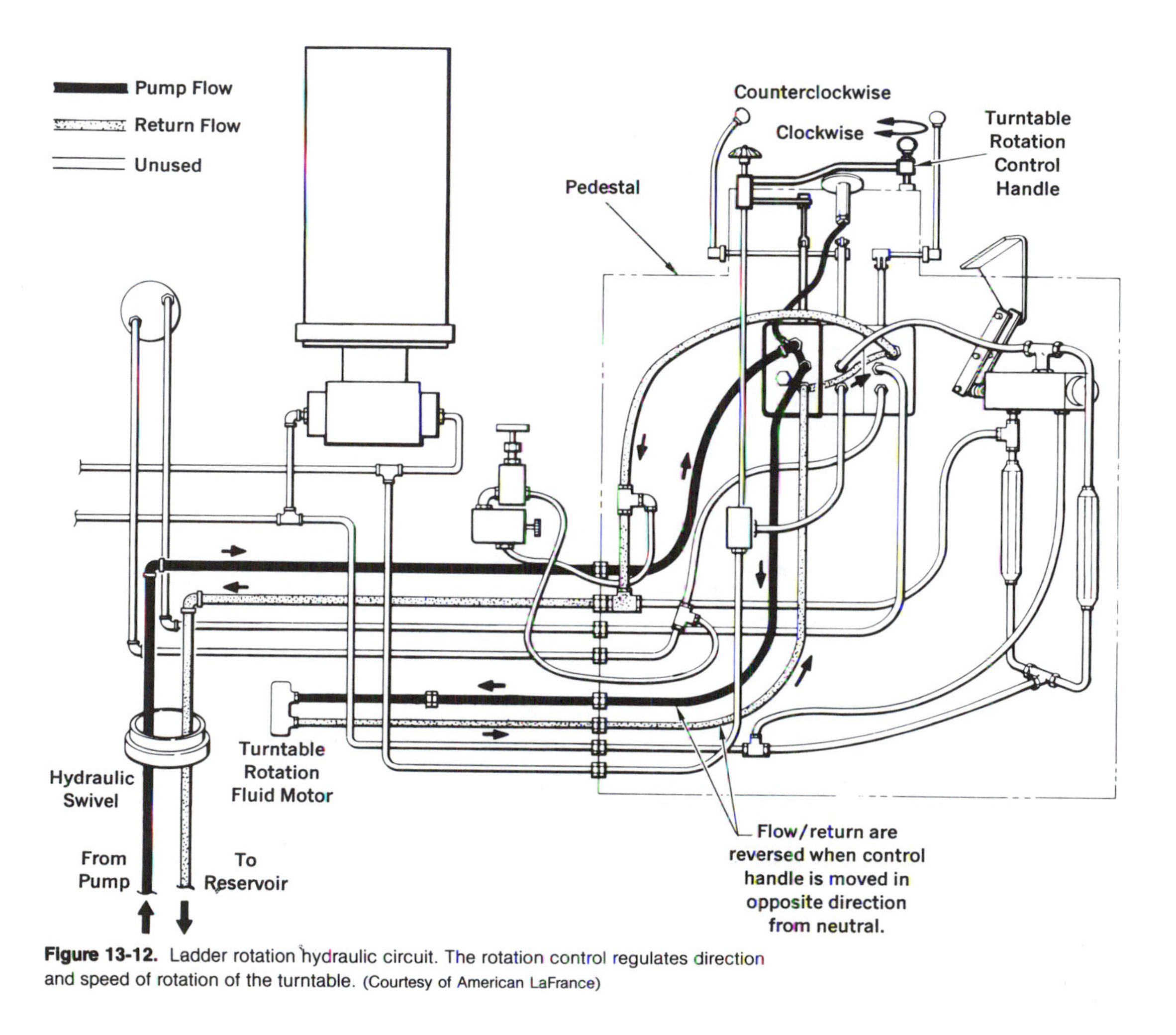

Figure 13-12. Ladder rotation hydraulic circuit. The rotation control regulates direction and speed of rotation of the turntable. (Courtesy of American LaFrance)

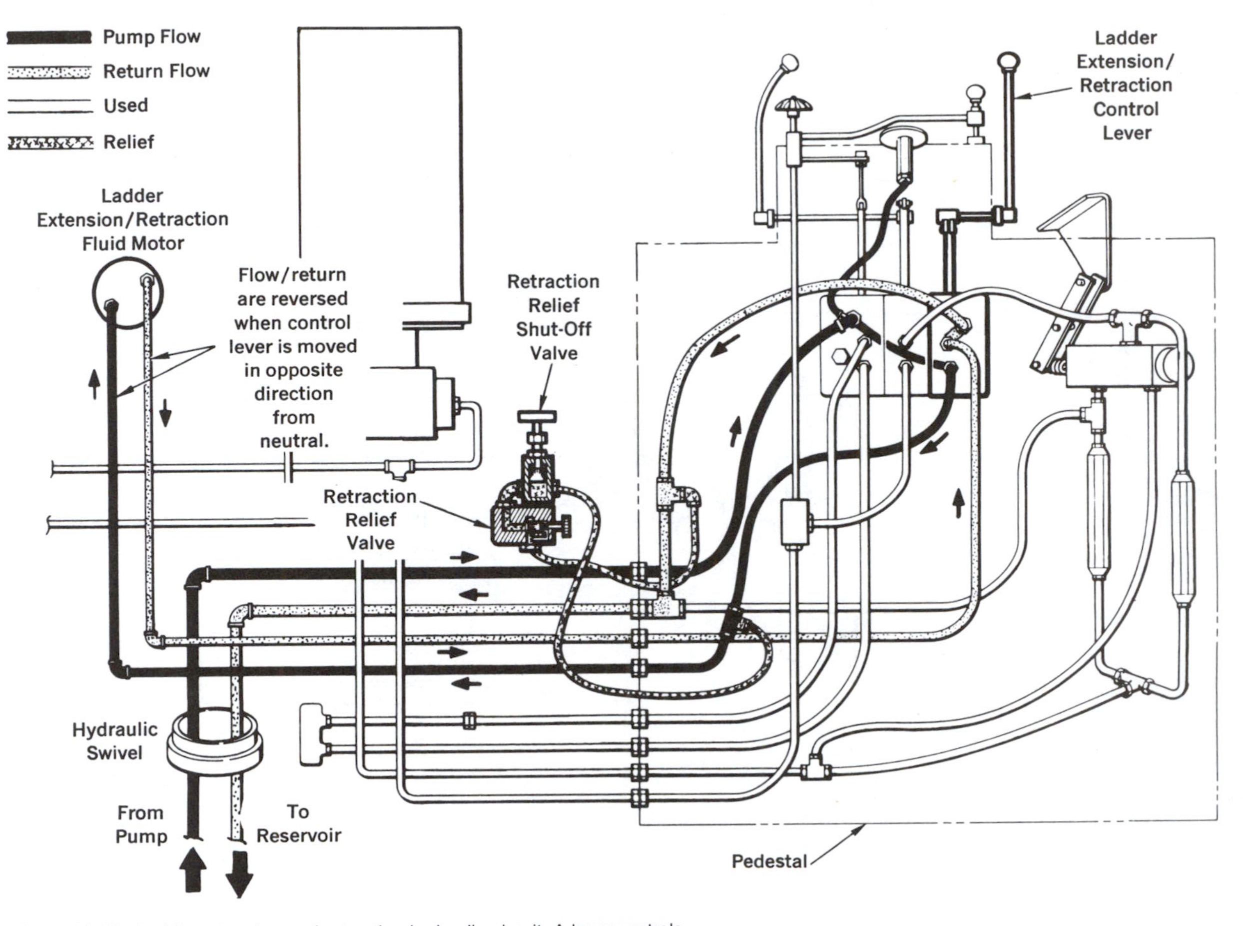

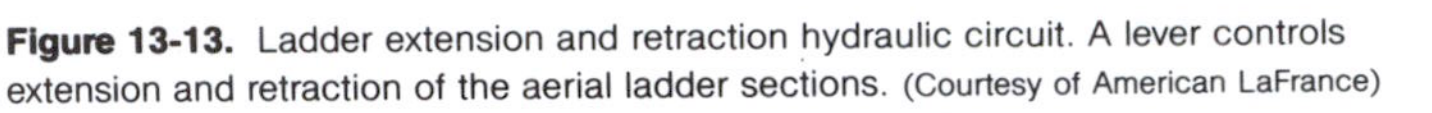

Figure 13-13. Ladder extension and retraction hydraulic circuit. A lever controls extension and retraction of the aerial ladder sections. (Courtesy of American LaFrance)

conditions of load and speed, and particularly if the hydraulic pump should be inadvertently disengaged or the engine stalled, the extension motor has a tendency to run away due to an unstable pressure condition of hydraulic fluid entering the motor. A stabilizing valve is installed in the hydraulic circuit between the extension control valve and the extension motor to overcome this condition. This valve contains both a check valve and a spring-loaded piston valve. The piston valve is designed to lift the check valve off its seat when pressure is applied to the piston valve. This pressure is supplied by an initial pressure in the hydraulic system through a small tubing line running from the control valve to the bottom of the stabilizing valve. The result is that unless the hydraulic pump is running to produce the initial pressure in the hydraulic system, the ladder will only drift down very slowly, regardless of its load, and under all conditions of operation control is perfectly stable and safe.

An automatic stop is provided to prevent overtravel of the ladder. This stop acts on the extension control lever so that the operator, upon feeling the reaction, should allow the control lever to return to neutral.

Caution. Do not resist the movement of the extension control lever as the automatic stop takes over. If the lever cannot move freely to the neutral position, damage to the stop linkage will occur.

An additional safety stop is provided at full extension and retraction. The hydraulic circuit contains a retraction relief valve factory set to approximately 750 psi. If retraction is accidentally attempted with the ladder locks engaged, the valve will divert fluid when the line pressure exceeds the valve setting, thereby preventing damage to the ladder or cables.

Outrigger hydraulic circuit. Figure 13-14 diagrams the outrigger hydraulic circuit. The position of the hydraulic selector control valve determines whether the flow of hydraulic fluid is directed to the turntable or to the outrigger circuits. When the selector valve control knob is pulled out, the hydraulic fluid is diverted from the turntable to the outrigger circuits.

With all the outrigger control levers in neutral, the outrigger control valve permits free passage of hydraulic fluid from the pump to the hydraulic reservoir.

When an outrigger control valve lever is pushed down, hydraulic fluid is directed to the upper end of the outrigger cylinder. The piston rod is moved out, which lowers the outrigger and forces the fluid between the piston and the lower cylinder head, back through the control valve to the hydraulic reservoir. Returning the outrigger control lever to neutral stops the fluid flow to the outrigger cylinder and locks the outrigger in position.

When an outrigger control valve lever is pushed up, hydraulic fluid is directed to the lower end of the outrigger cylinder. The piston rod is moved into the cylinder, raising the outrigger and forcing the fluid between the piston and the upper cylinder head back through the control valve to the reservoir.

Automatic cushioning stops are built into the outrigger cylinders to prevent overtravel in either direction. The stops shut off the flow of fluid in the cylinder as the piston approaches its limits of travel.

Spotting the Truck

The proper spotting of apparatus, while seemingly a simple operation, cannot be overemphasized in its importance. The position of the truck can determine the effectiveness, or lack of effectiveness, of an aerial ladder operation. Many things must be considered when selecting a spot for an aerial ladder to be raised: stability of the aerial ladder itself, overhead wires, steep grades, narrow roads, parked cars, and other apparatus.

In selecting a spot, before raising the aerial ladder, find the most level and solid footing available, consistent with the immediate requirements. The proper distance out

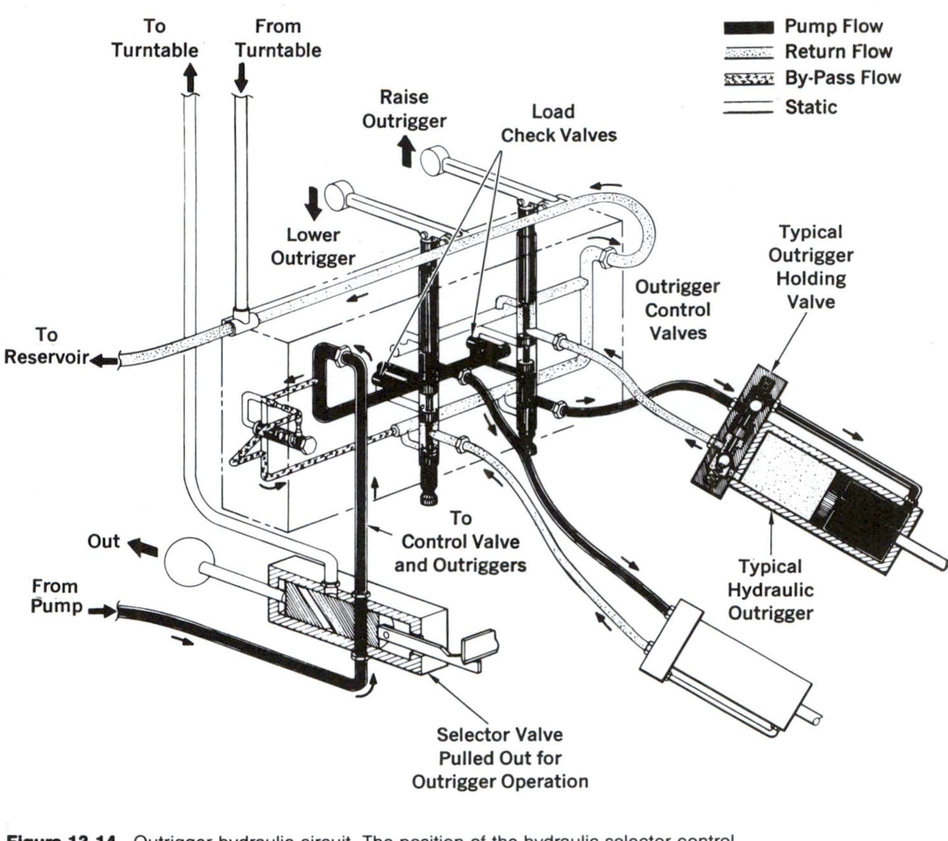

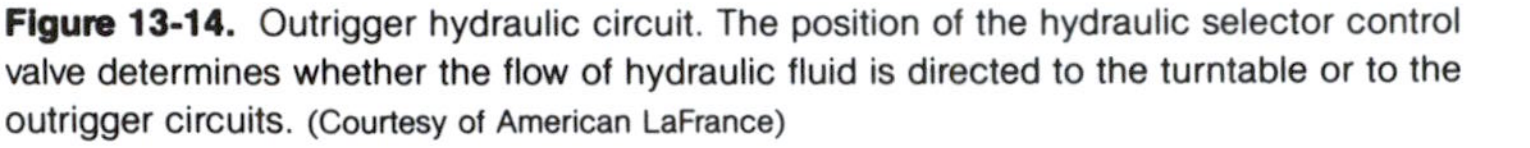

Figure 13-14. Outrigger hydraulic circuit. The position of the hydraulic selector control valve determines whether the flow of hydraulic fluid is directed to the turntable or to the outrigger circuits. (Courtesy of American LaFrance)

from a building will vary, and will depend on the height of extension required, the planned use of the ladder, and other conditions at the emergency. The best climbing angles are from 70 to 80° inclination from the horizontal. At these angles, with the tip supported, the ladder will carry its maximum load. The proper placement for maximum stability and the preferred climbing angle must be determined by the experience and good judgment of the aerial ladder operator.

In business districts, the curbs are usually 10 to 12 ft from the building line. When spotting can be made next to the curb, the aerial ladder operator knows that the reach will be between the fifth and eighth floors while maintaining a climbing angle of between 70 and 80°. Any operation below the fifth floor will result in climbing angles of less than 70°. When a spot is made outside of a line of parked cars, any operation below the seventh floor will be conducted at a climbing angle of less than 70°. As most spots have to be made outside of the line of parked cars, it is not unusual to see aerial ladders operate with climbing angles considerably less than 70°. It is considered good practice to locate the center of the turntable not more than 35 ft horizontally from the objective for any operation, regardless of extension. When it is necessary to exceed this distance, stay within the operating sphere for the particular apparatus.

Operating under Hazardous Conditions

Textbook operating procedures for apparatus often appear to make unrealistic recommendations: precise ladder angles, spotting the truck so the ladder is exactly at a right angle to the building, obtaining the maximum stability by angling or jackknifing the vehicle, laying the ladder into the building so that both ladder beams are bearing the weight, and other requirements that are important or vital to the safe and efficient utilization of an aerial truck. Even though these ideal conditions can seldom be achieved on the fire ground, they must be known and understood by all fire fighters, and adhered to whenever possible.

When manufacturers design an apparatus, they are concerned first with designing a strong vehicle that is safe and efficient; then a safety factor is added. Apparatus operators should never deliberately surpass the manufacturer's maximum load restrictions; the safety factors are intended to compensate for inadvertent excesses.

Operators should be governed by the loading recommendations because truck stability can be adversely affected. The stability of the ladder and the apparatus is most seriously jeopardized when the ladder is fully extended at a low angle in a cantilever position at a right angle to the truck chassis. The least force is exerted when the ladder is raised to its maximum angle and is supported by a building. Therefore, if less than ideal conditions are obtainable, the operator will be forced to compensate by altering the original plans. If the ground is soft and a close approach to the structure cannot be made, greater stability can be obtained by heading directly towards the building and extending the ladder over the cab. If a close approach to the building cannot be obtained, so that the ladder angle will be fairly flat, there are three alternatives: reduce the load on the ladder, support the tip of the ladder on a structure, or support the ladder with straight ladders or pike poles. When the ladder is used at low angles in a cantilever position, the strain can be reduced by lowering the load permitted on the ladder or restricting the ladder extension.

It has happened that an aerial ladder was excessively extended and then supported on a building at a very low angle because the truck could not approach the structure close enough. But when the operator attempted to raise the ladder off the building, the hydraulic pump and hoisting cylinders could not generate sufficient force. The fastest method of coping with this type of problem is to have one or two crew members climb the ladder, stand on the building, and help lift the ladder beams clear of the structure. After they descend, the ladder should be retracted to lessen the leverage on the hoisting mechanism. If this problem occurs, the hydraulic pump pressure and the relief valve adjustment should be checked.

High wind conditions. High-velocity or gusty winds will impose loads and strains that can adversely affect the ladder's strength and stability, especially if it is extended and unsupported. If possible, position the truck so the strain will not cause sidewise or twisting stresses. Use extra care in angling or jackknifing the truck and placing the ground jacks to achieve maximum stabilization of the vehicle. In case of gale strength winds of 35 mph or over, it is advisable to attach a guy rope to the top of the fly if it is to be extended very high. Tension should be maintained on the guy rope in the direction from which the wind is coming, especially if the wind is blowing sideways on the ladder. The guy line may be eased off when the ladder is in position against a building unless the wind is of an especially high velocity. If the wind is especially strong and the ladder is supported on a building, it will lessen the stress to secure the ladder tip to the structure.

Steep inclines. The strength of the cantilever bridge construction of an aerial ladder depends more upon design than upon the innate strength of materials. The cantilever design does not provide much support for side stress; in fact, all ladders give more support when the load is applied perpendicular to the rungs than when the load is applied from the side.

Rungs of an aerial ladder may be considered tilted any time they are not parallel with the horizontal plane of the earth's surface. If an aerial ladder is operated on level ground, the rungs will remain parallel with the earth's surface when raised to any position. When the truck is operated on inclines, the rungs will remain horizontal if the apparatus is in line (no jackknife) and positioned parallel with the curb, and if the ladder is raised either over the cab or over the trailer. When operating on an incline, with the truck in line and parallel with the curb, the rungs will be tilted any time that the ladder is raised while swung out at any angle from the centerline of the truck. The tilt will increase as the ladder is moved from an in-line position to a 90° angle from the in-line position. The tilt with a 45° angle from the in-line position will be one half as great as the tilt at a 90° angle. The crown of the road will also have an effect on the side stress.

Any time that the rungs of a ladder are tilted, a portion of the weight of the ladder and of the load (if there is one) acts sideways on the ladder. This effect is caused by a shift of gravity from the center of the rungs toward one side or the other. The center of gravity will shift toward the low side of the tilt. The greater the tilt, the greater the shift in the center of gravity and the greater the force applied sideways on the ladder. This side stress is immediately apparent to the aerial ladder operator because of the increasing difficulty of operation. When a truck is working on a 5% grade, with the ladder extended to a building directly in front, 5% of the ladder weight and load is converted to side stress. The amount of weight the ladder can support with safety is reduced under these conditions.

To compensate for excessive street grades, spot the apparatus down the slope from the area of operation. If headed downhill, it is necessary that the operator pass the spot where the aerial ladder is to be raised (Figure 13-15). If headed uphill, the operator should stop short of the intended point. There are two reasons for doing so: first, the rung tilt will be reduced and, second, the apparatus is more stable with the ladder extended uphill than downhill. The reason for this is that the lower the hoisted angle of the ladder (angle from the horizontal), the less load it will safely carry. An aerial ladder working on an incline is working at a lower angle from the horizontal when raised to the

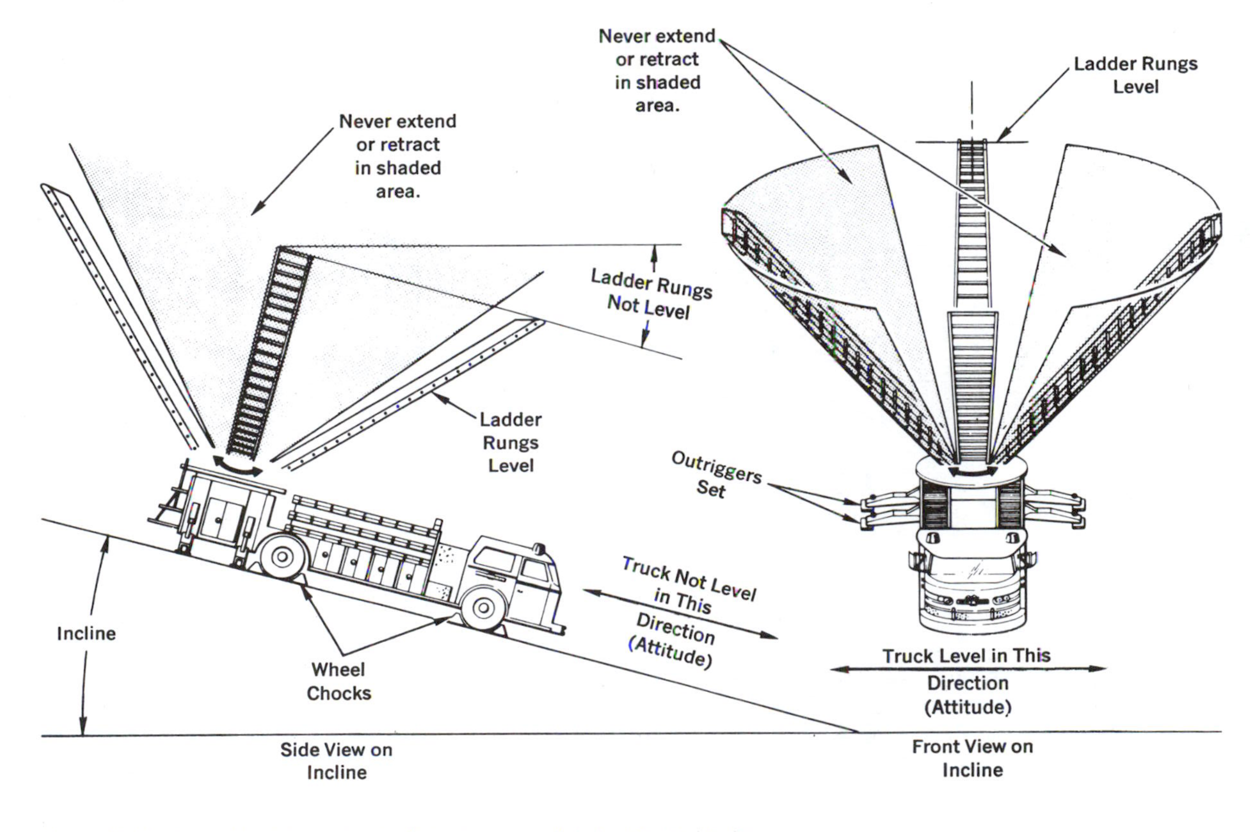

Figure 13-15. Safe ladder extension and retraction on an incline. Do not position the aerial ladder truck in such a manner as to jeopardize its stability. Avoid uneven or soft terrain. The maximum ladder capacity is achieved when ladder rungs are level.

sixth floor and spotted uphill from the area of operation, than when raised to the sixth floor and spotted downhill. When raising aerial ladders to upper floors while working on an incline, only one beam of the ladder will rest against the building. The ladder cannot then be considered to be supported, and should therefore be operated as if in an unsupported position.

Overhead obstructions. On some streets in the business and manufacturing districts, there are so many overhead wires that it is impossible to raise an aerial ladder; nevertheless, safe operations can be conducted with proper planning.

When approaching a spot with overhead obstructions, the operator should thoroughly size up the situation. It may be possible to select a spot, jackknife the apparatus, lower the ground jacks, and raise the ladder without any difficulty. In some cases it may be necessary to raise the main ladder out of the bed and then jockey the truck into position with the main ladder raised. In accomplishing this, engage the spring locks before raising the ladder. It is safe to move the apparatus with the main ladder raised, although this may take the coordinated effort of a driver, an aerial ladder operator, and a tiller operator. But do *not* move the truck with the ladder extended.

At times it may be necessary to select an alternative spot, such as an adjacent parking lot. When there is an unusually heavy concentration of overhead wires and other companies are already working in the most desirable spots, it may be necessary to position the truck on the sidewalk and raise the ladder at a particularly steep angle or to drive up on the sidewalk on the opposite side of the street and extend the ladder at a low angle. When working close to high-tension wires, be extremely careful that the ladder does not touch the wires or that it is not close enough for fire fighters to brush against the wires while using the ladder.

Railroad tracks. Occasionally an apparatus must drive down and over railroad tracks to reach a particularly strategic location. It is best to straddle the tracks, crossing them as little as possible to reduce the possibility of tire and wheel damage. If tracks must be crossed, it is best accomplished at an angle so only one wheel at a time is affected. It will greatly reduce the likelihood of tire damage if the crew places lumber between the tires and the tracks.

Jackknifing

Jackknife is a trade term in the fire service that means to turn the tractor of a truck at an angle to the trailer to secure good stability (Figure 13-16). In effect, an aerial ladder is a long lever that must be counterbalanced with a greater weight at the other end. The wheels and ground jacks on the fire side of the truck serve as the fulcrum; extending the stabilizing jacks as far as possible moves the fulcrum towards the ladder tip and changes the leverage ratio. Extending the ground jacks beyond the chassis dimensions will spread and distribute the truck weight and rotating stresses beyond the ground area covered by the vehicle wheelbase.

There are two methods of jackknifing: the inside and the outside. Theoretically, the inside jackknife is the most efficient because it moves the fulcrum point closer to the objective. Actually, however, the inside jackknife should be avoided, because it places the base of the ladder further from the building and longer side reaches are necessary. The advantages of the outside jackknife are that the base of the ladder can be placed closer to the building and the heavier portions of the apparatus are used most effectively for counterbalance.

The ideal angle is about 60° from the inline position and away from the direction of ladder extension; this is called an outside jackknife. Even when the tractor is jackknifed as much as 30°, the truck stability is more than twice as great as when in the inline position. Because of narrow streets, alleys, overhead obstructions, etc., jackknifing is not always possible. Good judgment

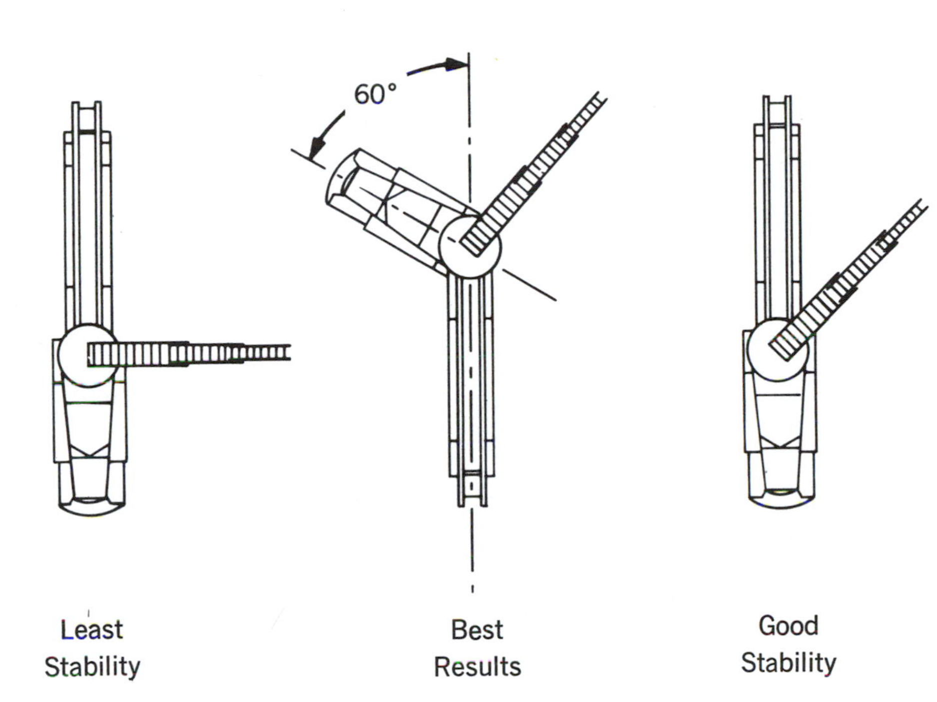

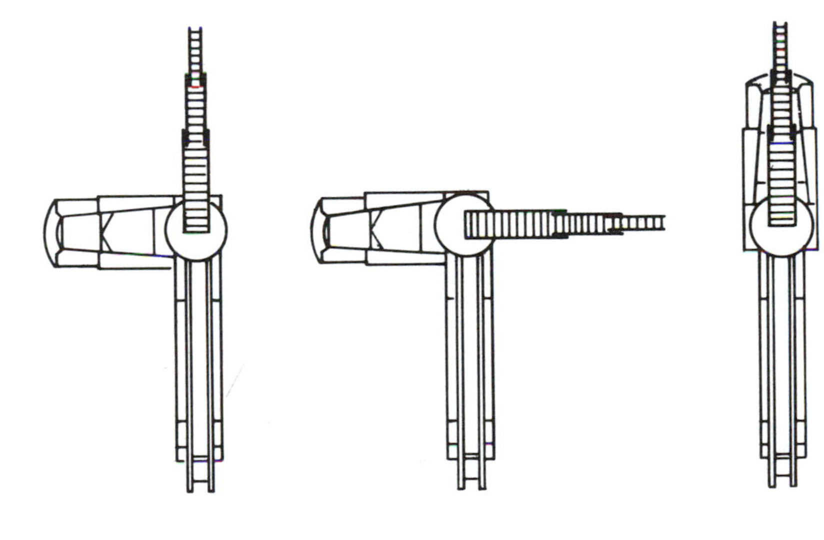

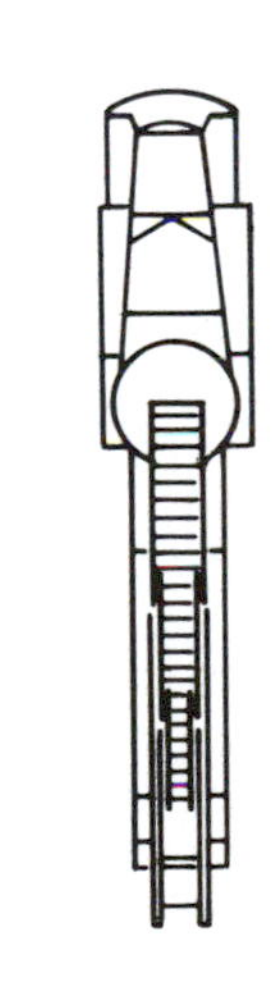

Figure 13-16. Jackknifing increases truck stability.

and a complete understanding of aerial trucks and their limitations will indicate the extent of safe deviation from recommended practices.

A 60° jackknife will provide excellent stability without excessively blocking most streets. The method of spotting gives all of the advantages that are required for effective aerial ladder operations: (1) a close-in spot, (2) good climbing angle, (3) maximum ladder strength, and (4) greatest ladder reach. A good method of achieving the jackknife can be practiced on a street with a tall building, average-width sidewalk, and no parked cars. Drive down the street as close to the curb as possible. Mentally drop a vertical line down the face of the building from the window or other location where the ladder tip is desired, then mentally place a line in the street at right angles to the objective point. As the front wheels cross the mental line in the street, cut sharply away from the curb with both the tractor and trailer until the rear tractor wheels are just past the desired spot. Stop, cut the front wheels of the tractor in the opposite direction, and back up until the rear tractor wheels are against or near the curb. After the initial turnout by the trailer, there will be practically no more movement of the tiller wheels; they will be capable of doing very little truck adjusting.

The first few times this method is attempted, it will probably be found that the tractor and trailer are each about 45° out of a straight line; this results in a jackknife of approximately 90°. This is excellent from a truck stability standpoint, but it really blocks the streets. Where traffic interference will become a problem, there may be an advantage to swinging the tractor and trailer each out 15° from a straight line; this will result in a 30° jackknife, which more than doubles the apparatus stability. Angles of more than 30° make the apparatus that much safer.

Stabilizing Single-Chassis Ladders

While single-chassis vehicles cannot be jackknifed to provide greater stability, they may be angled towards the building to improve conditions (Figure 13-17). The greatest stability is provided by extending the ladder directly over the front or rear of the truck, while the most hazardous extension is at right angles to the chassis. Any angling will improve the stability.

Aerial Ladder Maneuvers

When the apparatus has been positioned to start operating, after particular attention has

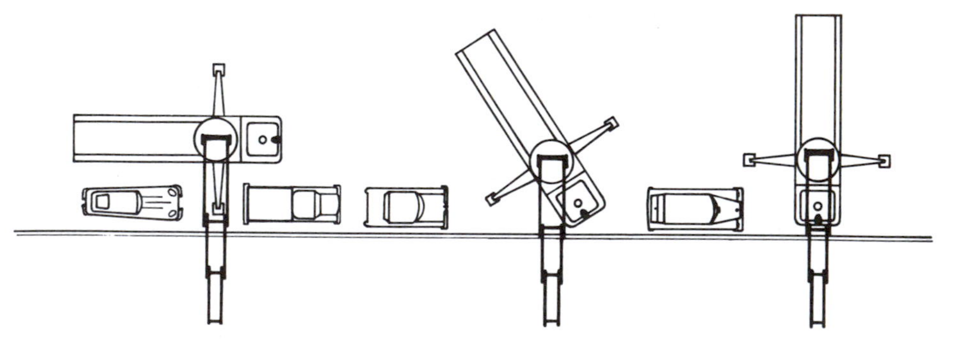

Figure 13-17. To improve the stabilization of a single-chassis truck, either head in or back in at an angle. Backing in a rear-mounted ladder will probably result in a better spot.

been paid to overhead obstructions, the engine should be left running; then place the road transmission in neutral, set the parking brake and the auxiliary braking system, if so equipped, and place the power takeoff in gear to drive the hydraulic pump. Throttle control on some models is transferred to the control pedestal by closing a switch on the dashboard.

On tractor-trailer aerials, the next step is to place the spring locks in operation, if so equipped; this is most easily accomplished just before the vehicle comes to its final stop. The tiller operator should swing the tiller wheel and seat clear of the ladder and properly stow the tiller post.

Next, set the stabilizing ground jacks to form a positive footing between the chassis and the ground for a stable ladder operation.

Before mounting the turntable, have a clear picture of what is to be accomplished with the ladder. Before touching the controls, check both sides of the truck to be certain that the chock blocks are correctly placed and the ground jacks are set. Then, glance along the apparatus to be sure that everything is clear and that the rest of the crew is ready.

Chock blocks should be placed as soon as the apparatus is properly spotted. On level terrain, the dual wheels of the tractor on the side on which the ladder is to be raised should be chocked at the front and rear. On an incline, place both chock blocks on the down grade side of the rear tractor wheels.

Some aerials require the energizing of a throttle switch on the control pedestal to raise the engine speed; this produces the proper hydraulic pressure in the extension, elevation, and rotation systems. Other aerials are equipped with an automatic throttle assembly that accelerates the engine to the proper speed when any of the three control levers is moved toward its operating position. The throttle is actuated for lowering or retracting only by a full depression of the levers, as idling speed is usually sufficient for these operations.

With the power takeoff in gear and all control levers in their neutral positions, fluid circulates through a selector valve to the oil reservoir; a small pressure reading will be indicated on the pressure gage. Movement of any control lever away from its neutral position will cause oil pressure to build up. The gage reading varies with the operation, but for any single motion it is generally close to that for which the relief valve is set. For lowering or retracting, the pressure reading is much less since gravity aids these two operations.

Ladder operations are carried out by one operator, and all levers and valves can be easily reached from operator's position at the control pedestal. Control levers are clearly marked to indicate their function and the direction of ladder movement; movement of the lever corresponds to the resulting motion of the ladder.

The speed of each operation may be regulated by the operator, depending on the amount the control lever is moved, from a very slow operation with the valve just barely open to full speed with the valve fully open. Very fine control may be exercised by using both hands on one control, one hand resting firmly on the pedestal and restraining the movement of the control, the other hand making the actual movement. If the control is the type that requires depression of a knob to unlock it, the only time the knob is touched is to release the catch; the movement is then made with the hands placed lower down on the handle. This prevents going past the lock in case of an emergency stop and, therefore, prevents reversing into the opposite movement. Move the control levers slowly at all times to avoid jerky motion and whip of the ladder.

It is recommended that each operation, such as raising the aerial ladder, extending sections or rotating the turntable, be accomplished separately. After the operator has become thoroughly familiar with the controls, these operations can be combined. It is much safer, however, to perform each operation independently of the others, so that the operator's attention is free to observe any obstructions that may interfere with placing the ladder in its proper position. Very little time is gained by raising the ladder and extending the sections simultaneously, because all the operations are slowed down when more than one is functioning at the same time. The usual sequence in placing the ladder is: elevate, rotate, extend.

Hoisting

The average opeator does not elevate the ladder up high enough when making the initial hoist out of the bed. The tendency is to raise the ladder to about 50°, then rotate towards the target, which necessitates having to raise the ladder to a steeper angle before being able to complete the rotation into a line with the building. This causes a slow, jerky operation. The initial elevation should be high enough so that when the ladder is rotated and extended to the desired location, the tip of the ladder is well above any supporting surface such as a roof or window ledge. Raising the ladder definitely above the necessary angle and then lowering the ladder into place and making any necessary corrections gives the entire operation continuity, saves time, and offers the bystanders and superior officers the impression of experienced know-how, rather than a fumbling, blundering, groping for position.

Rotating

Rotate the turntable to place the ladder in position. Sighting along the ladder beams will reveal whether or not the ladder has been elevated sufficiently, and proper adjustment can be made by means of the hoist control lever. The turntable should be gradually slowed down into a stop, particularly when the ladder is extended, rather than attempting to maintain one rate of movement until the ladder lines up and then suddenly stopping it. Sudden stops with the aerial extended can cause whipping, with consequent excessive stress on the ladder, rotation motor, and gears by lashing and jerking. Sudden reversal of rotation is also abusive and hazardous. The operator should never intentionally allow the ladder to strike any object during rotation.

Extending

Slowly and smoothly operate the extension control and extend the fly to the proper point. If the aerial ladder is to be raised to a roof coping, extend it several feet above the wall; this helps crew members to get on or off the ladder and makes it possible to avoid laying hose lines over the ladder tip. It also makes the ladder easier to find for fire fighters coming back to it in the dark or in smoke. For windows and fire escape balconies, raise the ladder 1 ft above the sill or railing. When the ladder has been extended to the proper position, operate the ladder lock lever to engage the locking pawls; then move the extension control lever to permit the extensions to drift back against the pawls.

Lowering into Building

Carefully release the hoist control and allow the ladder to drift gently into position. While the beams of metal ladders are not fragile, ordinary care should be exercised in handling. A blow to the bottom of the beam, which might occur if the ladder is permitted to strike the building violently, might dent the beam and affect its smooth action over the rollers or actually weaken the ladder structurally.

The hydraulic pump can be left running if the weather is extremely cold, or the ladder is to be used only for a short time. If it is necessary to leave the ladder in the same position for a considerable period of time, the hydraulic pump should be disengaged and the engine shut down.

Ladder Locks

When raising an aerial ladder and placing it in position, the hoist, extension, and rotation locks should be activated as each evolution is completed. As the last lock is closed, all three should be rechecked to be certain they are in the correct position. Many times at multistory fires, as the ladder is being lowered into position, truck personnel who are anxious to ventilate, or engine crew members with hose lines will start to climb the ladder so they may attack the fire. They should be detained until the ladder operator is certain that the extension locks are in place and the ladder retracted into the locks. Then as the ladder is finally positioned, all the locks should be checked again before allowing anyone on the ladder.

Drilling New Operators

Smooth handling and competent operation of an aerial ladder can only be obtained through understanding the apparatus mechanism, knowing its abilities and limitations, and extensive practice sessions at practical drills.

When teaching a new operator how to maneuver an aerial ladder, it helps build confidence to have the trainee raise the ladder to the limit of elevation and then extend it out to the maximum extension to demonstrate how the warning devices and limit controls work. Show the correct method of retracting and lowering the ladder into the bed; make a definite point that no rotation correction of the ladder should be made below the level of the highest point of any equipment or tiller seat; ladders have been damaged by not observing this precaution. Demonstrating the correct methods of operating the ladder, as well as the safety features, will produce a more knowledgeable operator.

Lowering the Ladder

When the ladder is to be lowered, first raise it away from the building a short distance; then extend the ladder a few inches to clear the ladder locking pawls, and then release the ladder lock lever. Move the extension control and let the ladder retract until it reaches the limit stop. Release the turntable lock and rotate the turntable until the arrow on the turntable lines up with the stationary marker on the frame of the truck. This indicates that the ladder is in the bedding position. While the ladder is retracting, the turntable may be rotated to the bedding position. Do not depend entirely on the arrows to indicate when the ladder is lined up; sight along the ladder beam, almost like sighting a rifle, to be certain that the ladder is correctly rotated. Move the hoist control and allow the ladder to settle into its bed.

Release the ground jacks and spring locks, remove and stow the chock blocks, and return the throttle control to the cab.

Safe Ladder Loading

Manufacturers specify different numbers of crew and amounts of equipment that can be safely supported by their aerial ladders, depending on the make, model, length of ladder, angle of inclination, support, and use. No general statement can be offered as a guide. Fortunately, all manufacturers incorporate an indicator of some type.

A pendulum-type inclinometer attached to the side of some makes of ladders indicates the angle of inclination and the maximum safe extension of the ladder. Other makes have a direction panel mounted under the control pedestal glass. Maximum safe ladder loadings are indicated where they are readily apparent.

With the ladder supported and fire fighters climbing or taking lines to an upper floor or roof, space the crew 12 to 15 ft apart (Figure 13-18); this will keep the ladder within safe loading limits. If the crew members stop to pass a hose up, or if the top

Figure 13-18. Space out crew when taking hose lines aloft. (Courtesy of the Los Angeles City Fire Department)

member is held up because of a difficult entry, this spacing should still be maintained.

When personnel are working on an unsupported ladder, loading must be kept to a minimum. When ladder pipes are being utilized, one crew member to an appliance is recommended.

Where a maximum number of people must be removed from upper floor windows in minimum time, an intimate knowledge of ladder loading and operational ability is required of the operator. If there are many people to be saved, one crew member attached to the ladder with a life belt should stay at the tip and get those trapped onto the ladder and started down. Another member should be positioned at an intermediate point below the fly to encourage and assist the people. One or two personnel on the turntable can help get the rescued people off the ladder and to the ground. When many people must be removed from various windows on several different floors, fire conditions might not allow time for each person to descend to the ground before moving the ladder. Experience has shown that it is rapid and safe to secure a fire fighter to the ladder tip with a life belt to help the jeopardized occupants, one at a time, onto the ladder, from which they can be transported to an adjacent roof.

Unusual Rescue Conditions

Some manufacturers recommend that their aerials never handle weights in derrick or crane fashion. Others not only sanction that use, but encourage it by incorporating a snatch block, anchored to the top of the fly section, and a hydraulically operated winch, mounted on the base section. Recommendations depend upon the strength of the ladder; they should be respected when deciding on proper procedures at a fire or rescue incident. A straight ladder can be used to help support the aerial ladder at low angles of inclination.

No matter what the rescue conditions, remember that a ladder is much stronger when the pull is vertical to the center of the rungs, not in a sidewise direction. Also, if possible, position the truck so the ladder extends over the cab or rear end; it is practically impossible to upset the vehicle when used in this position. Ladder load limitations must also be remembered and taken into account.

When removing victims from an excavation, canyon or cliff, fasten a hose roller to the ladder fly, attach a Stokes stretcher to a life line, and pass the line down the center of the ladder and over the hose roller. Extend the ladder so the stretcher will not drag on the sides of the excavation. After the victim is in the stretcher, heave on the line to raise the stretcher; retract the ladder fly, and remove the victim and stretcher. Use a guy rope to steady the stretcher.

When removing victims from roofs or upper floor windows, attach a Stokes stretcher to the ladder fly with a bridle so that it swings securely, or use a life line; elevate and extend the ladder so the stretcher is placed in the desired position. Place the victim in the stretcher and lower it to the ground. When raising and lowering patients in a stretcher, it is necessary to use restraints to prevent movement by the victim.

Ventilation With Aerial Ladder

Upper-story window ventilation may be accomplished with the aerial ladder. Because of the possibility of damage when an aerial ladder strikes a window sill, the glass should be broken by extending the fly into the panes rather than by dropping or lowering the ladder against the window frame. Beware of glass trailing down the ladder as well as falling to the ground below the windows.

Maintenance

Most of the problems encountered while operating aerial ladders are caused by inadequate maintenance. To keep the apparatus fully operational, routine lubrication policies should be formulated and followed.

Usually the most neglected part of the mechanism is the hydraulic system. These systems are composed of components with very close clearances. Control valve spools are lapped to an extremely close tolerance; internal leakage could cause the ladder to drift dangerously. A relief valve that is sticking, or is held open by a small particle, could prevent pressure buildup. Partially clogged lines can cause slow ladder operation. Abrasive contaminants circulating in the oil flow can cause the various components of the system to wear excessively, and thus fail, at an accelerated rate.

Mechanics who religiously change the oil and filters of the vehicle engine at frequent intervals rarely give the same attention to the hydraulic system. This fluid is similar to heavy-duty engine oil. It attains the same high temperatures and receives similar abuse to engine oils. The additives become depleted with use and age; sludge and other contaminants accumulate. Therefore, at least annually, and oftener if the apparatus has been subjected to excessive use, the entire hydraulic system should be drained and flushed. Refill the hydraulic tank with oil or fluid of the recommended viscosity. Clean the screens and replace the filters.

Many of the operating problems are caused because the oil becomes too hot. Excessive heating is usually the result of one or more of the following conditions: hydraulic tank oil level too low, prolonged operation at too high an engine speed (800 to 900 rpm is sufficient), oil viscosity too high, or the fluid contaminated by water or other impurities.

The ladder and all parts of the mechanism should be kept clean and inspected often for leaks or loose parts. Keep all lubrication fittings well greased. The slides, rollers, sheaves, and pulleys should be cleaned with a solvent and then lubricated. The cables should be thoroughly inspected and greased. Check for frayed strands, and then apply a wire rope lubricant or light oil with a brush or rag.

14 Elevating Platforms and Water Towers

Since the beginning of fire service history, fire fighting vehicles and equipment have been constantly improved; recently, fire apparatus have been developed and refined at an accelerated rate. The latest advancements have been made in the introduction of hydraulically operated elevating platforms and water towers.

Elevating Platforms

There are various types and models of elevating platforms, each with its advantages and disadvantages. Before any fire department can figure out which is the best for its purposes, it must decide what its basic duties will be; what tasks it will be expected to perform; what roadway and geographical restrictions may exist; what its crew assignments will be; whether it will respond on first-alarm assignments in a regular district or be available on special calls only. These and a multitude of other questions must be considered before any firm determination can be made as to which type and size of elevating platform should be purchased.

Functions

Elevating platforms, which are also called aerial platforms, are commonly used to supplement standard pumpers and ladder trucks. This type of apparatus is generally known in the fire service as Snorkels, though this is actually a trade name created by one manufacturer.

In general, it may be assumed that a fire department will be concerned first with the unit's potential as an elevated base for a large nozzle to be employed in water tower service and, secondly, as a means of rescue from upper floors. Other useful functions it serves are: as a portable elevator for transportation of fire fighting personnel, as a base for use by supervisory personnel or chief officers who wish to oversee or carry out a reconnaissance, as a platform for extending floodlights above an emergency operation, and as a standpipe connection to readily supply water for hose lines on the upper floors of a building. Figure 14-1 demonstrates the use of elevating platforms in fighting fires.

Figure 14-1. Master streams from a portable monitor and an elevating platform making a low-level attack on a supermarket fire (Courtesy of Mack Trucks, Inc.)

Types of Platforms

The types of elevating platform apparatus designed for the fire service consist of hydraulically operated booms of either telescopic, articulated, or a combination telescopic-articulated design (Figure 14-2). One advantage of the combination telescopic-articulated type is a saving in overall travel length; a 150-ft Firebird in the travel position is 47 ft long, as compared to many 85-ft articulated units. As the primary purpose of an elevating platform with most departments is to provide a fast, massive, and readily controlled heavy stream application, the monitor will occupy a prominent place; in fact, the vehicle is designed mainly to enhance the water delivery capabilities of the apparatus.

Other equipment regularly or occasionally maintained in the basket (platform) are breathing apparatus with a piped-in supply, 12 or 110 V receptacles for operating power tools and lights, flood lights, pike pole and axe for ventilation purposes, hose pack of 1½- or 2½-in. hose for making a hand line attack on a fire, winch for effecting rescues, and a protective curtain and spray nozzle on the underside of the platform.

Some fire departments operate elevating platforms as separate units, similar to water towers. Others have them mounted on a service ladder truck or a pumper. Aerial platforms have been constructed that carry a full complement of ground ladders, a main fire pump, a booster tank, and a full load of fire hose. When evaluating such trucks for municipal fire protection, consider the extent to which the elevating platform provides pumping capacity or needed ladder truck protection. In making such determinations, local structural conditions, existing fire apparatus in service, and individual characteristics of the particular truck in question must be considered.

Effective Reach and Height

The nominal height of an elevating platform assembly is measured by a plumb line from the top surface of the platform to the ground, with the platform raised to its position of maximum elevation. The longer elevating platforms will usually accomplish everything that the shorter ones will; in addition, they will have the added height and reach for additional duties. However, this added capacity will usually be accompanied by extra vehicular weight, length, and height that could prevent the apparatus from maneuvering through narrow, congested streets or over weak bridges. A compromise will usually decide the minimum size that will accomplish all the anticipated fire fighting duties and the maximum size that can be housed in the fire station and maneuvered through the city.

The effective height of an elevating platform governs not only the vertical height that the passenger-carrying platform assembly (basket) can attain, but the horizontal reach that can be achieved. Horizontal reach is about equally as essential as vertical height. Parked automobiles, sidewalks, building setback, power lines, street lights, and a large variety of other hindrances will help decide how close to the building the apparatus can be spotted. Rescue operations will probably require an ability to place the platform directly against the building; when a heavy stream operation is desired, the apparatus should be kept further back in case of a structural collapse. Whether the elevating platform is a telescoping or articulating type, or a combination of the two, will also affect the area of a building that can be serviced by a platform without moving the vehicle; the area of a structure that can be reached by an elevating platform is called the scrub surface.

Figure 14-2. Fifty-foot aerial platform with two articulating booms mounted on a triple combination pumper (Courtesy of Crown Fire Coach Corporation)

Apparatus Construction

There are similarities, and there are vast differences, in the construction of elevating platform apparatus. It would be impossible to describe the principal fabrication details of all the makes, models, and sizes of this type of apparatus. The following test explains the principles of a two-boom articulating type of apparatus, but many of the particulars will apply to some extent to every elevating platform.

An aerial platform is a completely self-contained, hydraulically operated unit mounted on a truck chassis. Its main frame, secured to the truck chassis, serves as a base for mounting platform components and accessory equipment. The main constituents are a turntable, elevating booms, and an operating platform. These units are assembled to form an integrated mechanical structure capable of both horizontal and vertical movement. A mechanical leveling system keeps the operating platform level at all times, regardless of boom movement. Automatic stops prevent overlifting either boom section, and keep the booms in a stable position. Retractable outrigger stabilizers, located on both sides of the main frame, enable the platform to be leveled and stabilized.

Power for operating the hydraulic system (Figure 14-3) is obtained from the truck power takeoff or from an optional power pack equipped with a gasoline engine. Either source of power operates a hydraulic pump that supplies fluid under pressure to actuate the turntable rotation motor, boom operating cylinders, and outrigger cylinders. The hydraulic fluid is distributed through the system in steel tubing and flexible hoses, and the fluid flow is controlled by a series of valves. Two control stations are provided; one is located on the operating platform for the aerial operation and the other at the base for ground control.

Hydraulic pressures vary from 1000 to 2000 psi, depending on the make, model, and requirements. This pressure, when introduced into a hydraulic cylinder, exerts its entire pressure on every square inch of the interior. Since the cylinder is built to withstand pressure, the force of the hydraulic pressure is exerted against the piston head. To give an example of the tremendous force that may be generated, in Figure 14-4 the piston head diameter is 9.5 in., an area of 70.88 sq. in. This area multiplied by the system pressure of 1500 psi, shows an available total force of 106,320 lb, or over 50 tons. The total push on a hydraulic ram is computed by the formula:

$$\text{Force} = 0.7854\, d^2 \times \text{psi of hydraulic pressure}$$

Booms

The several types of aerial platforms differ mainly in number of booms, in method of construction, and in the way the booms are controlled. Articulated aerial platforms generally utilize two booms, though three-boom apparatus have been constructed.

Telescoping-boom elevating platforms are of multiple-section construction, one base and three sliding sections, fabricated of high tensile strength aluminum alloy. The boom shell houses the telescopic waterway for the platform nozzle and for all lines, electrical and hydraulic, that lead to the platform. Mounted atop the telescoping boom is a permanent ladder. Boom sections are powered for extension and retraction by a series of hydraulic pistons.

Platform Operators

The most feasible operating arrangement is to have two operators on the platform and one operator stationed at the ground control. One operator on the platform handles the controls, while the other is responsible for positioning the monitor or other needed equipment. The operator standing by the ground controls near the turntable is primarily there to watch over the booms behind

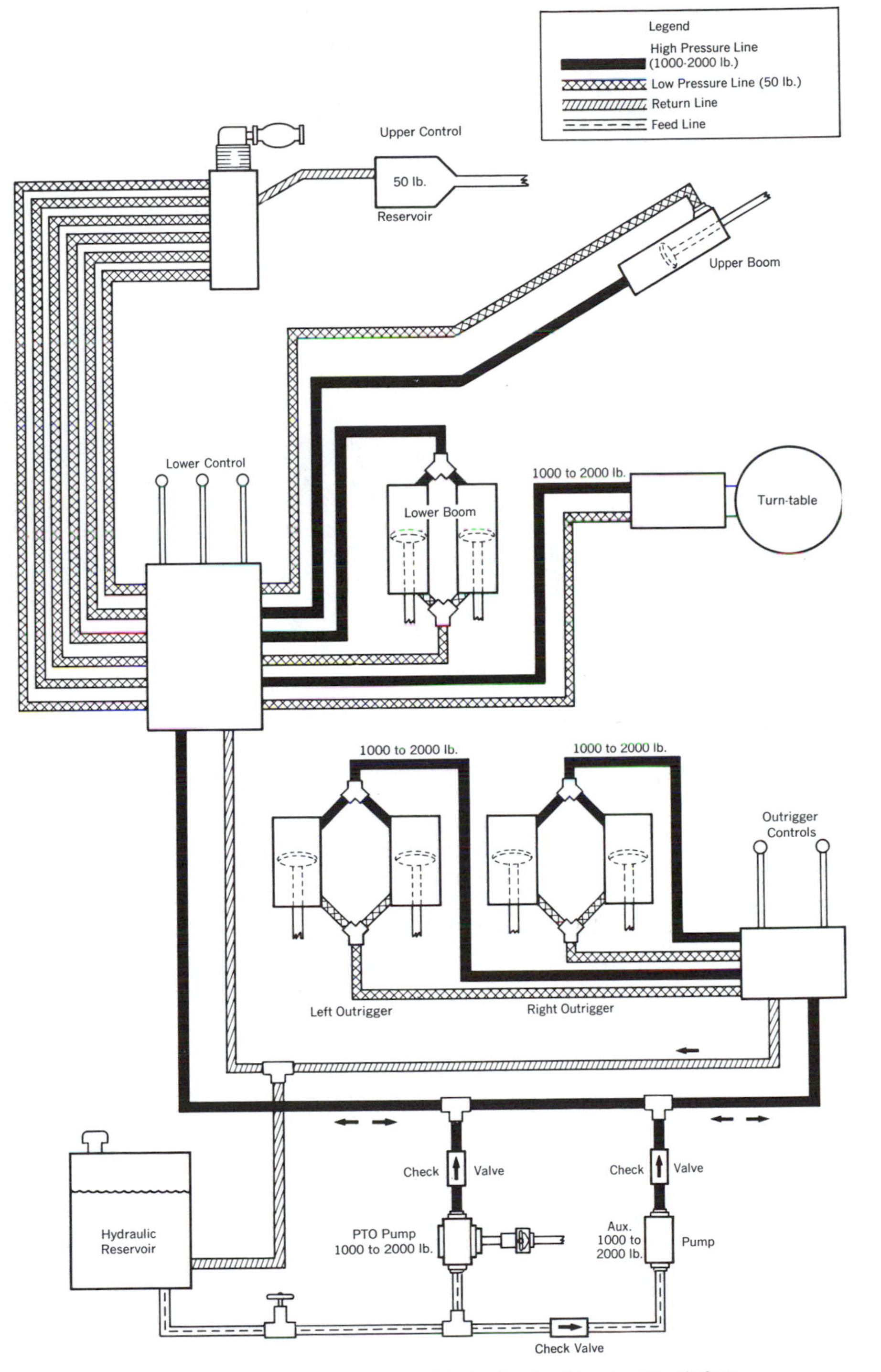

Figure 14-3. Schematic diagram of a typical hydraulic circuit for elevating platform

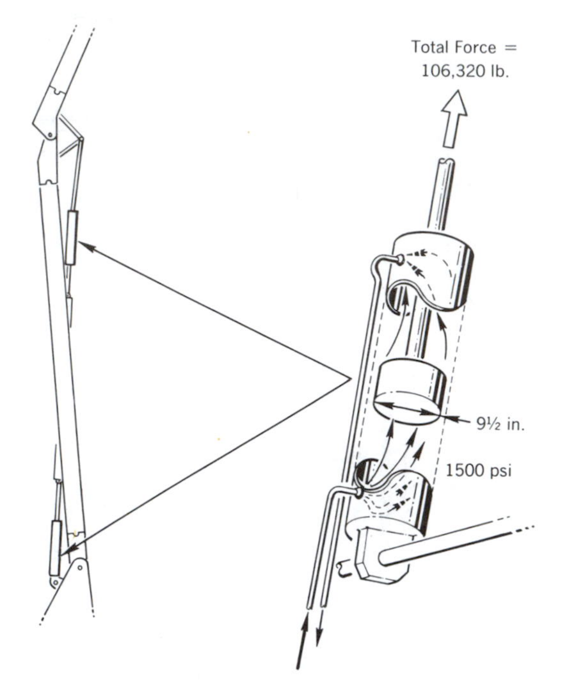

Figure 14-4. Hydraulic cylinder operation

the operator in the basket. The ground operator keeps in constant communication with the operators on the platform and alerts them of developments and orders.

The operator at the base constantly checks the stabilizers to make sure they do not sink into the pavement. When hose lines are being raised to upper floors by use of the platform, the ground operator maintains sufficient slack and sees that there is no possibility of snagging. On all models of elevating platforms, the lower controls mechanically override the controls in the basket in case the operators on the platform become incapacitated or an emergency situation develops.

Platform Operating Capacity

The platform's rated payload varies according to the manufacturer and the length of the booms. It is applicable only when the apparatus is located on firm, level ground and when the outriggers are extended and properly set.

The maximum load capacity of the platform of most makes of apparatus is decreased by the nozzle reaction force (measured in pounds) when the water system is activated. If the nozzle reaction is 200 lb, then only 500 lb can be loaded onto a platform with a carrying capacity of 700 lb.

Stabilization

Ground jacks and/or outriggers are provided on each side of the apparatus with the widest practicable spread between their feet when in the operating position (Figure 14-5). These are usually hydraulically operated with positive safety stops to prevent overextension and holding valves to lock the hydraulic cylinders at any up or down position. Some elevating platforms employ two or four outriggers to stabilize the vehicle, while others utilize four jacks and two out-

Figure 14-5. The hydraulic-powered outriggers and jacks lift the entire Firebird off the ground for better leveling and stability.

riggers to lift the vehicle off its springs or to raise the entire apparatus off the ground to level and balance the vehicle. Mechanical safety locks are usually incorporated in the devices to further protect against any possible failure.

The ground contact area of each stabilizing jack should be at least 140 sq. in. and, in addition, foot plates of ample strength should be provided that would increase the ground contact area to at least 575 sq. in. for use when the ground or pavement is soft.

Interlocks are provided to prevent any raising or movement of the retracted elevating platform booms or sections out of the bed until the ground jacks are placed in position to stabilize the vehicle. Also, this device will prevent the retraction of the outriggers if the boom is elevated. This is a safety requirement to prevent the possibility of overturning the apparatus when the elevating platform is raised before the vehicle is properly stabilized.

On some models, the outrigger interlock switches serve as an instability warning system. If sufficient weight is removed from one of the outriggers that there is a possibly unsafe condition, the vehicle's air horn will blow to warn the operator.

Factors Affecting Stability

The rated load capacity of an elevating platform varies according to model and manufacturer; it is figured for a maximum horizontal reach on firm, level ground and includes an adequate safety margin. Factors of stability and load-induced stresses are also based on the vehicle being level. When the truck is in other than a level configuration, its performance, stability, and safety factors are reduced in proportion to the degree of deviation from a level setup.

As an example, a rotation system was designed for level operation. When the chassis is on a slope from front to rear, the rotating functions now include the work of lifting the boom system uphill. If the booms are retracted to their shortest length and raised to the greatest elevation, this added load is at its minimum value. When the booms are extended and in a horizontal position, this added load is at its maximum. The degree of slope adds to the lifting force required in either operation. Therefore, under certain conditions, telescoping booms may have to be retracted and either articulating or telescoping booms may require elevation in order to rotate the turntable uphill. Because of this added load, the operator must never rotate at high speeds when on a slope and must always feather the controls when starting or stopping in order to avoid shock loads. A thorough understanding and appreciation of the factors described briefly in the above explanation should help the operator understand the need for careful, cautious operation when using the machine on other than level ground. When operating on a slope, all functions must be performed at reduced speed. Extra caution must be used on downhill rotation, which should be executed at one-half speed or less.

If the unit must be operated on a slope, the operation should be restricted to the rear and high side of the turntable if at all possible to avoid the risk of upsetting.

The tipping load includes horizontally applied loads from wind or nozzle reaction. Avoid extending the platform over the down wind side of the apparatus when operating in gales of 30 mph or over. When using the platform as a water tower, always keep the basket forward or outward from the vertical centerline of the turntable if the booms are articulating. Safe limits of operation are illustrated in Figure 14-6.

Articulating boom linkage is generally designed to operate normally with the lower boom in an approximately vertical position (75 to 85°). If the lower boom is placed in a near-horizontal position and the upper boom is elevated, a heavy overload is placed on the structure of the lower boom, turntable, and base. If the turntable has been rotated to place the booms at a considerable angle with the apparatus, this overload may overturn the rig. The lower boom should be elevated from the traveling position only with the upper boom folded, or within 30° from the lower boom, unless the turntable has been returned to the neutral position to align the booms parallel with the apparatus.

To be certain of avoiding an unsafe position of the lower boom, the operator should always work outward from the vertical centerline of the turntable. In other words, with the operator's back to the boom and right hand on the control handle, the operator should rotate the turntable until facing the working area to be reached, before moving the basket forward to the area.

When the aerial platform is operating near the maximum height position, any rapid motion of the lower boom on an articulating-type apparatus will cause a jerking or lashing movement at the basket. This is especially hazardous when operating with maximum load capacity. To prevent this hazard, the controls should always be operated smoothly and with extreme caution at the maximum height position.

Ground station hydraulic controls are more sensitive than the basket controls because of different metering characteristics of the two types of valves used in some models; other types may show no variations. If the ground controls are more sensitive, it may be safer to allow the ground control operator to

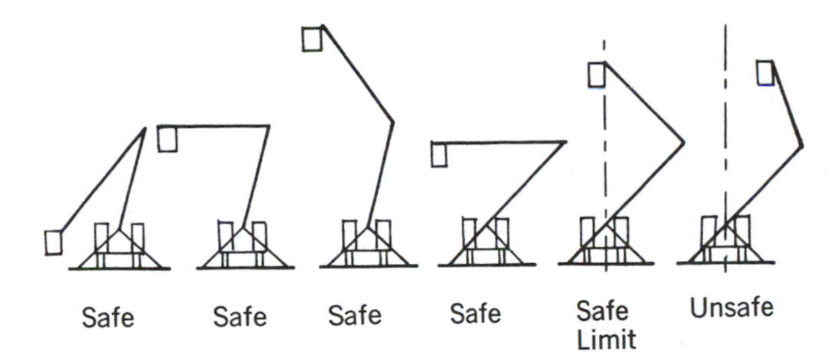

Figure 14-6. Safe limits of operation. At no time should the platform be maneuvered to a position that puts the elbow and the platform on the same side of the center of rotation.

position the booms when they are near their upper limits of travel or where extremely exact maneuvering is necessary.

Operating Controls

Platform controls consist of a single lever for actuation of the rotation assembly and upper and lower boom assemblies on some models of elevating platforms, while others utilize a separate lever for each function (Figure 14-7). On apparatus equipped with a single control handle, rotating the handle clockwise (right) causes the platform to move to the operator's right; rotating the handle counterclockwise (left) causes the platform to move left. The upper boom is raised by pulling up on the handle and lowered by pushing the handle down. The lower boom is actuated by pushing forward on the handle to raise the boom and pulling back on the handle to lower the boom. Any two or three combinations of movements can be accomplished simultaneously by a proper manipulation of the control handles.

Spotting Apparatus

The type of service to which the elevating platform is to be subjected is a determining factor in positioning the vehicle, since rescue operations frequently require contact with

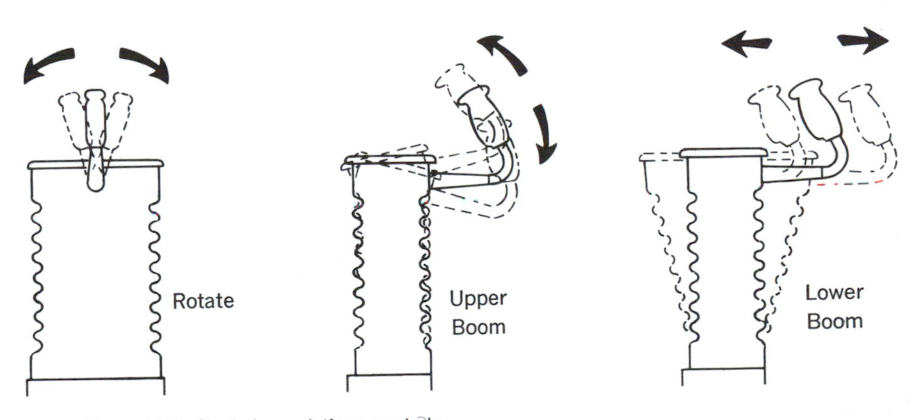

Figure 14-7. Single-lever platform controls

the building while water tower operations may dictate that the vehicle be placed a greater distance from the building (Figures 14-8 and 14-9). Whether an aerial platform is to be used to elevate and direct heavy streams or to perform a rescue, a size-up according to established fire fighting tactics is important. Consideration must be given to the extent that the fire has involved the building, the type of construction, the direction of fire travel, and the location of exposures. An analysis of these and other pertinent factors is needed to determine a good location for spotting aerial apparatus. Other factors that affect the placement of apparatus are signs, overhead wires, streets, space for outriggers, space for other fire apparatus, and the possibility of falling walls. Unlike an aerial ladder, the booms of an articulating-type elevating platform may require unobstructed space on the opposite side of the turntable. Such space is needed to provide complete freedom of operation for the elbows of the articulated booms.

Whenever operation on a slope is mandatory, the chassis should be facing either uphill or downhill, preferably downhill; it is far better to back up a hill to secure a proper spot than to drive up the hill. The most stable operation is to work over the rear of the apparatus. It is recommended that operation be limited within 45° of rotation to either side of the fall line of the particular grade on which the apparatus is being used. The fall line of a hill is the theoretical line straight down which a ball would follow if allowed to roll free. When operating as a water tower on an incline, keep the direction of the water stream pointed downhill where possible. Further, it is recommended that the platform load be reduced below rated capacity and the lower boom of articulated types be positioned below a maximum elevation or about 75° when operating on an incline. If the hill is approached from the top, drive straight down and stop at the required spot. If the hill is approached from the bottom, back up the hill to the designated point. When in position, it is extremely important that all wheels be chocked on the downhill side after setting the brakes. A positive anchorage in the form of one or more cables or lines connecting the vehicle to a tree or some solid structure might be desirable if the slope is slick or icy.

There are many times when an aerial platform must be used in alleys that can be reached only after making a sharp turn. If the turn cannot be negotiated while moving forward, it may be advantageous to back in. The other end of the alley may provide a better access.

It is important to remember, whenever spotting any type of fire fighting apparatus, that a portion of the street should be left clear for the passage of later arriving equipment.

When it is evident that heavy stream appliances will be placed in operation, lines should be laid by engine companies before the arrival of the elevating platform.

Water Towers

Water towers, designed to inject heavy streams of water into the upper floors of buildings, have been utilized in the fire service since before the turn of the century (Figures 14-10 and 14-11). The early models were large, heavy, and unwieldy; they were slow and required the services of many fire fighters to place them into operation, so officers were reluctant to use them in the early offensive phases of fire fighting. Gradually, they were phased out and replaced with other types of apparatus.

Lately some very modern and sophisticated models of water towers are being offered by manufacturers; some feature an articulated boom, while others are telescopic. They are very mobile and versatile. Operating with these water towers, a small crew can attack a large fire in a minimum amount of time. One operator, who has fingertip control of boom movements, nozzle elevation, sweep, and pattern, is usually in absolute command. The remote control of the monitor and nozzle eliminates the

Figure 14-8. Apparatus is spotted adjacent to curb so that street is not blocked, but there is ample maneuvering room for the elevating platform.

Figure 14-9. Using monitor through upper-story window (Courtesy of the Chicago Fire Department)

Figure 14-10. In this photo of a 1913 fire at Third and Broadway in Los Angeles, the water tower has been shut down, but wagon batteries and handlines are still being used. (Courtesy of the Los Angeles City Fire Department)

Figure 14-11. This water tower, which was manufactured in the fire department shops by the mechanics, utilized water pressure to raise the telescoping sections. (Courtesy of the Los Angeles City Fire Department)

need for a fire fighter at the tip of an aerial ladder or in the elevating platform basket to direct the stream.

These appliances are usually installed on a large-capacity pumper. Permanently installed and preconnected waterways eliminate the need for lugging cumbersome fire hose up a ladder and attaching a ladder pipe.

Construction

Modern water towers consist of a metal elevating tower of two or more booms or sections equipped with a nozzle capable of providing a large capacity mobile and elevated stream of water. The height of the water tower is measured as the vertical height from the discharge end of the nozzle to the ground. The horizontal reach of the tower when it is fully extended at right angles to the apparatus is measured from the center of the turntable to the end of the nozzle. The raising, extending, and rotating devices are similar to the elevating platform systems; they will allow multiple movements of the tower simultaneously, all controlled from the ground. To avoid possible injury in case of apparatus contact with live electric wires, the controls should be located so the operator does not stand on the ground when operating the tower (Figure 14-12).

Figure 14-12. Rear view of a Squrt water tower showing the ground controls (Courtesy of Crown Fire Coach Corporation)

Modern specifications dictate that the permanently installed nozzle be supplied by water supply piping that should be capable of delivering at least 750 gpm to the tower nozzle with a pressure loss of not more than 75 psi with the tower at maximum elevation. Nozzles are equipped so the operator can control the rotation and elevation. The nozzle should be sufficiently mobile that it will rotate at least 45° to either side of center and will raise and lower at least 45° above and below horizontal; this flexibility will allow a large segment of the fire to be deluged by the fire stream without moving the tower.

Water towers are excellent for applying foam on a fire because the extinguishing agent can be applied gently, instead of projecting it from a remote location.

Water towers should be equipped with at least three 2½-in. hose inlets, or other sizes of equivalent capacity, to assure that the apparatus can be adequately supplied. The inlets should be equipped with clapper valves to prevent water discharge from any unused hose connections. A suitable pressure gage should be installed at the control pedestal to indicate the water pressure at the base of the tower or lower boom section.

It is suggested that nozzle tips of 1⅜, 1½, 1¾ in., and a spray nozzle of at least 500-gpm capacity, be carried on every water tower. Where a variable flow spray nozzle capable of a discharge range of between 500 and 1,000 gpm is provided, these additional tips need not be furnished.

A spotlight of not less than 100,000 candlepower should be provided on the apparatus so the operator may observe the effects of the fire stream.

A control station is provided for the operator at or adjacent to the turntable. It is desirable to have the controls so located that the operator will not be standing on the ground while operating the tower in order to avoid possible injury in case the apparatus accidentally contacts energized electrical wires.

The vehicle is stabilized at the driver's position by a device that locks the driving wheels of the chassis at the wheel brake drums. Ground jacks are provided on each side of the vehicle to provide the widest practicable spread.

To prevent damage to the mechanism at full extension or full retraction, positive stops are provided to limit boom or section travel to those positions of operation determined as safe. Provisions are made that, in the event of failure of any extending mechanism, the gravity descent of the water tower will be kept at a speed that will prevent damage to the equipment or danger to the personnel. Should the normal power source fail to operate, an auxiliary source of power should be available to permit operation of the water tower in any of its normal operating positions.

A visual and/or audible signalling device is provided at the driver's position to indicate when the power takeoff (PTO) mechanism is engaged.

Fire Fighting Operations

One of the most valuable advantages of using a modern water tower for combating a blaze is the speed with which the apparatus can be placed in operation; this makes it an important asset when making a blitz attack on an incipient fire. Properly spotted, a water tower mounted on a triple combination pumper can use the water in the booster tank to launch an immediate attack on the fire while other apparatus and the remainder of the crew are laying supply lines and readying hand lines. However, caution must be exercised not to drive the flames and heat into uninvolved sections of the structure or to injure occupants and other fire fighters by directing high-pressure streams of water into occupied areas of the building. Figures 14-13 through 14-18 demonstrate the many ways water towers are put to work in fighting fires.

Figure 14-13. An articulated-boom Squrt easily applies a heavy stream of water into a fifth-floor window. (Courtesy of Snorkel Fire Equipment Company)

Figure 14-14. Horizontal reaches are possible. (Courtesy of Snorkel Fire Equipment Company)

Figure 14-15. Walls and fences are no deterrent. (Courtesy of Snorkel Fire Equipment Company)

Figure 14-16. Maneuverable Squrt avoids electrical wires as it combats a fire. (Courtesy of Snorkel Fire Equipment Company)

Figure 14-17. Aerial platforms and Squrts with articulated booms can fight low-level fires. (Courtesy of Snorkel Fire Equipment Company)

Figure 14-18. Snorkels, Squrts, and other types of aerial platforms are useful for applying foam to aircraft or flammable liquid fires. (Courtesy of Snorkel Fire Equipment Company)

15 Truck Company Procedures

Truck companies are just as essential as engine companies at most fires; the truck company's duties of forcible entry, rescue, and ventilation, as well as a myriad of other tasks, must be accomplished without delay if life and property losses are to be held to an absolute minimum.

Truck personnel must frequently make major decisions and use their initiative without any supervision. Each crew member must know the duties of truck companies; the best method of ensuring that the crew will be thoroughly efficient, other than through experience, is by constant prefire planning and training.

Truck Company Apparatus

Aerial ladder apparatus are commonly called trucks, and the fire department units manning them are referred to as truck companies (Figure 15-1), because their principal function is to provide the ladders and a large assortment of tools and appliances needed at fires and other emergency incidents. A modern truck company may be equipped with an elevating platform or a water tower, instead of an aerial ladder; when this occurs, the apparatus will still be equipped with a full complement of ground ladders and equipment. Ladders are necessary for life saving and to facilitate fire fighting operations when all normal means of entrance and exit are cut off by the fire.

In addition to rescue work, members of truck companies let smoke, heat, and gases out of buildings by ventilation, assist engine company personnel in getting lines into position to direct fire streams where needed, and help to protect contents from smoke and water damage. The ladder truck, therefore, is equipped with a large assortment of special equipment and appliances for these purposes. Truck company service is required at almost all structure fires; much of their service must be performed simultaneously with the advancement and application of hose streams. If no truck company is assigned to the fire, other personnel then must be delegated to this work.

Truck Stabilization

Regardless of whether a truck company is supplied with an aerial ladder, elevating platform, or water tower, the vehicle must be

Figure 15-1. A truck company in the background is raising its ladders while the engine company is stretching its lines; together the companies will launch a coordinated attack on the fire. (Currier and Ives Print; courtesy of the Library of Congress)

stabilized before the aerial equipment is raised (Figure 15-2). An aerial apparatus is very narrow in comparison with its length. When the ladder or boom is projected out to one side, the narrow width must provide the counterbalance for the long, heavy weight. Therefore, it is necessary to widen the supporting structure of the vehicle when the aerial device will be operated at a right angle to the vehicle chassis.

Spreading the ground jacks to their full width, jackknifing a tractor-trailer, or angling a truck will increase the stability. After the apparatus has been spotted, the brakes applied, and the chock blocks placed to immobilize the wheels, the ground jacks and outriggers should be spread out to their full width and firmly set. When there is a line of parked cars at the curb, it is possible to position the vehicle closer to the building by spotting the truck so that outriggers may be extended between the parked automobiles. Once a ladder truck has been positioned, leave space behind it so ladders may be pulled out of the rack.

The safest and most stable method of operating any aerial device is the one most often overlooked by the operators, and that is to use the ladder or boom directly over the cab or rear of the vehicle. With tractor-trailer aerial ladders, the most stable position is directly over the cab with the trailer in line with the tractor. It can readily be seen that the ground support is evenly distributed with the fulcrum point moved to the front or rear wheels and the counterbalancing accomplished by the large mass of the vehicle. This is the preferred method of operation when soft ground is encountered and extended reaches at low elevations are necessary, such as combating fires in schools, hospitals, and other structures which are set back from the street. It must be remembered, however, that extended side reaches must be avoided when operating in this manner.

Figure 15-2. Retractable outriggers and jacks on both sides of elevating platforms enable unit to be leveled and stabilized. Overall rear outrigger spread of this 90-ft Calavar Firebird is 16 ft. The lower control panel is shown swung out to the side.

Spotting the Truck

Truck company apparatus, whether a ladder, platform, or water tower, must be positioned in exactly the right spot if it is expected to deliver its full potential effectiveness. The spot will vary according to what duty will be assigned; rescue and ventilation activities may require a closer positioning than heavy stream use. Because truck spotting is so critical, many fire departments run the truck company ahead of the engine company to avoid being blocked out of the correct position. Although it is desirable to jackknife tractor-trailer aerials or angle park single-chassis trucks to achieve maximum stability, this sometimes cannot be done in narrow streets because the roadway would be blocked to other apparatus.

When it is necessary or desirable to spot a truck on a slope, always position the apparatus with the front of the vehicle headed downhill. If the hill is approached from the top, drive straight down and stop at the most advantageous position. If the hill is approached from the bottom, back up the hill to the desired point. When in position, place all parking brakes on and place the chock blocks on the downhill side of the wheels. If the slope is very slick or icy, it may be wise to make a positive anchorage by cabling or chaining the apparatus to a convenient tree, building column, or other solid structure.

Ladder pipe or elevating platforms should be positioned at a good vantage point to give the operators an overall picture of the fire situation, as well as the positions of fire fighters and building occupants. Aerial fire streams may be used to break windows for ventilation while simultaneously combating the fire. Fire fighters can rapidly remove windows on the upper floors with pike poles, thus saving truck crews from the punishment of ventilating from the interior of the structure. The ladder or basket should be positioned to one side and slightly above the window to protect the occupants from a backdraft explosion or from being cut by falling glass.

Choosing the correct location for a heavy stream application on the fire ground rests on the following requisites: accurately interpreting conditions on the fire ground; knowing the capacity of the available equipment; and having a thorough knowledge of the involved property and its environs. Knowledge of equipment and property should be obtained from drills, fire department inspections, and preplanning. Lack of such knowledge can lead to unnecessary loss of life and property.

When a large fire is threatening to spread to adjacent structures, the first heavy streams should be positioned downwind where they can protect the threatened exposures and simultaneously attack the main body of the fire (Figure 15-3). The apparatus should not be spotted where the crew and equipment could be jeopardized and overwhelmed by a wind change or a sudden spread of the fire.

Spotting an apparatus so that a heavy stream device may be utilized means estimating the length of time it will require to stretch the hose lines, obtain a sufficient flow of water for an adequate fire stream, and raise the nozzle to the correct elevation. With this information in mind, the officer will then have to estimate where the fire will be when the aerial stream is placed in operation. One of the most common errors is to position the heavy stream device where the fire is burning when the company arrives on the fire ground, instead of where the fire stream can accomplish the most good when the nozzle is positioned and an adequate fire stream has been developed. One of the pitfalls to avoid is spotting the apparatus where it can be effective only if the fire does not advance after the order is given to set up an aerial stream operation.

Sometimes handlines must be used to protect exposures and prevent the fire from spreading while master stream operations are being organized. There are other times when handlines are a waste of time because of the enormous amount of heat being generated. For this reason, all efforts should

Figure 15-3. The first heavy streams should be positioned downwind where they can protect the threatened exposures and simultaneously attack the main body of the fire.

be made to set up the master streams. The officer in charge must make this decision during the size-up.

So that maximum use of the monitor may be realized at a fire of large area, such as a lumber yard or a massive frame structure, the apparatus should be spotted in a position to sweep the fire area or structure. It should be placed where it is not in the direct line of exposure, yet can still protect adjacent structures and uninvolved materials. The apparatus should be positioned to safely reach the greatest volume of fire.

It is equally important to consider the extent of the fire in the building, the building's construction, the direction of fire travel, and exposures that may necessitate the use of a water curtain. When projecting heavy streams of water, care should be taken to prevent unnecessary water damage.

An aerial platform should not be placed in areas where overhead or adjacent obstructions may be encountered, or where the terrain is sloped, broken up, or marshy. If it is necessary to operate in a heavily obstructed area, care must be taken in placement to avoid contact with the electric current in such installations as signs, street lights, high-tension wires, elevated railroads, and electric current supply wires of all kinds. While it is sometimes necessary to operate in heavily obstructed areas and under adverse conditions, it is obvious that the normal range of efficiency will be limited to a greater or lesser degree by each obstruction lying in the path of the movement. The quantity of water being projected by the nozzle is important because the water load on floors and provisions for drainage need to be considered.

The operation of an elevated nozzle differs greatly from other forms of fire stream application in that the position of the operator and the nozzle may be varied to a tremendous degree. This three-dimensional flexibility permits the operator to aim the stream anywhere in a hemispheric pattern from ground level with some aerial devices to a full extended height and to each side of the vehicle's position. In addition to the up, down, and sideways motion, the nozzle may be moved close to, or away from, the building.

Elevating Platforms

In the application of fire streams, two operators can utilize the flexibility of an aerial platform to its greatest advantage (Figure 15-4), one operator aiming the monitor nozzle and the other positioning the basket as needed. Coordination between the two operators is essential; when sweeping the face of a building from window to window, a running conversation between them keeps each appraised of the other's intentions. The operators in the basket should keep in mind that when a constant pressure is maintained at the base of the platform, nozzle pressure will vary proportionately with the operating height. In the case of an 85-ft aerial platform, nozzle pressure will vary over 35 psi when the elevation is changed from ground level to a fully extended height. There will be 0.434 psi change in nozzle pressure for each foot the platform is raised or lowered.

Ladder Pipe

A ladder pipe is usually placed in position with the ladder still in the bedded position. When the ladder pipe and hose are attached and connected up, the ladder should be raised and extended to the approximate location of use before the lines are loaded. This keeps unnecessary weight off the ladder during its most difficult period of adjustment. When the lines are loaded and the ladder pipe is ready for use, the ladder should be rotated into place. If the fire is very hot and the ladder is close, the fire stream should be flowing as the ladder is rotated into position to prevent possible damage to the ladder.

When the ladder is unsupported, it is desirable to use the ladder pipe at a 70°

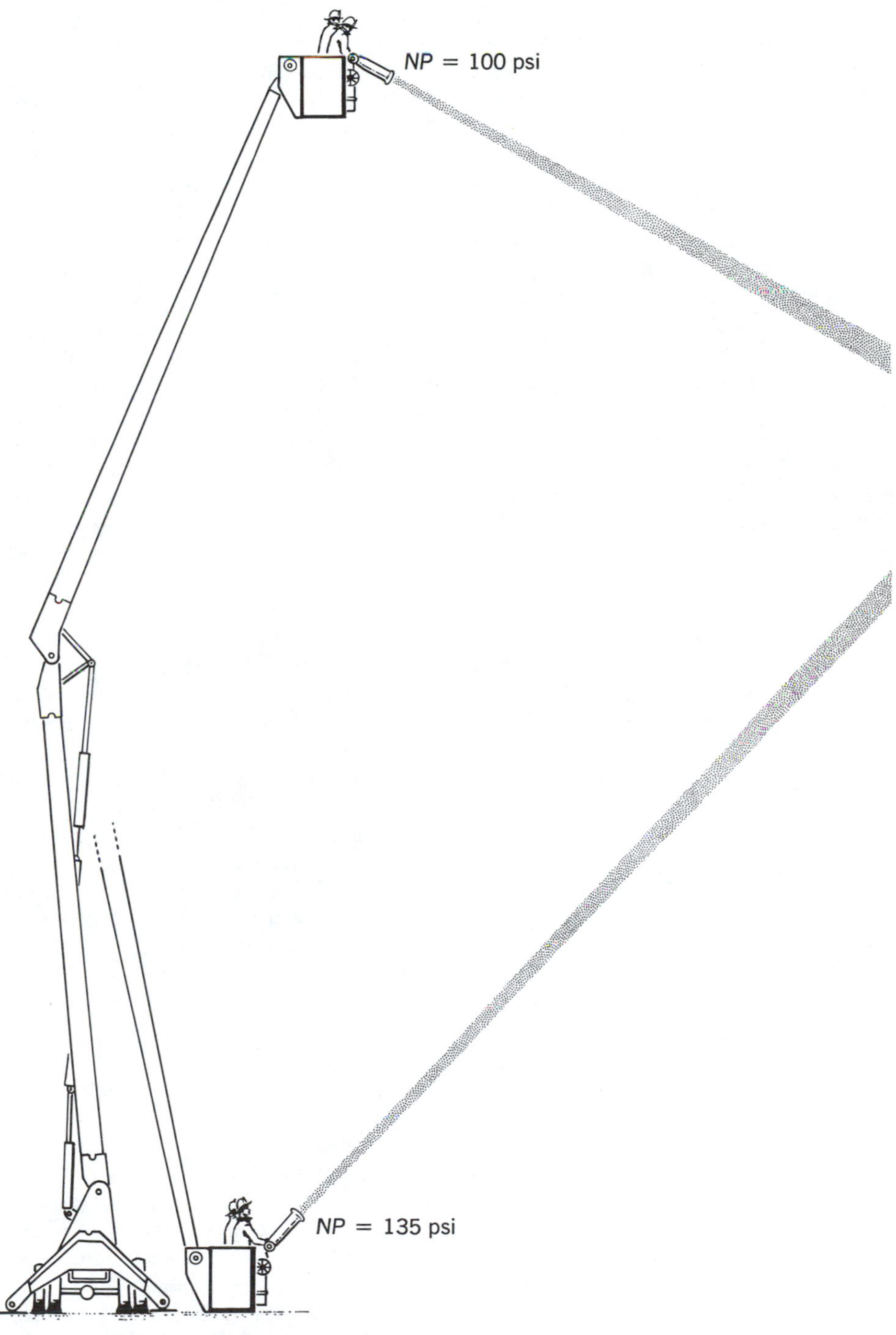

Figure 15-4. Effect of platform elevation on nozzle pressure

angle of inclination. At 70° and above, the ladder can handle the weight of the hose, appliances, and water, and operate in good fashion; and at 75° and under, the nozzle reaction is usually not sufficient to force the ladder back to where control is lost. To clarify this last statement: to lift the ladder, hydraulic oil under pressure is allowed to flow into cylinders near the base of the ladder. In these cylinders are closely fitting pistons which are directly connected to the ladder. As the oil enters, the pistons are forced upward and the ladder will be pushed upward. When the flow of oil is stopped, the ladder is supported through its connections to the pistons by the oil remaining in the cylinders. In the normal manner of lowering the ladder, a valve is opened and the oil is allowed to escape back to its reservoir; this permits the ladder to be lowered. However, a nozzle reaction pushing back on the ladder may remove the weight of the pistons from the hydraulic oil, and they may actually pull a suction on the hydraulic cylinders so that oil from the reservoir is pulled into the cylinders and the ladder moves backward. This can be expected anytime a ladder pipe is used at 80° inclination or above. In other words, the force generated by the nozzle reaction exceeds the force exerted on the ladder by gravity.

To regain control of the ladder if this happens, the flow of water to the ladder pipe can be shut off to stop the nozzle reaction; or if the ladder pipe is being controlled by an operator on the ladder, the stream can be pointed upward to cause the nozzle reaction to be directed longitudinally along the beams of the ladder. Aerial ladders are provided with mechanical force-down devices, but to activate and attempt to use them against a nozzle reaction is not recommended.

Some ladder pipes have a hydraulic cylinder back-pressure gage that shows the static oil pressure at the bottom of each cylinder; this gage indicates the degree of control the operator can exert over the ladder. With the ladder raised between 70 and 75°and the ladder pipe not operating, the gage will probably read somewhere between 450 and 600 psi pressure. As the stream is turned on, the nozzle reaction will start to reduce this figure. It is good practice to try to retain 100 psi on this gage so that control over the ladder can still be exerted by the operator. If the nozzle reaction makes this gage reading go below 100 psi, the ladder inclination can be lowered slightly to cause the weight of the ladder to come to a more favorable position and to combat the nozzle reaction; pressure to the ladder pipe can also be lowered to cause less nozzle reaction. Keeping 100 psi on the cylinder back-pressure gage is a safety measure that has been found to work very well. There are several factors, however, that can wipe out this safety margin in a hurry, such as a pressure surge or a rapid change in the vertical direction of the ladder pipe stream.

Inclinometer and load indicator readings are conservative and provide for perfectly safe operation under all conditions. They may be exceeded only with care and good judgment.

Twisting has a weakening effect on the ladder sections. Actually, twisting is the first step to ladder collapse and is liable to occur in a high wind, when the turntable is not directly opposite the objective (one beam engaging a roof or balcony first), or under any other undue lateral stress. The lead hose to the ladder pipe should be laid near the middle of the ladder rungs, between the beams; it should never be hung over the side of the ladder.

Limit the sideways movement of the ladder pipe to not more than 15° either way from the perpendicular (straight ahead position). Rotate the turntable if more than a 15°swing is required; use the hand crank if rotation cannot be accomplished smoothly with the control lever. Move the ladder pipe slowly and steadily without jerking. The water should be turned on and off gradually to avoid causing a violent whip of the aerial ladder. A fire fighter on the ladder directing a ladder pipe stream should use a safety belt attached to one of the ladder rungs.

Heavy Fire Streams

Most fires are brought under control and extinguished by streams from small handlines. When the quantity of water being discharged through handlines is insufficient to absorb the heat being generated by a fire, the use of large volume streams is indicated. Very few blazes develop to the extent that large streams are required to extinguish them. Occasionally a delayed alarm, structural conditions, explosion, weather, or other circumstances will result in the rapid and uncontrolled spread of fire. The control and extinguishment of such an advanced fire and the protection of exposures can require the use of large volumes of water at high pressures.

The terms *master, heavy,* and *large caliber* fire streams usually refer to a water discharge, either as a spray or solid stream, in excess of 400 gpm. Heavy stream appliances include portable monitors, mounted turrets, deluge sets, wagon batteries, ladder pipes, water towers, and monitors mounted on elevating platforms.

Water

Extinguishing a fire with water commonly involves cooling the burning material to a temperature lower than its ignition temperature; or, in other words, the removal of heat from the burning material. Water is the best extinguishing agent for general fire fighting operations because it is chemically stable under ordinary fire conditions, it is one of the best cooling agents available, and it is inexpensive. As water readily absorbs heat, it is therefore a good heat-transfer medium.

To be effective, the water must reach the burning material itself and impinge on the burning surface; the material must be cooled to a point where it no longer glows or continues to emit flammable gases. Heat absorbed by water at a fire will be carried away from the fire either in a cloud of vapor or in a hot liquid runoff. The physical state of the water, whether it is steam or liquid, will depend on the volume of heat absorbed.

Fire Streams

A fire stream is the water discharged from a nozzle. The shape of a fire stream as it leaves a nozzle is influenced by the design of the nozzle, the discharge pressure from the tip, and the condition of the nozzle. The course of a fire stream is affected by gravity, friction due to air resistance, wind velocity, and obstacles encountered. In general, a fire stream that penetrates into a fire and cools the burning material below its ignition temperature is considered to be an effective fire stream.

Of prime importance to the development of a good fire stream is an adequate quantity of water passing through the nozzle at the correct pressure. Unless the quantity of water and the pressure are appropriate for the fire ground situation, the fire stream so developed may not be effective. The quantity of water depends on the available supply. Nozzle pressure is influenced by water supply, pump pressure, nozzle size, hose size, and hose layout.

Spray or fog streams, in contrast to solid streams, generally absorb heat from hot atmospheres more rapidly, cover a greater area with water, and use less water. However, they require a higher discharge pressure, have a shorter reach, have less penetration, and are less effective in cooling and extinguishing fires in subsurface areas that have reached the incandescent stage, such as deeply charred wood. Where a fire seriously exposes other buildings or structures, spray streams played on the exposed structure will tend to keep it cool and prevent windows from being broken by the heat.

Fog and spray streams can be used to screen and protect personnel and equipment because they absorb and reflect the heat from the fire, and act as a thermal shield. When used on attack, this rolling fog will penetrate where solid streams cannot. Upon contact with intense heat or fire, the water particles rapidly absorb the heat being generated by the fire and are converted into steam. The steam is then carried upward by the fire's thermal currents and additional particles of water are drawn in to continue the attack; this rapid steam generation absorbs tremendous amounts of heat energy. The steam also blankets the fire as the water vapor replaces the air; this will often snuff out the fire when the oxygen content of the atmosphere in the structure is lowered below 16%. It is possible to obtain maximum efficiencies with limited supplies of water when using this technique.

Heavy Stream Application

Large fire streams controlled by a monitor or deck pipe (Figure 15-5) have been favorites of both municipal and volunteer fire departments since their introduction in the 1800s. Today, heavy stream appliances are capable of controlling gallonages from 400 to 2000 gpm. Their major advantages, in addition to high gallonage, are velocity, reach, superior fog pattern, and the ability to be left unattended or abandoned while operating if necessary. This feature is particularly important when structural exposures need to be protected but additional personnel are not available.

Unless the structure is too engulfed with fire to permit the use of handlines, the initial attack should be made by stretching hose lines into the uninvolved sections of the building (Figure 15-6) to prevent further extension of the blaze and to protect the occupants' means of exit. If the initial attack is made from outside of the structure, the force of the fire stream could spread the heat, smoke, and fire throughout the building.

The penetration of a heavy stream discharged from a turret at ground level and directed into a building will be influenced by the angle the stream makes with the horizontal (ground). For example, a heavy stream discharge from a portable monitor at ground level that is aimed at a second story window will penetrate some distance into the room before striking the ceiling. Were the same stream aimed at an eighth story window, the stream would strike the ceiling almost immediately upon entering the room; it would, therefore, have a penetration of only a few feet, making the stream ineffective.

The third floor is generally considered to be the highest story into which a heavy stream appliance at ground level can throw an effectively penetrating fire stream. For this reason, some heavy stream appliances are mounted on devices that can be elevated, such as aerial ladders, water towers, and elevating platforms. Other appliances are designed to be portable so they can be readily positioned in locations from which an effectively penetrating fire stream can be aimed, such as the roof of an adjacent building.

Another factor affecting the penetration of heavy streams are obstructions in the path of the stream such as partitions, closed doors, furniture, or stock. Unless it is readily apparent that a heavy stream is being effectively applied, it is usually better to redirect the stream, to move the nozzle to a more advantageous position, or to shut down the appliance. Because it is frequently difficult to reposition heavy stream appliances on the fire ground, they should be initially spotted in a location from which an effective stream can be directed at the fire.

A vital factor in operations involving the use of larger streams is to shift the point of impact; that is, to keep the stream moving. Several hundreds of gallons per minute corresponds to a ton or more of water being delivered every minute; it can be readily seen that one or two minutes of application in any one spot is sufficient to wet down and cool nearly any amount of fuel to a point

Figure 15-5. Elevating platform and wagon batteries applying heavy streams of water on a burning structure (Courtesy of Mack Trucks, Inc.)

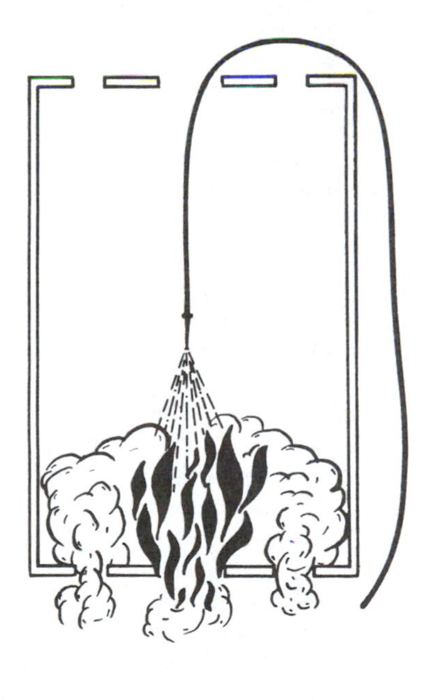

Figure 15-6. The initial attack should be made by stretching hose lines into the uninvolved sections of the building to prevent further extension of the fire and to protect the occupants' means of exit.

where it will no longer burn. If after a short application at any spot there is no appreciable reduction of flames and heat, it should be evident the stream is not reaching the heart of the fire and some other point should be selected. Steam being emitted from the fire area is an indication that water is striking hot or burning materials.

The operation of high-pressure fire streams into any building should be under the direction of the officer in charge of the fire. It should be ascertained that no civilians or fire fighters will be jeopardized by the force of these streams projected into a window or other opening (Figure 15-7). A powerful fire stream has the ability to reverse the thermal rise of heat and gases caused by a fire, thus forcing them against persons struggling down stairways and corridors and making it difficult or impossible for fire fighters to enter or ventilate the building. Solid streams issued under high pressure are extremely hazardous to personnel in the building; they can also quickly place such a weight of water into a building that a danger of structural collapse is created.

As a general practice, when a building is totally or greatly involved, extinguishment should start at the lowest point and work upward. If the fire is in a room just inside the window, a close application of the 90° fog pattern is desirable. In a direct attack on a fire from outside of a building, care must be taken that the fire is not driven into the untouched portions of the structure.

Taking hose lines to upper floors of a building presents a laborious and time-consuming job. With the turret nozzle charged and ready, the nozzle can be raised and positioned in a window on the floor below the fire, or the fire floor, and hose lines can be supplied by removing the nozzle tip. A rapid attack on the fire by means of hose lines may be made in this manner. Hose lines may be connected to pumpers on the ground, and the nozzles raised to the fire floor, when height does not create too much strain on the hose couplings and pull them apart.

Operators in the basket of an elevating platform should be constantly aware of the location of the valve that activates the downward spray nozzle under the platform. They should be concerned that it is unobstructed and operable at all times, since use of this device could save their lives in an extreme emergency. When operating over a fire, an explosion, upsurge of flames, or change of wind direction may carry smoke, heat, and flames up under the basket; this would necessitate prompt and continued operation of the downward spray until the basket could be moved to safety.

It is extremely important to watch for signs of the fire's spreading. If the fire is extending, streams should be concentrated primarily to prevent the spread, whether it is vertical or horizontal.

Shut down heavy stream devices as soon as it is evident that the building may be entered with hose lines; this reduces the water damage and danger to hose operators.

Operating Procedures

Before starting master stream operations, several basic facts should be kept in mind:

1. Safety precautions must be constantly observed when using any heavy stream appliance. Increasing numbers of small fire departments are either purchasing articulating boom vehicles and extension ladder trucks that were designed for the use of utility companies, tree trimmers, etc., or they are making contingency plans for the dispatch of these units from other city departments in case of a fire. These practices should be discouraged unless the vehicles were engineered to withstand the forces exerted by pressure reactions and loads peculiar to fire department use. Because of leverage and ladder construction, the effects of nozzle reaction of ladder pipes must be given more consideration than that usually necessary with other heavy stream appliances (Figure 15-8).

Figure 15-7. Before directing any fire stream into a building, operators should be positive that no occupants or fire fighters will be jeopardized.

2. Unless the need is clearly apparent, large-caliber streams should be placed in operation only on the orders of the officer in charge of the fire.

3. A ladder pipe operation will demand the total effort of both a truck and an engine company. Consideration should be given to the calling of extra companies to be utilized for other fire duties.

4. A ladder pipe, elevating platform, or water tower is a highly mobile heavy stream appliance that can be used where other streams are not practicable (Figure 15-9). They have a particular value in the control of fires in upper stories, and in supplying water curtains at high elevations. Fires in extensive roof areas can often be readily confined and controlled by major streams.

Figure 15-8. The effects of nozzle reaction on ladder pipes must be given more consideration than is usually necessary with other heavy stream appliances.

Figure 15-9. Some types of elevation platforms and water towers are sufficiently maneuverable that they can project a fire stream into a structure at any height from below ground level to their full extension.

To attack fires in the upper levels of multistory buildings, a large stream should be operated only long enough to effect a knockdown of the main body of the fire. It should then be shut down and handlines advanced to complete extinguishment. Until the heavy stream ceases operation, areas surrounding the fire are usually untenable and personnel cannot safely and effectively bring handlines to bear on the fire.

It is possible to utilize the elevated nozzle as a standpipe by simply removing the tip and extending hose lines from the appliance. The nozzle should be thrust into a convenient window, the tip removed, and the hose or adapter fitting attached. Some departments carry a hose pack in the platform baskets in readiness for a fast handline attack on a small fire.

Roof areas that would not otherwise be accessible by other means are within the range of elevated fire streams. Although the use of hose streams over roofs is seldom justified because of the difficulties created for fire fighters working below, spray streams are sometimes valuable in effecting containment and quick knockdown of fires that have vented through roofs.

Wet drills will quickly demonstrate the extremely long reaches possible with master streams that make them especially effective in combating fires covering large open areas.

Electrical Hazards

Truck companies usually carry heavy safety rubber gloves and insulated cutters because fire fighting operations often jeopardize fire crews by exposing them to electrical hazards. These dangers may be present anywhere in the country because rural areas are becoming as electrified as the cities. Fire fighters should study the problems and know how to protect themselves when they could become exposed to a possible electrical contact during fire fighting or rescue operations. A standard operating procedure that has been approved by the local power utility should be adopted.

Electrical Utilities

Circuits carrying electricity from power companies to the consumers are divided into transmission, distribution, and service. Transmission and distribution circuits carry high voltages; service circuits carry low voltages.

Transmission circuits are the lines from the generating plants to the distribution points. Line voltages are usually above 66,000 V.

Distribution circuits refer to the wiring from the distribution points to the transformers that service consumers. Line voltages vary from 600 to 66,000 V.

Service circuits contain the wiring from the low-voltage side of the transformers to consumers' homes or places of business. Line voltages are 600 or less. These low voltages should not be interpreted to mean that they are safe, because voltage as low as 110 can be lethal to a well-grounded person.

Fire Stream Hazards

Where a fire in a large structure is extensive, heavy streams should be positioned where they can reach the greatest volume of fire (Figure 15-10). Consideration must be given to the involvement of the building, its construction, direction of fire travel, and exposures which would require protection. Extreme caution must be exercised where electrical wires, street lights, large overhead signs, and other hazards and hindrances could create problems. Often an adjacent parking lot will offer a strategic and safe location.

Fire fighters on hose lines or directing heavy stream applications could become injured or killed if they should inadvertently direct a fire stream into or at energized electrical wires or equipment. The amount of current that may be conducted back to the nozzle will depend on the following factors: (1) the voltage involved, (2) the distance from the nozzle to the electrically charged line or equipment, (3) the purity of the water

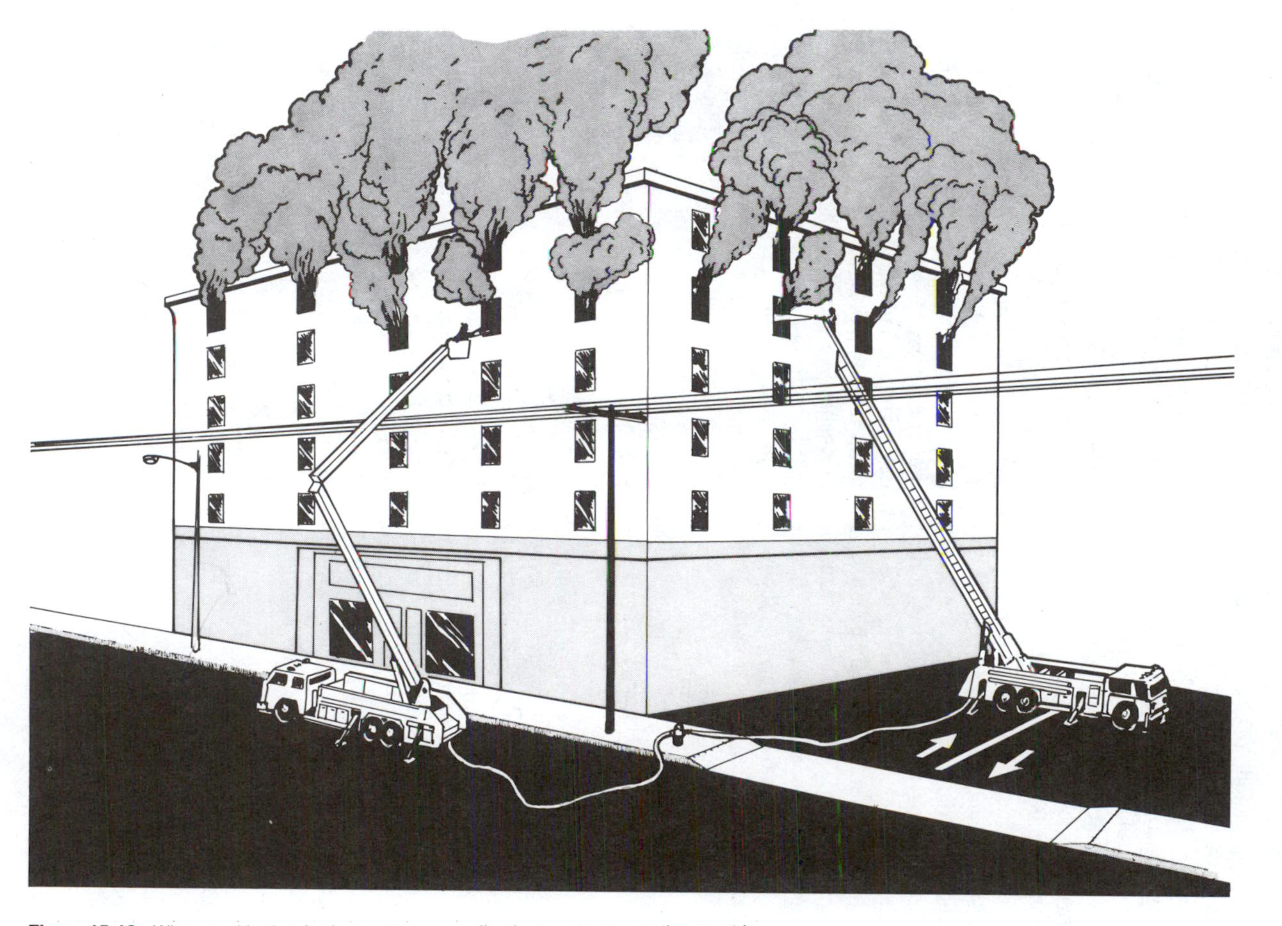

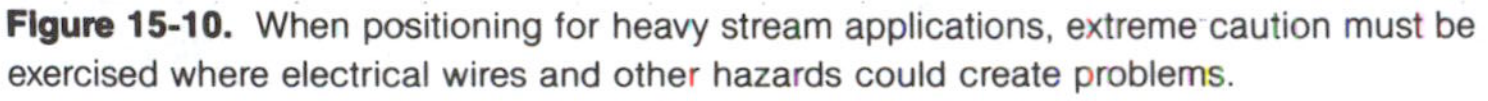

Figure 15-10. When positioning for heavy stream applications, extreme caution must be exercised where electrical wires and other hazards could create problems.

in the fire stream, (4) the size of the fire stream, and (5) whether the fire stream is solid or broken.

The application of hose streams on charged wires or equipment should be avoided whenever possible. However, during fire fighting operations, fire streams often must be utilized in close proximity to charged circuits, so all fire fighters must be aware of the possible dangers, as well as the permissible procedures.

Remember, the only nonconducting extinguishing agents considered safe are carbon dioxide and dry powder types that have a nonconducting applicator. Salty or dirty water, or the discharge from a soda acid or a foam extinguisher, may have such a high conductivity that no safe rule can be adopted. If it becomes necessary to use water known to have a high conductivity, such as ocean water, then a spray or fog stream should be employed.

Minimizing Electrical Hazards

Certain procedures are suggested to minimize electrical hazards. For ordinary household circuits of 120 V to ground, any type of nozzle can be held within a few inches of the charged wire without danger. The distance at which a straight-stream nozzle can be positioned without discomfort is 3 to 4 ft from a wire carrying 550 V. It is safest for fire fighters to consider every wire as one carrying high voltage because of the possibility of contact with high-voltage wiring at some remote point.

Spray and fog streams are the safest for all fire fighting activities as there is little chance of a hazardous amount of current being conducted back to the nozzle because of the air spaces between the water droplets. Therefore, whenever possible, fog nozzles should be employed to cool down the equipment and extinguish the fire.

It is recommended that the spray nozzle not be equipped with a long applicator, which could touch the energized equipment. Also, the spray nozzle should not be adjusted while it is being directed on the electrical equipment because of the danger of accidentally changing from a spray to a straight stream when the nozzle is too close for safety with a solid stream.

When directing a heavy stream through high-tension wires, a fog pattern should be used because the force of a solid stream could push the wires together and cause a short circuit; a spray stream is less likely to do so.

When handling ladders or long tools around electrical wires, particularly if they are constructed of metal, extreme caution must be observed. Also, when climbing or descending a ladder, death or injury can result if a fire fighter barely brushes against or pushes live wires together.

When dropping a rope from an upper floor as a life line or to pull up hose or equipment, make certain that the rope falls close to the face of the building. Throwing out a coil of rope could cause it to go over and through high-tension wires; this would probably drag them together, causing them to arc, spark, and possibly fall to the ground.

Shutting Off Power

Whenever a building has been damaged by a fire, explosion, or other incident sufficiently to render the electrical wiring and other devices unsafe, the power should be shut off. Leave the power turned on as long as possible so that lights, elevators, and other electrical conveniences can be utilized. When it becomes necessary to interrupt the electrical service, it is best to shut off the power only in the hazardous section of the structure.

The best method of disconnecting the electrical service is to open the main switch. If the switch panel or box has been damaged or is wet, or if it is necessary to stand in water while opening the switch, under no condition should the switch handle be grasped manually. Utilize a pike pole, piece of rope, or some other nonconducting object or wooden-handled tool to operate the switch.

When available, use tested and approved rubber safety gloves when working around energized electrical equipment.

Another method of disconnecting the electrical service is to cut the wires on the outside of the building. It is best to adopt the rule that fire fighters should never cut electrical wires unless human life is endangered. Trained power company personnel are usually available for a quick response at any time, day or night; it is considered bad practice to unnecessarily jeopardize fire fighters and expose them to the dangers of electrocution.

Most fire departments do allow their personnel to cut the low voltage (600 V or less) service wires into a structure if utility line personnel are not readily available. The best place to cut the electrical service into a building is at the drip loops, which are the portions of the wires between the anchor insulators and the point where the wires either enter the building or descend down the wall. If the wires are cut at the drip loops, the wires between the pole and the building will not drop to the ground and jeopardize other persons.

Fallen High-Tension Wires

Fire fighters are commonly dispatched to a location where high-tension wires that should be suspended on poles and crossarms are sagging or fallen. Even when no one is injured or immediately jeopardized, this is an extremely hazardous situation that constitutes a menace to the public and everyone else in the vicinity.

Notify the power company immediately. Unless a victim who is still alive must be rescued, no attempt should be made by fire department personnel to touch, move, or cut any wires. The fact that they may appear totally inactive and do not sputter or spark is no indication that they are dead. Assume that every wire is energized and dangerous. Whenever any wire must be cut, use the proper tools, wear safety rubber gloves, and observe all safety precautions. Reach up as high as possible to cut the wires so the dangling ends will not remain a hazard; clip the wires on both ends, then coil and remove the section of wire so it will not continue to cause apprehension.

Wires may fall on wire fences, automobiles, railroad tracks, metal buildings, or other conductive material; if this happens, a lethal charge of electricity could cause death or injury at a considerable distance from the downed wire. A fallen wire lying in a puddle could cause the entire ground area to be energized. Electric arcs and flashes can cause burns and eye injuries, even when there is no direct contact with the wire. Keep everyone away from under the power wires for at least one pole length on each side of the break because the adjacent spans of wire may have been weakened and could drop without any warning.

When a fire company is called to the scene of an electrical emergency, approach the area with caution. Stop the apparatus well away from the fallen wires as they are difficult to see, particularly at night or during stormy weather. Be very careful when descending from the apparatus as a wire lying in a pool of water may have rendered the entire area lethal. If any portion of a vehicle is touched by a live wire, the entire vehicle will likely be energized; it is improbable that the contact between the vehicle and the earth will be efficient enough to cause the electrical charge to form a circuit to ground. No one should touch any vehicle, fence, or structure that could be energized. If an automobile accident victim is still in the car, tell him to remain there until the current is turned off. If it is necessary to get out of or off an apparatus or other vehicle, jump clear so contact is not made simultaneously with the apparatus and the ground. Positively make sure that both hands and feet are clear of the vehicle before touching the ground.

In one accident, an automobile with two persons in it struck a power pole carrying high-tension wires. One wire broke and dropped on the car. The two victims, still dazed, were still in the automobile when responding fire crews arrived, recognized the danger, and advised the two persons to remain where they were until the utility company could respond. The driver panicked, opened the door, placed one foot on the ground, and immediately received a lethal charge of current. The passenger remained in the car and was later removed relatively unharmed.

If the conditions are hazardous, prevent any persons from approaching too close to the wires so that they will not inadvertently become harmed. It is important that sufficient guards be placed around the scene to restrict unauthorized persons from entering until the utility line personnel declare the area safe.

Ladder and Boom Hazards

All aerial ladders, elevating platforms, and water towers are constructed of metal, so they can be considered excellent conductors of electricity. Although vehicle tires are made of rubber, they are not very good conductors of electricity; therefore, if a ladder or boom contacts a charged wire, it will not become automatically grounded. Electricity follows the path of least resistance; if someone standing on the ground touches the apparatus, the current will probably flow through the person's body.

Extreme caution must be exercised whenever operating any boom or ladder around electrical wires. Consider all wires as being charged, even if the utility company claims that they have been deenergized; fallen wires in the vicinity may have reenergized them. The essential point is never to provide a path to ground for an electrical circuit. If the ladder or boom actually makes electrical contact, the safest course for the operator to follow is to get away from the controls and touch nothing until the power is turned off. Personnel on the ladder should keep their hands and feet on the rubber-covered rungs until the circuit is cut. It is possible that a ladder or boom could swing into a high-voltage wire with such force as to cause the wires to come together. The covering on these wires is not of sufficient insulation strength to prevent a short circuit. The temperature of an electric arc is about 3300° F; at this temperature, the wires may burn through, break, and drop to the ground; this causes additional hazards. Also, the apparatus and personnel are in danger of damage or injury from the arcs and sparks.

If it becomes necessary to get off the apparatus while it is electrically charged, jump so that contact with the truck and ground is not made simultaneously. Crew members on the ground must keep clear of the charged apparatus.

There is very little difference from the standpoint of electrical conductivity whether ground jacks are in place or not. In dry weather they make a poor grounding contact. An exception would be if the jacks were resting on a rail, in a puddle of water, or any other similar electrical ground.

Ventilation Strategy

Ventilation, as applied to the fire service, refers to the planned and systematic clearing of objectionable smoke, heat, and noxious gases from a structure, ship, or other area.

If a davenport were burning in an open field, extinguishment would be simple because the heat and smoke would rapidly dissipate into the atmosphere. The same furniture burning in a building will cause large amounts of heat and smoke to accumulate and build up because they cannot escape to the outside atmosphere rapidly enough. Consequently, fire fighters would have to advance to the fire under very unpleasant conditions caused by the confined heat and smoke. Even if the fire fighters are protected by breathing apparatus, the high heat will be punishing and the toxic gases and smoke will be hazardous.

Difficulties caused by the interior fire can be greatly reduced by making the conditions as similar as possible to those of the outdoor fire; that is, by thoroughly ventilating the building. However, ventilation should not be performed in a haphazard manner. It is second in importance only to the application of water in fire fighting strategy and tactics. In many cases ventilation is essential to permit the proper application of water. Ventilation is one of a truck company's primary responsibilities; every fire fighter should be thoroughly acquainted with it.

Ventilation is concerned with accomplishing any or all of the following:

1. Facilitating rescue operations by removing smoke and gases that are endangering occupants who may be trapped or unconscious.

2. Making the area more tenable for fire fighting operations. Fire fighters can enter the structure to search for victims and to approach close enough to the fire to extinguish it with a minimum of delay, smoke, water, damage, and hazards to firemen.

3. Removing toxic and/or explosive gases and vapors from the structure.

4. Controlling the spread of the fire.

5. Minimizing backdraft conditions.

Rules and Procedures

It would take an entire book just to comprehensively cover the subject of fire fighting ventilation. Officers should teach and drill their crews in all aspects of this subject, because effective ventilation is one of the most important factors leading to efficient structural fire fighting. In fact, often the success or failure of a fire ground operation depends on the degree of attention devoted to ventilating a building (Figures 15-11 and 15-12).

A

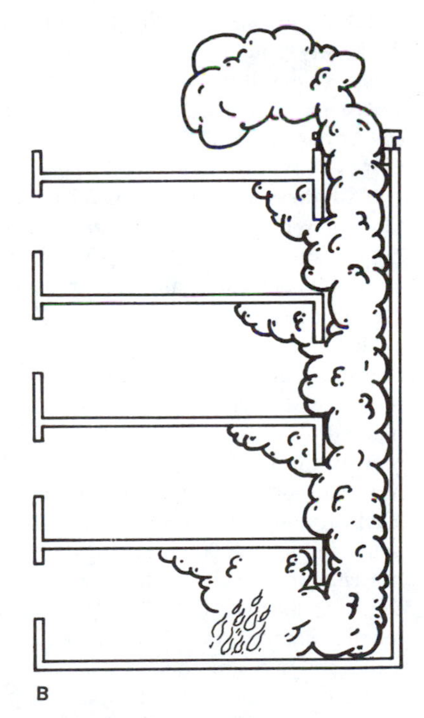

B

Figure 15-11. (A) Heat and smoke confined in the building. (B) After the roof has been opened over the stairs or elevator shaft, the structure becomes tenable as the heat and smoke escape.

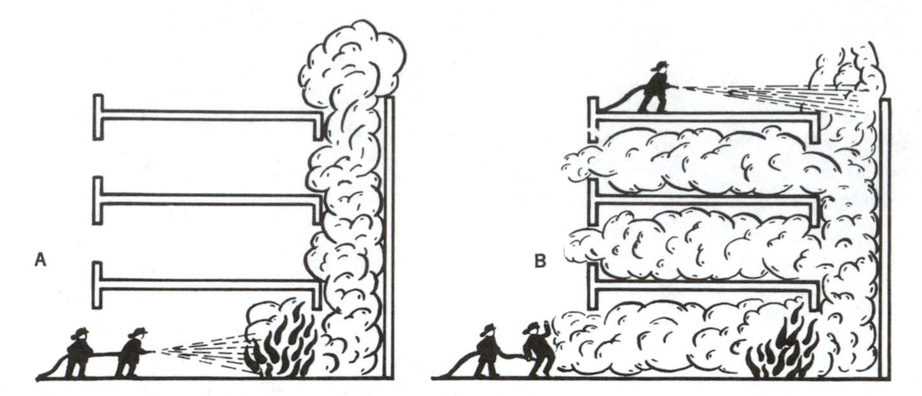

Figure 15-12. (A) Strategic ventilation will allow prompt fire fighting. (B) A fire stream projected down through a ventilation opening in the roof cools the column of heated smoke and gases and reverses their orderly movement from the structure.

The building and the type and location of the fire dictate the method of ventilation. The roof of a multistory building must be opened above the fire at a top floor blaze, and over the stairwell if the fire is on a lower floor. Any ventilation hole must present a complete passageway to the outside for the smoke and heated gases to escape. Push the ceiling down, too, after opening a hole in the roof.

If windows are broken out for cross ventilation, remove all of the glass, screens, drapes, and curtains. A window screen can hold sufficient smoke and heat in a room to prevent entry.

Skylights are often placed on the roofs over stairwells and elevator shafts; removing these would be the easiest and most practical method of ventilating a structure. After removing the skylight, place it upside down near the hole. Its presence will alert other truck crew members about the possibility of a hole in the roof, and inverting it intact will prevent other crew members wasting time removing the glass. If the skylight cannot be readily removed, the glass may be slid out unbroken or cleared out with an axe.

When a pitched roof is opened up, the hole should be made at the peak. Heat and gases will accumulate at the highest point.

Fire fighters going to the roof should carry adequate tools to do the job. Chain saws, axes, roof ladders, and other cutting and ripping tools may be necessary. A long pike pole can often be used to break out the top floor windows from the roof. Many companies routinely carry a life line every time they go to the roof at a fire because it will offer them a means of escape.

Rapid ventilation is possible in several ways. In case the whole building is burning, a sideward sweep with a nozzle will remove glass from the windows without causing excessive water damage. Caution must be exercised, however, if the building is occupied as the flying shards of glass could cause wholesale injuries. A slower method, accomplished with no water damage or injuries is to utilize an elevating platform or aerial ladder; one operator on the tip or in the basket can stop at each window and open it by hand or break out the glass with an axe or pike pole. Personnel on the ground should be alerted before glass or debris is dropped. Precautions must be taken that fire fighters are slightly above and to one side for protection from falling glass and, possibly, from a backdraft explosion.

When the roof beams and sheeting have been subjected to fire, they are probably weakened; they could be close to collapse (Figure 15-13). If sponginess is detected, it means an unsafe condition. An elevating platform can often place truck personnel in the correct position for cutting a ventilating hole. If the roof is deemed to be hazardous, a safety line should be secured between the men and the basket. If the roof is obviously totally unsafe and broken through, the hole may be enlarged with a pike pole while working from the safety of the basket. A pike pole should be poked down through the hole to be certain that the ceiling has also been opened up so the smoke and heat can escape.

Backdraft Explosions

Because of the personnel hazards involved, it is important to be informed about backdrafts, or smoke explosions. Backdrafts are a constant threat to fire fighters; every fire

Figure 15-13. Ventilation of spongy roof is safe with aerial ladder or basket of elevating platform close at hand for exit of fire fighters.

fighter should understand their nature, causes, forewarnings, and the measures that can be taken to prevent them and to protect personnel against them. They are not uncommon at fires; in many instances it is only a matter of good fortune that more fire fighters are not killed or injured by backdraft hazards. When it is necessary to enter an atmosphere with an oxygen deficiency (less than 16%), or when the building is heavily charged with smoke, gases, and suspended solids, a self-contained breathing apparatus should always be worn.

When common carbonaceous materials, such as wood and paper, are burning in an atmosphere that contains sufficient oxygen, carbon dioxide is generated. But when a fire occurs in a tightly closed structure, the fire gradually depletes the normal 21% of atmospheric oxygen; when the oxygen content is lowered below 16%, the combustion of a free-burning fire will slow down and smoulder. Incomplete combustion produces carbon monoxide, which is a highly toxic and flammable gas. A mixture of carbon monoxide and air has an explosive range of 12.5 to 74% by volume and an ignition temperature of 1204° F. When these fuel gases are heated to their ignition temperature, they will burn if enough oxygen is present; they require only the admission of sufficient fresh air to cause rapid burning, the expansion of which may be enough to cause an explosion.

When a fire is confined within a building, conditions are comparable to a furnace burning with doors closed and damper shut. To ventilate at a lower floor before providing top ventilation would be like opening the door of the furnace while the damper is closed. Additional oxygen would be added with a corresponding increase in the intensity of the fire and in the generation of gases; this could turn the building into a raging inferno with flames engulfing all floors. Moreover, if sufficient oxygen is admitted from a lower floor before top ventilation is provided, there is always the possibility of instantaneous combustion of the confined and heated carbon monoxide mixture. Proper ventilation by opening the building at the highest point will permit the heated gases and smoke to be expelled and make entrance at the lower floors safer.

Backdraft dangers are not restricted to buildings, floors, or rooms. Any area that has an overheated mass of flammable gases can present a hazard, even though the entire structure has been ventilated. Explosions have occurred when a closet, ceiling void, automobile trunk, truck cargo area, box car, packing crate, or other closed void was opened to allow oxygen to enter.

Careful consideration must be given to the fact that where there is human occupancy on the upper floors of a building, the mere presence of smoke in any quantities on those floors will cause apprehension, and may create a panic condition. For this reason alone, it is imperative that fire fighters should be sent to the involved floors at the earliest possible moment to comfort the occupants and ease their fears. In addition, of course, they will be concerned with their normal fire fighting and ventilation duties.

Because of the increase in intensity of the fire when supplied with additional oxygen, charged hose lines should always be in place and fully manned before extensive ventilation is effected. To accomplish good, effective fire fighting, a great number of different operations must be conducted simultaneously; proper timing is the result of dedicated drilling and good teamwork.

Warning Signs of a Backdraft

A backdraft is a rapid, almost instantaneous, combustion of flammable gases, carbon particles, and tar balloons emitted by burning materials under conditions of insufficient oxygen. Therefore, backdraft conditions do not normally exist except where a fire is burning in a confined space, but the dangers must be recognized and considered when ventilating and providing for forcible entry.

Fortunately, backdrafts do not occur without forewarning; their very nature creates certain conditions that are usually discernible to knowledgeable firemen.

Everyone knows the appearance of a free-burning fire that has sufficient oxygen for complete combustion; there are few persons who have never gazed into a bonfire, campfire, or fireplace. One indication of incomplete combustion is unusually thick, black smoke at an ordinary structure fire where common materials such as wood, paper, cloth, and other similar substances are burning. The smoke is thick and black because of the extreme lack of oxygen; this causes carbon particles to be released into the atmosphere in great quantity. This is also an indication that generous amounts of flammable gases are being emitted.

Exterior Indications of a Backdraft

Anytime fire fighters are approaching a structure where there is a suspected fire, they should remain alert to the possibility of a smouldering fire. Feel the doors and windows before attempting to open them; backdraft conditions are possibly present if the glass or door panels are hot and there is very little or no fire immediately adjacent. If upon attempting to look into a building through a window it is found that visibility is poor, and yet no fire is in evidence or there is just a faint glow through the smoke, it is reasonable to assume that a fire has been smouldering for awhile in order to produce sufficient smoke to cause such poor visibility. Warm smoke may be emitted rather forcefully, as distinguished from oozing, from around doors, windows, and other small openings. In these cases, the smoke is usually forced out intermittently in puffs.

When opening a door or window where a backdraft may be encountered, stand to one side of the opening, remain low, and have charged hose lines available and ready to attack the flames. Any time there is a chance of a backdraft, personnel should not stand in, nor directly in front of, any door, window, or other opening. It is safest if no one remains within the V-shaped force pattern that will emanate from such openings. The gaseous products of the explosion will expand with some force as they are driven through the openings due to the lesser pressure of the atmosphere outside the building.

Signs of an Imminent Backdraft

There are some definite indications of an imminent backdraft, but, unfortunately, visibility inside a structure usually precludes such observations. However, fire fighters who have lived through a smoke explosion have shared their experiences. If the flames assume a pale or sickly yellow hue and seem to lose their liveliness, it is an indication that the fire is becoming oxygen starved and that flammable gases are being generated. The presence of a heavy volume of hot, dense smoke swirling around with great force near the ceiling, or at all levels, is a very hazardous sign; the smoke is usually grayish-yellow in color and the swirling atmosphere is generally accompanied by a peculiar whistling sound as fresh air is drawn inward.

If fire fighters are inside the building when backdraft conditions develop, they should leave as quickly as possible. Those who have been caught in an explosion state that they could see it coming; under hazardous conditions, drop to the floor and crawl out. A backdraft starts with what appears like a reddish rolling mass of fire that rapidly increases in size and spreads throughout the entire structure, overtaking the fire fighters in the course of travel. While the force of the backdraft is evident from floor to ceiling, the fire itself usually stays high and comes only to within a few feet of the floor. Those who drop prone are usually not burned, nor are they thrown about by the force of the explosion. However, those who remain standing are not only burned on exposed skin surfaces, but are thrown considerable distances.

Backdraft Prevention

Backdraft conditions are not always readily apparent upon arrival at a structure fire, or they may develop after fire fighting operations are under way. Assuming that hazardous conditions are recognized upon arrival, every effort should be exerted to prevent the backdraft from occurring. Prompt and proper ventilation will be the correct precaution, but even this cannot be relied on to absolutely prevent an explosion. Ventilate in accordance with prescribed methods, and attempt to cool down the atmosphere and drive out the flammable gases and vapors by the use of fog streams.

The example of good fire fighting and ventilation in Figure 15-14 contrasts with the example of poor fire fighting in Figure 15-15. In Figure 15-14 truck crew members ladder the building and cut holes in roof to permit the heat and smoke to escape. A hole cut in roof at eaves allows a direct handline attack on the attic fire. A 1½-in. line was taken into the floor below the fire so an attack could be made through the attic scuttlehole. Notice that in Figure 15-15, there is no breathing apparatus on the crew members, so an interior attack would be punishing; no holes have yet been cut in the roof to allow heat and smoke to escape.

The fire fighting attack on an attic fire should either be from below through the scuttlehole or by pulling the ceiling. On a steep high roof, an effective attack may be made by cutting a hole in the roof at the eaves. *Never* direct a fire stream down through an opening made for ventilation purposes; with very few exceptions, a hose line should be taken to the roof only to protect adjacent property.

An example of the vagaries of smoke explosions occurred when two engine companies and one truck company responded to a reported fire in a large single-story nightclub in the early morning hours. The doors and windows, both front and rear, were very hot, so there were no doubts that conditions were right for a backdraft. The first engine company laid hose lines to the front door, charged them, and had their crew ready to advance the fire streams. The second engine company laid lines to the rear door, charged them, but stood by; it is seldom advisable to advance fire streams simultaneously from two different directions. All crew members were equipped with self-contained breathing apparatus. When the hose lines were in position, the truck company used power saws to cut a large hole in the hottest portion of the roof; immediately, flames leaped 20 ft into the air to indicate that the hole had indeed been cut over the main body of the fire.

Upon being advised that the structure had been ventilated, the second engine company opened the rear door and windows; then they took a position to one side of the doorway. The first engine company opened the front doors and windows and waited to see if a smoke explosion would occur. A dense mass of thick black smoke poured from the front and back doors and windows; the roof opening still emitted flames very forcefully.

As the fire appeared to be confined to the rear half of the structure, it was decided to advance the hose lines from the front. All of the fire fighters except one were crawling on their hands and knees while working the spray streams to drive the heat and smoke out of the rear of the structure; one crew member was standing while helping to advance a backup line in the rear. The smoke was so thick inside the building that visibility was absolutely zero. Suddenly a backdraft occurred; flames and smoke were propelled with tremendous force from the front and rear of the structure. The crew members who were prone were not injured; the one who had been standing was very seriously burned, and would have been killed had it not been for the breathing apparatus. The fire fighters at the rear door were still standing to one side, so they were not hurt; flames were driven the full width

Figure 15-14. Example of good fire fighting, with ventilation holes cut in roof to allow heat and smoke to escape (Courtesy of the San Francisco Fire Department)

Figure 15-15. Example of poor fire fighting: no breathing apparatus on crew members; no holes yet cut in roof for ventilation

of the rear alley and seared a distant wall. After the smoke explosion developed, the atmosphere rapidly cleared; the fire was quickly confined and extinguished.

Investigations showed that the correct fire fighting and ventilation procedures had been complied with, and that all precautions had been taken. The nightclub had been redecorated many times and two separate and complete false ceilings had been installed under the original structure. The fire originated in a rear storeroom, spread up into the area between the false ceilings, and burned there until sufficient oxygen in the building had been consumed that the blaze subsided into a mass of smouldering embers. The hole in the roof did ventilate the main body of the fire, which was above the false ceilings, but not the dense atmosphere of flammable gases inside the night club. Sufficient oxygen entered through the front and rear openings to create an atmosphere that was within the explosive range of the mixture, glowing embers were present, ignition of the gaseous mass occurred, and a backdraft explosion was inadvertently created. Fire fighters should learn by this true experience that even though all precautions are taken, at the present time there is no method of being absolutely certain that a backdraft explosion will not happen.

At a fire in a carpet warehouse, fire fighters were killed because the truck crew did not ventilate correctly. A large hole was cut in the roof before the fire fighters advanced their lines. But, the heat, smoke, and flammable gases could not escape through the hole in the roof because no one poked a hole through the ceiling with a long pike pole. In this case, a ceiling was constructed several feet below the roof structure. The truck crew cut a large hole through the roof, felt down through the hole with an axe and could not detect a ceiling below, and moved on to other duties. Because the heat and flammable gases could not escape, they mushroomed down and filled the lower floors. Oxygen entered the structure with the entry of fire fighters with the hose; a violently devastating smoke explosion occurred after the fire fighters were within the structure.

Remember: When opening up a roof to ventilate a potentially explosive atmosphere from a structure, check to be certain that the heated smoke and gases are escaping through the hole. There may be an underlying ceiling to prevent the passage; firedoors or an enclosed stairway may be interfering with the circulation; or there could be a variety of other causes for an opening in the roof not allowing the smoke and gases to escape. High-rise buildings often develop a condition referred to as a "stack effect," which prevents the smoke from rising to the underside of the roof; the smoke from a fire in the basement or the lower floors will rise until it cools sufficiently to be the same density as the surrounding atmosphere. This plug of cold smoke, usually 10 to 15 floors above the fire, will effectively prevent the heat and smoke from rising any higher. Whenever vertical ventilation through the roof is prevented, horizontal ventilation through windows and the use of power blowers will have to be depended upon.

(Courtesy of the New York City Fire Department)

16 Cold Weather Operations

Fire fighters usually view the approach of the winter season with no great degree of enthusiasm. A heavy snow storm and extremely cold temperatures can be a deadly enemy, causing fires and other incidents that demand the fire department's attention, resulting in response delays, and adversely affecting the fire fighting operations. This is the season of the heaviest work load, most difficult operations, and greatest personal hardships. The majority of a fire department's cold weather problems are associated with the operation of apparatus and equipment. Preplanning and preparedness go a long way in minimizing these difficulties (Figure 16-1).

Predicting Weather and Road Conditions

Regardless of the section of the country, some winter operating and driving problems can be expected to cause problems. The element of risk can be minimized when potential problems are understood and respected. When driving in a cold climate during the winter season, be extra cautious and alert; a heavy snow storm and extremely cold temperatures can be a fire fighter's worst enemy.

Certain measures can be taken to alleviate some of the problems before or during a heavy storm. Weather advisories should be used to the fullest extent; it is extremely beneficial to determine as accurately as possible in advance what difficulties the storm might bring, how much snow is expected to fall, how low a temperature may be anticipated, the duration of the storm, and whether it may be accompanied by sleet or ice.

Emergency Preparedness and Handling

In communities where volunteers supply most or all of the fire fighting labor, the response of the volunteer members to an emergency could be delayed because of snow or ice on the roadways. Volunteer departments should consider the advisability of maintaining some personnel on duty continuously whenever delayed response could occur.

Where deep snows are common, the locations of hydrants should be marked before

Figure 16-1. Fire fighters view approach of winter season without enthusiasm. (Courtesy of the New York City Fire Department)

the first snowfall. A good method is to fasten a pole to the hydrant or paint a marker on nearby trees or telephone poles.

Extra shovels should be carried by all apparatus to uncover hydrants or to provide access to snowed-in structures.

Where streets are normally blocked during a snow storm, the normal assignment of companies may have to be adjusted because the closest station may not be accessible. It may be advantageous to assign extra personnel to reserve apparatus so the dispatching of companies to a fire could be more flexible and reinforced.

Hose Lines

Although roads may be cleared immediately after a snowstorm, many secondary streets wind up with a usable width that is too narrow to assure the maneuvering of apparatus after the automobiles of volunteers and spectators have entered the area. Therefore, even if the need for a fire stream is not immediately apparent, it is a good practice to lay at least one line when the second-due company is not on the scene. It is far better to pick up an unused line than to attempt to maneuver an apparatus around obstacles on a snow narrowed roadway.

Extra efforts should be expended in winter months to see that the hose lengths are tightly coupled to prevent leaking. Once a hose line is charged at freezing temperatures, it must not be shut down until it is ready to be picked up. If it is not practicable immediately to drain and pick up the hose lines, some water should be kept running through the hose by slightly opening the nozzle; this will prevent the line from freezing. When closing the nozzle, slowly nudge the shutoff handle towards the closed position until a small stream is still flowing. If the nozzle is completely closed, it could freeze shut and be rendered immovable until it is thawed out.

When lines are to be picked up in frigid weather, and there are several lines in the street, take time to select the hose that is mostly on top of the others as the first one to pick up. Remember, this has to be a fast operation, so the method likely to tangle the hose least should be used. A crew member should be stationed at each coupling; all personnel should concentrate on picking up one line at a time. The moment the line is shut down at the pump, as each crew member uncouples a length of hose, it should be walked out to its length, with the hose draped over one shoulder to drain it.

Do not take time to drain the hose completely as it is crucial that the hose rolling start immediately. If the hose is dry on the outside and it is rolled rapidly, even though the weather is extremely cold, the hose can probably be rolled before a crunching feeling indicates that ice is starting to form. Avoid unnecessary bends; haste in getting the hose rolled is more important than neatness.

Where hose is frozen into thick sheets of ice, it may be necessary to spread salt over the area; this action will thaw the mass sufficiently so the hose may be dug out. Care must be taken to thoroughly wash off the salt on the hose as soon as possible thereafter.

Handling fire hose in a frozen condition is a common cause of damage. With enough time, water will permeate fabric hose jackets to various degrees and will freeze in the winter months. Fibers in the jacket will become weakened if the hose is treated roughly; threads are likely to be broken, seriously damaging the hose. A good rule is to handle frozen hose as little as possible. When the hose is frozen solid, the best transportation is a flat bed truck. If many lengths are involved, and if the temperature can be expected to rise well above freezing the next day, it may be the wisest decision to barricade the street and post a guard to protect the hose lines until nature thaws out the area; this will be much easier on both the hose and the personnel.

If the hose is not frozen solid, several crew members working together can make two or three bends in the hose and place it over the pumper hose bed for transportation back to the station. In quarters, thaw the hose before scrubbing and straightening.

Frozen Hydrant

Occasionally, fire ground operations are complicated because the most strategic hydrant is frozen. It is poor practice to spend more than a minimum amount of time trying to cope with a difficult hydrant. The best procedure would be to move on to the next available hydrant or to radio to the next-arriving engine company to choose another hydrant and supply the water.

Only the valve may be frozen. Placing a wrench on the stem and tapping smartly may release the valve stem. The blows should be moderate to prevent bending or breaking the rod. A 1½-in. pipe telescoped over the end of the hydrant wrench can sometimes provide the added torque needed to break the valve loose. Care must be exercised to prevent injury in case the pipe slips off when pressure is applied.

A hydrant may be checked to determine if ice is blocking the barrel by lowering a weight on a strong string into the barrel. If the weight cannot be lowered the full depth to the valve, then ice is present. A frozen hydrant may sometimes be detected by striking the hand over an open outlet. Ice in the barrel will cause a higher tone to be generated than if the hydrant were completely empty.

Probably the most satisfactory method of thawing a hydrant is with a steam hose from a thawing device. Sometimes chemicals are utilized, but the hydrant must be thoroughly flushed afterwards.

Pumpers

Preplanning and preparedness go a long way toward minimizing cold weather difficulties in pumper operation. Some pumpers circulate an engine coolant through the fire pumps to prevent freezing and have heated booster reels and compartments. On others, it is essential to keep the pumps and piping drained whenever pumps are not being used. Booster reel piping is particularly vulnerable to damage from freezing. Booster hose and piping can usually be adequately drained in quarters by placing the fire pump under a vacuum, using the dry test method. After 15 to 20 in. of vacuum is observed on the suction gage, the booster nozzle and pump-to-booster valve are opened. This permits water in the booster line and piping to be drawn into the fire pump, which in turn can be drained. Pressurized air cylinders with adapters can be utilized to blow the water out of the hose and piping. During extreme temperatures when a particularly long worker is anticipated, to prevent freezing of the tank it may be necessary to drain the booster tank, employ a tank heater element, or circulate the water from the pump, through the hose, and back to the tank filler.

Soda-acid pump tanks, pressurized water, and foam extinguishers must be protected against freezing. They should be moved to an area containing some heat. In many cases they can be placed alongside a running engine, which will provide enough heat to prevent freezing and expansion of the contents. Expansion could rupture or weaken the container. Extinguisher hoses should be drained by gravity before the tank is refilled during freezing weather.

Aerial Ladders and Aerial Platforms

Accumulation of ice on raised ladders and platforms presents great difficulties and has severe implications (Figure 16-2). It is possible for an unsupported ladder at a normally safe angle to become loaded with ice to the point where stability is lost. A combination of a heavily iced condition and a moderate wind has been known to result in the twisting and eventual collapse of aerials. Some aerial platforms are equipped with heated booms to prevent icing.

A light-to-moderate coating of ice will often render it impossible to either raise the ladder away from a building or to retract the

Figure 16-2. Cold weather may interfere with apparatus operations. (Courtesy of the New York City Fire Department; photo by M. Gallagher)

ladder sections. Exercise great care when the ladder is in this condition as the ladder locks may stick in a locked position. Efforts to retract sections can then damage rungs or cause injury or a weakening of the cables. It is especially desirable for the contact surfaces of the sliding ladder sections, rollers, and pulleys to be liberally lubricated during freezing weather; this will often help resist the formation of ice at critical points and facilitate the removal of ice that may form when in service under conditions involving spray or rain.

Raised aerial ladders and platforms should be returned to bedded position as soon as conditions permit. After it is determined that the ladder locks are released, and there is still difficulty in retracting the sections, raise the ladder to a nearly vertical position to reduce friction. Some ladders depend solely on gravity for retraction; under iced conditions, bedding the ladder will be a formidable, if not impossible task. Other ladders use a fraction of their available hydraulic pressure for retraction and, on some, high pressure is available. Retracting an ice coated ladder must be accomplished with extreme care, for full hydraulic pressure exerted in the retracting operation could cause damage.

If ice should accumulate on a ladder to the point where it cannot be elevated away from a building, or be raised vertically, it may be necessary to place a portable ladder under the extended sections and assign crew members to give the hydraulic mechanism an assist in raising or retracting the ladder. Care must be exercised to avoid any damage to either the aerial or the portable ladder. Where the ladder is raised at such a steep angle that it is difficult to assist by using a portable ladder, it is often possible to tie a rope to the top of the ladder and pull from an adjacent roof top. Guy ropes attached to the top rungs and given an upward pull may ease off the weight load.

Exercise great care when a ladder is coated with ice, as this may cause a failure of the ladder if it is moved before defrosting. To remove excessive ice, the best results will be obtained with a steam hose.

If all other efforts to lower a heavily iced aerial ladder fail, and it is impossible to chip or steam off the ice, a crane may be utilized to take the excess weight off the hoisting mechanism and base. Departments exposed to freezing conditions should preplan their operations should the necessity occur. They should have an up-to-date list of steamers, cranes, and other equipment that would be available on a 24-hr basis.

Conditions permitting, efforts should be made to position apparatus where they will not be exposed to spray from fire streams, ladder pipes of other aerials, and wagon batteries. It is good practice to occasionally check heavy stream appliances and couplings to see that no leaks are spraying water over the ladder sections.

On some ladders, the cross shaft carrying the cable pulley located at the base of the ladder must be well lubricated and carefully watched to see that the pulley rides freely across the shaft. Improper tracking on the cable drum causes a shortening of the cable and will severely damage the ladder. The ladder locks must receive careful attention for proper release whenever sections are to be retracted.

In cold weather, keep the hydraulic circuits operating to prevent sluggishness or freezing. Use sand or salt under the stabilizing jacks and tires when operating on ice and snow.

Use special care to avoid needlessly stationing fire fighters on aerial ladders or platforms during cold and windy weather.

Winter Driving

Winter is the season of special driving problems; it is the period of snow, sleet, and skids in most of the country.

Visibility

There is one rule of safe driving that cannot be ignored: danger must be seen before it can be avoided. Although winter driving conditions cannot be controlled, hazards on the road can be spotted, and averted in time, with good visibility. Statistics indicate that a great number of accidents can be traced to impaired vision during bad weather.

Brush or scrape off all the windows; do not be a peephole pilot. If the apparatus is equipped with a heater or defroster, and there is time, warm up the engine and turn on the fan before starting out; be sure that the inside air is warm enough to prevent condensation on the glass.

Keep the wiper blades in top shape; never attempt to operate the wipers before scraping away ice and snow. If the wipers are immobilized, the internal drive gears may be damaged.

Turn on the low headlight beams at dusk, in rain or snow, or in just gloomy weather. Remember that glare is aggravated by a dirty windshield.

Traction

Next to good vision, effective traction is the most important factor in winter driving. Traction is the ability of the tires to grip surfaces of varying friction: dry, wet, slippery, or rough. The problems of hydroplaning, skidding, and sliding are caused by a loss of traction and, therefore, a lowering of friction between the tires and the pavement.

Tire chains are extremely efficient in mud, good in snow, and fairly effective on ice-coated streets. However, there is no tire tread or chain that is completely dependable. Tests show that reinforced tire chains provide more than seven times the pulling ability of regular street tires on glare ice and better than four times the pulling ability of regular tread on loosely packed snow. To be effective, tire chains must be kept tight; therefore, tension springs are recommended.

Remember to remove the chains when dry pavement is reached as chains wear down fast on dry roads. Tire chains should be checked and the necessary repairs accomplished well in advance of the first snowfall. Driving with defective chains sets up an unsafe condition; it is possible for faulty chains to rupture a brake line or other vulnerable component of the vehicle.

Proper tire pressure will allow the tread to grip the pavement in the most efficient manner. If the air pressure is too high, traction may be reduced. It is necessary for both tires of dual wheels to be correctly inflated to ensure proper load distribution and traction.

Snow tires without studs offer no advantage over regular tires for stopping on ice, but they will stop a little better than standard tires on most other surfaces; however, they improve pulling ability 28% on ice and 51% on snow.

Studded snow tires develop three times the pulling ability of regular tires on glare ice. Placed on the rear wheels, they reduce braking distance on ice by 19% as compared to regular tires.

It is better and safest to place the best gripping treads on the rear wheels to avoid front-end spinouts.

Shoulders

The shoulder of the highway may be firm and frozen on the surface; slushy and boggy beneath the cover. A great number of heavy vehicles have pulled off a winter highway and found themselves trapped in axle-deep mud under a solid covering; remember that fire apparatus weighs far more than passenger cars.

Icy Roads

Ice is the deadliest enemy of winter driving, particularly the "black ice" that may strike anywhere from the southern deserts to the frigid northern mountains. Wind blows snow and slush into the roadway and cold winds freeze it suddenly. Although the highway surface looks normal, it becomes completely covered over with an almost invisible glaze that will cause a vehicle to slide off the road at the slightest excuse. Icy roads are particularly treacherous because they are difficult to spot. The most obvious indication of an icy road is the behavior of other vehicles. Make it a habit to check the road for a long distance ahead, where visibility allows.

If there are indications of an icy pavement, slow down. Do not stop unless the vehicle gives indications that it is becoming unmanageable. When slowing down or stopping, apply the brakes gently and gradually; you cannot make a panic stop on ice. Tests prove that pumping the brakes gives the quickest stop that still keeps steering control. On a vehicle equipped with a manual transmission, leave the transmission in gear until just before motion ceases.

Winter Maintenance

The rigorous weather of winter greatly increases the preventative maintenance problems for apparatus operators.

Instruments and Gages

In winter operations, where fire streams are being used or there is a heavy rain or snowfall, efforts should be made to cover the open cabs, windshields, and other control areas to protect the gages and instruments from the elements. With open cab apparatus, a canvas or salvage cover can be used over the entire windshield and cab to protect the dashboard instruments and other vulnerable equipment. These measures also make it far more comfortable to drive back to quarters after the fire has been extinguished. Whenever an aerial ladder or elevating platform has been raised, the controls should be covered when they are not being operated, to prevent freezing.

Cleaning the Apparatus

Apparatus maintenance is substantially increased by the salt and other materials that are spread on the streets to allow better traction. Before the apparatus is housed, the entire undercarriage and surface should be flushed with clean water. Sanding and touching up the paint is essential to protect against corrosion as the salt road spray enters wherever paint is loose or worn away.

Door Locks

Moisture blows into locks, hinges, and crank handles where it then freezes. When temperatures fall low enough, oil and grease will begin to stiffen, making operation difficult. Eliminate such problems by using powdered graphite or other dry lubricant instead of a petroleum substance.

Cooling Systems

Whether or not the fire apparatus is provided with antifreeze for the cooling systems depends on department policy and climatic conditions. These factors will play in important part in establishing procedures for operating at fires.

Many departments do not use antifreeze, but depend on relatively warm fire houses to maintain vehicles in an operable condition. If the cooling system does not have antifreeze, with the weather below freezing, engines are kept running while they are out of quarters.

To aid the engine in warming up quickly to its operating temperature, a shield over the lower part of the radiator is advantageous. A piece of heavy cardboard or corrugated paper is adequate. During periods of mild temperatures, drivers should be alert for any signs of engine overheating and remove the shield.

During periods of subfreezing temperatures, engines of all apparatus parked near the fire scene and not being used should be kept running. However, they should be kept running only at a speed necessary to maintain the battery charge; the ammeter is the guide and there should be no discharge indicated. When weather conditions are extreme and it may reasonably be anticipated that the company will be operating over a long period, efforts should be made to remove apparatus not being used to adjacent fire stations as conditions permit.

Fuel

A close watch should be maintained to prevent the accumulation of water in the fuel tank, filter bowls and fuel lines. When the freezing of water in fuel systems becomes a problem, frequent servicing of the filter will usually remedy the situation.

Air Brake System

A common source of trouble during cold weather is the freezing of water that accumulates in the air brake system. Prevention is simple: remove the moisture from all air tanks daily by cracking the air pet cocks until the air discharge is dry.

17 Apparatus Maintenance

It is mandatory that all apparatus be fully operational for every response. To attain this degree of reliability, with the subsequent saving of costly repair bills, planned preventative maintenance with an assignment of responsibility is required.

Proper maintenance of their apparatus is one of the most important functions of fire department personnel. All equipment assigned to a company should be maintained in an exemplary manner, motivated by the fire crew's knowledge that such maintenance will provide greater personal safety for them and dependable service for their apparatus. In a maintenance program, the operator should not dismantle the apparatus to the point where it would be considered out of service; it must be able to respond with reasonable promptness to any alarm.

The knowledge, skill, and dedication of the apparatus operator are undoubtedly the most important factors of a good preventative maintenance program. Preventative maintenance calls for inspection, adjustment, and repair before failure occurs. If a carefully predetermined maintenance program is followed, with its schedules of systematic, periodic inspections, and if the necessary repairs and replacements are performed, the frequency of costly repairs and inoperative vehicles is lessened.

A well-organized preventative maintenance program has a schedule of daily, weekly, monthly, bimonthly or quarterly, semiannual, and yearly inspections of departmental automotive equipment. Typical maintenance schedules are shown in Figures 17-1 through 17-3. Much of this routine inspection and maintenance is the responsibility of drivers and operators; the more comprehensive scheduled maintenance should be accomplished by trained fire apparatus mechanics while the vehicle is out of service. Check sheets for the different inspection schedules should be made up so that no point is overlooked.

In the performance of proper maintenance, an accurate check can be made only if the part to be inspected is free from dirt or excessive accumulations of oil and grease. Clean apparatus can be more effectively maintained. Possible defects may be detected with less margin for error and before they cause any malfunctions. So, the first step in a proper maintenance program is: *keep apparatus clean!* Also, by cleaning the

PREVENTIVE MAINTENANCE is a periodic lubrication, inspection and adjustment according to schedule. Its purpose is to preserve new-vehicle efficiency and to obtain maximum life of the components. USE THIS SCHEDULE IN CONJUNCTION WITH THE LUBRICATION CHART.

DAILY SCHEDULE
(or while pumping*)

Inspect and Report:

Cooling System ---------------- Check for leaks. Add rust inhibited water to correct level. In winter check specific gravity of anti-freeze.

Fuel ------------------------- Fill tank before stand-over. Check operation of gage.

Hand Brake Lever ------------- Set hand brake in full release position. Turn adjusting knob on top of handle clockwise until lever pulls hard over center with a distinct click. Report if clevis pin is at top of slot.

Tires ------------------------ Check pressure when tires are cold. Check tightness of rim lugs. Do not bleed if tires are hot.

Battery ---------------------- Check specific gravity. If 1.225 or below, report. Add distilled water to 3/8" above plates.

Lights ----------------------- Report any defective parking, running or emergency lights. Check operation of all flashing lights.

Starting --------------------- Engine should fire promptly. Failure to do so warrants fuel and electrical system check.

Air Gage -------------------- 60 lbs. min. before moving vehicle.
90 lbs. min. to 120 lbs. max. -operating range

Ammeter --------------------- Should indicate charge with engine running at part-throttle.

Temperature Gage ------------- Should indicate 160° to 180° F. *If temperature exceeds 180°F., cut in auxiliary cooler to maintain 160° to 180°F.

Oil Pressure ----------------- Check for 10-20 lbs. @idle; 45-75 lb. maximum.

Windshield Wipers ------------- Check operation speed control and condition of wiper blades.

Engine, Transmission,
Rear Axle -------------- Check for oil and fuel leaks.

Air Tanks -------------------- Open drain cock to remove moisture.

Fire Pump (Main) ------------- *Check gland leakage. (10-25 drops per minute)

Priming Pump ---------------- Check air vent opening at top of U-tube at primer oil reservoir.

Figure 17-1. Preventative maintenance daily schedule (Courtesy of Mack Trucks, Inc.)

1-MONTH SCHEDULE
Perform Daily Schedule

Inspect and Report:

Main Fire Pump ----------------	Vacuum Test. Pressure Test. Check pump shift mechanism. Check priming system. Check transfer valve operation. Check all operating valves. Check all gages. Check auxiliary cooler operation. Check relief valve and pilot valve operation. Check all electrical connections. Check all manual overides. Check gland leakage (10 drops per minute, to one or two drops per second). Check pump gear box for oil leakage.
Booster Pump -----------------	Pressure Test. Check P.T.O. and Linkage. Check all operating valves. Check pump packing (10 drops per minute, to one or two drops per second).
Booster Hose and Reels---------	Check condition of hose. Check electric and manual rewind. Check reel and roller condition. Check all electrical connections.
Booster Tank ------------------	Check for leakage. Check vent, filler neck, cover and tubing connections.
Aerial Ladder -----------------	Check angle drive & P.T.O. Check spring locks. Check mechanical or hydraulic jack operation. Operate ladder. Record____rpm, ____psi. Raise, extend, rotate and lower. Check all hydraulic connections. Inspect cables, rung locks and all indicating lights. Check all electrical connections.
Aerialscope -------------------	Check rotation gear assembly and P.T.O. operation. Check jack and outrigger. Check hydraulic connections. Check aerialscope operation from platform and console. Raise - Extend - Rotate - Lower - Check electrical system. Check reservoir system. Check portable hydraulic pump system. Check platform leveling. Check boom bracket bearing assemblies. Check interlock system.

Figure 17-2. Preventative maintenance monthly schedule (Courtesy of Mack Trucks, Inc.)

FIRE APPARATUS PREVENTATIVE MAINTENANCE INSPECTION FORM

	1st	2nd
UNDER HOOD (Inspection)		
Coolant level & Radiator Core		
Fan and Generator belts		
Hood catches		
Engine Mounts		
Oil, fuel, and water leaks		
Carburetor and choke		
Generator, regulator & alternator		
Motor lights - Compartment		
DRIVER COMPARTMENT		
Master cylinder		
Brake pedal		
Clutch pedal		
Steering wheel play		
Siren and/or horn		
Batteries and Connections		
Seat upholstery		
Windshield wipers		
Lights: Warning and others		
Instruments		
Parking brake adjustment		
Vacuum clutch		
Cab doors, windows, etc.		
OUTSIDE VEHICLE		
Wheel nuts		
Front end		
General body inspection		
Oil pan & front & rear seals for leaks		
Spring mountings		
Brake system for leaks		
Tail pipe and muffler		

	1st	2nd
FUEL SYSTEM		
Throttle adjustment		
Carburetor		
Gasoline leaks		
Gasoline tank		
Gasoline lines		
Fuel pump, electric		
Air cleaner		
FIRE PUMP (main)		
Packing		
Transfer valve		
Working valves, disch., intake		
Leaks: water, air		
Gaskets		
Shift		
Gauges		
Relief valve or regulator		
Piping		
Priming pump and oil reservoir		
Booster reel, brake, and gland		
Pump transmission		
Pump performance, vacuum - pressure		
AERIALS & SNORKELS		
Hydraulic System - Leaks, relief valve, working pressures, etc.		
Ladders - Lubrication, operation, cable adjustments, etc.		
Turntable - Rotation, gears		
Jacks - Hydraulic, mechanical, operation		
Control Stand - Operating levers, gauges, etc.		

	1st	2nd
Radius rods		
Rear axle oil leaks and level		
Tie rods and knuckle stops		
Transmission oil level		
Drive shafts and universals		
BOOSTER TANK		
Check inside surface		
Water level gauge		
Leaks		
TESTING - ROAD		
Engine perform. & acceleration		
Gear shift action		
Overdrive performance		
Brake perform. (service - parking)		
Steering action		
TESTING - ROAD (Inspection)		
Front-end action		
General ride		
ENGINE OILING SYSTEM		
Oil pressure		
Oil lines		
Oil filter		
Oil level & condition of oil		
IGNITION SYSTEM		
Spark plugs		
Wiring		
Timing		
Magneto		
Distributor		

ALL BLANKS ANSWERED BY OK OR IF (√) SEE REMARKS BELOW.

Remarks ______________________________

Serviceman ....

Date ....

Figure 17-3. Check sheet for semiannual and annual scheduled inspections (Courtesy of Ward LaFrance Truck Corporation)

vehicle, company members will become more familiar with the components of the apparatus and know where the lubrication points are located.

A complete dissertation on the subject of the preventative maintenance of fire department apparatus would require an entire book by itself. This chapter is mainly concerned with inspections and maintenance as practiced by fire department personnel; the more complicated semiannual and yearly inspections are usually performed by trained and knowledgeable mechanics; they follow the normal preventative maintenance recommendations of vehicle manufacturers. Any questions that may arise concerning the maintenance of fire apparatus should be referred to a responsible fire department mechanic.

Daily Schedule

Immediately upon reporting on duty, apparatus operators should determine that the equipment is ready for instantaneous response and fire fighting duty. Discuss the previous shift's activities and the condition of the apparatus with the relieved operator. Did they have a fire? What hose and equipment was used? Was the hose changed and the equipment serviced? Did everything perform well, or were there problems? Exchanging information and suggestions will keep all personnel better informed and cooperating.

Keeping the fuel tanks of pumpers full is especially important, since the engine often has to operate at maximum power for several hours after the apparatus reaches the fire. Unless the fuel tank is checked when the operator first comes on duty, and after each run, there is a danger that the tank has been partly depleted and the fuel supply is inadequate for a long pumping operation. Few officers, if any, will allow a driver to blame the other shift when the apparatus runs out of fuel on the way to a fire or has an inadequate amount of water in the booster tank just after a change of platoons. Do not trust the other crew to leave the apparatus and equipment in correct condition.

The National Fire Protection Association publication 1901, *Standard for Automotive Fire Apparatus,* requires that the fuel capacity for apparatus with pumping equipment shall be of a size that will permit the operation of the pumping engine for not less than 2 hr when operating at rated pump capacity. While most manufacturers install sufficiently large fuel tanks that this will not create an immediate problem, some pumpers barely exceed the minimum fuel capacity requirement. There have been instances where pumpers have run out of fuel during the 2-hr capacity tests; apparatus being placed out of service at a critical time during fire ground action is not too uncommon. In some localities, it may require over 2 hr for a fuel tanker to respond to the fire ground after it has been called. Each fire department must have arrangements for refueling apparatus at all times, day or night.

Spend a few minutes every day to place turnout clothing on the apparatus, adjust the seat and mirrors, determine that all valves and controls are in the proper position for an immediate response to a fire, close and check all compartment doors, check the ladders for security, and visually inspect to determine that all fire fighting equipment is in place and that there is no body damage.

Even though the daily check of an apparatus may appear to be a routine matter, it is extremely important; a planned procedure should be devised and adopted so that nothing is overlooked. A good method is to begin in the driver's seat; place the gearshift in neutral, check the clutch and brake pedals, put the parking brake on and the pump shift lever in road position, check the gages and instruments, test all lights and the radio, check the windshield wipers, and ascertain that everything in the cab is ready for an immediate response. Then check the operator's panel or position; determine that all the valves and controls are in the correct position for fire fighting duty; be certain that someone has not inadvertently left the pump transmission in the wrong position. Proceed around the vehicle, inspecting all equipment to be certain that nothing is missing. Inspect all tires. Close inspection will determine if any unreported damage to the vehicle has occurred. Until this thorough check of the fire fighting apparatus has been carried out by the operator reporting on duty, the vehicle cannot be considered fully available for an immediate response to a fire.

Starting Engine at Change of Shift

The practice of starting the engine at change of shift is a survival of the early days when automotive apparatus was replacing horse-drawn steamers (Figures 17-4 and 17-5); gasoline engines in those days were not very dependable. This test is an unnecessary precaution of no practical value with modern engines and only causes needless wear and abuse. It does not prove that the engine will start the next time it is needed.

Automotive engineers and lubrication experts agree that a large portion of engine wear takes place during the first few seconds after starting, before the normal flow of oil starts and an oil film is established. Moreover, when the engine is cold, water and fuel condense on the working parts inside of the engine; this results in the corrosion of valve lifters, springs, and other internal parts. After the engine is warmed up, this condensation vaporizes and is drawn out through

Figure 17-4. Early gasoline engines were not very dependable, so they were started every change of shift as a test. This 1913 Seagrave hose wagon was powered by a Frayer-Miller 6-cylinder air-cooled engine.

the crankcase ventilating system, doing no further harm.

Water vapor resulting from the burning of fuel and air is condensed in the crankcase and will mix with carbon and dirt particles in the oil, forming an emulsion that will settle out and form sludge. This sludge will deposit in the oil pan, on the walls of the crankcase, on the valve chamber, in oil passages, and on other internal parts of the engine. Sludge may interfere with the proper functioning of the lubrication system.

Oil contamination can be minimized by running the engines long enough to get them thoroughly warm every time they are started; this helps to vaporize the contaminating products in the crankcase and will allow the crankcase ventilating system to perform. Engines should not be operated in confined quarters without adequate ventilation because of the hazards of carbon monoxide asphyxiation.

Body Maintenance

Frequent and regular washing, waxing, and polishing will lengthen the life of the painted finish and bright metal trim on the apparatus. Wash with warm or cold water to remove dirt, dust, and road grime; this will help preserve the original luster of the paint. Do not use hot water or strong soaps and detergents. Never wash or wax in the direct rays of the sun, or where metal is hot to the touch; this may cause a streaking of the

Figure 17-5. This 1917 American LaFrance chemical and hose wagon had six cylinders cast in groups of two with intake valves on one side and exhaust valves on the other; some of these engines had four spark plugs per cylinder and two ignition systems—battery and magneto. Even then, starting the engine was often a challenge.

finish. When salt water has been sprayed on the paint, or when calcium chloride, salt, or other spray from icy roads is deposited on the finish, the apparatus should be washed as soon as possible.

Sponge off the wheels, tires, under fenders, and other contaminated portions of the apparatus. Keep the water off of the hot engine, manifolds, brake drums, and other heated parts of the vehicle as the sudden cooling could cause cracking or distortion of the metal.

Tires

Tires should be kept clean; they should be checked every morning and after each run for any readily apparent softness or carcass defects. At least once a week a pressure gage should be used to determine if the air pressure is correct. All air-pressure testing should be done on a cold tire, not right after a run or after the tire heats up in the summer sun.

Study the tread to detect any unusual wear patterns that would indicate a need for wheel alignment. Check the tread for foreign objects and remove them. Valve caps should be kept on the valve stems to protect the valve cores. In localities that require chains in the winter, these should be examined to determine if they are ready for use.

When tires are underinflated, they tend to wear on both edges of the tread; where overinflation exists, tires will show wear in the center of the tread. If tires show undue wear on one edge, they may be out of line. Improper balancing of a tire and wheel is indicated by flat spots, and on the steering wheels it may cause cupping. Flat spots and tread abrasion can also be caused by heavy braking, out-of-round brake drums, or locked-wheel stops. Feathering at the edges of the tread grooves, depending upon the direction of the feathering, denotes excessive toe-in, excessive toe-out, aggressive cornering, or fast turns.

If the apparatus has dual wheels, pay close attention to the air pressure, because one tire may be low in pressure but it may not be apparent. When checking the tires, test the wheel lugs for tightness.

Fuel Tank

The fuel tank should be kept full to provide the maximum operating time, prevent the accumulation of moisture and condensation in the tank, and reduce the likelihood of drawing sediment into the fuel lines.

Caution: When filling the fuel tank, keep the filler spout or nozzle in contact with the metal of the tank to prevent the accumulation of any static electricity from the flowing liquid; avoid the possibility of an electric spark igniting the flammable vapors. If a battery charger is operating in the vicinity of the fuel tank, shut it off before refueling. Do not have the vehicle inside a building when filling the tank if it can be avoided. If it cannot be avoided, provide the building with complete ventilation to prevent the dangerous accumulation of flammable vapors.

Response Schedule

Just as important as the routine daily check schedule is the check schedule immediately after response to a fire.

At Scene, after Fire

Load, secure, and inventory all equipment.

Unless the hose is frozen, load it all back into the hose compartments ready for an immediate response to another fire. Some departments roll wet hose and carry it on the tailboard back to the station where they get a dry, fresh load. There is no guarantee that the next fire will wait for the apparatus to leisurely be placed back in service. If another fire is encountered on the way back to the station, this practice could cause delays and confusion.

Place all valves and controls in the proper position to insure instantaneous operation at the next fire.

Close and secure all compartment doors. Secure all ladders.

Refill the booster tank at the scene or nearest hydrant so that the apparatus is fully operational.

In Quarters, after Fire

The engine, transmission, and other operating parts of a vehicle should be examined at least weekly, and after every time it has been subjected to hard usage at a fire or drill. All components of a vehicle should be maintained clean and well painted so any leaks, looseness, missing bolts and nuts, or other defects or points of potential failure may be detected. Fuel, oil, or coolant leaks should be traced and repaired. Check the condition of all lines, hoses, and other components. Carefully check all of the drive belts. If the adjustment limit has been reached, if belts are cracked, frayed, or oil-soaked, they should be replaced. A well-maintained apparatus seldom has any sudden breakdowns.

Service as needed: fuel, radiator coolant, engine oil, primer oil tank, hydraulic fluid, extinguishing systems tanks, etc.

Test all lights.

Inspect for damage all fire hose and fittings that were used.

Clean the windshield, mirrors, and entire vehicle.

Inspect the tires for damage.

Inspect and service all equipment used: breathing apparatus, fire extinguishers, portable pumps, generators, blowers, stretchers, salvage covers, chain saws, life lines, etc.

Periodic Testing and Maintenance Procedures

To keep fire fighting apparatus in top condition, it is important to keep up periodic testing and maintenance check ups. The following test and maintenance procedures should be regularly scheduled and practiced.

Road Test

At least weekly, each piece of automotive fire apparatus should be started and given a road test. Some departments merely start the engines once a week, if they have not been run meanwhile, and operate them in quarters at idle speed until warm. It is harmful to an internal combustion engine to allow it to run for a period of time with no load on it. A road test will allow the engine to warm up properly, permits a close observation of its operation, and lets the driver obtain needed practice in handling the vehicle. The operator should be alert to detect any unusual noise or vibration in the engine, transmission, differential, or any other component of the vehicle.

The road test will check the vehicle for power and engine performance. Clutch operation will show slip, ease of engagement, and grabbing. Any excessive emission of smoke should also be noted. The riding and handling characteristics of the steering, springs, shock absorbers, seat, body motion, and cab motion may be carefully evaluated.

Fire Fighting Appliances

Pumpers, foam trucks, crash apparatus, aerial ladders, aerial platforms, and every other type of fire fighting vehicle should be operated at least weekly under realistic conditions to be certain that they will perform efficiently on the fire ground when needed; this will allow the operators an opportunity to practice. There have been many examples of fires lost when an apparatus would not operate correctly because of deteriorated or corroded components, or undetected defects.

Pumpers

With the engine at operating temperature and water flowing through the pump, check the operation of the pump shift controls, panel throttle, pressure control device, changeover valve, and discharge gates.

Priming devices should be operated periodically to determine their ability to develop and maintain an adequate vacuum and to discover if any air leaks have developed.

It has happened that transfer valves have frozen in a neutral position, rendering the pump inoperable, or it has stuck in one position. Whenever the pump is operated, maintain a steady pump pressure and operate the transfer valve to the full extent of its travel several times from series to pressure and back; this will keep the valve working freely. Carry the transfer valve in the position that is most likely to be used. Most pumpers carry this valve in the series (pressure) position as the largest number of fires, especially in the early stages, require small hose streams.

Relief valves and governors should be placed in operation weekly to keep them free. Pump at 100 to 150 psi discharge pressure and cause the regulating device to operate by turning the nozzle on and off several times. This will help prevent pilot valves, pistons, and other parts from sticking. Lubricate according to the manufacturer's instructions.

Lubricate the pump shift linkage pivots and pins; throttle linkage pivots and pins; pressure control linkage pivots, pins, and pistons; discharge gate linkage; and changeover valve linkage. Check the pump shaft packing for excessive leakage; approximately 30 to 60 drops per minute is satisfactory to keep the packing cool while pumping. Mechanical seal types should not drip.

The suction strainers in the pump inlets should be inspected to be certain that they are in place, intact, and not damaged or clogged. If one is broken, it could allow stones and other foreign material to enter the pump and damage the impellers and other components.

Pump Maintenance

Except for lubrication, fire pumps require very little attention. However, it is important that the recommended preventative maintenance be performed at the proper intervals. Following is the suggested maintenance for Hale Fire Pumps; other manufacturers have similar programs:

Weekly. Test the priming system, transfer valve, and relief valve or governor. Check the controlled leakage of the packing gland and the oil level in the priming tank. Lubricate the ball discharge valves.

Monthly. Complete the weekly test and dry vacuum test. Lubricate all remote controls. Check the drive line flange bolts and the oil level in the pump drive unit. Grease the suction tube threads and inspect the gaskets for deterioration. Inspect and clean the strainers. Visually inspect the drive line oil seals.

Semiannually. Complete the weekly and monthly tests. Drain sample oil from the drive unit and check for contamination. If contaminated, drain, flush, and replace with S.A.E. 90 oil.

Annually. Complete all previous tests. Check gages for proper calibration. Test suction hoses and check for collapsed inner liners. Drain drive unit oil and replace with S.A.E. 90 oil. Complete annual service test on the pump.

Note: If the pump is used extensively, it should be repacked every two years. Good preventative maintenance lengthens the life of a pump.

During freezing weather, be sure to drain all water out of the pump. This can be done by opening the discharge valves, removing the suction inlet and discharge valve caps, and then opening all drain valves and cocks for gage lines and cooling lines. After the pump is completely drained, all the caps should be replaced and the valves closed. Do not put off closing the drain cocks or valves until later as forgetting to close them will result in failure to get water when attempting to work at draft.

After pumping salty or dirty water, connect the pump to a hydrant or other source of fresh, clean water and pump for a few minutes to flush and clean out the pump, discharge valves, relief valves, gages, and cooling lines.

Service Charts for Hale Pumps

The following check points for maintenance servicing of Hale pumps are keyed to the diagrams in Figures 17-6 and 17-7.

1. *Air vent hole.* (See under (14).)

2. *Bearing and packing cooling lines.* Once a year check the drilled holes in the pump body and packing housing for blockage.

3. *Bearing, sleeve.* Grease every 3 to 6 months, depending upon severity of pump usage. Use 3% molybdenum disulfide grease.

4. *Bearing, sealed oil reservoir.* This bearing is lubricated and sealed for the normal life of the pump. For severe service, check the oil level once a year; if required, fill to the top with S.A.E. 90 oil.

5. *Check valves.* On parallel-series pumps, frequently remove the suction inlet strainers and reach in with the hand or a rod to make sure that the valves are free to swing and that no foreign matter is caught between the valve and the seat.

6. *Discharge valves.* Using a brush, lubricate the face of the ball with a waterproof grease once a week; work the valve a few times and then regrease. If a drain valve is installed, insert an oil can tip in the open end of the discharge valve and allow some oil to run down the connecting pipe to the closed drain valve. The remote control linkage connecting the valves should be lubricated monthly.

7. *Drain.* Drain the pump after using. Open the pump body drain cocks. On two stage pumps, move the transfer valve to both the volume and pressure positions. If the pump is equipped with a power operated transfer valve, open the drain valves so the pilot and actuating piston will empty. Drain the governor or relief valve. After the pump is completely emptied, all caps should be replaced and the valves closed.

8. *Gaskets.* Inspect suction hose and inlet cap rubber gaskets frequently. Foreign matter under these gaskets, or faulty gaskets, will cause air leaks that may prevent obtaining water when working from draft. Even if water is obtained, faulty gaskets could cause an irregular, pulsating fire stream. Replace the gasket if the rubber is cracked or has lost its resiliency.

9. *Gear case drive unit.* The pump bearings, drive unit bearings, and all gears are supplied with oil from the drive unit housing.

10. *Gear case dip stick.* Keep the oil level between the high and the low marks on the bayonet gage.

11. *Gear case magnetic drain plug.* Flush out the old oil and replace every 3 to 12 months, depending on model and amount of pump usage.

12. *Gear case oil level and fill plug.*

13. *Priming control valve.* (With rotary gear priming only). If the valve has a tendency to stick, oil the path of the button or disassemble to clean and lubricate.

14. *Priming pump.* Keep the primer oil tank filled with S.A.E. 30 engine oil. Always lubricate the priming pump after drafting; this can be done by running the priming pump after the main pump has been drained. Gearshift type: place the priming pump in gear and run the engine at 800 to 1000 rpm. Clutch type: place the main pump in gear and press the priming button when the engine is running at 800 to 1000 rpm. Electric priming pump: press the priming button.

Continue to operate the priming pump until oil sprays out of the priming pump discharge. Should oil continue to leak from the primer, check the small air vent hole (1)

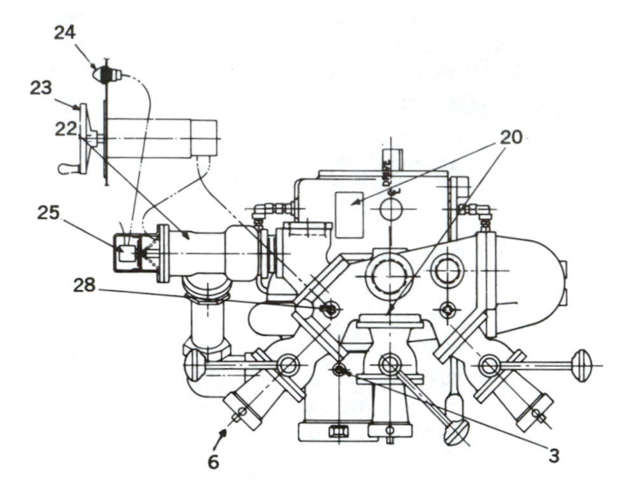

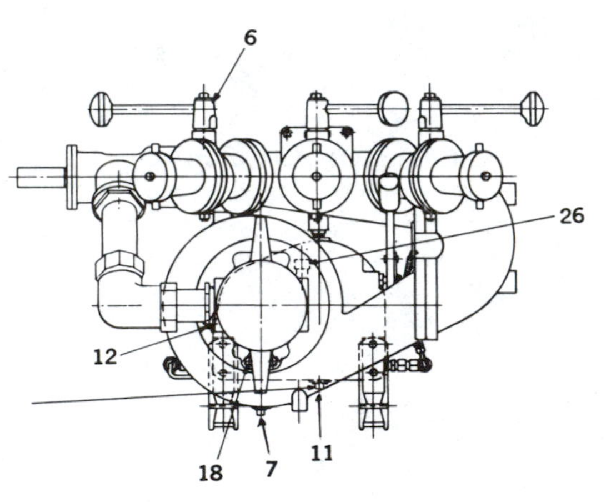

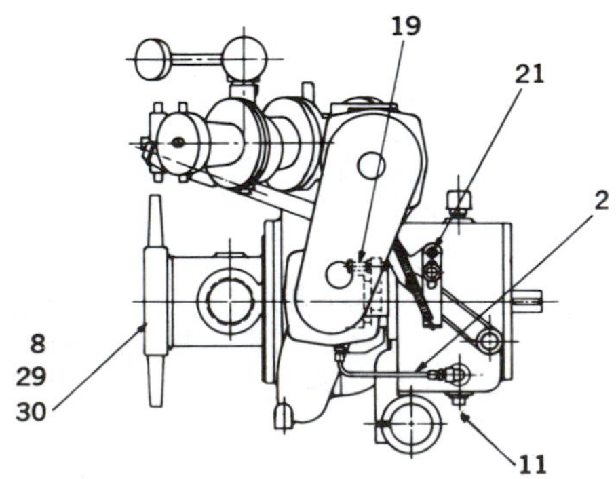

Figure 17-6. Service chart for Hale front-mount pumps (Courtesy of Hale Fire Pump Company)

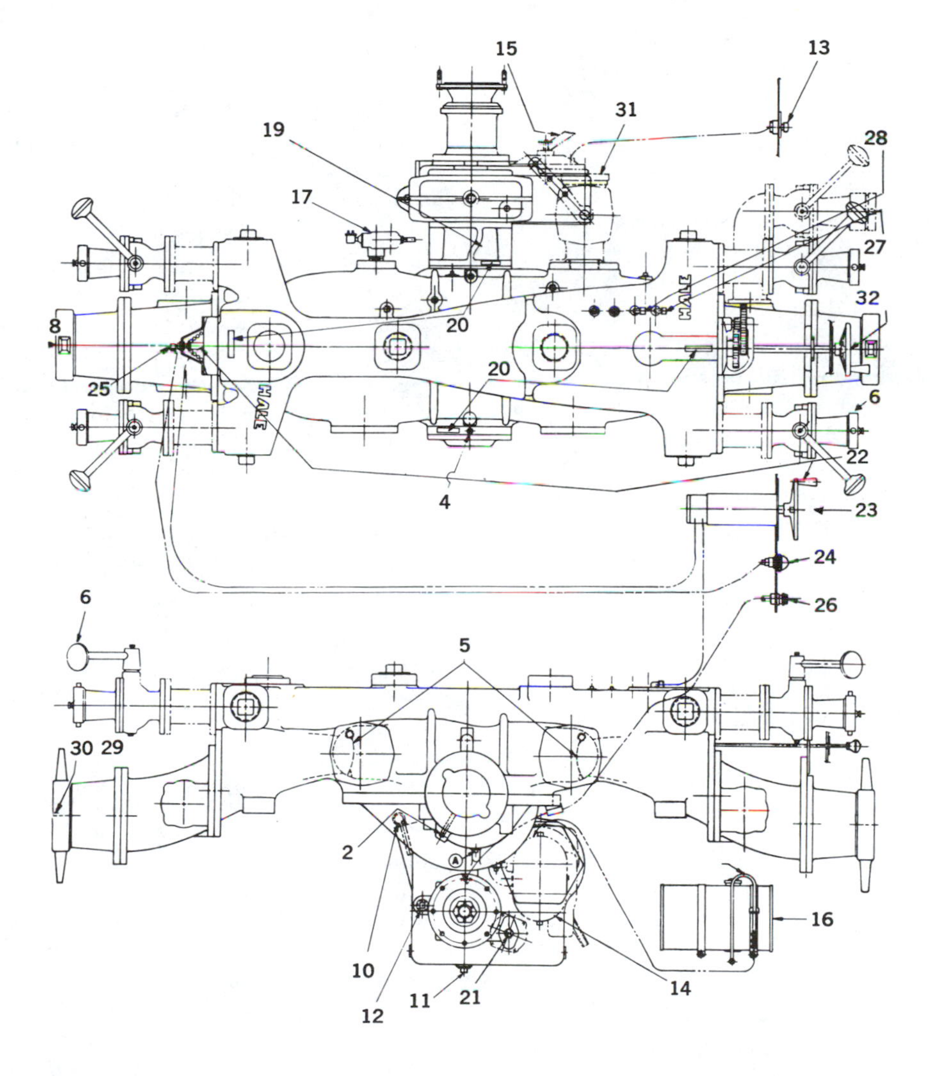

Figure 17-7. Service chart for Hale midship pumps (Courtesy of Hale Fire Pump Company)

in the top of the tubing fitting at the primer oil tank. This hole should always be open so air can flow into the hole; this prevents the oil in the tank from syphoning over into the priming pump.

If a priming operation is not using oil, water is entering the tank through the priming valve from the main pump.

15. *Priming pump manual control*

16. *Priming pump oil tank*

17. *Priming valve*

18. *Pump heater connections.* This system circulates warm water from the engine cooling system through the pump to prevent freezing during cold weather.

19. *Pump packing.* A slight trickle of water, about 10 drops per minute, should leak past the shaft packing to lubricate and cool the shaft and packing, thus preventing the packing from burning and scoring the shaft. Do not tighten the packing unless air enters while priming or considerable water leaks out.

20. *Pump serial number.* Check here to ascertain pump number.

21. *Pump shift.* Lubricate the linkage. If apparatus is equipped with a power shift, lubricate the shifting cylinder and piston yearly with a few drops of vacuum cylinder oil.

22. *Relief valve.* Refer to specific relief valve or governor maintenance instructions.

23. *Relief valve control*

24. *Relief valve pilot light.* A bright pilot light glows when the relief valve is open for flow.

25. *Relief valve pilot light switch*

26. *Speed counter.* This connection, joined to a similar fitting on the pump operator's panel, is driven at 1/10th of the engine speed. This fitting is used to obtain the true engine and pump speeds.

27. *Strainer*

28. *Strainer, relief valve control*

29. *Strainer, suction tube.* These strainers are sprung into position; they can be easily removed for cleaning.

30. *Suction tube threads.* If made of cast iron, lightly coat the threads with grease once a month.

31. *Tank valve and check valve.* Inspect the check valve to be sure that it swings freely and seats properly. If the pump is equipped with a strainer between the tank valve and the pump body, remove the strainer housing and clean periodically.

32. *Transfer valve.* (Not required on single stage pumps). On handwheel geared types, remove the old grease every six months and apply multipurpose grease. On power cylinder types, lubricate the power cylinder once a year by removing the vacuum line at each end of the cylinder and adding one ounce of vacuum cylinder oil. If either the pilot valve or the hydraulic power piston has a tendency to stick, disassemble and clean thoroughly.

33. *Transfer valve grease fitting.* If so equipped, fill the grease fitting with multipurpose grease once a month.

34. *Yoke.* If pump has a motor yoke, grease once a month with multipurpose grease.

Pump Flushing

All equipment, including the pump, should be flushed out with clean water after each pumping operation in salty, dirty, or sandy water. Connect the pump to a hydrant and flow water through all components that were exposed to the contaminants. When used, the priming device, engine cooler, and other accessory equipment must also be flushed until clear water is discharged. If the booster tank has been filled or the radiator fill has been used, drain and refill the tank or cooling system.

Tasting water flowing from the pump is the most dependable method of determining when all traces of salt have been flushed out.

Aerial Ladders and Aerial Platforms

Run the apparatus through a full cycle of operations. Test the operation of the stabilizer jacks, spring locks, and parking brakes. Observe for hydraulic leaks, proper operation of all controls, ladder locks, section rollers, limit stops, and other components. Carefully check the hoisting cables for broken strands. Inspect the rotation pinion and gear under the turntable for wear and lubrication. Inspect the collector rings under the turntable for corrosion, the brushes for wear, and the wires for secure connections.

Lubricate by applying one or two drops of oil to the control linkage pivots and pins, cable pulleys, hoisting cylinder pivots, and bed lock pivots. Lubricate the limit control mechanism. Wipe the extension cables and spring compensator lightly with an oiled rag.

Aerial Hydraulic Systems

Oil performs three functions in the hydraulic systems of elevating platforms, aerial ladders, and water towers:

1. Transmission of power from the hydraulic pump to the operating units, cylinders, and fluid motors.
2. Lubrication of all moving parts in the system.
3. Cooling of operating pump and motors.

Heat radiation from the lines and supply tank cools the pump and fluid motors. The work done by the pump to raise the fluid pressure produces heat, which in short periods of operation is satisfactorily dissipated through the oil without high oil temperature. Under conditions of prolonged operation, such as a training session of two hours or more, the oil temperature may rise to a point where it thins out sufficiently to make the motors nonoperational, and the cylinders will act "spongy" with a poor response to the controls. Therefore, it is a good practice at a lengthy drill or a prolonged fire to take the pump out of gear when the mechanism is not being operated. One of the prime requirements for oil used in the hydraulic system is the ability to retain a working viscosity at elevated temperatures. To ensure this quality in a hydraulic oil, it should have a minimum viscosity index of 105.

Hydraulic oil which will be used under high temperature conditions of 200° F or more require an oxidation inhibitor. Such an additive reduces the effect of heat and acts to prevent breakdown of the oil and formation of sludge and varnish, which causes faulty relief valve operation and plugging of lines. Other important qualifications are: antiwear additives, proper viscosity to maintain a lubricating film on all moving parts at every operating temperature, a corrosion inhibitor, an antifoaming agent, a pour point depressant, and demulsibility.

Preventative maintenance of the hydraulic system has two principal objectives:

1. To keep the equipment fully operational at all times.
2. To keep repair costs at a minimum and avoid out of service time.

Both of these objectives are achieved by the use of the recommended oil in the system; a regular schedule of oil change at least once a year, or as recommended by the manufacturer; and cleaning the strainers and replacing line filters, if equipped, at least every six months.

Hydraulic oil becomes contaminated from several sources. Dirt enters the tank through the vent, water forms from condensation with temperature changes, metallic particles are worn from the pump and fluid motors, and the packings and seals contribute particles. All mix into an emulsified sludge accumulation when left too long. Contaminants in the system lead to malfunction of the relief and other valves. Lines become plugged, causing possible pump cavitation because oil will not flow sufficiently free to supply the requirements. Pump pressure will be reduced to less than the required amount. These contaminants, together with the oxidized oil, can only be disposed of by an oil change.

Some manufacturers supply screens or filters in the system; other manufacturers depend upon the larger particles settling out of the supply in the tank. This condition makes it imperative that the oil be changed regularly to avoid excess wear and malfunction.

Electrical Circuits

The vast majority of operating difficulties that befall any automotive vehicle are caused by electrical malfunctions. A few may be related to fuel or mechanical components, but these are relatively rare. Automotive electrical circuits appear complicated. This is really not true, and a basic knowledge of electricity will allow an apparatus operator to maintain apparatus in a more knowledgeable manner.

Electricity may be defined as the flow of electrons through a conductor at the speed of light. A battery, generator, or alternator creates the necessary electromotive force (EMF); this pressure is measured in volts (V). The quantity of electrons passing a given point is indicated in amperes (A), and the resistance to flow is stated in ohms (Ω). As the electrons are displaced from their orbits in the conductor atoms in a domino fashion, an electrical circuit must be complete and unbroken so electrons will be able to progress back to the original atoms.

This is analogous to fire department hydraulics, and the laws of electricity may be closely compared to those governing the flow of liquids through pipes and hoses. A battery or alternator creates volts of electrical pressure sufficient to force a quantity of electrons (stated in amperes) through a conductor while overcoming resistance measured in ohms. In hydraulics, a pump creates sufficient pressure in pounds per square inch to force water measured in gallons per minute through a fire hose to overcome friction loss resistance stated in pounds per square inch. A complete circuit would be similar to a pump drawing water from the bottom of a tank and discharging it through a hose back into the top of the tank so the water level would remain constant.

Ohm's Law is the basic rule of electrical circuits. If this is understood, everything else is simplified:

$$\text{Amperes} = \text{volts} \div \text{ohms, or}$$
$$\text{Volts} = \text{amperes} \times \text{ohms, or}$$
$$\text{Ohms} = \text{volts} \div \text{amperes}$$

An increase or decrease in any of the three factors in Ohm's Law will affect either or both of the others. If the voltage in a circuit drops because of a low battery, and the ohms resistance in the conductor does not change, the amperage available to light a lamp or drive a motor will drop a corresponding amount. If the resistance in a circuit increases, due to corroded terminals or a damaged wire, the available current will decrease. Keeping this simple basic rule in mind while maintaining and servicing a vehicle, the operator should check all connections and terminals for cleanliness and tightness, all wires for damaged insulation and broken strands, and the wiring should be securely clamped in place so excessive movement will not cause undue stress.

All circuits should be protected against amperage overload in case a short circuit occurs. This can be achieved by the use of fuses or circuit breakers. Usually, automatic reset circuit breakers are preferred over fuses; their advantage is that when a short circuit or other overload is corrected, the circuit breaker automatically reactivates the system for use. When troubleshooting, an automatic reset circuit breaker will continually click on and off to indicate trouble; this clicking will continue until the trouble in the circuit can be found and corrected.

When a vehicle is designed, engineers carefully calculate what size wires and switches should be utilized in each circuit; they seldom specify any larger conductor than is necessary. The actual selection of wire size is primarily dependent upon the amperage load and the length of the conductor; the voltage and mechanical strength may be considered to some extent. High amperage must be transmitted with a minimum amount of loss by resistance. When a 12-V battery is used to supply current to a hose reel rewind motor which draws 10 amperes through 20 ft of wire, the engineer will be concerned with calculating what gauge of wire will not create excessive resistance. If a larger motor which requires more amperage is later substituted, if broken strands of wire or corroded connections increase the resistance, if another device or lamp is added to cause additional demands on the circuit, or if a partly discharged battery will not supply the necessary electromotive force (volts), the motor can not possibly operate satisfactorily. This example can be similarly applied to every other electrical circuit.

A complete electrical circuit is necessary before the moving current can accomplish any work; in fact, before it can even move. The electrons flow from negative to positive, though some books state the opposite. Batteries are normally grounded by attaching a cable from the negative battery terminal to the engine block; the frame and body are then grounded to the engine by mounting bolts and grounding straps. All lamps, motors, relays, spark plugs, and other electrical components are also grounded to the engine, frame, and body. This is referred to as a common ground and it allows a complete circuit to be made with only one wire to each component. Electrical fixtures mounted to a nonconducting material, such as a fiberglass body or a plastic surface, require two wires to conduct the current.

If a battery is used to light a lamp, the current flows from the negative battery terminal to the engine block, through the frame and body to the lamp socket, through the lamp filament to heat it up and make it glow, along the wire, through a closed switch, and back to the positive battery terminal. If the switch has not been closed, or if the wire is broken, there is an open circuit and the lamp does not light; if there is excessive resistance due to corroded terminals, loose connections, partly broken wires, or a poorly grounded lamp socket, the lamp will glow dimly or not at all; and if the current is short circuited because damaged insulation allows the conductor to touch a ground, a fuse or circuit breaker will probably open because of the increased flow of electricity.

Corroded battery terminals can cause defects in the operation of every part of the electrical system; check both ends of each battery cable for corrosion and tightness. If a battery cannot be kept fully charged, regardless of the amount of driving, first check the drive belt tension of the alternator, condition of the battery cables, and for wiring and terminal defects in the charging circuit before assuming that the battery is defective. A slipping drive belt can not rotate a generator or alternator satisfactorily under load. A squeal during starting of the engine or while suddenly accelerating the vehicle indicates a loose belt. However, excessive slippage is not always announced by a squeal. The voltage regulator controls the output of a generator or alternator, but it cannot detect if the increased battery voltage is due to a fully charged battery or a partly charged battery and increased resistance in the wiring and terminals of the charging circuit.

Lights

The operator should check all lights and warning devices daily to be certain that they are all operating at peak efficiency. Sometimes lamps gradually dim because of gradually increasing resistance in the wiring circuit due to loose connections and corrosion. Compare the brightness of the lights; if one appears to be dimmer than the others, the trouble is usually a loose or corroded bulb or ground connection. Clean the surfaces with fine sandpaper and reassemble; tighten all connections thoroughly.

Check for loose connections if the lamp flickers when it is tapped or when the vehicle is moving. If the lights are dim when the engine is stopped or idling, but they become brighter when the engine is speeded up, the cause may be a low battery charge or corroded battery terminals.

If an unusually large number of lamps burn out, a voltage regulator which is set too high is usually the cause. This excessively high voltage output will also cause the battery to require an extraordinarily large amount of water to be added to the cells when they are checked.

Batteries

Modern fire apparatus are equipped with alternators which produce a fairly high charging rate. However, the newer equipment is also equipped with a large number of radios, lights, and other power-consuming devices that place a greater demand on the battery and generating circuit. The engine seldom runs long enough at a sufficient speed to replace the current being used during normal fire department operations. At fires, give the battery every possible opportunity to maintain its charge. Turn off the radio if it is not needed for fire ground operations; when lights are not required, turn them off; hoard electrical power so an ample supply will be stored in the battery when needed. One of the most detrimental operations for an engine is to allow it to run at an idle speed with no load. Shut down the engine when it is not being utilized. If it is necessary to allow the engine to be left running in cold weather to prevent freezing, operate it at a sufficiently high speed so the ammeter will indicate that the battery is being charged.

The *Standard for Automotive Fire Apparatus,* NFPA 1901, recommends that the capacity of a battery should be commensurate with the size of the engine and the anticipated electrical load. The capacity should not be less than 120 A-hr rating at a 20-hr discharge rate for gasoline engines and 200 A-hr rating at a 20-hr discharge rate for diesel engines using a 12-V starting system. When a dual battery system is supplied, each battery should be of the capacity required for a single battery system. One or more polarized receptacles should be provided for charging all batteries when in quarters.

Dual batteries are recommended, with suitable switching arrangements, by which either battery may be used for ignition and lights alone, for starting alone, or for starting, ignition, and lighting. The wiring and switches should be arranged so that both batteries may be used simultaneously in starting, neither the lighting wiring nor the ignition wiring shall be connected to both batteries simultaneously except when both starter switches are operated, and the generator or alternator shall be connected to supply the battery which is being used for ignition and lights. Most departments alternate batteries weekly.

To provide maximum service and efficiency, a battery must be kept in a charged and clean condition. The battery case should be checked to see that no cracks or leaks develop and that vents are not blocked. The battery carrier and hold-down should be secure and free from corrosion. Clean the top surface of the battery with ammonia water or a baking soda solution to neutralize the acid, rinse with clean water, and wipe dry; accumulations of dirt and battery acid form a conductor between the terminal posts, causing a constant drain on the battery charge. The terminal posts and cables must be bright, free of corrosion, and tight enough to make a good clean and firm contact. To retard any buildup of corrosive deposits, coat the terminal connections with a thin layer of vaseline or grease after the connections are tightened.

Lead-acid storage batteries are manufactured by alternating positive plates of lead peroxide with negative plates of sponge lead; the positive plates and the negative plates are separately connected to each other, with a nonconducting separator between each plate to prevent internal short circuits. The plate assembly is then placed in a cell and covered with a dilute sulfuric acid electrolyte. Depending upon the manufacturer, the specific gravity of the electrolyte in a fully charged battery will range between 1.260 and 1.280 when the temperature of the electrolyte is 80°F. A battery cell made from the above materials will generate an electrical pressure of 2.2 V, regardless of the size of the plates or the number of plates in the cell; a 12-V battery will require 6 cells. The capacity of a battery is rated in ampere-hours; it is determined by three components: weight of positive material or number and size of positive plates, weight of negative material or number and size of negative plates, and weight of sulfuric acid or percentage of sulfuric acid in the electrolyte.

As a battery is used, it becomes discharged. Both the positive and the negative plates gradually change to lead sulphate when the lead peroxide and the sponge lead chemically react with the sulfuric acid in the electrolyte; the sulfuric acid concentration slowly becomes depleted and the specific gravity of the electrolyte drops. The condition of a battery should be checked by measuring the specific gravity of the electrolyte with a hydrometer. If the fully charged battery originally had a specific gravity of 1.280, a measurement of 1.250 would indicate a charge of about 75%; a reading of 1.225 would show that the battery is only 50% charged. Electrolyte is drawn up into a glass tube and a weighted float will indicate the strength of the solution. If the liquid is either hot or cold, the specific gravity will require compensation as most batteries are rated with an electrolyte temperature of 80°F. When testing the electrolyte, avoid spilling any of the solution as the sulfuric acid will damage clothing, the paint on a vehicle, or sensitive body areas such as the eyes.

Sufficient distilled water should be added after checking the cells to bring the level of the electrolyte to the full mark; if there is no split-ring indicator, the liquid level should be ⅜ in. to ½ in. above the plates. Do not overfill; expansion of the electrolyte could cause it to be discharged through the cap vents, thus lowering the sulfuric acid concentration and corroding the terminals and battery case. Failure to maintain a sufficient level of water in the battery can cause permanent damage, by loss of capacity, as the exposed plate areas will become hard and dense; the plates will lose their conductivity.

When a battery becomes discharged, the sulfuric acid concentration is lowered; in extreme cases, the electrolyte may be close to the specific gravity of distilled water. A discharged battery may freeze in extremely cold weather. If a charger or booster cables are to be attached to a battery that may be frozen, do not attempt a boost until all of the cells have been checked. If ice can be detected, or if the electrolyte cannot be seen, do not attempt to charge the battery until it has been thawed out. Even a fully charged battery will not put out its rated voltage or amperage at low temperatures. Remember, a storage battery is a chemical device; the speed of a chemical reaction slows down at low temperatures and speeds up when the temperature rises. A battery may be fully satisfactory during warm weather, but may fail when the temperature drops.

A battery should not be allowed to remain discharged any longer than absolutely necessary. The chemical reaction by which the sponge lead and lead peroxide on the plates gradually change to lead sulphate as they generate electricity can be reversed by recharging the battery with direct current of the same voltage. The lead sulphate reverts back to the original plate chemicals and the sulfuric acid concentration in the electrolyte becomes greater. If a battery is kept active and fully charged, it will probably operate satisfactorily far beyond its warranty period. When a battery is allowed to remain discharged for a lengthy time, the lead sulphate on the plates becomes stabilized and it is increasingly more difficult to revert it back to the original active chemicals. Thus, the active area on the plates becomes reduced; this results in a lowering of the battery capacity.

Batteries should be constantly monitored when they are being charged in the station because there is no voltage regulator, as there is on a vehicle, to prevent them from being overcharged. The maximum charge rate, unless circumstances demand a quick-charge, should be 7% of the rated ampere-hour capacity of the battery. A 200 A-hr battery would be limited to a maximum charge rate of 14 A. Many departments have a regulation that the specific gravity of a battery should be taken at intervals of not over one hour. When the rated specific gravity of the battery is attained, the charger should be promptly turned off and the charging cable detached. More fire department batteries are ruined by overcharging than by any other single cause. Many years ago batteries were made so a specific gravity of 1.300 indicated when they were fully charged; occasionally, either obsolete manuals or inadequately trained operators still believe that a battery is not fully charged until a specific gravity of 1.300 is attained. Manufacturers lowered the sulfuric acid concentration in the electrolyte of modern batteries to achieve a longer life. It is impossible to recharge a battery to a higher specific gravity than it had when new.

During charging, the water (H_2O) in the electrolyte is changed into its component elements, hydrogen and oxygen. These gases are released through the battery cap vent holes. Hydrogen gas has a wide explosive range; an accumulation of the released hydrogen and oxygen gases in the atmosphere could generate an explosive force sufficient to kill or injure anyone in the close vicinity. If the conditions are just right, and a source of ignition such as a spark or a glowing cigarette is present, the battery can be blown apart like a bomb. The acid and debris will be thrown in all directions. To prevent an explosion during charging, sparks must be kept from getting into the battery or near its top. The hazards are similar to working around gasoline; with a source of ignition in proximity when the vapor-air concentration is just right, an explosion is inevitable.

Fire apparatus are usually provided with polarized receptacles so the batteries may be recharged with the least inconvenience. It may not be necessary to open the battery compartment to attach the charger cables. Caution will have to be maintained if the batteries are checked while they are being charged when the charger clips are attached directly to the battery terminals or if a booster charge is used to start a stalled vehicle. Battery compartments should be well ventilated to prevent an explosive accumulation of gases.

If the charging clips are to be attached to a battery, connect the positive clamp first and then attach the negative clip to a good ground at least a foot away from the battery; attach the negative wire to the engine for best results. The charger should not be turned on until all connections are made.

If booster cables and another vehicle are to be used to start an apparatus with a dead battery, the exactly correct procedure must be adhered to if expensive damage to the charging circuit components or an exploded battery is to be prevented. In this case, two batteries can double the risk and the charger cannot be switched off. Be certain that the two vehicles are not touching each other. Connect the booster cables in the following sequence. First, take the red positive clamp and attach it to the positive post of the booster battery; be careful that the other end of the cable does not touch the vehicle. Place the other red clip on the positive post of the dead battery. Take the negative clamp of the other cable and attach it to the negative post of the booster battery; the remaining clamp is attached to a good ground on the engine of the troubled vehicle. The ground connection should not be attached close to the battery; if a spark is generated, it should be far enough away from the battery that it will not ignite any flammable gases. If the clamps are incorrectly attached so the polarity of the batteries is reversed, which would occur if the positive terminal of one battery were connected to the negative terminal of the other, and vice versa, current would be forced through the voltage regulator, alternator, and other components in a reverse direction. Expensive damage would instantaneously and inevitably occur.

Cooling System

During the last 20 to 30 years the heat load on the engine cooling system has more than doubled, though the physical size of the radiator, which provides the principal means of heat transfer to the atmosphere, has only increased about 10%. The efficiency of the cooling system must be maintained continuously if proper engine operation and anticipated service life are to be realized.

Through burning, an internal combustion engine releases the energy stored in fuel; a portion of this energy is converted to mechanical force for turning the crankshaft. The cooling system is just as important to an engine as the lubricating system or the fuel system. The cooling system is very often neglected until some serious defect appears. It is important to keep the cooling system in perfect condition because neglect could result in expensive repair bills and complete failure will ruin an engine within a few minutes.

A 300-horsepower diesel engine operating under a 70% load burns enough fuel to keep 25 five-room houses comfortably warm in weather of ten degrees below 0°F. This heat must leave the combustion areas by three routes: approximately one-third is converted to useful horsepower; one-third is removed with the exhaust gases; and one-third is dissipated by lubricating oil, air flowing over the engine's exterior, and the cooling system. In addition, most automatic transmissions also use the coolant in the radiator to cool the transmission fluid.

During the time fuel is burning in the combustion chamber, flame temperatures are approaching 3000 to 3500° F. The cylinder head, valves, pistons, and cylinder walls are exposed to this heat, though only for a short time period, but the average temperature reached by these components exposed to the heat of combustion is relatively high. Exhaust valves will reach 1200° F, piston crown temperature will attain 500° F, and the cylinder wall temperature (upper half) will reach 250° F on the water side. The coolant flow in these areas must be in relatively high volume and at high velocity to swiftly carry away the heat. Any condition adversely affecting the rate of heat transfer to the coolant, or to slow or to stop coolant circulation, can be critical.

Sludge can restrict the rate of coolant flow in the system. With reduced circulation, local boiling of the coolant can occur. This will raise the temperature of metal in the high heat areas in cylinder walls and cylinder heads, further increasing coolant temperature and reducing or destroying the lubricating oil film in the cylinder. Any restriction in the coolant flow reduces the rate of heat dissipation by the radiator, thus keeping coolant temperatures high, cutting power output, reducing lubricating oil viscosity, and increasing the rate of oil oxidation to create varnish and sludge.

The purpose of the cooling system is to control engine operating temperatures within specified limits. If the temperature is allowed to get too high, engine oil will break down and metal-to-metal contact will exist. Also, parts such as pistons and rings can expand to the point where scuffing and scoring occurs.

On the other side of the fence, if an engine is allowed to operate at too low a temperature, acids and sludge can form in the oil which will cause corrosion damage. Engine efficiency will be greatly reduced and fuel wasted.

The optimum operating temperature is just below the point at which damage would begin to occur to internal engine parts. The manual for the particular engine should be consulted to ascertain the recommended maximum operating temperature.

Water has long been the basic fluid used in engine cooling systems, but water by itself will no longer satisfy cooling system requirements. The basic requirements for a coolant in any internal combustion engine are to:

1. Provide an adequate medium for transfer of heat from the combustion area and cylinders to the atmosphere.
2. Prevent corrosion in the cooling system.
3. Prevent the formation of scale and sludge.
4. Control the pH condition.
5. Be compatible with all metals, hose, and other materials that it could contact.
6. Provide adequate protection against freezing when required by climatic conditions.

The coolant most often recommended by manufacturers today is a 50% ethylene-glycol/water solution the year around. This solution not only provides freeze protection to approximately –20° F, but it contains rust and corrosion inhibitors that protect the system from efficiency robbing deposits.

Most modern cooling systems have a special radiator cap that permits the development of pressure within the cooling system. This pressurization further increases the boiling point of the coolant and, consequently, the heat-transfer capabilities of the cooling system. There is an approximate 2.5°F increase in the boiling point of the coolant for each additional pound of pressure. At sea level, water boils at 212°F; with a 15-psi cap, the boiling point will be raised to about 250°F. A 50% ethylene-glycol/water solution at sea level boils at 227° F and at 265°F under 15 psi pressure. The additional 53° possible temperature differential between the ambient air and the coolant greatly increases the heat-transfer capability of the radiator and, thus, the efficiency of the cooling system.

To control engine operating temperatures, the fluid in the cooling system must circulate constantly and the walls surrounding the coolant must be good heat conductors. Operators have reduced the effectiveness of the cooling system, and caused a consequent heating problem, by painting a radiator with the wrong type of paint or by applying an excessive thickness of nonheat-conducting enamel to the engine block, heads, and oil pan. Anything that interferes with heat conduction, water seal, or temperature control will reduce the efficiency of the cooling system and will increase engine operating defects and inefficiency.

It is essential that these systems be kept clean and leakfree, that filler caps, thermostats, and pressure relief mechanisms be correctly installed and maintained at all times, and the coolant be kept at the proper level. *Warning:* Use extreme care when removing the pressure cap from a radiator. The sudden release of pressure from a heated cooling system can result in a violent eruption of coolant from the radiator fill opening. If this boiling liquid is splashed on a person, it could cause serious scalding injuries.

Most vehicles depend upon air flowing through the radiator to maintain the engine at its most efficient temperature. While the fan does circulate some air when the vehicle is standing still, a large amount of the cooling is derived from the forward motion of the truck. Pumpers must have an auxiliary method of cooling the engine when the apparatus is pumping at a fire. Modern pumpers incorporate a heat exchanger to prevent overheating while stationary. This system diverts water from the discharge side of the fire pump to circulate through the heat exchanger; the water than returns to the suction side of the pump. Thus, the cooling water does not dilute nor contaminate the engine coolant.

Operators should periodically observe the engine temperature gage and adjust the auxiliary cooling valve to maintain the engine at its most efficient operating temperature. If the auxiliary cooling was used while pumping salty or dirty water, the heat exchanger and the connecting lines should be flushed out.

Some of the older apparatus utilize a radiator fill valve that diverts water from the discharge side of the fire pump to flow water directly into the engine cooling system. Do not use this method unless absolutely necessary. If used, first remove the radiator fill cap so that the radiator and hoses will not be subjected to possible bursting from overpressure. Just barely open the radiator fill valve to flow a small stream because it is possible to crack hot metal by cooling it suddenly. If the radiator fill valve is used while pumping salty or dirty water, the entire engine cooling system should be thoroughly flushed and refilled with clean coolant.

Check the radiator, engine, and hose connections for evidence of leaks. Inspect the hoses to be sure that they are not excessively soft, as deteriorated hoses may partially collapse when the engine is operating and thus restrict the coolant flow.

When antifreeze solutions are maintained in the cooling system, periodically make certain that the protection has not been lessened by a loss of fluid. Cooling systems that have a conditioner installed should be maintained according to the manufacturer's instructions.

Drive belts should be checked for tightness; about ½-in. play is correct. If a belt tension gage is available, adjust the tension to the manufacturer's instructions. Belts should be clean and free from oil or grease, with no cuts or visible wear. Dual or triple belts should all be replaced at the same time in matched sets. Check the temperature gage to make sure that it is operating. Particular attention should be paid to the radiators on cab-ahead-type apparatus to make certain that they do not become blocked by leaves, insects, paper, etc., which prevent proper air circulation through the core.

Crankcase Ventilation

The function of the crankcase ventilation system is to remove the blow-by gases that pass the piston rings, condensation vapors, and fumes from the crankcase. When permitted to remain, these contaminants produce sludge and varnish in the oil, which interfere with proper engine operation, permit rust formation, and develop highly corrosive acids that etch polished engine surfaces.

Oil Fill Cap

In some engines, the oil fill cap also serves as the breather unit for the crankcase ventilation system. It generally contains a wire mesh filter that serves to cleanse the air entering the system. Since the proper functioning of the crankcase ventilation system is dependent on the correct amount of clean air entering the crankcase, it is essential that the oil fill cap be serviced at the interval specified on the lubrication chart.

Conventional Ventilation System

The conventional ventilation system utilizes the forward motion of the vehicle to develop a low-pressure area at the lower end of the road draft tube from which crankcase fumes are exhausted. This action permits air to enter the oil fill cap and to pass down into the crankcase; there it picks up crankcase gases and fumes which are then carried into the atmosphere through the draft tube. This system has two shortcomings. First, its ability to function is governed by the forward motion of the vehicle. Therefore, at low road speeds, or when the vehicle is stopped, the system is quite ineffective. Second, crankcase vapors that contain unburned hydrocarbons are released into the atmosphere and add to the air pollution. Since the advent of air pollution standards, this type of system is not being used.

Positive Crankcase Ventilation System

The positive crankcase ventilation system (PCV) uses the vacuum created in the engine intake manifold to draw clean air through the crankcase and into the combustion chambers. The air stream draws with it the fumes and vapors from the crankcase and directs them into the engine cylinders where they are burned. By using the positive action of engine vacuum and by burning the crankcase hydrocarbons, the positive crankcase ventilation system overcomes the two basic faults of the conventional ventilation system. Most PCV systems employ a check or air control valve which limits the amount of air flowing through the system according to the ventilation requirements of the engine at varying loads and speeds. In time, depending on the mileage driven and the mechanical condition of the engine, the sludge and carbon accumulation on the valve causes it to stick or to become completely plugged. When the valve becomes inoperative, fumes remain in the crankcase; this could cause dangerous or premature wear of engine parts and serious engine damage due to the development of sludge, varnish, rust, corrosion, and acids. Further, normal engine operation is interfered with since a stuck valve can upset the fuel-air ratio. The PCV valve also performs the important function of preventing a crankcase explosion by automatically closing in the event of an engine backfire.

Servicing the PCV check valve is quite simple. In some systems the valve is a crimped assembly that is washed in solvent and then blown dry with compressed air. To determine if it has been properly cleaned by soaking, shake it and listen for the clicking sound of the valve free in the housing. If the valve has not been freed up, replace the unit. Other PCV systems have valves that can be disassembled. Some valves should not be serviced, but merely replaced. Refer to the manufacturer's service instructions.

The normal wear of piston rings and cylinder walls results in increased blow-by. This condition in turn places a greater burden on the ventilating system. Therefore, as engine wear increases, more frequent servicing of the crankcase ventilation system is advisable.

Crankcase Oil

Crankcase lubricating oil becomes contaminated by unburned fuel, water, soot, and similar combustion by-products. In time, contamination will reduce the lubricating quality of the engine oil to a point where severe damage to the engine may result. Because of the number of cold starts, short runs, and frequent excessive choking of gasoline engines, the contamination of the oil and the condensation of water vapor causes a higher rate of dilution in the fire service than in other use. Oil change intervals are planned to avoid this needless engine wear. Preferably, engine oil should be changed as soon as possible after the engine is stopped and is still warm so that any foreign materials in the oil will be drained out with it. Check the oil condition and level on all inspection operations; this is the most important item on this schedule and should be done regardless of mileage since the last change.

The oil pressure gage should be observed immediately after starting the engine and periodically during the entire time of operation. The reading on the oil gage does not indicate the quantity of oil in the engine; it only shows that there is some oil in the crankcase and that it is being pumped under some pressure. A large number of vehicles have indicator lights instead of gages to show that at least a minimum amount of pressure exists. A shortage of oil would be indicated by an inoperative or a flashing indicator light, or by the oil gage needle fluctuating spasmodically or registering zero. When the oil pressure indicator fluctuates or is inoperative, the engine should be stopped immediately and the cause determined and remedied. Every apparatus operator should be thoroughly familiar with the normal pressure gage readings for the particular vehicle.

Check the crankcase oil level frequently. Do not measure it while the engine is running as the reading will be false. When checking, wipe the dipstick clean before taking a reading; reinsert the dipstick fully and then remove it to determine the oil level. The check should be taken before the engine has been started, or else a few minutes after is has stopped, allowing time for the oil to drain down into the oil pan.

If no oil has been added and the dipstick shows a rise in the oil level from the previous reading, it can indicate trouble. Check the dipstick carefully for indications of fuel or water in the crankcase. Fuel in the crankcase oil possibly indicates a ruptured diaphragm in the mechanical fuel pump. A leaking head gasket, cracked block or head, or other engine defect may allow coolant to leak into the oil pan; this would also be indicated if the coolant level keeps dropping without any signs of an external leak.

Engine lubricating oils perform four functions:

1. Prevent metal-to-metal contact of moving parts.

2. Assist in carrying heat from the bearings and other heat producing components of the engine.

3. Clean the engine parts and hold sludge and other contaminates in suspension until they are filtered out.

4. Form a seal between the piston rings and the cylinder walls to prevent blow-by of the combustion gases.

The piston rings must ride on a film of oil to minimize wear and prevent cylinder seizure. At normal rates of consumption, oil reaches a high-temperature zone at the upper end of the piston where rapid oxidation and carbonization can occur. In addition, as oil circulates through the engine, it is continuously contaminated by soot, acids, and water originating from combustion. Until the oil detergent and dispersant additives are exhausted, they aid in keeping sludge and varnish from depositing on engine parts.

Oil that is carried up the cylinder wall into the combustion chamber is normally consumed during engine operation. Excessive oil consumption will leave carbonaceous deposits in the combustion chamber that could foul spark plugs, cause valve sticking, or result in detonation or preignition.

Oil does not wear out. The purpose of changing oil and oil filters is to protect the engine from oil that has deteriorated because the detergent and dispersant additives have been used up, decomposition products have formed, and the oil contains contaminants. There is a definite limit as to the amount of lubrication and protection any lubricating oil can provide to an engine, regardless of the quality of the oil, and additives not only have a capability limit too, but they are consumed during engine operation and can only be replaced safely by draining and replacing with fresh oil. The condition of the oil does not necessarily relate to the miles or hours of engine operation, or to the length of time the lubricant has been used; nor is it unduly influenced by the age or make of engine in which it is being used. The dominant factors that control the condition of the oil in engines are operating temperatures, engine adjustments, crankcase breathers, air intakes oil filters, and the amount of stop-and-go driving.

The reduction of friction and wear by maintaining an oil film between moving parts is the primary requisite for any lubricant, and this is especially true of crankcase oil. The film thickness and its ability to prevent metal-to-metal contact of moving parts is related to oil viscosity, which is a measurement of the oil's resistance to flow. The better oils will maintain this viscosity over a wide temperature range; lower-grade oils are usually viscous at low temperatures and excessively thin when heated. Temperature is the most important factor in determining the rate at which deterioration or oxidation of the lubricating oil will occur. The oil should have adequate thermal stability at elevated temperatures, thereby precluding formation of harmful carbonaceous and ash deposits.

Oil Change

The subject of oil changes and the proper time for changing the filter can be simply resolved by a laboratory analysis of a crankcase oil sample drawn from the engine. There are laboratories that specialize in this service, and it is generally recognized that they save the departments more money than they cost because they allow the lubricating oil and filters to be utilized for the maximum effective time period; good oil is not thrown away just because a time period has elapsed. Through experience a laboratory can tell what oil conditions are normal for different types and makes of engines. They can advise whether the oil and filter should be changed, or if they may be used longer. However, the most important benefit derived from periodic analysis of the engine's crankcase oil by a competent laboratory is discovering mechanical and/or operational defects that require correcting. The viscosity and fuel dilution of the oil is checked. The quantity and types of metal particles and other foreign matter is examined. The results of an analysis can be the basis for immediately requiring a complete overhaul of an engine, or it could determine that there is no basis for fears about the condition of an engine as there are no serious defects.

Oil changes should be made according to department policy. As a rule, replacements are made on a calendar basis, instead of operating hours or miles driven, because fire fighting apparatus is subjected to far more rigorous conditions than any other type of vehicle. The engines are not allowed to thoroughly warm up before they are subjected to high-speed operation and expected to deliver their maximum horsepower. In addition, pumpers, aerial ladders, and other apparatus often operate at fires and drills where the mileage on the odometer would not offer any clear indication of the engine's performance. Most departments change the engine lubricating oil on active apparatus every two or three months and after every lengthy pumping operation. Relief apparatus, which is operated infrequently, should receive an oil change at least every six months.

Oil Filters

Engines are provided with a replaceable element oil filter to remove the solid contaminants. Normally, oil filters should be changed on alternate oil changes or annually, depending on their size and capacity. Accumulations of sludge in the filter housing should be removed before replacing the filter. Be certain of a complete and tight seal on the filter housing after replacing the filter; wipe it clean and closely observe for leaks immediately after starting the engine. If heavy sludge is found on the filter at the time of change, it indicates that the detergent additive in the oil was exhausted and the oil change interval should be shortened.

If the oil feels gritty or contains metallic particles, the filter is plugged and should be changed more often. Also, metallic particles indicate rapid wear that will lead to premature major overhaul of the engine.

Most diesel engines with turbochargers also have a filter that must be changed at each oil change; this is very important. The turbocharger is lubricated under pressure as part of the engine lubricating system. Turbochargers rotate at speeds to 48,000 rpm at an engine speed of about 2100 rpm; this high speed shows that correct lubrication is essential. When changing the replaceable element filter, be sure the shell is filled with clean oil before the filter and shell are reassembled.

Air Filters

All internal combustion engines require plenty of clean air of the correct temperature. Dirt is an engine's worse enemy; it has been proved many times that the best way to obtain extra service from an engine is to keep dirt out of it. Even though it is usually invisible, atmospheric air has a large amount of finely divided dust in suspension, which is extremely abrasive when drawn into the engine. No engine has ever been designed that will give satisfactory service if it is allowed to breathe dirty, unfiltered air.

A filter or cleaner is used to remove dirt and dust from the air the engine breathes. As dirty air is the main cause of wear on pistons, rings, cylinder liners, valves, and bearings, it is important to check and service the air cleaner regularly. A small hole in the tubing or hose between the cleaner and the engine can admit a surprising quantity of abrasive material. The service instructions given on manufacturers' charts are for normal conditions. If operating under dusty or off-road conditions, service two or three times as often as regularly recommended.

With this fine filtering, the pores or openings do become clogged, and as the degree of clogging increases, there is a decrease in the amount of air entering the engine. This restriction changes the fuel to air ratio and results in an enriched fuel mixture. This not only produces objectionable black smoking of the exhaust, but causes fouling of spark plugs in a gasoline engine and fouled injector nozzles in a diesel engine. The excess fuel is not all burned and it dilutes the engine oil, cutting lubrication on the cylinder walls, causing sticking of the piston rings and inlet valves, and accelerating bearing wear. Fuel consumption increases and what is worse, engine power decreases because the engine is starving for air.

Avoid overtightening the mounting clamp when replacing the air filters on carburetors; tightening too much may distort the shape of the carburetor air horn, causing the butterfly choke valve to stick.

Dry Type

A dry-type air cleaner is a positive filtration type consisting of treated paper as a medium that air passes through to reach the engine. This type increases in efficiency as the dirt load builds up a cake or bed in the filtering medium; the efficiency increases upwards from a minimum of 99.5% and is constant at all engine speeds. Remove the air cleaner cover and element. Gently shake and tap the paper-pleated element to remove loose dust and dirt. Some instructions recommend cleaning the element by directing a flow of compressed air outwards from the center of the filter; other instructions advise against the use of compressed air as it may perforate the element; comply with the manufacturers' recommendations. Wipe out the housing.

Replace the element if it becomes oily, cannot readily be cleaned, has holes through the paper, or if the rubber sealing surfaces are distorted. The price of a new filter is trivial when compared to the potential damage that could occur.

Composite Dry Type

Composite dry-type filters utilize centrifugal force and a paper filter for efficient two-stage cleaning. Deflector vanes or tubes create a cyclonic twist to the entering air. Dirt particles are separated by this action through centrifugal force and are thrown outwards where they drop into a cup or bin. The cleaned air then passes through a paper filter where the remaining dust is removed. Empty the collecting traps and clean the paper element as instructed for the dry types of filters.

Oil Bath Type

When servicing an oil bath filter, remove the air cleaner from the carburetor or air manifold, using care to avoid spilling oil. Remove the element. Wash the filter element in kerosene or solvent, shake to remove the surplus solvent, allow the element to dry, but do not use compressed air as the pressure may blow holes in the mesh. Wipe the housing clean and remove old oil. Refill the reservoir to the level mark with fresh engine oil; do not ever utilize old crankcase drainings. Reassemble and replace on the engine.

Polyurethane, Oil-Wetted Type

For the polyurethane oil-wetted filter, remove the element from the air cleaner housing, carefully remove it from its mesh support and wash the element in solvent. Squeeze out surplus solvent but do not wring out as the element may tear. Dip in engine oil, squeeze to distribute the oil, and replace it in the cleaner.

Wire Gauze, Oil-Wetted Type

To service, remove the wire gauze oil-wetted air cleaner from the engine, remove the element, and wash it in solvent. Shake the element to remove surplus solvent, allow it to dry, but do not use compressed air to dry it. Reoil with engine oil, drain, and replace on the engine.

Power Brake System Filters

Most air and vacuum power brake systems have air filters of some type; they may be small and hard to find or service.

Hydrovacs

Clean the Hydrovac vacuum cylinder and remove the pipe plug from the air inlet end of the cylinder. Add oil to the level of the fill hole as specified on the lubrication chart and replace the pipe plug. Inspect, clean with solvent, and reoil the Hydrovac air cleaner.

Air Brake Filters

Remove the air filter on the compressor, clean it in solvent, and replace.

Brakes

All drivers should have a knowledge and feel of the vehicle's steering, brakes, clutch, and all other operating components. Each time the apparatus is moved, the brakes should be tested before they are needed. Try the brakes by making a test stop; the distance traveled may be as little as 10 ft. By the action and feel of the stop, an evaluation of the condition of the brakes can be made if the driver is familiar with the vehicle.

Hydraulic Brakes

With hydraulic brakes, check the travel of the brake pedal; if excessive travel is evident before firm resistance is met, it indicates the need for a brake adjustment. A spongy brake, as contrasted with a firm resistance when brakes are applied, indicates the presence of air in the brake lines; bleeding the brakes should correct this problem. Test to determine if fluid is by-passing the cup in the master cylinder by applying the brake with about 50 psi pressure on the pedal. If the pedal moves to the floor, or if there is a notable increase of travel toward the floor as compared to normal braking travel, the master cylinder is defective and should be repaired. All vehicles equipped with hydraulic brakes have a reservoir for holding brake fluid; this should be checked to make sure that the level is maintained ½ in. below the fill opening. If the vehicle is equipped with disk brakes, be certain that the fluid used is a heavy-duty type with a boiling point of over 450° F, as disk brakes generate a greater amount of heat more rapidly than other types. If the fluid boils, gas bubbles will be formed; braking power will either disappear or be greatly diminished.

Air Brakes

With air brakes, if compressor is self-lubricated, the oil level should be checked at the same time the oil level in the engine is checked, and replenished if necessary. The oil should also be changed when the engine oil is changed. Other compressors are attached to the engine lubricating system and the oil does not require any separate maintenance.

Should it be necessary to drain the engine cooling system to prevent damage from freezing, water-cooled compressors must also be drained.

There should be at least 60 psi air pressure in the air brake system before the vehicle is moved. Periodically observe the dash gage while driving; if the air pressure drops to a low point or if the warning buzzer or light signifies low air pressure, stop the vehicle and correct the problem.

Pressure build-up test. Run the engine at fast idle. The time requred to raise the air pressure from 50 to 90 psi will vary with different vehicles, but should not exceed five minutes.

Leakage tests. Run the engine until the governor cuts out. Wait until air pressure stabilizes. With brakes released, dash gage pressure drop should not exceed 2 psi per minute for single vehicles and 3 psi per minute for combination vehicles. Make a full brake application and allow pressure to stabilize. With the brakes fully applied, pressure drop should not exceed 3 psi per minute for single vehicles or 4 psi per minute for combinations. If the air pressure leakage is excessive, check the system connections and applicable units with a soapy water solution; a leak will blow soap bubbles. Most departments try to adhere to a more strict avoidance of air leakage because an apparatus cannot stay in the station for 5 or 10 minutes after they have been dispatched to a fire while they wait for the compressor to build up the air pressure to the minimum safe driving pressure. A good practical standard is to avoid any leakage that would cause the warning buzzer or light to signify low air pressure when the vehicle has been driven within the last 24 hours.

Reservoirs store compressed air until needed for brake operation and to provide sufficient air for several brake applications, even after the engine stops. Bleed the air storage tanks weekly; daily during freezing weather. The following procedure should be used:

1. Run the engine to build up air pressure in the storage tanks.
2. Shut off the engine.
3. Crack open the petcock at the bottom of the reservoir; hold the palm of one hand in the path of the air discharge. When the air gives no more indication of moisture, shut off the petcock completely.
4. After bleeding, start the engine and build up the air pressure to the maximum. Check the petcocks with soapy water to detect any air leakage.

Determine the need for a possible brake adjustment by measuring the travel of the brake chamber push rod, as indicated by the manufacturer's recommendations, every time the vehicle is lubricated. This check is especially important with air brakes as the need for a brake adjustment cannot be detected by the pedal operation.

Parking Brakes

Parking brakes, which should be used only when the apparatus is not in motion, need to be checked occasionally to make sure they will hold the vehicle on a hill. On some types of parking brakes it is possible for the driver to make adjustments while in the cab. These types are tightened by turning the knurled adjustment knob, located on top of the handle, in a clockwise direction. The pressure of the brake lining on the drum is directly related to the degree of effort required to pull the brake handle over the center; this should be a definite effort. When parking on a steep hill, take an extra turn on the adjusting knob and then pull the handle up fully. Remember to take off this extra turn when ready to move the vehicle again or the partially engaged brake will cause drag and unnecessary wear on the brake linings.

Fuel Pumps

When the apparatus is equipped with both mechanical and electrical fuel pumps, the electrical pump should be operated at least once a week to check its condition. Bypass valves on the mechanical pump, if so equipped, should be closed down to allow the electrical pump to operate independently. This practice reduces the possibility of overlooking a defective pump.

Vehicles equipped with switches for operating the electrical pumps are not always wired so that the ignition switch will control the pump. Therefore, always make sure that pump switches are turned on and off when starting and stopping the engine.

Fuel Filters

As engines became more efficient and powerful, the clearances in the engine decreased. To keep engines operating properly and to obtain maximum service, more efficient filters had to be developed. It is easy to overlook the fuel line filter because it is usually relatively small, sometimes hidden from view, and does not give frequent trouble. But when from neglect or oversight the filter becomes restrictive to fuel flow, trouble usually happens suddenly with little or no advance warning. Generally the engine suddenly lacks power, runs erratically, balks, and perhaps backfires or spits just when it is needed most. Good maintenance requires that all fuel line filters be cleaned or replaced, depending upon the size and type, at least as often as the oil filters.

Some fuel filters can prevent water from passing through; moisture in the fuel, usually caused by condensation, causes a slime or jelly to form. If this is not filtered out, it will cause a gum to build up and affect the flow through the carburetor jets, fuel pumps, lines, and injector nozzles.

Gasoline filters are utilized in the fuel line between the pump and carburetor to prevent dirt, rust, and scale from clogging the carburetor jets and passages. A neglected fuel filter may restrict the flow and cause the engine to starve for fuel. This may happen suddenly with no advance warning just when power is needed most.

The number of diesel engines in the fire service is increasing and the maintenance of fuel filters is, if anything, much more important than with a gasoline engine. All diesel engine manuals stress the importance of this part of engine maintenance.

The operating clearances in a diesel engine fuel-injection pump and the openings in the nozzle are so small, and so critical to continued good operation, that fuel filtering must be carefully and regularly maintained. Dirt in diesel fuel oil, even microsized particles, cannot be tolerated. Injection pumps wear and score; nozzles plug and burn. To preserve the economy and reliability of a diesel engine, dirt, dust, rust, and grit must be filtered from the cleanest fuel supply.

Routine cleaning or replacement of these filters is one of the most important maintenance operations for diesel engines. Service the filters at the recommended intervals and whenever the fuel pressure gage indicates an abnormally high pressure.

Diesel fuel systems have a strainer between the fuel tank and the fuel pump. In some cases, it may be incorporated in the fuel pump housing as an integral part of the pump assembly. The second and most important filter in a diesel fuel system is between the primary fuel pump and the fuel injection pump or the fuel inlet manifold. A fuel-injection pump is on 4-cycle engines and an inlet manifold is on 2-cycle engines. Or, there may be two filters in series in the fuel line between the transfer pump and the injection pump. Filters of the metal-edge type are also provided in each injection nozzle housing or in the manifold for each nozzle.

There are thus two fuel line filters plus the filters for each nozzle. The maintenance manual for the particular diesel engine will give detailed instructions covering the maintenance of fuel filters and these should be faithfully observed. The maintenance periods are given as optional schedules, one on a mileage basis, the other on a time basis. For fire service, use the time schedule.

Ignition Systems

The performance of a gasoline engine ignition system is dependent on the efficient performance of the battery, wiring, ignition switch, and all components of the charging and ignition circuits of the vehicle's electrical system. Poor performance by any part could result in an inoperable engine. The starter could demand such a large percentage of the available current from a weak battery that the ignition system would be deprived of sufficient energy to provide enough voltage to the spark plugs to ignite the fuel-air mixture in the combustion chambers. The entire electrical system of the vehicle must be maintained to the highest degree if the engine is expected to operate efficiently and effectively.

Spark Plugs

Remove the plugs, clean, file the electrodes flat, and regap according to specifications. The ceramic insulators should be clean and dry. Any accumulation of dirt, grease, or moisture can cause the plug to ground itself instead of firing properly. Inspect the ceramic portion for any chips or cracks that might cause misfiring. Plugs should be checked for tightness so there is no compression leakage. All high-voltage wire connections should be clean and secure so as to provide conductivity.

Distributor

The distributor should be inspected to make sure that it is free from oil and dirt accumulations, and that all connections are clean and tight. Wires should be checked for fraying and cracking of their insulating covering; defective wiring can cause short circuits, loss of high-voltage spark through leakage, misfiring, rough engine performance, or difficult starting. The distributor cap should be checked for carbonization and pitting. Any of these could be contributing factors to poor engine performance.

The distributor should not be overlubricated. Wipe any excess lubricant from all parts. The distributor shaft lubrication will vary with the distributor design; an oil cup, grease cup, lubrication fitting, or an oil fill hole with a plug are some of the most common used. A wick inserted into the top of a hollow distributor shaft carries the oil to the automatic spark advance mechanism. Remove the cap, remove the rotor, and then apply one or two drops of light engine oil to the wick; replace the rotor and cap. Apply a trace of high-melting-point grease to the breaker arm cam to lubricate the cam faces and the breaker arm rubbing block.

Check the breaker points for burning and pitting.

Chassis Lubrication

A resistance to motion is always present; this influence exerts a definite drag on every form of movement known to science, and is called friction. The word friction is defined as that force which acts between two bodies at their surfaces of contact so as to resist their moving upon each other. The strength of this frictional force depends upon the pressure existing at their contacting surfaces and the character of the surfaces. The cause of friction in both solids and fluids is their tendency to stick together and resist being moved.

Friction is usually considered to fall in one of three categories:

1. Sliding friction: that resistance to motion developed when two surfaces slide over and rub against each other under pressure.

2. Rolling friction: encountered when a body with a curved surface rolls over the surface of another body.

3. Fluid friction: the resistance to flow offered by any fluid material, a physical phenomenon usually measured as viscosity.

All metals, regardless of how highly polished they may be, show rough surfaces under microscopic examination. When two pieces of metal are brought together, the contacting surfaces tend to grip one another, or it may be said that they attract one another, because of the toothlike surfaces. All fluids, regardless of their cohesive and adhesive qualities, tend to keep the toothed surfaces of such solids apart when placed between two solid surfaces such as metals. When two surfaces are kept apart by a fluid film, they are said to be lubricated.

Lubrication is the process of reducing or combating friction. One function of a lubricant is that of making bearing surfaces slippery so that the least possible amount of effort is required to overcome the resistance to motion caused by friction. In most cases, this is accomplished by interposing a layer of fluid material between the contacting surfaces so that fluid friction is substituted for sliding friction. This procedure is highly effective because fluid friction is only a small fraction of what the unlubricated sliding friction would be.

The functions of any lubricant are (1) to maintain an unbroken oil film between the moving surfaces, which prevents destructive wear on the surfaces of the bearings; (2) to prevent the excessive generation of heat that would otherwise be created by friction; and (3) to carry away the heat normally developed during bearing operation, in order that the bearings may be operated at the lowest possible temperature. In addition to the above functions, chassis lubrication performs another important service because the grease prevents water, dirt, and other abrasive materials from entering the bearing spaces. No lubricant ever entirely overcomes friction–it only lessens or minimizes it.

Chassis lubricants are compounded when relatively minor percentages of certain metallic soaps are mixed with a petroleum oil to form a material that has a semisolid, plastic, fibrous physical structure. Chemical additives are used to govern consistency, to improve resistance to heat, to improve resistance to oxidation, to give extreme pressure characteristics, and to improve rust preventative qualities.

For complete chassis lubrication, consult the applicable chart and directions furnished by the vehicle manufacturer. This guide will show the location of every lubrication point, the lubricant to be applied, and the interval at which the service should be performed.

When lubricating the front suspension points, a more efficient job can be done if the vehicle is raised with a jack or hoist so that the joints have no weight on them. This will allow grease to enter deep into the joints. Apply the lubricant intermittently, turning the wheels from side to side to distribute the grease.

Avoid any excessive greasing of the brake cams, the clutch release bearing, or the universal joints. On aerial ladders and aerial platforms, lubricate all points thoroughly; a grease fitting for lubricating the ball races is located on top of the turntable; rotate the turntable slowly while lubricating to properly distribute the grease.

Transmission and Differential

Check the fluid level in the transmission and differential every time the vehicle is lubricated. If fluid must be added, make certain that the correct lubricant for that specific component is used. The manufacturer's warranty could be voided, and expensive damage may occur, if the incorrect lubricant is not used.

Clutch Release Bearing

Some clutch throwout bearings are sealed for life, and others require periodic lubrication. Remove the lower plate or cover under the flywheel to expose the service point. A fitting or plug may be provided at the service point. Wipe the plug or fitting clean before removal or lubrication. Lubricate sparingly at low pressure, wipe off excess grease, and replace the plug if one was removed. Be careful not to apply excessive lubrication as it could get onto the frictional surfaces of the clutch plates. Replace the plate or cover.

Pilot Bearing

Check the lubrication chart carefully to determine if there is a pilot bearing in the rear center of the flywheel that requires periodic lubrication. On some engines, it is necesssary to remove the flywheel bottom cover, rotate the flywheel until the plug appears in the opening, replace the plug with a grease fitting, lubricate sparingly with low pressure, and replace the plug and cover.

Clutch Adjustment

Usually, poor driving practices are the number one cause of too frequent clutch adjustments. The driver, perhaps without realizing it, rides with one foot on the clutch pedal; other drivers accelerate brake lining wear by riding with one foot on the brake pedal. The driver is not intentionally putting any pressure on the pedal, but just the weight of a foot takes up all the free travel and any involuntary reflex, such as hitting bumps on a rough street, causes a small or partial release of the clutch pressure; this starts an increased rate of wear and heating which reduces the capacity of the clutch to transmit full torque. Also, this practice causes accelerated throwout bearing wear.

The clutch free travel should be checked on the routine inspections. The proper method of clutch adjustment is shown in Figure 17-8. When the travel is down to ½ in. on most makes of clutches, readjust the clearance to 1½ in. of free travel. Consult the manufacturer's instructions; if the clearance is different from above, comply with the service manual. This free travel check should be made by moving the clutch pedal pad by finger application, not foot, as the contact point is sensitive.

Some unsynchronized transmissions have a clutch brake to stop the pilot shaft from rotating when placing the transmission in first or reverse gear when the vehicle is stopped. When driving a vehicle with this type of transmission, depress the clutch enough to actuate the clutch brake only when the vehicle is stationary. Stopping the gears being rotated by the pilot shaft while the vehicle is in motion will make it difficult to mesh them with other gears being driven by the drive shaft.

Rules of Good Driving

The driver can help to reduce maintenance work by observing a few on-the-road rules of good driving:

1. Do not ride the clutch. Keep your left foot on the floor except when shifting gears.

2. Do not skip gears. When leaving the station or accelerating in traffic, make a progressive shift. Shock loading adds to the problems of the clutch and the entire drive train.

3. Do not shift into the next higher gear until the engine speed is near the governed speed. A lugging engine shock-loads the clutch and the entire drive train back to the tires.

4. Do not ever coast with the clutch disengaged. This can tear out a complete drive train on sudden engagement.

5. Do not place the transmission into gear until ready to move the vehicle. When waiting at a traffic signal or other time delay, place the transmission in neutral and release the clutch pedal.

6. Do not slip the clutch to hold the vehicle steady on a hill. This practice will wear out a clutch faster than any other single cause.

7. Adjust your gear selection to traffic conditions and grade or road conditions.

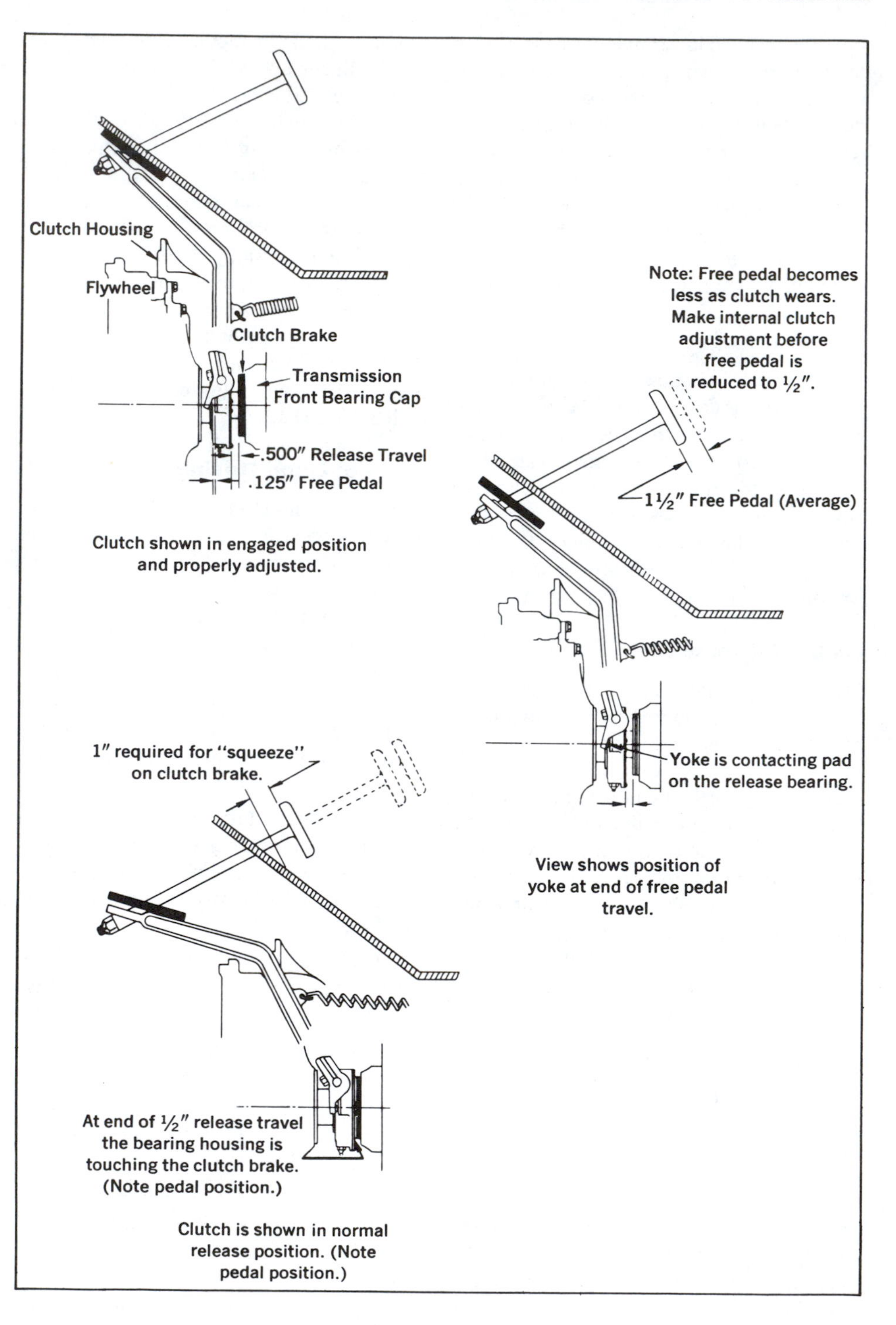

Figure 17-8. Proper method of checking the clutch adjustment

18 Driving Fire Apparatus

If accidents are to be avoided, modern traffic conditions require that fire apparatus drivers be skillful, mature, and responsible. Good drivers are made, not born. They are made through a training program that includes initial instruction, driving training, the development of proper attitudes and driving habits, and training in the exercise of restraint and good judgment. Some of the factors that govern skillful driving are:

1. The condition and limitations of the vehicle
2. The physical features of roads, streets and terrain
3. The attitudes, behavior and reactions of other highway users
4. Varying light and weather conditions
5. The personality and makeup of the driver

Despite all the mechanical improvements and automotive safety advances, the driver is still the key to traffic safety. It is essential to be in good physical condition, have sound driving skills and habits, and develop and maintain proper attitudes. The requirements to be a skilled driver include:

1. Attitude
2. Skills
3. Knowledge
4. Judgment
5. Habits
6. Physical fitness
7. Mental fitness

Driving a heavy apparatus loaded with crew and fire equipment through congested traffic leaves no room for error. To qualifv for this responsibility, the driver must be thoroughly trained to acquire the skills necessary for emergency driving operations. In addition, the driver must be given basic mechanical information relevant to fire apparatus, must understand all pertinent traffic regulations, must establish good traffic habits, and must develop proper attitudes toward traffic responsibilities.

Every portion of the fire department vehicle, including its load, should enter into the driver's thinking. The very weight and balance of the apparatus are important. The engine with its high power, the transmission with its intricate mechanism, the brakes, which must always be effective for the safety of all concerned, the tires which provide the frictional surface for traction and for stopping or turning, the steering mechanism, the starter and the various switches are all under the sole control of the driver, who must be skillfully trained as to their proper use.

Any street intersection, to the driver of a fire apparatus responding to an alarm, brings the hazard of cross traffic, which is seldom controlled for the benefit of the driver by the signal light or the stop sign: in fact, quite the opposite is true. These signal controls are often a hazard. However, the driver does have some traffic control on the vehicle in the form of a warning red light and siren. Unfortunately, these devices do not always control traffic instantly, and when they fail, the emergency apparatus driver must know how to avoid a dangerous situation by alternative actions.

Emergency right-of-way does not ever permit a driver to lose sight of the responsibility for the safety of all users of the streets. The law gives certain exemptions to emergency apparatus drivers when responding under lawful use of a red light and siren, but it does not give any valid cognizance to an arbitrary use of these emergency rights which unduly endangers life or property.

Defensive Driving

A defensive driver is one who makes allowances for one's own deficiencies and for the lack of skill and knowledge on the part of others, and who recognizes there is no control over unpredictable actions of drivers and pedestrians or over weather. The defensive driver concedes right-of-way and makes other concessions to avoid collision, looks ahead, and watches situations develop.

In defensive driving, a driver can expect others to act in one of two ways. They may do what is expected of them, or they may do things other than what they are expected to do. Consequently, defensive driving involves making two types of predictions:

1. Predict the predictable.
2. Predict the unpredictable.

It is the personal responsibility of the driver to operate without accident. Alertness is the key to avoid that last-moment futile effort to escape an accident.

Safe driving is made up of: 20% safe vehicle: 20% good physical condition of the driver: 60% good attitude.

Expressed differently, it could be said that safe driving is very much a mental process requiring thought, attention, courtesy, wisdom, and a sense of alert responsibility, rather than an unthinking physical process involving just fast reactions and skill in manipulating the vehicle controls. Good attitude prompts interest and an effort to develop both knowledge and skill so that defensive and safe driving practices become habit.

A preventable accident is one in which you failed to do everything you could have done to prevent it (Figure 18-1).

Attitude

A driver's attitude, which is a crucial determinant of safe driving, includes mental or emotional regard for oneself, for others, for the vehicle and for surrounding conditions. A driver with a poor attitude usually has some excuse for anything that happens–the other guy was wrong, the street was poor, the intersection was blind, etc. Attitudes are not inborn, but are learned and therefore can be improved. We have all known the type of person, who, acquiring the authority of red lights, siren, and the power of several tons of fire engine, is suddenly a big shot and loses all considerations of safety and courtesy. This attitude is wrong! To be a *good* driver, the temperament *must* change.

Other Emotional Factors Affecting Driving

Some people, in order to compensate for failure, do absurd things. The unrecognized

Figure 18-1. Inattentive drivers may not relinquish the right-of-way, even to a fire engine, with red lights and siren, going to a fire.

person looks for a chance to appear brave and powerful; the really important person has a sense of modesty and normally appears humble. The normal person obtains satisfaction from life itself, but the thwarted person may obtain satisfaction by stepping into a car and use the power in an attempt to get recognition. Such a person may be recognized by some of the following acts:

1. Insisting on the right-of-way
2. Trying to "get even" with drivers who pass
3. Edging in to cheat someone out of a parking place
4. Making pedestrians scramble to safety

Focus Fixation

There is a strong tendency to steer toward the spot to which your attention is focused. The muscles of your body tend to adjust toward the goal of your attention. It therefore follows that a driver whose attention is poorly controlled and easily distracted is likely to steer the car in that direction. Some situations that might tend to distract the uncontrolled attention of drivers are:

1. The scene of an accident
2. A member of the opposite sex, in or out of the car
3. Sharp light reflections
4. A back-seat driver

The Top-Notch Driver

Some of the characteristics of a good driver are not only motor skill but also emotional balance and control. Adjustment and maturity are evidenced by such traits as:

1. Self-control
2. Acceptance of responsibility
3. Good judgment
4. Good sportsmanship

The driver of mature psychological makeup has a quality that is very important in driving. This quality is controlled attention. Attention is focused constantly on the path the car should take while considering all factors in the situation. Such a driver is aware of all that is happening in the field of vision that could possibly affect the driving situation. Control, whether of attitudes, emotions or attention, is a distinguishing characteristic of the person who is psychologically mature.

Vehicle Operational Practices

The following driving procedures and practices must be understood and followed by all fire company drivers.

Prescribed Response Routes

Drivers of companies must follow, as closely as possible, their prescribed fire alarm response routes of travel. Officers and drivers must be on the alert for other responding companies, particularly at street intersections, and should make due allowance. Normally all fire department vehicles responding from a multiple company station should take the same route to an alarm. If more than one vehicle is using the same street when responding to, or returning from an alarm, they should do so in a single file, and no vehicle should pass another unless the latter is disabled.

Signalling and Correct Position

Good signalling is far more than a wave of a hand or a flip of a switch. Coordination of every facility is required to make intentions clear. Signalling includes:

1. Switching on turn signals 100 or more feet ahead of the turn. For safety, exceed the Vehicle Code minimum requirements.
2. Using both the inside and outside rear view mirrors and getting into a proper posi-

tion where the vehicle, by its very position, becomes a forceful commanding signal.

3. Slowing down to turning speed. This is a signal too, and is accentuated by the red flash of the stop lights.

4. Sounding the horn, should it be necessary.

Correct position. The vehicle position serves, day or night, as the most visible forceful method of signalling. For example, if the driver is preparing to turn left, the vehicle's position both invites and ensures that following traffic will pass on the side away from the planned turn.

Turn signals. With their bright flashing lights, turn signals command attention day and night, and they permit the driver to keep both hands on the wheel. At times it may be an additional safety factor, however, to use hand signals in conjunction with the turn signals. This may be necessary when the glare of the sun causes the indicator lights to be indistinguishable or, in a passenger car, when the hand and arm would protrude outside of the automobile body.

Do not signal an overtaking vehicle that it is safe to pass. To do so transfers a part of the responsibility for safe passing from the overtaking driver to the fire department driver. In the event of an accident, the driver and the department may be held liable. Similarly, do not signal a pedestrian that it is safe to cross in front of a vehicle which may be stopped; another vehicle may pass on either side and hit the pedestrian. In this case, it was the pedestrian's responsibility to cross only when conditions permit a safe crossing.

Wrong position. Avoid the hazards of a wrong or sloppy position:

1. Make no "question mark" or "buttonhook" turns–that is, start to turn to the right at the beginning of a left turn.

2. Start no right turn so far from the curb as to invite some "curb cruiser" to crowd alongside and then be "mousetrapped."

3. Avoid the dangerous habit of cutting left-turn corners.

4. Do not straddle lanes of travel.

5. Avoid "blind-spot" driving. Stay out of the *other driver's* blind spot.

Safe Backing Practices

Avoid backing wherever possible. When obliged to back, get out and walk back or around the apparatus to make certain there is nothing behind it. Then back immediately, watching sharply.

In backing a large vehicle, have someone as a guide (Figure 18-2). Have the guide stand to one side and signal–not call. Regardless of being guided, the driver is not relieved of the responsibility to back safely. The guide does not have control of the apparatus and can neither start it nor stop it.

Never back a vehicle around a corner at an intersection in order to turn around; drive down the street to a side street or driveway and turn around. If it is only a short block, drive around it.

Limitations of Siren Audibility

Sound waves set up by sirens are directional. Tests have shown that these sound waves have a greater intensity ahead than to the sides or rear of the emergency vehicle. Buildings deflect sound waves. Motorists approaching the intersection from right angles to the path of the emergency vehicle do not receive a signal of the same strength as those motorists directly ahead. Tests have indicated that the audibility of a siren around the corner at an intersection where buildings are located is as much as two-thirds less than on a straightaway. Hills also affect the direction of siren sound waves. When an emergency vehicle is approaching the crest of a hill, the perception of siren sound waves by motorists on the other side is greatly reduced.

Figure 18-2. Safe backing practice

Interference Factors

Areas containing greater background noises such as those made by streetcars, buses and trucks, reduce audibility. Furthermore, personal hearing limitations vary considerably. Some persons are more perceptive to high tones while others can hear low tones best. This is a reason for operating the siren throughout its entire tone scale, fluctuating from a high to a low pitch. A person whose hearing is impaired may be given a restricted operator's license which merely requires that his vehicle be equipped with adequate rear and sideview mirrors.

In cold and rainy weather, most motorists drive with all windows closed. One test has pointed out that audibility is diminished by one-third inside a vehicle with all windows closed, compared to audibility when the driver's window is open.

Noise inside a vehicle or truck, created by its engine, and noise created by a radio or conversation will greatly reduce a driver's perception to the sound emitted by an approaching siren.

Freeways

It is better not to operate a siren while responding on a freeway or throughway. As a rule, the traffic is traveling faster than the fire apparatus. The tendency of most freeway drivers is to slow down and attempt to move over to the right when a siren is heard, and this aggravates the traffic conditions.

Limits of Reliance on Siren

The effectiveness of the siren under varying field conditions is not as great as might be expected. As an added safety factor the driver of an emergency vehicle should assume that all motorists are partially deaf, that they are inattentive to their driving, that the windows of their vehicle are closed, that a radio is playing and that conversation is taking place within their vehicle, and that they will become confused if and when they hear a siren or see a responding emergency vehicle. These conditions, the background noise of the district and conditions at each intersection should determine the extent to which a driver depends on the siren to clear traffic. Many officers and drivers consider the air horn an effective device to clear traffic at intersections; however, the air horn is not a legal emergency vehicle warning device and should not be used to the exclusion of the siren.

Potentially, the most serious accidents involving emergency apparatus can be expected at intersections. Fire department drivers must give adequate warning of their approach and should be prepared to make safe emergency stops, particularly at blind and congested intersections and when entering against traffic signals.

Total Stopping Distance

This is determined by three measurements: perception time, reaction time, and vehicle braking distance. These three conditions depend on several factors, such as: the type and condition of vehicle, the street surface conditions, the driver's physical and mental limitations and the speed of the vehicle.

The type and weight of a vehicle, type and condition of its braking system and tire conditions will affect braking distance.

Street surface conditions. Traction decreases as speed increases. This decrease in traction develops much more rapidly on wet surfaces than on dry, and is most severe at the beginning of a rain before the oil and dust on a street is washed away. The type of street surfacing is also a factor.

Driver limitations. In addition to vehicle braking distance, there is perception distance and reaction distance. The perception distance is the distance traveled while the driver recognizes the need to stop and decides to do something about it. The reaction distance is the distance traveled while the driver acts to apply the brakes. Of these two factors, perception time is more variable.

Speed is the primary factor that governs all other items mentioned. If speed is doubled, the perception-reaction distance will be doubled and the vehicle's braking distance will be about four times as great. For example: a motorist who decides to stop while traveling at 20 mph may have a perception-reaction distance of 44 ft and a braking distance of 20 ft, for a total vehicle stopping distance of 64 ft. At 40 mph these factors would increase to a perception-reaction distance of 88 ft and a vehicle braking distance of about 80 ft, a total vehicle stopping distance of 168 ft.

Perception-reaction time is not a factor in the braking distance requirements of the Vehicle Codes. However, in stopping vehicles under normal driving conditions, perception-reaction distances are important factors. In estimating the average total stopping distances, most agencies allow ¾ of a sec. for each, perception time and reaction time; this is a total time lapse of 1½ sec.

Laboratory tests of drivers' stopping distances will frequently show results far better than those indicated by charts. Under such controlled test conditions, drivers know that an emergency situation is to be encountered and consequently perception-reaction time is considerably shortened. This alertness can also be applied to drivers of emergency vehicles whose perception-reaction time will probably be much less than the accepted average of 1½ sec for motorists.

Fire department drivers should take into consideration the various factors that affect total stopping distances. They should not expect the average motorist to stop as quickly as they themselves can, as the motorist's perception-reaction time is greater.

Speed of Apparatus

Instead of being essential to the response of fire apparatus, speed is positively objectionable. Speed often results in a serious and unnecessary accident which not only prevents the apparatus from reaching the fire, but may cause death or injury to fire fighters and to civilians.

Of even greater importance is the psychological effect of speed on the crew. The natural result of speeding to a fire is that it incites a lack of logical judgment when the fire is reached. The most effective approach to the control of the fire is often missed, and windows and doors are unnecessarily smashed in the effort to apply fire streams speedily. Speed in ventilation, before hose lines are in position, has proven disastrous in many fires. Speed in discharging hose streams before the fire is located, or before salvage covers are spread, has caused thousands of dollars of unnecessary water damage. The officer who makes certain of an alarm location and responds with sane and safe driving will arrive with a company prepared for action. If the driver of the apparatus has the vehicle under control at all times, especially at street intersections, the excitement incident to dangerous speed will be lacking, and the officer and crew will be in a state of mind to control the fire in a cool and deliberate manner.

Drivers should govern their own choice of speed according to a basic speed rule, which can be stated in a very simple way: "Never exceed a speed that is reasonable and proper for existing conditions, *even* where the law permits a speed higher than that at which you are driving." Safe, prudent speeds vary with such factors as the driver's reaction time, brake efficiency, the condition of the pavement, the weather and traffic congestion. A speed that is reasonable when there are few persons or vehicles on the street may be excessive in heavy traffic, or at hours when school children must cross the streets. Over the same stretch of roadway, speeds that are safe at certain times are unsafe at other times.

Under all circumstances, the speed of fire department vehicles turning at any street intersection should be reduced to a minimum consistent with safety and with due consideration of traffic.

On return from an emergency call, and in all other nonemergency driving, all traffic laws must be observed by drivers of fire department vehicles.

Whenever any department vehicle is required to respond to a nonemergency alarm and the company officer and/or driver is informed by the dispatcher that a fire is not involved or that human life is not in immediate danger, the driver should drive the vehicle in a safe and cautious manner, observing all traffic laws and rules. DO NOT USE THE SIREN OR RED LIGHTS DURING THE RESPONSE.

Crossing Controlled Intersections

When responding to an alarm and confronted with a traffic control red light, the apparatus should be slowed down and stopped, if necessary, until all traffic proceeding on the green light realizes that the fire department vehicle is signalling for the right-of-way. The apparatus should remain halted and out of the intersection until the traffic comes to a complete stop. Upon approaching a stop sign, fire department vehicles should slow down and should not proceed until all other vehicles yield the right-of-way. If the siren and red light do not get the right-of-way, it should not be taken arbitrarily under any condition.

Entering Area of Emergency

Drivers approaching the scene of a fire or other incident should exercise judgment in parking their vehicles so as to prevent an unnecessary concentration of fire apparatus in one location. Companies that are not first-in on a fire should hold back at the nearest intersection without committing themselves, unless it is clearly apparent that they are needed in front of the building. Often all responding companies will crowd into the immediate location of the response, creating a traffic jam and reducing the flexibility of the operation.

Positioning Vehicles at Fires

Drivers of pumpers must use good judgment in parking their vehicles so as not to interfere with the positioning of other fire department vehicles. Particular attention should be given to the positioning of pumpers for the use of hose reel lines, ready lines, or when acting as a manifold so as not to obstruct aerial ladder operation. Aerial ladders usually operate from a position directly in front of the fire scene.

At fires, unless otherwise ordered, first-due squads should be positioned on the side of the street opposite the premises involved so that ladders, tools and equipment are readily available. Aerial trucks should be positioned as directed by the officer-in-charge.

At all times, operators of fire department apparatus must exercise good judgment in parking their equipment so as not to interfere with the operations of responding companies. They should particularly avoid blocking the removal of ladders from the rear of truck company apparatus.

Aerial Truck and Elevating Platform Drivers

The driver of an aerial ladder or elevating platform truck should remain with the apparatus until it is obvious that the immediate necessity for movement and operation of the apparatus and aerial ladder has been passed. Thereafter, the driver should report to the officer at the scene, who will direct further movements.

Engine Drivers

The driver of a pumper that has responded to the fire scene should also remain with the apparatus until it is obvious that the immediate necessity for movement and operation is over. Thereafter, the driver should report to the commander, who will direct future movements.

Disposition of Equipment

All equipment that is not needed at the fire scene should be returned to quarters or parked in a safe location with all lights and engines turned off. Only one unit need be manned for radio contact with communications.

Skidding

Every driver is familiar with the way the apparatus tends to skid outward on a curve and with the tendency of the body to roll over in the same direction. The centrifugal force responsible for these effects varies with the differences in bodies, loads, roads, curves and speeds. Speed is the most important of these factors because of its squaring effect; i.e., as the speed doubles, the centrifugal force increases four times. REMEMBER, A SKIDDING VEHICLE IS OUT OF CONTROL (Figure 18-3).

A curve should always be approached at a speed that will not require braking while in the turn and which will permit the application of power. Tires do not have as much of a tendency to skid sideways, when power is being applied to the wheels, as they do when a turn is being made under coasting or braking conditions. If the vehicle starts to skid, turn the steering wheel in the direction of the skid. Do not disengage the clutch or

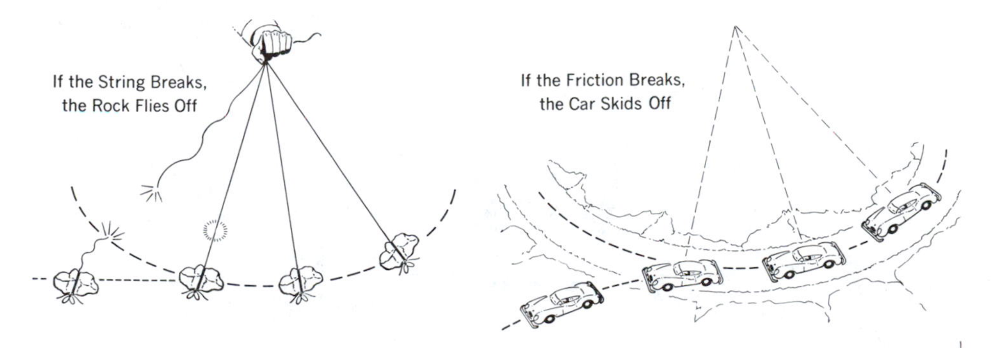

Figure 18-3. Centrifugal force and skidding

brake sharply. Apply the brakes intermittently, letting the wheels revolve.

Crossing Railroad Tracks

When crossing railroad tracks or a wide bump which would normally hit both front wheels at the same time, the shock to the front and rear axles may be softened, if clearances permit, by taking the bump one wheel at a time. Approach at a slight angle and do not apply power while making the crossing. These steps are needed to avoid excessive strain on the drive train.

Chock Blocks

To avoid any possibility of a rollaway from a secured position on a grade, place one chock against a left rear tire and one against a right rear tire–in front of the tire if headed down grade, in back if headed up grade (Figure 18-4). If on level terrain, place one chock in front of left rear tire and one chock in back of right rear tire. When vehicles are equipped with a special chock block, only one is used. IT IS THE DRIVER'S RESPONSIBILITY TO CHECK THE CHOCKS BEFORE DEPARTURE, AND TO ALWAYS USE THEM WHEN SECURING. It has been found that the safest method of placing a chock block is to set the block on the ground close to the tire and shove it into position with a foot. Chock blocks must be placed square and snug against the tires to be effective. On a grade, place all blocks on the downhill side of the wheels. If four chock blocks are available, such as on an aerial truck, the placement of two chocks under the rear trailer wheels as well as under the rear tractor wheels provides added safety.

The Use Of Flares

These useful safety devices can turn killer if they are misused in an area with flammable liquid spills or suspected explosive compounds. *Do NOT use to leeward* if such conditions should exist. This cannot be overemphasized. In fact, it would be well to delete their use altogether in such situations

Figure 18-4. Prevent rollaways with chock blocks

and to resort to battery-operated or safety lights for traffic control.

When fire department vehicles respond to an emergency on a freeway, throughway or other busy thoroughfare, they should attempt to park directly behind or ahead of cars involved in the emergency. This position will minimize the interruption of traffic flow as well as the exposure of crew and apparatus to danger from collision. Avoid, as much as possible, the direction of lighted headlights and spot lights into flowing traffic. Revolving warning lights, tail-lights and emergency red lights should be kept in an *on* position. Department vehicles equipped with four-way hazard warning signals should have these lights turned on and flashing to alert the traffic. A series of three or four flares should be placed around the scene of the emergency, starting at least one hundred feet behind the emergency and ending at the immediate side of the emergency scene. If the emergency occurs on a curve or at night or during the early morning hours when traffic often flows faster, flares should be spotted from a point several hundred feet to the rear of the emergency. Needless to say, the person spotting the flares should use extreme care in anticipating the actions of on-coming cars.

Safety When Driving in Fog

Visibility is at its worst in fog. In dense fog, drive slowly using the low-beam headlights, which will throw the light down on the road where needed, rather than out into the fog where it will be reflected back. Avoid sudden stops. Signal stops by tapping on the brake pedal; make stop lights blink to alert the following driver. Never assume that there is a clear road ahead when driving in fog, except for the distance that can actually be seen.

Rear-View Mirrors

Before passing another vehicle, always carefully check traffic conditions both to the front and rear. Check in the rear-view mirror and the side mirrors. Be sure to check the blind spot (Figure 18-5), the narrow area to the rear left where a car can be entirely concealed by the part of the apparatus between the mirror or mirrors and an approaching car. Never create an additional or

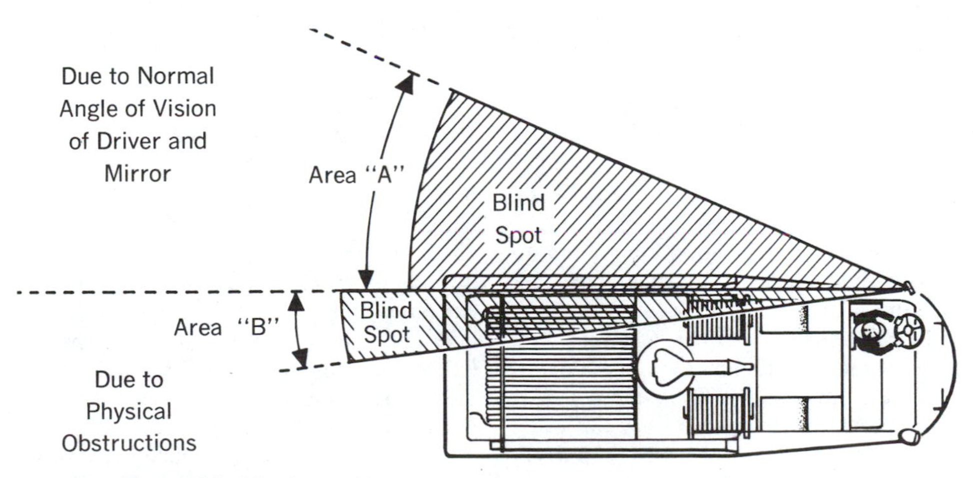

Figure 18-5. Driver's rear vision

broader blind spot by hanging a coat or helmet in such a way that it reduces the view to the side or rear.

On cab-ahead types of apparatus, utilize the side mirrors to center the apparatus between the lines in the street. Until drivers become familiar with this type of apparatus, they have a tendency to drift too far to the right and may strike a parked vehicle.

Safe Driving Reminders

Skid marks on the pavement are an indication that a vehicle has not been under proper control.

The lives and welfare of fire fighters depend upon the driver and the officer-in-charge every time a department vehicle leaves the station.

Officers should operate the siren so as to produce a regular high-and-low pitched sound. *Caution:* Never ride the siren button.

Do not permit the promiscuous use of sirens. Do not use sirens when converging with other fire department vehicles at the scene of an emergency and do not fail to reduce speed and to proceed cautiously, as this is one of the most dangerous parts of a response.

Remember that car windows of other vehicles may be closed and their drivers may not hear sirens or bells.

A GOOD DRIVER takes his foot OFF the gas pedal and places it ON the brake pedal at each intersection. This practice reduces reaction time.

Better time can be made by getting out of the station quickly, than ever can be made up by speeding to alarms.

Traffic lights and signs are an aid to driving, but they are not a guarantee that safe driving conditions exist.

Inspect the vehicle often—and keep it in safe operating condition. Do not accept an "Off Shift" driver's report that the vehicle is in proper condition to drive.

THE BEST PIECE OF FIRE FIGHTING APPARATUS IN THE WORLD IS USELESS UNLESS IT ARRIVES AT A FIRE SAFELY.

Driving Techniques

The following driving techniques must be thoroughly understood and practiced at all times and under all conditions.

Seat

Adjust the driver's seat so that the right foot rests easily and fully on the accelerator pedal. If the driver sits too far back, the foot rests near the base of the pedal, requiring heavy foot pressures. Any slight foot movement at the pedal base will give excessive sudden action at the top of the pedal which is almost certain to result in undesirable, jumpy, wheel-spinning action.

Mirrors

Adjust the mirrors for best visibility to the rear.

Seat Belts

Fasten the seat belt BEFORE starting the engine.

Gages

Purpose of gages. The dialed gages on instrument panels provide necessary information for efficient road operation. The location and type of gages varies depending on the make of the apparatus. In general, information can be found concerning the engine oil pressure, water temperature, intake manifold vacuum and operating speed. Other gages provide information indicating the air brake reservoir pressure, the amount of gasoline in the fuel tank, the charging or discharging of the battery in the electrical system, the apparatus road speed, etc.

Proper use of gages. The proper use of apparatus gages requires three things of a driver: (1) becoming thoroughly familiar with the type and location of all gages provided; (2) learning the operational norms for engine oil pressure, engine temperature, air brake reservoir pressure, etc.; and (3) developing a systematic procedure for checking the gages provided. A suggested procedure is to start in the upper left-hand corner of the instrument panel and work to the right; in the event more than one row of gages is provided, read each row separately.

After starting the engine, a quick check of all gages, with a particular emphasis on oil pressure and brake reservoir pressure, will indicate whether it is safe to operate the apparatus. Operating gages should be checked from time to time while the apparatus is on the road to determine whether the apparatus is functioning properly or whether some condition has arisen which makes it unsafe to continue operation.

Parking Brake and Transmission

The parking brake should be set and the transmission placed in neutral before starting the engine.

Starting Gasoline Engine

Turn on the ignition switches, open the throttle slightly, depress the clutch (this takes some load off the starting motor by disengaging the transmission) and engage the starter. If the engine does not start promptly, close the choke; after the engine starts, gradually open the choke until the engine is running smoothly.

Occasionally, the engine may fail to start because the battery is partly discharged and the draw of current by the starter is causing a low voltage in the ignition circuit. In this case, depress both starter switches (if the vehicle is so equipped).

On apparatus equipped with dual ignition systems, the following check should be made often with the engine at operating temperature. Set the hand throttle at a fast idle (600 to 800 rpm). Turn off one ignition switch and observe the decrease in engine speed: 100 rpm drop, or more, indicates that the ignition system which is being used is not fully efficient. Then check the other system.

Starting Diesel Engine

Because of the many different ways of starting and operating diesel engines, refer to the manufacturer's guide for the exact model of engine involved.

Starting Off

After starting the engine, allow the engine to warm up until it will take the throttle without hesitation. Then, depress the clutch fully and shift into the proper starting gear. Depress the throttle slightly and increase the engine speed to prevent stalling the engine when the clutch is engaged. Reengage the clutch smoothly and do not further increase the engine speed until the clutch is fully engaged. When starting from a dead stop, the interval of clutch slippage should not exceed 1½ seconds.

The clutch pedal should be fully depressed only when the apparatus is at a standstill preparatory to placing the transmission in gear. Full depression of the pedal actuates the clutch brake on constant-mesh transmissions, thus stopping the pilot shaft and the countershaft in the transmission. Once the apparatus is in motion, the clutch pedal should be depressed just far enough to disengage the clutch and to allow movement of the gear shift lever into or out of gear.

Proper Gear for Starting Off

Naturally, an empty truck can be started satisfactorily in a higher transmission gear ratio than when partially or fully loaded. If auxiliary transmissions or multi-speed axles are used, they must be in the lower ratios for satisfactory starts.

A good rule of thumb for the driver to follow is: empty or loaded, select the gear combination that lets one take up the slack and start moving out with an idling engine or, if necessary, just enough throttle to prevent stalling the engine. After the clutch is fully engaged, the engine should be accelerated to approximately the governed speed for the upshift into the next gear.

Clutch Action

The major cause of clutch failures could be summarized in two words: EXCESSIVE HEAT. Excessive heat is not the amount of heat a clutch can normally absorb and dissipate, but the amount of heat a clutch is *forced* to absorb and which it attempts to dissipate. Most clutches are designed to absorb and throw off more heat than encountered in normal clutch operation without damage or breakdown of the friction surfaces. Clutch installations are engineered to last many thousands of miles under normal operating temperatures. If properly used and maintained, they will give satisfactory service for that length of time.

However, if a clutch is slipped excessively or required to do the job of a fluid coupling, high heat quickly develops to destroy the clutch. Temperatures generated between the flywheel, driven discs and pressure plates can be high enough to cause the metal to flow and the friction facing material to char and burn. Heat or wear is practically nonexistent when a clutch is fully engaged but during the moment of engagement, when the clutch is picking up the load, it generates considerable heat. An improperly adjusted or slipping clutch will rapidly generate sufficient heat to destroy itself.

Proper training of drivers as well as mechanics is essential for long and satisfactory clutch life. Drivers should be taught how to operate the apparatus properly and not left to experiment for themselves. Some of the more important points to cover in a driver-training program are: teaching the trainee to start in the right gear, to recognize clutch malfunctions, and to know when to write up a clutch for readjustment.

While most clutch pedals work the same, they are not always adjusted the same. Some clutches begin to function just as the clutch pedal is moved downward, some not until the pedal is depressed 2 or 3 in. Generally a clutch pedal has approximately 1½ in. of free travel before it affects clutch action. As the pedal continues downward, the clutch is completely uncoupled in about 1 in. more of travel. The pedal still has approximately 3 in. of travel before it is all the way in. When releasing the pedal, it is obvious that the pedal will come out approximately 3 in. before affecting clutch action, another inch will complete the coupling, and the last 1½ in. will be free travel. The distance of pedal travel may not be the same on every vehicle, but the principle is the same.

Riding the Clutch Pedal

This is very destructive to the clutch since a partial clutch engagement permits slippage and excessive heat. Riding the clutch pedal will also put a constant thrust load on the release bearing, which can thin out the lubricant. Release bearing failures can be attributed to this type of operation.

Holding the Vehicle on an Incline with a Slipping Clutch

This is asking the clutch to do the job normally expected of a fluid coupling. A slipping clutch accumulates heat faster than it can be dissipated, resulting in early failures.

Coasting with the Clutch Released and Transmission in Gear

This procedure can cause a high driven-disc rpm due to the multiplication of ratios from the final drive and transmission. It can result in throwing the facing off the clutch discs. Driven-disc speeds of over 10,000 rpm have been encountered in such a simple procedure as coasting a vehicle down a hill. While an ample safety factor is provided for normal operation, the burst strength of the facing is limited.

Engaging the Clutch While Coasting

This can result in tremendous shock loads and possible damage to the clutch, as well as to the entire drive train.

Automatic Transmissions

Several types of automatic transmissions are used on fire apparatus, but their functions are similar. For the purposes of instruction, one model of an Allison fully automatic transmission (HT 740D) will be discussed; facts covered here may be applied with slight adaptation to other makes and models.

These transmissions offer completely automatic shifting. After the driver selects the drive range, the vehicle accelerates rapidly through the gears. All gear changes are automatically timed and applied correctly, without great loss of engine speed, during the shifts. In all normal driving situations, the pressure of the driver's foot on the accelerator pedal controls the automatic shifts in the transmission, providing the performance desired. When the accelerator is depressed to full throttle, the transmission will automatically upshift at a speed just below the maximum governed engine speed. With less pedal pressure, the transmission will upshift at a lower engine speed. Automatic selection of the best gear ratio for performance, as well as fuel economy, is assured.

For added control, the driver can move the control lever to a lower range position. This permits the driver to restrict upshifts, matching power output and speed to road, load, and traffic conditions. For example, the driver may wish to keep the transmission from upshifting to the direct drive range in stop-and-go traffic. This is done by simply selecting the next lower range. Shifting in all lower ranges remains the same, but the transmission will not make the final upshift at what might be an inopportune time for the traffic flow.

The Allison HT 740D automatic transmission is hydraulically operated with four forward gear ratios and one reverse. Basically, it consists of a hydraulic torque converter, a planetary gear train, and a control system to automatically change gear ratios and supply oil under pressure to the converter and lubrication circuits. The converter multiplies the engine torque and also does the work done by the clutch on a stick transmission. Control is exercised by means of a range selector lever, adjacent to the driver's hands.

Selection of the correct range provides better control; it also avoids undue hunting by the transmission for the required gear. Here are the range selections and when to use them:

R (Reverse). The vehicle should be completely stopped before moving the selector lever from a forward gear to Reverse, or from Reverse to a forward gear. Reverse has only one range.

N (Neutral). Neutral is the position for starting the engine. It is also the position to use during stationary operation of the accessories.

D (Drive). The Drive position is used with normal loads on level roads with an open highway ahead. The transmission starts in first gear, then automatically upshifts to second, third, and fourth gears, as speed and load dictate.

3 (Third). Third position is used when in the city or suburbs, or any place where traffic and load conditions will not permit top speed. The transmission starts in first gear, then automatically upshifts to second and third gears.

2 (Second). Second position is selected when top speed is severely limited, such as in congested city traffic. The transmission starts in first gear and automatically upshifts to second as speed increases.

1 (First). First position is "low." It is selected when starting with an extra heavy load, pulling through mud and snow, or driving up or down steep grades; use it for slow road speed control.

The transmission can be upshifted or downshifted at full throttle. Although there are no speed limitations on upshifting, there is on downshifting and reverse. Good driving practices indicate that downshifting should be avoided when the vehicle is above the maximum speed attainable in the next lower gear. Therefore, the good driving habits have been designed into the Allison transmission shift pattern. The downshift inhibitors within the valve body prevent those harmful shifts when the vehicle is going too fast for the next lower gear. If downshifts are attempted at excessive speeds, the inhibitors prevent the selected downshift until the vehicle reaches an acceptable speed.

To use the engine as a braking force, shift the range selector to the next lower range. If the vehicle is exceeding the maximum speed for a lower gear, use the service brakes to slow the vehicle to an acceptable speed where the transmission may be downshifted safely. An automatic, compared to a manual shift transmission, has a longer coast-down time. Until a driver becomes accustomed to this characteristic, it may be necessary to manually downshift to reduce speed. With a little experience in driving with an automatic transmission, a driver will learn to decelerate a bit sooner, or brake until the automatic downshifts occur; this will reduce the need for manual downshifting.

Rocking out. If the vehicle is stuck in deep sand, snow, or mud, it may be possible to "rock" it out. Shift to low gear and apply a slight steady throttle (never full throttle). Then by moving the range selector between low and reverse, rock the vehicle free. Time the shifts to take advantage of the forward and backward movements of the vehicle. If the driving wheels break traction or spin, apply less throttle.

Towing or pushing. Before towing or pushing a disabled vehicle which is equipped with an automatic transmission, the driveline should be disconnected or the drive wheels lifted off the road. The engine cannot be started by pushing or towing.

Parking brake. There is no "Park" position in the transmission shift pattern. Therefore, always apply the parking brake to hold the vehicle when it is unattended. Be sure the selector is at Neutral.

Driving on ice or snow. Automatic transmissions continually provide proper balance between required power and good traction; the driver can have better control of the vehicle because of this smooth, constant flow of power through the drive train. However, when driving on ice or snow, any acceleration or deceleration should be made gradually.

Temperatures. The transmission oil temperature is indicated in some vehicles by a gage specifically designed for this purpose and in some vehicles by the engine coolant temperature indicator. Extended operations at low vehicle speeds with the engine at full throttle can cause excessively high oil temperatures in the transmission. These temperatures may tend to overheat the engine cooling system, as well as possibly causing transmission damage.

Caution: The engine should never be operated for more than 30 seconds at full throttle with the transmission in gear and the vehicle not moving. Prolonged operation of this type will cause the transmission oil temperature to become excessively high and will result in severe overheat damage to the transmission.

Normal oil temperature for both on and off the highway operation is 160° to 200°F. Oil temperatures should never exceed 250°F for on-highway operation or 300°F for off-highway operation.

If an excessive temperature is indicated by the engine coolant temperature gage, stop the vehicle and determine the cause. If the cooling system appears to be functioning properly, the transmission is probably overheated. Shift to Neutral and accelerate the engine to 1200 to 1500 rpm; this should reduce the oil sump temperature to the operating level within a short time. If high temperatures persist, stop the engine and have the overheated condition investigated by a mechanic.

If an excessive temperature is indicated by the transmission oil temperature gage, stop the vehicle and shift to Neutral. Accelerate the engine to 1200 to 1500 rpm. The temperature should return to normal within two or three minutes; do not operate the vehicle unless the oil is cooled.

If the transmission overheats during normal operation, check the oil level in the transmission.

Oil check procedure. Since the transmission oil cools, lubricates, and transmits power, it is important that the proper oil level be maintained at all times. If the level is too low, the converter and clutches will not receive an adequate supply. This can result in poor performance or transmission failure. If the level is too high, the oil will foam; this will cause the transmission to overheat.

Before checking the oil level, clean around the end of the fill pipe before removing the dipstick. Dirt or foreign matter must not be permitted to enter the oil system because it can cause valves to stick, cause undue wear of the transmission parts, or clog passages. Check the oil level by the following procedure:

1. Operate the transmission in a drive range until the normal operating temperature (160° to 200°F) is reached. The oil must be warm to insure an accurate check; the oil level rises as the temperature increases.
2. Shift through all drive ranges to fill the clutches and oil passages.
3. Park the vehicle on a level spot, shift to Neutral (N) and apply the parking brake. Let the engine run at idle speed.
4. Check the oil level after wiping the dipstick clean. The safe operating level is between FULL and ADD marks on the dipstick.
5. If the oil level is not within this range, add or drain oil as necessary to bring the level to the FULL mark.

Hydraulic retarder. Some models of Allison transmissions are equipped with a retarder. When fully applied, the retarder has a braking effect much greater than engine braking alone. However, it is not a brake and it will not bring the vehicle to a complete stop. It can be applied in any gear, but it is primarily used on downgrades in the lower ranges.

The hydraulic retarder saves wear on both the driver and the brakes. Use the retarder when descending steep grades, snubbing in stop-and-go traffic, slowing down on curves, and maneuvering on icy or slick roads.

The hydraulic retarder has only one moving part, a vaned rotor which is driven by the vehicle wheels through the drive line. When the retarder pedal is depressed, a surge of oil is released. The rotor throws this oil against the fixed stator blades, which resist the flow. This makes it harder for the truck wheels to turn; the vehicle is thus slowed down. The oil does all of the retarding work, absorbing the heat generated by the retarder action. When retarder action is not called for, the valve is closed and the oil is evacuated from the cavity; there is no more power absorption in the retarder.

Proper Throttle Usage

The secret of proper throttle application is to use a sensitive but relaxed foot on the gas pedal. This is generally referred to as "feath-erfooting." The sooner a driver is able to apply "just a little more" or "just a little less" throttle pressure, the sooner the shifting will be smooth. Too much or too little throttle pressure during the shift will result in roughness. The degree of roughness will depend on the degree of error in throttle usage. Probably the most significant sign of a really good driver is judgment of how much throttle should be applied and the ability to apply just that amount. (This ability is especially noticeable in shifting but applies to all driving situations.) Although it is basically true that the throttle is pushed down for power and speed, and that pressure is removed from the throttle for slowing and stopping, a driver can improve performance considerably by considering the throttle as a precision instrument. Used properly it can correct many ills of sloppy operation.

Lugging

When an engine is being operated properly, the momentum of its moving parts (pistons, connecting rods, crankshaft, etc.), and the pressure exerted on the top of the pistons (by the burning fuel in compression chambers) are in relative balance. If the engine is performing an operation which requires more horsepower than the engine is producing, such as pulling a steep grade in higher gear, the momentum of the moving parts (the speed of the engine) is reduced. When this occurs the engine runs in an unbalanced condition because the pressure exerted on the top of pistons exceeds the momentum of moving parts. This unbalanced condition is called lugging. Lugging most often occurs when the engine rpm's are at or near the bottom of the operating range, producing only a small portion of the total brake horsepower available (brake horsepower is the power developed by an engine at any given speed) while the engine is under excessive load.

Continued lugging of apparatus engines will greatly shorten engine life by causing severe damage. This damage results from the engine being subjected to a severe shock, similar to striking the top of the pistons with a sledge hammer, each time a cylinder fires. Some apparatus are equipped with a vacuum gage showing the intake manifold vacuum in inches of mercury. When the engine is being operated under a load and the vacuum gage shows less than 5 in. of mercury, the engine should be considered lugging regardless of the engine speed.

Overspeeding

The engine of an apparatus should be considered as overspeeding any time the engine speed exceeds 90% of the peak engine speed. Overspeeding most often occurs when the apparatus is being rapidly accelerated when the transmission is in a low or intermediate gear, and when the apparatus is descending a steep grade.

When an apparatus engine is overspeeded it is again running in an unbalanced condition; however, in this instance the momentum of moving parts exceeds the pressure exerted on the top of the pistons. This unbalanced condition is especially severe when the engine is overspeeding with little or no load as when the vehicle is descending a grade with the throttle closed. When this occurs there is very little force exerted on the top of the pistons to retard the tremendous inertia developed as they travel upward in the cylinders. The result is that a severe strain is placed on the pistons, wrist pins, connecting rods and the lower half of the connecting rod bearings as the engine tries to pull itself apart, so to speak.

Torque

Webster defines torque as "that which produces or tends to produce rotation." In everyday language, torque is a twisting effort which tries to make things turn. An engine will give its most efficient operation if the rpm can be maintained somewhere between speeds at which maximum torque is developed and 90% of peak engine speed. If the engine is operating under a heavy load and the speed falls below the point at which maximum torque is developed, momentum is rapidly lost and lugging often results.

Engine Speeds

When driving the apparatus, every effort should be made to keep the engine speed within the limitations of the operating range by the use of proper gear selection. If the engine rpm approach 90% of peak speed, shift into a higher gear; if the engine is under load and the rpm's fall to the bottom of the operating range, shift to a lower gear. Under ordinary driving conditions it is not necessary to raise the engine speed to the upper limit of the operating range in each gear when shifting up through the gears. The engine speed need only be raised high enough to allow the shifting operation into the next higher gear to be completed with the engine speed still above the lower limit of the operating range. If maximum acceleration is desired or required, engine speed may then be raised to the upper limit of the operating range in each gear.

While operating in any single gear, brake horsepower produced by an engine increases as the rpm's are raised; thus it is desirable to keep the engine speed in the upper portion of the operating range while ascending a grade during mountain driving operations. When descending grades, with the braking effect of engine compression utilized, it is good practice to hold engine speeds in the lower two-thirds of the operating range.

Terrain Affects Shifting

Since it is necessary to keep engine speeds within the proper range, it is a must for the driver to consider the immediate terrain and momentum of the vehicle when contemplating a shift. It should be remembered that as soon as the clutch is depressed at the start of a shift, the momentum of the vehicle changes in relation to the immediate terrain. As the clutch is depressed, road speed will:

1. Change very little on a level roadway.
2. Decrease going upgrade.
3. Increase going downgrade.

How much road speed will change during the shift depends on:

1. The momentum at the start of the shift (the greater the momentum, the less noticeable the change).

2. The steepness of the grade where the shift is made (the steeper the grade, the more speed is affected).

3. How long it takes the driver to make the shift (the longer it takes, the longer the vehicle is uncoupled and the more momentum changes).

Obviously a knowledge of proper throttle regulation is needed to adjust engine speeds to match road speeds under varying road conditions. When a driver understands how much terrain affects shifting, he should be able to change gears with less difficulty. The position of the gear lever before the shift has no bearing on the amount of throttle needed during the shift into the selected gear nor on the releasing of the clutch at the completion of the shift.

To determine how much throttle is needed during a shift consider:

1. Which gear is being selected.
2. How fast the vehicle is traveling.
3. How fast the engine will be going in that gear at that speed.

The driver must determine how much engine speed will be needed in that gear to match the road speed of the vehicle after the clutch is released at the completion of the shift. The driver must determine, then regulate, then hold that amount of throttle until the clutch is released. If the driver's judgment has been reasonably correct, a smooth entry into the selected gear will be accomplished and, equally important, it will result in a smooth clutch release. If a shift is made in a smooth manner with no noticeable change of engine or road speed when the clutch is released, it is obvious that no vehicle abuse has occurred.

Gear Shifting

The mark of a good driver is the manner of shifting gears, probably the most important of all driving operations. Correct use of the various gears will greatly increase apparatus performance, give longer engine and transmission life and add to the safety of driving operations.

Calculating Gear Split

The split between gears refers to the difference between ratios of gears in the road transmission, expressed in engine rpm. In other words, split is the number of rpm engine speed must be decreased, when shifting up, or increased, when shifting down, to properly complete the shifting operation desired. Knowing the split between gears takes a great deal of guesswork out of gear shifting and will often greatly improve driving performance. For example, the split between third and fourth gears is 700 rpm; the apparatus is operating in fourth gear at an engine speed of 1100 rpm, and the driver wishes to shift into third. Knowing that the split is 700 rpm, the driver can shift into neutral, increase the engine speed until the tachometer reads 1800 rpm, and complete the shift into third.

Gear split can be calculated with the following formula:

$$\text{Split} = 90\% \text{ of peak engine speed} - \left(\frac{90\% \text{ of peak engine speed} \times \text{gear ratio}}{\text{next lower gear ratio}}\right)$$

The peak engine speed and the gear ratios can be found in the apparatus manual.

For example, suppose an apparatus has a peak engine speed of 2400 rpm and is equipped with a five-speed transmission. The ratio in fifth gear is 0.78 to 1.00, and the ratio in fourth gear is 1.00 to 1.00. 90% of 2400 is 2160 rpm. Application of the formula is:

$$2160 - \left(\frac{2160 \times 0.78}{1.00}\right) \text{ or } 2160 - 1684.8$$

In subtracting 1684.8 from 2160, we have a difference of 475.2; thus, the split between fourth and fifth gears in this road transmission is 475.2 engine rpm. Therefore, when shifting between fourth and fifth gears, engine speed must be decreased 475 rpm when shifting up and increased 475 rpm when shifting down. This formula determines the gear split when engine rpm is increased to the governed speed. The split at lower engine speeds will be correspondingly reduced. Thus, if the governed speed is 2000 rpm, at 1500 rpm the split will be reduced by 25%, and at 1000 rpm the difference in engine speed will be reduced by one-half.

Double-Clutching

Shifting of gears in heavy apparatus, either up or down, should always be accomplished using the double-clutching technique regardless of the type of road transmission. Double-clutching provides an opportunity for the engine speed and the road speed of the apparatus to synchronize, preventing the undue strain on the drive train.

Heavy duty engines, because of the weight of their pistons, connecting rods, crankshafts, and other parts, are slow to accelerate and decelerate. Engine speed must be increased or decreased so it will be correct for the gear into which the transmission is being placed, and this relationship will vary according to the road speed of the vehicle. If the clutch on the apparatus is disengaged long enough to allow the engine to reach a correct speed, the counter shaft in the transmission will completely stop; the gears on the main shaft, being driven by rear wheels, will still be rotating. Thus, the moving gears will be attempting to mesh with the gears at a standstill and clashing will result.

Double-clutching is a two-part operation:

1. Depress the clutch; shift the transmission into neutral; then reengage the clutch. This will allow the transmission counter shaft to rotate at the correct speed.

2. When the engine has accelerated or decelerated to the correct speed for the next gear, disengage the clutch; shift the transmission into gear; then release the clutch.

Shifting Up

When the apparatus is accelerating and the engine speed reaches the point at which a shift can be made to the next highest gear without allowing the engine speed to fall below the lugging point, depress the clutch and shift to neutral position. At the same instant the clutch is depressed, release the throttle. With the transmission in neutral, release the clutch to allow the engine to sychronize the gearing in the transmission. When the engine speed falls to the lower limit of the gear split, depress the clutch, shift into the next higher gear, release the clutch, and depress the throttle to pick up the load and provide a smooth flow of power.

Shifting Down

Shifting down, or backshifting as it is sometimes called, can be accomplished much more easily by using the following technique:

When ascending a grade and the engine rpm are near the lower limit of the operating range, depress the clutch and shift into neutral; hold the throttle where it is; *do not let off on the throttle.* The engine speed will immediately begin to increase when the engine is relieved of its load. When the shift lever is in neutral position, reengage the clutch. When the engine rpm reach the upper limit of the split between the gears, complete the shifting by depressing the clutch and shifting into the next lower gear. As the clutch is released, the engine will pick up the load smoothly and evenly with little or no strain placed on the road transmission and drive train. When using this technique on level terrain, it will probably be necessary to depress the throttle slightly while the road transmission is in neutral (to allow the engine rpm to build up) because the engine will be operating under a light load.

Fire apparatus transmissions are of a selective type; therefore, it is not necessary to progress up or down through all of the gears. It is only necessary to have the transmission in the proper gear so that the engine will be in the correct operating range for that road speed. Therefore, at times gears may be skipped when shifting down in order to prevent lugging of the engine. An example of this would occur if an apparatus were traveling down the street in fourth gear at 40 mph and traffic conditions suddenly caused the driver to slow down to 15 mph. The engine would be below the operating range if the driver shifts into third gear, so the correct procedure would be to downshift directly from fourth to second gear.

Two-Speed Rear Axle

Some apparatus are equipped with an extra cluster of gears located in the rear axle which is commonly referred to as a two-speed axle. Drivers are bewildered when first exposed to this device, but understanding and application develop appreciation of its value. Its purpose is to change the gears in the rear axle to either a high or low range, thus allowing a choice of two speeds in any transmission gear. Apparatus having a four-speed forward transmission will have a possible eight speeds forward if equipped with this device.

Gears in the two-speed axle are changed by a motor, operated either by vacuum or by electric current. The motor is most commonly operated by a button, mounted near the top of the gear shift lever, which moves up or down approximately one inch. If the button is pushed down to "low side," the motor changes the range to low in the rear axle. If the button is lifted up to "high side," the motor changes the range to high in the rear axle. The "low side" is a lower gear; the "high side" is a higher gear.

The driver must correctly regulate the throttle and operate the clutch to get the most satisfactory results from the two-speed axle. If a driver understands the two-speed, it will be a great help in maintaining momentum, whether uphill or on level ground, and whether the vehicle is loaded or empty. But its main purpose is often overlooked; it is most valuable to better engine care because it enables the driver to keep the tach within its range with less variation of road speed. For example, the tach range between transmission gears is usually about 1000 rpm, whereas the tach range between the high side and the low side of the two speed is approximately 500 rpm.

While the principles of the two-speed rear axle operation and of road transmission operations are very similar, there is one slight variation of technique–clutch timing. In two-speed operations, the button and throttle are moved at the same time in the same direction and the clutch is depressed a split second *afterward.* The clutch must not be depressed until the button is moved, but should follow immediately.

Operating Range

The operating range of an engine can be defined as the engine speeds between the point where the engine is considered lugging under load and 90% of peak engine speed. The engine's lugging point and the peak engine speed are both usually predetermined by the manufacturer in the interest of good performance and long engine life.

The ideal operation of a gasoline engine (from the combined standpoint of engine efficiency, economy and long trouble-free mileage) is attained at speeds ranging from about two-thirds of the top rpm engine speed to within about 200 rpm under the top rpm engine speed. The accelerator must also be in a position corresponding to about two-thirds to three-quarters torque. For an engine with a top speed of 2600 rpm, the range described above would be from 1700 to 2400 rpm. At this rpm speed range, the engine is usually most responsive to the accelerator, and its torque build-up will be achieved more readily with the opening of the throttle.

Where a tachometer is provided in the driver's compartment, it is a simple matter to control engine speeds by knowing what the engine speed operating limits are and by then watching the tachometer. Where only a speedometer is provided in the driver's compartment, engine speeds are transposed to the speedometer in terms of allowable speeds in each gear.

Steering

The art of steering is more than just turning the wheel. Safely and correctly controlling the horsepower of a fire fighting apparatus is a challenging problem that requires skill and attention (Figure 18-6).

Use of Hands

Except for those times when it is necessary to shift gears, both hands should be kept on the steering wheel. The suggested position for hands is about ten and two o'clock on the wheel. With the hands in this position, there is less possibility of losing control should the apparatus inadvertently strike a chuck hole or rut in the road. Turning movements should be made smoothly using the hand-

Figure 18-6. Safely and correctly controlling the horsepower of a fire fighting apparatus was as much a challenging problem one hundred years ago as it is today. (Courtesy of the Los Angeles City Fire Department)

over-hand technique (Figure 18-7). The hands should be kept palm down with the fingers gripping the outside rim of the steering wheel.

Tendency to Straighten Out

When an apparatus is turning, there is a tendency for the front wheels to straighten out because of wheel alignment. This tendency can be used to recover from a turn by relaxing the grip on the steering wheel, allowing it to slide freely through the grip and then regripping the wheel as the front wheels reach the desired position.

Power steering units are designed to retain the tendency of the front wheels to return to the straight-ahead position when making a turn. However, with this type of steering, do not rely on the front wheels to return to the center automatically. It is always wise to steer in both directions, turning and straightening, when using power steering.

Shifting in Turn

If over half a turn of the steering wheel is required while turning, do not shift during the turn as both hands are needed on the wheel. Gear changes may be made while turning if the turns are slight, and it is not necessary to change the position of the hands on the wheel.

Speed and Steering

Drivers should be alert to foresee situations far enough in advance that the apparatus

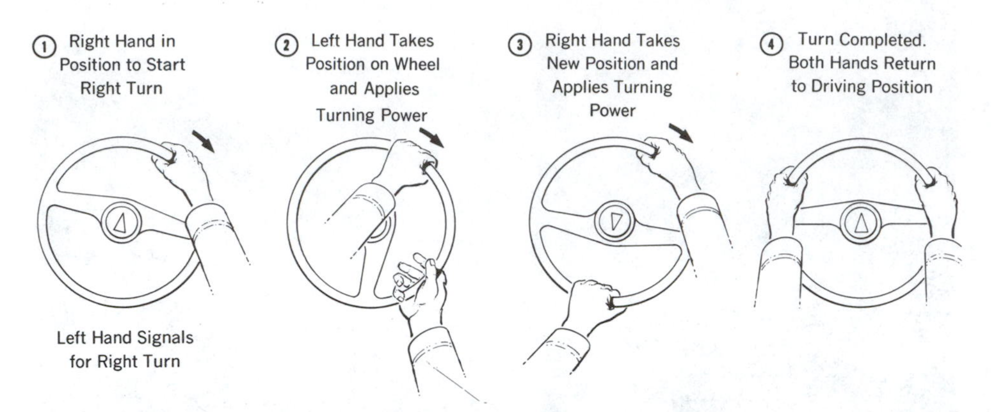

Figure 18-7. Hand-over-hand steering on curves

need not be jerked into a turn, but can be let into it smoothly. If the steering movements cannot be made fast enough during a turn, it is not the fault of the steering mechanism; the apparatus is traveling at a road speed too fast for the conditions encountered.

Brakes

A hindrance to reducing the number of accidents involving emergency vehicles is a lack of understanding and appreciation of the actual stopping distances that are necessary at various speeds and under unusual conditions. The total distance a vehicle travels between the time that a hazard appears in the driver's vision and the time that a vehicle can be brought to a total stop depends mainly on three factors: (1) the driver's preception–reaction time, (2) the driving speed, and (3) the braking distance. Perception–reaction time is not a factor in the braking distance requirements of the vehicle codes; when estimating the average total stopping distances, most agencies allow ¾ of a sec for each, making a total time lapse of 1½ sec. As the perception–reaction distance varies directly with the speed, if the speed is doubled, the perception–reaction distance is doubled.

The braking distance is the distance traveled between the point of first retardation by the brakes and the point where the vehicle actually stops. Under the same conditions, braking distance increases with the square. If the road speed is doubled, braking distance becomes four times as great.

Braking Principles

A moving vehicle possesses energy that must be dissipated before the vehicle can be brought to rest. Occasionally this is done by crushing and bending metal in a collision; more usually it is accomplished by wearing and heating the brake shoes, drums, tires, and the road when the brakes are applied.

The key points in the braked stop are two friction points at which the wearing and heating occur. These are the area of contact between the brake shoes and the brake drums and the area of contact between the tires and the road surface. As the brake shoes press harder and harder against the brake drums, more and more energy is used up in generating heat and wearing the shoes; thus, the wheels turn slower and slower. If the pressure is great enough, the wheel may stop turning before the vehicle stops, and the tire will then start to slide on the roadway. This is called "locking the wheels" or making a "locked wheel stop."

Once the wheel starts to slide, additional pressure on the brake shoes makes no difference in the rate of stopping. If the roadway surface is very slippery, such as glare ice, very little pressure on the shoes will be required to lock the wheels. If the surface has very good gripping ability, such as roughly textured concrete, it may be nearly impossible to lock the wheels.

If the wheels do lock, the friction between the tire and the road determines the length of the stop. Thus, it would be quite useless to test brakes on ice; the poorest brakes could lock the wheels and the stopping distance would be determined by the gripping power of the ice. It would be equally foolish to proclaim that a vehicle would stop in a specific number of feet without knowing the gripping power of the road surface.

Road Surface

One of the important factors that make some braking distances greater than others is the gripping ability of the road surface, usually referred to as the *coefficient of friction.* It is measured by the amount of drag or retarding force it puts on the vehicle, expressed as a proportion of the vehicle's weight. A heavy vehicle bears down harder than a light one, and the drag is correspondingly greater. A road with a coefficient of friction of 0.50 (or 50%) will produce a drag of 1500 lb on a vehicle weighing 3000 lb, but the same road will produce a drag of 5000 lb on a truck weighing 10,000 lb; these figures presume that the wheels lock in both cases.

Coefficient of Friction

The coefficient of friction is low for slippery surfaces; it may be as low as 0.05 on glare ice and 0.20 on packed snow; the best road surfaces may range as high as 0.90. It is not possible to give figures for particular kinds of roadways, for many factors affect the coefficient of friction, such as the following:

Water on the surface is the most important factor. On concrete, particularly with oil drippings on the surface, water may increase braking distances by one-third or more. On gravel or cinders, water may sometimes shorten stopping distances.

Speed is another important factor. At 40 or 50 mph some pavements, especially when wet, have only half the coefficient of friction which they have at 10 mph.

Tires do not have as much effect as one might think. Smooth treads decrease gripping ability slightly on ordinary pavements, with the greatest decrease on wet surfaces.

Chains on ice or snow will roughly double the coefficient of friction.

Temperature has a slight effect on the gripping ability of dry road surfaces, but this is not important except when bituminous surfaces begin to melt and bleed, thereby reducing the coefficient of friction. On snow or ice, however, temperature has an extremely important effect. Stopping distance tests from 20 mph on glare ice, for example, have shown an average braking distance of 120 ft at temperatures near zero, increasing to an average of 225 ft near 32°F.

Foreign material on the surface may change the coefficient. Loose gravel on a hard surface may act like ball bearings and cut the gripping ability in half. Dirt on the pavement may have no effect when dry, but the first sprinkle of rain may cause it to act like a film of soap on the road. Oil drippings on concrete pavements carrying heavy traffic may, when wet, make the roadway more slippery than many wet asphalt pavements. On the other hand, sand or cinders on ice will increase the coefficient of friction.

Texture of the pavement is important. Gritty concrete made with sharp edged aggregate is much better than a very smooth surface. It is unfortunate that the texture of pavements changes with age and wear. Traffic gradually polishes the sharp edges; it also adds oil drippings to reduce the friction.

Stopping without Sliding the Wheels

When the wheels do not slide, friction in the brakes absorbs the energy of the moving vehicle. In this case, there is an important difference from the locked wheel stop; the pressure between the sliding surfaces has nothing to do with the weight of the vehicle. In the locked wheel stop, there is a constant relationship between the weight and the retarding force; a heavy vehicle produces a greater retarding force so that the stopping distance is the same. When the friction is within the brakes, however, the drag depends on the number of brakes, the size of the lining surfaces, the kind of lining, and the force applied to the shoes to push against the drums. Each individual brake may produce a different drag, but added together they represent the total retarding force on the vehicle.

Slope of the Road

If the vehicle is going downhill, braking distances are increased; stopping distances are decreased if the vehicle is traveling uphill.

Applying the Brakes

So far the stopping of a vehicle has been considered simply as whether the brakes locked the wheels or not. Actually, the wheels will not all work exactly together; one may slide while the others do not, or one may slide before the others. It is commonly stated that the best and quickest stop is made when the wheels are all at the point of impending skid. The various coefficients are believed to be highest at this point.

Effect of Speed

The most important aspect of stopping a vehicle is the effect of increased speed. The principle involved is a fundamental one of physics, which says that the energy of a moving body is in proportion to the square of its speed. Since when braking a motor vehicle the major problem is to dissipate the energy of the moving vehicle, a higher speed means more energy to dissipate; thus, it takes longer to stop. Since the energy is in proportion to the square of the speed, however, the braking distance also is in proportion to the square of the speed. In other words, twice the speed means four times as far to stop; three times the speed results in nine times as far to stop.

Driver Perception Time

Perception time is the period it takes the driver to perceive a dangerous situation after it has appeared. If a driver is intently watching the stop light of the car ahead and it turns bright red, it will be perceived almost instantly and perception time is zero. If, however, the driver is looking at the scenery off to the side of the road when the stop light comes on and sees the light two seconds later, when looking ahead again, the perception time is 2 sec.

Perception time varies tremendously with circumstances. It is very closely connected to attention. Alert drivers have shorter perception times than those who let their attention wander. A driver who is usually quick to perceive may sometimes be distracted by something especially interesting to him and experience a dangerously long perception time. It is not practical, therefore, to embody the distance traveled during perception time in tables or charts of stopping distances for vehicles at various speeds. As the average perception time is ¾ sec, the same as attributed to a driver's reaction time, these figures will provide some idea of how much distance might have to be added for perception time to the total stopping distances.

Driver Reaction Time

Unfortunately, no driver can apply the brake instantly upon seeing a hazard; time is required to see or hear, to think, and to act. This reaction time varies greatly in different individuals, and for the same person under different conditions, but most of the reactions in driving require between ½ sec and 1 sec. A figure of ¾ sec is commonly used. The distance traveled by the vehicle during this reaction time must be added to the braking distance to arrive at the total stopping distance.

Stopping Distance Table

It is quite impossible to develop any simple table or chart which will tell how far it will take a vehicle to stop. So many factors enter into any particular stop that the chart could not possibly fit all possible combinations. A stopping distance chart serves a useful purpose only as an educational tool to help explain the general problems of speed and stopping distance to drivers. As such, a chart usually shows average figures. Such a chart is of no value to an engineer or a police officer trying to study a particular accident, for the chance that a specific case would be *average* in all aspects is extremely remote. Therefore, Table 18-1 is not under any circumstances to be taken as indicating the stopping distances for a certain vehicle under specific circumstances.

Types of Brakes

Brakes may use any one of several different types of operating energy: mechanical, vacuum, hydraulic, or air. All of these operate similarly with little difficulty.

However, air brakes do present a few operating characteristics that will be discussed further.

Operating the brakes of an air-braked vehicle differs very little from operating the brakes of a passenger car. Because the operation of the brake pedal requires very little physical effort, proper control of the brakes is easily accomplished. The distance the brake pedal is depressed determines the amount of air pressure delivered to the brake chambers, and the brake chamber pressure determines the braking force. Thus, the driver should keep in mind that he is operating a brake valve capable of giving finely graduated brake control and should make full use of this control.

An air-braked vehicle should not be moved unless the air gage shows at least 60 psi of air pressure in the air brake system, as the brakes are not safe at lower pressures. Some makes of apparatus are equipped with a dashboard warning light that stays lit until

TABLE 18-1. STOPPING DISTANCES FOR HEAVY TWO-AXLE TRUCKS ON DRY, CLEAN, LEVEL PAVEMENT

SPEED		DRIVER	VEHICLE	TOTAL
MILES PER HOUR	FEET PER SECOND	REACTION DISTANCE, ft	BRAKING DISTANCE, ft	STOPPING DISTANCE, ft
10	15	11	10	21
15	22	17	22	39
20	29	22	40	62
25	37	28	64	92
30	44	33	92	125
35	51	39	125	164
40	59	44	165	209
45	66	50	210	260
50	73	55	255	310
55	81	61	310	371
60	88	66	370	436

the pressure reaches 60 psi. A vehicle should not be moved while this light is on.

While operating the vehicle, the driver should periodically observe the air pressure registered by the dash gage to be sure that pressure is being maintained properly. If the air pressure drops to a low point, or if the warning buzzer or light signifies that pressure is low, the vehicle should be stopped and the trouble corrected.

Application of Air Brakes

The following braking techniques should be practiced when driving air-braked vehicles to ensure the public safety and the ability to stop safely in emergencies.

Regular stops. Whenever possible, brake applications should be started far enough in advance so that moderate air pressure can be used. In making normal stops, the brake application should be hard at first and then graduated off as speed is reduced so that at the end of the stop, little pressure remains in the brake chambers. Never apply air brakes lightly at first and then increase the pressure as the speed diminishes. This not only makes a longer stop but also makes a rough stop because of the final high pressure.

With any braking system, the fastest stop is made when enough resistance is applied to the rotation of the wheels to produce faint tire marks on the roadway. Once the wheels are locked, braking efficiency decreases. Power-applied brakes do not necessarily stop the apparatus quicker, but they replace muscular effort with power. Do not lock the wheels and put the vehicle into a skid because it will not be under control. Release the brakes to the point where the wheels are rotating in order to regain control of the vehicle. If in a skid, do not apply the brakes; keep the wheels rotating; let up on the throttle gradually and turn the wheels in the direction of the skid. Do not disengage the clutch until just before stopping.

Fanning. The fanning of air brakes is the repeated rapid application and releasing of brakes during a stop. This practice is to be avoided because it gives poor brake performance. Fanning does not increase the brake chamber pressure, but rather, lowers the reservoir and line pressure by wasting air.

Descending grades. When descending light grades, the compression of the engine, through proper gear reduction in the road transmission, can often maintain the apparatus at a safe speed. The brakes can also be applied intermittently to keep the apparatus well under control.

In descending steep grades, the speed of the apparatus should be maintained at a point which allows a full and complete stop with a safe distance if it becomes necessary. To prevent excessive road and engine speed, the road transmission should be in the gear which will keep the engine speed in the lower two-thirds of the operating range when the brakes are constantly applied with light to moderate pressure. This type of application can be maintained by just slightly depressing the brake pedal. When brake application pressure gages are provided on the apparatus, maintain an application pressure of approximately 10 psi. A constant brake application, when descending steep grades, will generate less heat than will hard intermittent applications. This is important, for it is the production of excessive brake drum temperatures which causes brakes to fade. Internal expanding brakes fade because as the drums heat, they expand away from the brake shoes, greatly reducing the effectiveness of the brakes; to a lesser degree, the coefficient of friction of the lining also decreases.

Air-Brake Locking Device

Some compressed-air brake systems provide a *secondary air-brake locking device,* controlled by a small toggle installed on the instrument panel of the driver's cab.

The toggle is operated in two positions. *Note:* THIS IS NOT TO BE CONFUSED WITH TOGGLE-CONTROLLED SAFETY BRAKES. With the toggle in the ON position, air is locked in the braking system and the brake shoes will remain in contact with the brake drums. With the toggle in OFF position, the brake shoes are retracted from the brake drums. The escutcheon plate further indicates, *"Do not use this lever to park."* This means basically that the toggle would not be engaged while in quarters or while parked on the street for other than emergency duty. It also means that the ON position should never be used if the normal air pressure is not being maintained. DO NOT USE THIS TOGGLE AT ANY TIME WHILE THE VEHICLE IS MOVING.

The toggle can be placed in the ON position when the apparatus is stopped and the engine is running. It may, in addition, be used with the engine shut off, providing the driver is assured that adequate air pressure is being maintained. Air brakes will assist the parking brake in holding the vehicle on a hill or in position for fire duty. Since this air lock system provides assistance only, it must be considered as a secondary system and used as such.

Parking Brakes

The term parking brake (emergency brake, commonly known as the hand brake) indicates a secondary means of stopping a vehicle. Should the normal foot brake fail, the parking brake (emergency) can be used to bring the vehicle to a safe stop; but an understanding of the braking system will point out the disadvantages of using parking brakes indiscriminately.

The parking brake on most modern heavy fire apparatus is on the final drive shaft and is separate and distinct from the foot brakes. This is necessary in order to provide two means for stopping the apparatus. Some apparatus have a double-face brake on the drive shaft while others have a single-face brake. In all cases, direct pressure is applied to the facing by a cable, a rod and/or a combination of the two. A ratchet is provided to allow the brake handle to engage in the ON position. Normally, the parking brake should be applied only when the vehicle has been brought to a full stop. Engaging the parking brake while the apparatus is underway will cause:

1. Glazed surfaces to build up on the lining surface which will result in brake failure
2. Burning of the brake surface due to friction
3. Terrific stresses to be placed upon the drive train, as well as upon rear-end assemblies
4. Stretching of cables, rods, linkages, etc.

The foot brake should be used to stop the vehicle and then the parking brake should be applied (followed by placing chock blocks under the rear wheels). When a serious emergency exists, the parking brake may be pulled on rapidly, but the driver should realize that this is submitting the running gear to a severe strain. Whenever possible, avoid applying the parking brake while under way. Since the parking brake affects only the driving wheels of the tractor, there is a possibility of jack-knifing a tractor-trailer rig if the brakes are applied while the vehicle is in motion.

Hand Air Brake

The hand air brake lever on the right side of the steering column on a late model apparatus accomplishes two purposes:

1. The emergency braking system, which has a separate reservoir, separate air lines and double-diaphragm rear brake chambers, is actuated by a hand air brake.

2. It allows the driver to apply the brakes with the right hand. Thus, it permits him to leave his right foot on the accelerator pedal when starting up a hill.

Under ordinary circumstances the hand air brake is not to be used as a parking brake for two reasons:

1. There is often a small amount of air leakage when brakes are applied. If air brakes are depended on to hold the apparatus when the engine is stopped, the vehicle may start moving when the reservoir air pressure drops.

2. Friction causes heat. When brakes are applied, heat is generated; this heat causes the brake drums to expand. If the brakes have been utilized to the extent that the drums are expanded, and the air brakes are set, the brake shoes will be held against the drums with great force, and the drums are likely to warp when they contract due to cooling.

If the limitations on the use of the hand air brake are understood, then this system should be used whenever it may be beneficial. Do not apply full reservoir pressure to brake chambers; the further the lever is pulled, the greater the application pressure. When parked on a hill, apply the parking brake and chock blocks to prevent the apparatus from moving; if there is the slightest doubt that this is sufficient, and it is not practical to re-park the apparatus, then a moderate pressure may be applied to the air brake system by means of the hand air brake.

Check occasionally to be certain that the hand air brake is not partially on. Dragging the brakes will result when the brakes are not completely released.

Pulling Steep Grades

In case the apparatus cannot pull a steep grade and stalls in lowest gear and the brakes refuse to hold, shift immediately to reverse and let the rig back down in gear. The only alternative is to let the engine stall in low gear, and then the dead engine can help hold the apparatus while the wheels are being chocked, after which the gearshift lever can be returned to neutral, the engine started and another try made.

Safety Brakes

Safety brakes are a combination air pressure–coil spring system that provides automatic braking, should a failure occur in the service air brake system. The system also provides a manual control of a positive parking brake. Automatic brake application occurs when there is abnormal leakage of air or a complete failure in the service brake system.

Should an unexpected emergency application of the safety brakes occur, the stop will not be too abrupt because the application of safety brakes is approximately 70% of the braking force of the service air brake systems. For a given speed, service brakes would be capable of stopping an apparatus in a shorter distance than would safety brakes. The safety brakes may be applied as a parking brake by means of the manual control whenever the apparatus is parked. Normally, the safety brake is not used to bring a moving vehicle to a stop by using mechanical application, although it is possible to do so.

Emergency Air Reserve Tank

All apparatus equipped with safety brake systems should be outfitted with an emergency air reserve tank. The purpose of this installation is to supply emergency air to release the spring-loaded tension on brake diaphragms once the normal air brake system has failed and the fail-safe feature of the air brake system has activated. The objective is to provide an emergency means of moving disabled vehicles out of the path of normal traffic. This emergency air reserve system is controlled by a push-pull button which is mounted on the instrument panel of the cab; on aerial ladders, an additional push-pull button is mounted at the rear of the trailer. *Cauton:* Do not confuse this button with the normal operating push-pull button of the safety brakes.

Techniques of Driving

Steering

Shift before the turn. When approaching a turn and it is obvious that a shift of gears will be advisable, SHIFT BEFORE THE TURN so that both hands will be on the wheel throughout the turn. Also, apply the throttle leaving the turn, since the transmission will already be in the proper gear. Avoid shifting in a turn.

Remember the wheelbase. When turning a corner, be sure to know approximately where the rear wheels are located. They are probably farther back than those on a car due to the difference in the wheelbase. Check the rear wheel location before driving. Avoid hitting curbs.

Recovery while maneuvering. When maneuvering in limited quarters (such as a turn-around), recover the steering before the vehicle stops. When stopping and changing direction, turn the steering wheel quickly just before stopping so that the vehicle will turn in the desired direction the instant motion is resumed. This applies to forward stopping before backing or to a backing stop before moving forward. Avoid turning the wheel while the vehicle is stationary.

Backing–wheels not at the corners. Check the location of the tires in relation to the ends of the vehicle before driving. The overhang (the distance the rear end protrudes beyond the rear wheels) varies greatly on different types of vehicles. Be aware of this overhang and govern your steering accordingly. Do not become so involved in maneuvering the rear end of the vehicle that the rear wheels are forgotten, or vice versa.

A known, but often overlooked, fact of backing is that the back wheels are leading and the front end is following. To change the direction of travel when backing, the following end must swing to the side in order to point the vehicle towards its goal. The sharper the turn, the more the following end must swing. While looking in the direction of backing, do not forget this swing; be fully aware that backward steering is different.

Clutch

Delay depressing when stopping. Use engine compression to help the brakes slow a vehicle when stopping. Do not depress the clutch as long as it is helping to hold back but DO depress it when it ceases to help–usually this is near the engine idle speed. Avoid depressing the clutch and coasting to a stop–this requires excessive brake usage.

Do not ride the clutch. Because of the numerous clutch applications necessary, the driver often keeps a foot ready at the pedal anticipating its use. This is proper, but there is one common fault found in many drivers of which they are entirely unaware; quite often the foot is depressing the pedal slightly, causing a small, unnoticeable, but damaging amount of slippage. This is known as riding the clutch. If not actually using the clutch, be sure the foot is off the pedal–not resting on it. AVOID RIDING THE CLUTCH.

Extended stop–clutch out. The mechanics of a clutch are such that the pedal should not be depressed for extended periods when the engine is running. This does not apply to shifting, but refers to stops such as waiting at a railroad crossing for a train to pass or standing at a stop signal. Any time it is evident that the pause will be prolonged, the gear shift lever should be placed in neutral and the clutch pedal released. The driver should anticipate the moment safe progress can be resumed and place the transmission in gear just before that moment so the move can be made without delay.

AVOID holding the clutch in for extended periods.

AVOID waiting so long in preparing to move that traffic behind is unnecessarily delayed or the opportunity to proceed is lost.

Anticipate Your Move and Prepare

What gear to select to start? This is a debatable subject with many drivers and depends partially on the physical makeup of the gears, which may differ considerably from one type of vehicle to another. Some vehicles are meant always to be started in the lowest gear, while in others, the lowest gear is to be used only for the extremely hard pull—terrain and load have a strong effect upon proper gear selection.

To select the proper gear for starting, apply the general principles of engine and clutch usage. If the vehicle moves with a minimum of clutch slippage and engine strain, the gear is obviously satisfactory. The driver will soon discover whether the lowest or the next lowest gear is better.

Gears Grind When Shifting While Stopped

Quite often gears will grind when a driver attempts to shift into the starting gear after a stop. This problem, more pronounced in some vehicles than in others, has three possible driver solutions:

1. Be sure the clutch is fully depressed.
2. Be sure the engine is idling and pause to allow time for the gears to stop turning.
3. Select another higher gear, and then try the lower gear again.

Driving and Tillering Aerial Ladder Trucks

There is no activity in the fire service which catches the interest of observers more quickly than the work of the driver and tiller operator, when a ladder truck is responding to an alarm. The public is always thrilled by an aerial ladder truck weaving in and out of traffic with the tiller operator negotiating turns that appear impossible. There is, however, much more to the job than steering and maneuvering the apparatus.

Driver–Tiller Operator

The driver and tiller operator of a tractor-trailer ladder truck have a dual responsibility that is unique among fire fighting apparatus drivers. The driver is responsible for the proper maneuvering of the tractor, and the tiller operator controls the lateral movement of the trailer. However, both must work together as a team and coordinate their actions to provide a safe and efficient operation of the tractor and trailer as a unit.

Truck company officers should be fully aware of the fact that the driver and tiller operator who work together will soon learn each other's operating techniques and develop into an efficient driver–tiller operator team. In this respect, company officers must exercise good judgment in choosing the team. Officers must enforce strict adherence to good driving practices and compliance with traffic laws. It is equally important that company officers provide their relief drivers and tiller operators with driving practice, whenever possible, in order that they may maintain their driving and tillering skills; this allows them an opportunity to thoroughly learn the techniques used by the regular apparatus operators. It is also good practice for the regular driver of a truck company apparatus to practice tillering in order to become more familiar with joint driver and tiller operator responsibilities.

Driver–Tiller Operator Signals

A system of signals between the driver and tiller operator is essential to the safe operation of a tractor-trailer ladder truck. Either a horn or buzzer is provided for the driver and tiller operator. Use of the following signal code by the crew is mandatory:

One blast . . . STOP IMMEDIATELY
Two blasts . . . PROCEED FORWARD
Three blasts . . . REVERSE

Signals should be given distinctly and may originate with either the driver or tiller operator. The signal for FORWARD or REVERSE *MUST* be acknowledged by a repeat of the signal before action is taken by the driver. The STOP signal is intended primarily for use by the tiller operator to signal an emergency stop or a necessary stop during a backing operation. A series of short signal blasts may be utilized by the tiller operator to gain the attention of the driver or to request a nonemergency halt of the vehicle.

The forward signal may be used by the tiller operator to indicate that the tiller end of the truck is in the proper position and ready to proceed forward. It is the company officer's responsibility to see that crews are safely positioned on the apparatus before directing the driver to roll; however, the order should not be given until the tiller operator has signalled that all is properly secure and ready to go. There is no excuse for an accident due to a truck leaving the station or proceeding forward before the tiller operator is properly seated and ready.

The reverse signal signifies that a backing operation is necessary. When given by the tiller operator, it shows that the way is clear and that all is in readiness to maneuver the trailer backward.

Drivers of a ladder truck can always aid the tiller operator by clearly indicating their turns ahead of time. This can be done by giving standard arm signals well in advance of legal requirements so that the tiller operator may have ample time to anticipate the tractor movement. Duplication of this arm signal by the tiller operator will also warn other vehicles and pedestrians of the truck's direction of travel. Because of elevated position, the tiller operator may often find it advantageous to the prevention of accidents to supplement the standard arm signals by additional arm movements to attract attention.

The Tiller Operator

The tiller operator must be thoroughly competent. Driving assignments include not only straight-line driving, turning and backing, but also the proper placement of the trailer at fires so as to ensure the ready removal of ladders and the safe operation of the aerial ladder. Tiller operators of aerial trucks must be particularly aware of such factors as the distance of the trailer from the base of the building involved, the angle of trailer placement in regard to the position of the tractor, proper clearance from overhead, side and rear obstructions, and grades and slopes of the working area. Tiller operators must also be qualified to operate the aerial ladder, as well as be familiar with the duties assigned to all truck company personnel. They share the driver's responsibility for the safety of the public and of other vehicles when responding to or returning from alarms. A thorough knowledge of prescribed routes to first-alarm assignments and of conditions to be anticipated are also essential to an efficient tiller operation. Finally, the tiller operator has the responsibility for the proper care and maintenance of the trailer's apparatus and of ladders, tools and other equipment carried.

Tiller Control When Leaving Fire Station

When preparing to drive a tractor-trailer ladder truck from a fire station, the tiller operator has several items to check before giving or returning the signal to proceed forward. On approaching the vehicle, the position of the trailer wheels for alignment with the tractor should be checked. If they are not in line, they must be brought into line as soon as the apparatus starts to move. The elevated seat position provides the tiller operator with a vantage point from which to check the readiness and safety of the crew and the proper security of ladders and equipment. The tiller operator must be secured in the bucket seat by the provided safety belt before giving the signal to proceed.

As the truck starts to move, the tiller operator should center the trailer in the doorway (Figure 18-8). In leaving some stations it is necessary, because of a narrow street or obstruction, for the tractor to make a sharp turn. Assuming a sharp right turn is started by the driver, it may be necessary to turn the tiller wheel slightly to the left to keep centered in the doorway. Clearance must be maintained on the right side of the trailer ahead of the tiller operator. As the trailer wheels clear the doorway and the tractor continues to turn further to the right, the trailer pivots on its wheels and moves the ladder overhang towards the left side of the door. This pivoting action must be corrected by steering to the right to keep the ladder overhang in the center of the doorway. If the reverse of the above situation occurs and the ladder truck is turned sharply to the left out of the doorway, then compensating tillering precisely the reverse of that described will be necessary.

Tiller Control When Traveling Forward

Tiller operators should, whenever possible, keep the trailer wheels parallel with the frame of the trailer and in a direct line with the tractor. This is especially important

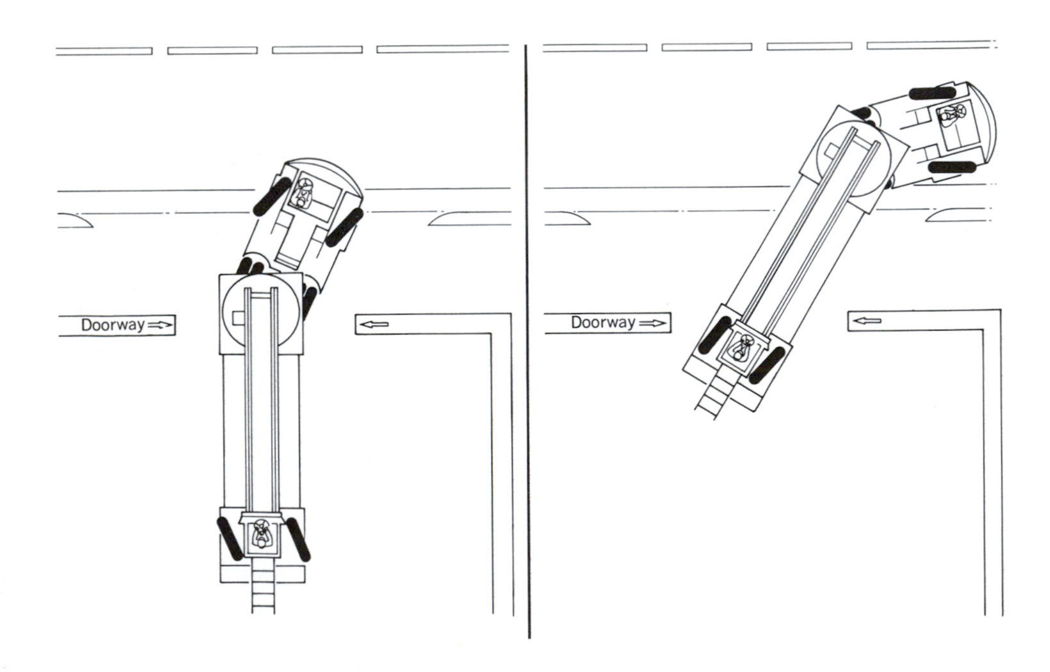

Figure 18-8. Tiller control on leaving station

when streets are wet or icy, and the danger of skidding is more probable. When the tractor and trailer are not in line, the width of the apparatus is greatly increased; this also increases the chance of being involved in an accident.

When traveling in a forward direction, turning the tiller wheel counterclockwise turns the trailer wheels to the left and moves the rear of the trailer to the left. Turning the tiller wheel clockwise moves the trailer wheels and the rear of the trailer to the right. When the tractor and trailer are in line, little actual movement of the tiller wheel will be necessary while traveling forward. Under this condition, it is normally necessary for tiller operators to control only the drift to one side or the other, similar to that which occurs when driving a passenger car forward. When making a turn, or when required to weave in and out of traffic on an emergency response, the tiller operator should keep the trailer in line with the tractor as much as possible; the tiller wheel should be turned only enough to maintain adequate clearances. In most turns where traffic is congested by stopped or parked vehicles, or where narrow streets are involved, the tiller operator may have to turn the wheels in order to keep the trailer in alignment with the tractor and to maintain adequate clearance for the trailer and ladder overhang. This particular maneuver must also be compensated for by a smooth recovery turn of the trailer wheels to maintain a proper tractor-trailer alignment.

It is essential that the tiller operator be alert to conditions on streets ahead whenever an apparatus is traveling. Particular attention should be given to trucks, buses and streetcars because their height presents an added hazard to the ladder overhang and may require the tiller operator to take precautionary maneuvers. Tiller operators who are alert and who anticipate situations that may develop, can plan compensating measures sufficiently in advance to prevent an accident. It is the duty of all of the truck company, as well as that of the driver and tiller operator, to keep a sharp lookout for any overhanging obstructions which could injure the tiller operator or damage the apparatus.

Tiller Control at Intersections

Particular care and alertness must be exercised by the driver and tiller operator at intersections (Figure 18-9). If the intersection is clear and wide, a right or left turn should not be difficult if the driver steers the tractor into a wide turn. If traffic is heavy, however, or streets are narrow, extra precautions and good judgment must be used; the tiller operator must keep constantly aware of vehicles on each side and to the rear of the apparatus. Even after starting into a turn, developments may occur which will require the tiller operator to maneuver the overhang of the trailer into the clear.

There are several methods of turning at an intersection, depending on the conditions and the immediate requirements. On most corners the trailer will track without an excessive amount of tillering. When approaching a turn at an intersection, the tiller operator should anticipate whether additional traffic lane space will be required to make the turn. It may be necessary to partially block a line of traffic from the rear by steering the trailer slightly into whichever lane it is necessary to keep clear. When it is necessary to move out wide with the trailer on making a turn at an intersection, the tiller operator should start turning away from the corner at about the time that the front tractor wheels first enter the intersection. This maneuver will prevent overrunning the in-line position of the tractor-trailer and will compensate for the centrifugal force of the turn.

If it is necessary for the driver to turn sharply at a point near an intersection curb line, the tiller operator will have to turn the tiller wheel quickly in the opposite direction to avoid striking or overrunning the curb. It

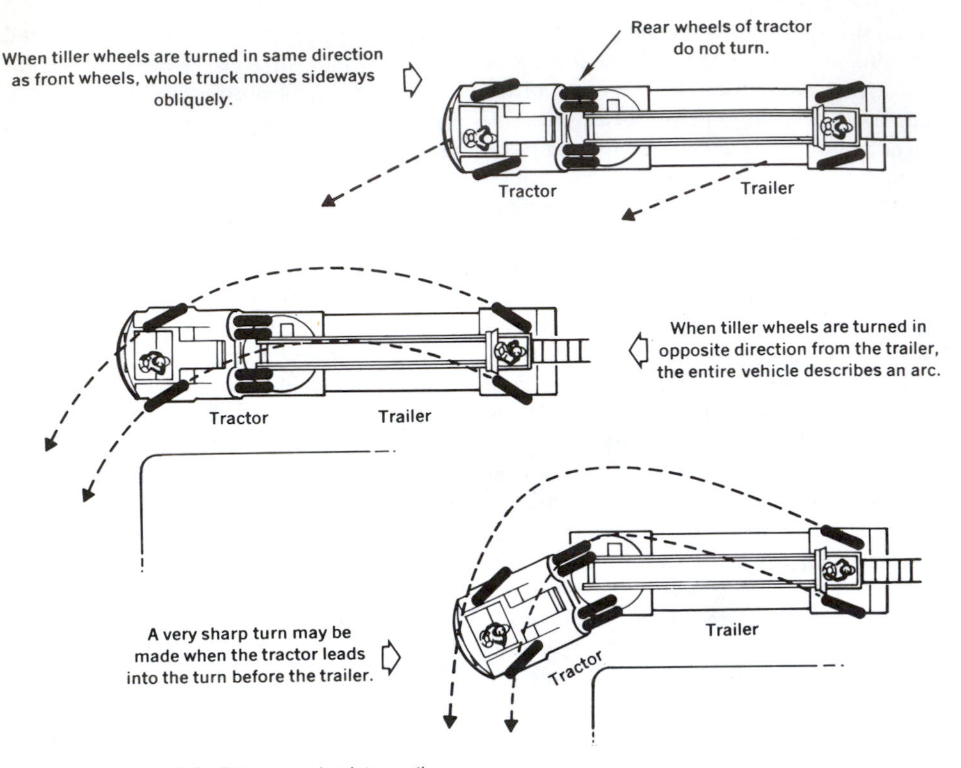

Figure 18-9. Tiller control at intersections

is essential that the tiller operator should properly judge and maintain side clearances on any turn sufficient to permit the safe passage of the rear overhang of the trailer. If at any time there is doubt regarding safe clearance, the tiller operator should immediately signal the driver to stop.

Do not over-tiller on turns. Avoid making a sudden swing-out that will require a sharp counter swing-in unless it becomes necessary to avoid a collison. After a turn, the tiller operator should bring the trailer in line with the tractor quickly and smoothly.

Maneuvering Ladder Trucks Backward

In any backing maneuver of an aerial ladder truck, close coordinated action must be maintained between the driver and the tiller operator, particularly in the use of signals. It is very easy to inadvertently jackknife the apparatus when backing if close cooperation is not maintained. It is also essential that fire crews be stationed as guides to warn the driver and the tiller operator of any inadequate clearances and to control traffic during the backing of the apparatus.

In backing operations, the driver must control the movement of the apparatus at a slow and smooth speed, and be in the position to stop the apparatus immediately and to steer the tractor so that it tracks the trailer in proper alignment. The tiller operator must pay particular attention to the position of the tractor, as well as to clearance on both sides and to the rear of the trailer. The trailer wheels must be smoothly guided to the parking spot and the wheels must be aligned with the trailer frame when the apparatus is stopped. Steering trailer wheels of a tractor-trailer apparatus, *when backing*, requires the opposite control of that used to steer a single unit vehicle. For example, if the tiller wheel is turned to the right, the trailer will travel sideways to the left; if the tiller wheel is turned to the left, the trailer will move to the right. This deviation from normal apparatus control is one of the reasons why company officers should allow relief drivers and tiller operators every chance to practice.

A good rule to remember is that when the truck is moving forward, with the hand at the top or front of the tiller steering wheel, the rear of the trailer will move in the direction the hand is moved. When backing, with the hand at the bottom or rear of the tiller steering wheel, the trailer will move in the direction the hand is moved.

Backing into the Station

Unless a fire fighter is lucky enough to belong to a company whose engine house has front and rear doors, the apparatus will have to be backed into the station (Figure 18-10). The procedure here will vary with the width of the street, the set-back of company quarters and traffic conditions. Of equal concern during the entire maneuver will be the securing of coodinated action between the driver and the tiller operator through the use of signals.

Where the street in front of quarters is wide with no obstructions, the driver should stop the apparatus short of the station on the right side of the street and have company members control traffic. Then the driver should make the required turn from a point fairly close to the front of the station. The

Figure 18-10. Maneuvering ladder truck backward

tiller operator should turn in the opposite direction from that of the tractor until the trailer is in line with the doorway. The trailer wheels should be straightened for backing just prior to stopping the truck. Check for obstructions and traffic and, if clear at this point, prepare for backing. If the steering wheel is held at a point near the bottom, it is a simple operation to push the wheel toward the direction of travel that is required. If this method is used, it will eliminate the possibility of turning the wheels in the wrong direction. In backing operations, tiller operators should pay particular attention to the position of the tractor, as well as to clearance on both sides and the rear. When parking apparatus in quarters, guide the trailer wheels to the predetermined parking spot and align them with the truck.

When the street in front of the engine house is narrow, post fire crew to control traffic. The truck should then be driven to the side of the street opposite the station. The tiller operator should steer the trailer wheels so that the rear of the trailer is drawn to about the center of street. The apparatus should be stopped when the rear of the trailer is a few feet past the door of the station.

The driver must back the tractor in the same arc as that used by the trailer to prevent any possibility of a jackknife. Particular attention must be paid to maintain adequate clearance on either side of the tractor and at the front of the trailer as the apparatus enters the doorway. The tractor must be properly aligned with the trailer, and, as the apparatus is parked, any signal to stop must be immediately obeyed when given by the company officer, the tiller operator or fire crew stationed as guides.

Jackknifing

Jackknifing is a type of accidental tractor-trailer skid in which the tractor or trailer skids sideways due to poor road conditions, or to improper braking or steering techniques (Figure 18-11). There are two distinct kinds of jackknifing:

1. Tractor jackknife, in which the rear of the tractor skids sideways and comes around;
2. Trailer jackknife, in which the rear of the trailer skids sideways and comes around.

It is very difficult to correct a jackknife that has gone beyond a 15° angle. Furthermore, the faster this critical 15° angle develops, the greater the force and severity of the jackknife. On the other hand, if recovery of a jackknife is started before it reaches a 15° angle, control is more easily regained and far more certain.

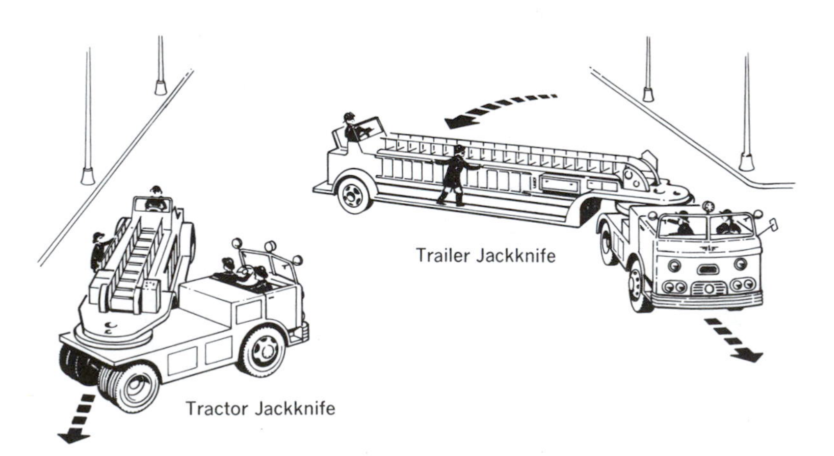

Figure 18-11. Tractor-trailer jackknife

Promptness in starting corrective action is the most important point in recovering from a jackknife. Not only must the driver know what to do if the truck begins to jackknife, but it must be done immediately. When pavements are wet or icy, steering with little or no use of the throttle or the brakes is the most effective driving technique for recovery from a jackknife. If pavements are dry, moderate acceleration is sometimes recommended.

If a tractor jackknife develops, immediately turn the front wheels toward the same direction toward which the rear of the tractor is sliding, similar to the method of recovering from a skid in an automobile.

If a trailer jackknife develops, it is probably due to locked trailer wheels because of overbraking by the driver. When these wheels are locked, they have no directional power and will be subject to any external side force; this is completely independent of other forces acting on the vehicle. Recovery from this situation calls for quickly getting all wheels rolling again by partially releasing the brakes; steering is very important to bring the tractor and trailer into line.

Positioning an Aerial Ladder Truck

The proper positioning or parking of a ladder truck is the responsibility of the company officer in charge of the apparatus. In no case should any truck be parked where it can obstruct access to the fire area for other apparatus. This precaution is particularly important in the case of fires in narrow alleys; in that case, the truck should be parked near the street and alley intersections closest to the fire, unless otherwise ordered by the officer in charge of fire.

When parking a ladder truck, it is the responsibility of the tiller operator to make certain that the area directly to the rear of the trailer is clear so that bed ladders can be removed easily. It may be advantageous to stop the trailer so that the rear is farther from the curb than the front; holding the rear of the trailer out in this manner will allow bed ladders to be removed even if another vehicle parks behind the truck. To facilitate operation of the aerial and removal of bed ladders if the company is ordered to work, the tiller operator should move the tiller seat and post out of the way. (With some makes of apparatus this is not necessary.) When a ladder truck is on a grade, the vehicle should be placed with the tractor in the downhill position whenever possible. This precaution is essential, particularly in the case of ladder trucks on which the removal of a tiller shaft will free all bed ladders.

The tiller operator should be alerted immediately if an aerial is to be raised. Because of the height of the position at the rear of the trailer, the operator can be of considerable assistance to the driver in spotting conditions that may affect the placement and operation of the aerial ladder. Overhead obstructions, high-tension wires, street clearances, distance from building, and other factors that affect the spotting of the apparatus are best analyzed by the tiller operator.

Tillering, like other specialized operations, requires considerable practice and training. A beginner must be closely supervised by an experienced tiller operator. Supervisors should be in a position to take over the tillering instantly if necessary.

They should also stress the importance of the following:

1. Good signal practice
2. Trailing in-line on a straight roadway
3. Bringing the trailer into line as the turn is completed
4. Adequately observing the ladder overhang on turns
5. Avoiding rough and jerky maneuvers
6. Over-tillering

If the tiller operator observes these precautions and has the alert cooperation of all members of the company, the aerial truck will be skillfully maneuvered to the emergency.

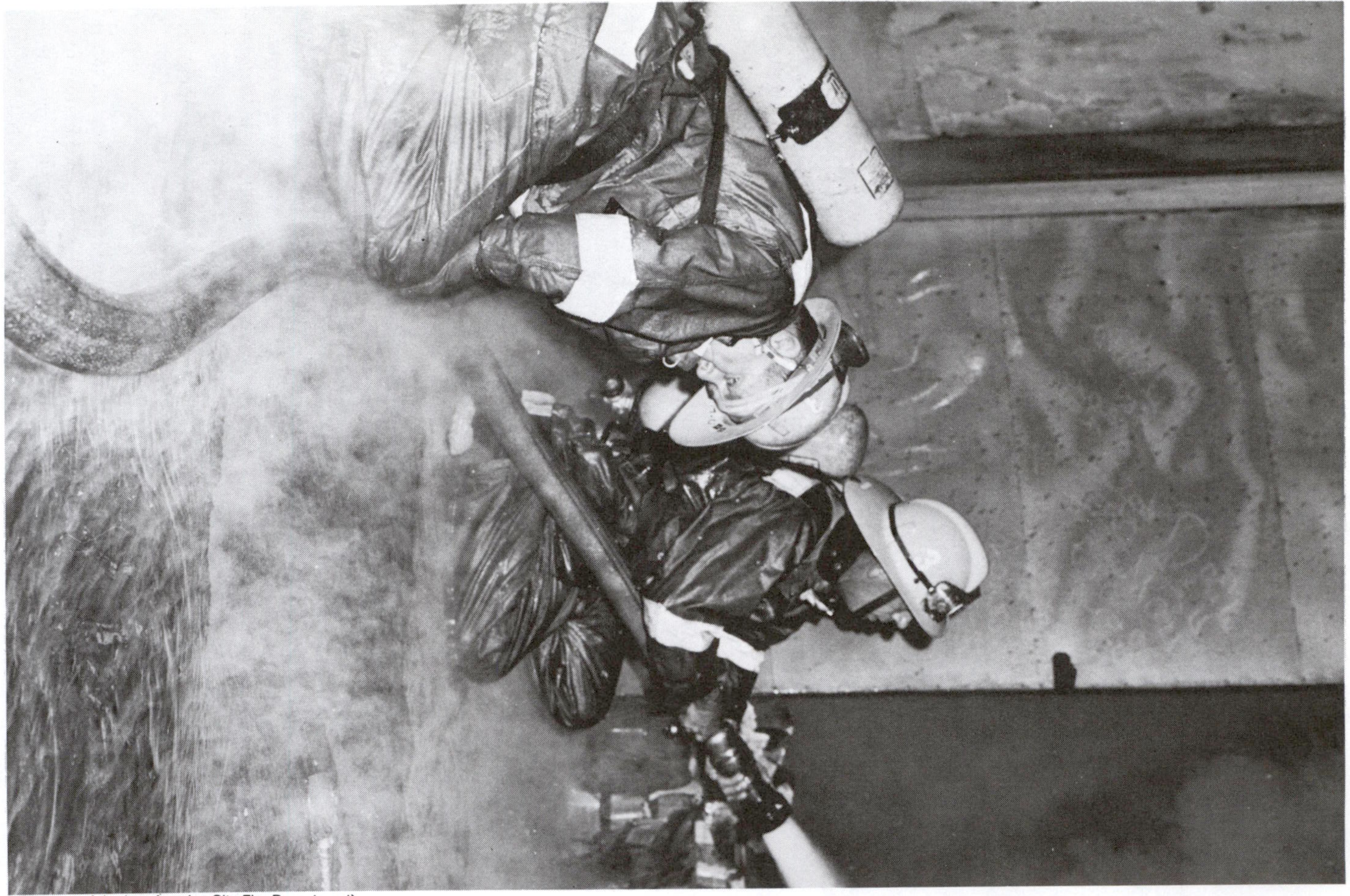

(Courtesy of the Los Angeles City Fire Department)

Glossary of Terms

Acceptance test Test conducted to determine whether new fire apparatus, especially pumping engines, meet the specified pressure and discharge requirements.

Adapter A hose coupling device for connecting hose threads of the same nominal hose size but having different pitch and diameters. *See also* **Reducer.**

Advance a line An order to remove line of hose toward a given area from point where the hose-carrying apparatus has stopped.

Advancing (line) Moving a line of hose by hand toward a fire or toward a position that is to be covered.

Aerial A hydraulically or mechanically operated turntable ladder attached to a ladder truck and manufactured in various lengths, such as 65 ft, 75 ft, 85 ft, 100 ft or higher. Also refers to the complete aerial ladder truck and its ancillary equipment. Aerial ladders are usually built in 3 or 4 sections of aluminum alloy or steel. Earlier types were of wood with two ladder sections.

Aerial ladder truck *See* **Aerial.**

Aerial platform apparatus A mechanically raised platform mounted on fire apparatus and designed for rescue and fire fighting service. The platform may be an articulated (folding) boom, a telescopic boom, or a combination of the two types. This apparatus can operate fire streams into upper floors and provide a means of access to and exit from a building. (Also called *elevating platform apparatus.*)

Aerate Combine water with air.

Air chamber A pump chamber filled with air, used to cushion pulsations caused by operation of pistons or gears displacing water in positive displacement pumps.

Aircraft crash truck *See* **Crash truck.**

Air foam (Mechanical foam) A type of foam for smothering Class B (flammable liquids) fires. It is produced by adding a liquid foaming agent to water to make it capable of foaming in the presence of air introduced by mechanical action in a foam maker or playpipe. The foaming agent is termed a stabilizer.

Air foam nozzles A special playpipe or nozzle incorporating a foam maker to aspirate air into a solution, thus producing air foam.

Air foam pumping A method of producing foam under pressure, employing self-contained positive displacement pumps to automatically proportion and mix air, water, and stabilizer. Some fire department pumping engines are also equipped to make air or mechanical foam, usually by means of an auxiliary tank containing foam stabilizer and a proportioning device to provide required percentage of the stabilizer to mix with water discharged from the fire pump. Where foam liquid or stabilizer does not pass through the pump, it is known as an *around the pump proportioner.*

Air foam stabilizer A liquid which, depending upon its concentration and nature, is used in water in a proportion of 3% or 6% to produce foam. Two fundamental types are: *low expansion* and *high expansion.* Special *alcohol* type, low-expansion foam stabilizer is available for use on fires in liquids that act as foam breakers to destroy foam produced by ordinary foam chemicals or ordinary air foam stabilizers.

Air lock Trapped air in a pump or piping which prevents complete priming of the pump.

Alternator An alternating current generator provided on automotive vehicles. It has the advantage of maintaining battery charge at relatively low engine speeds. Where a suitable transformer is provided, it may also provide current for small power tools and floodlights.

Angle of departure Referring to the rear overhang of fire apparatus, the angle between the road surface and a line from the rear point of contact of the rear tire to any unsprung projection of the vehicle behind the rear axle.

Apparatus A term referring to mobile vehicles, including fire boats, which are used to transport fire department personnel, equipment and appliances to fires or other emergency incidents.

Apparatus floor The main floor of a fire department station on which the fire apparatus is quartered; it also includes the watch desk, patrol room and other facilities. In a restricted sense, the actual room where the apparatus is housed.

Appliance Any of a variety of tools or devices usually carried on fire fighting apparatus. The specific item should be designated as *heavy stream appliances, lighting appliance,* etc.

Applicator A special pipe or nozzle attachment for applying foam or water fog to fires. Designated *foam applicator* or *fog applicator* and usually also identified by the size of the hose employed, as *"2½-in. fog applicator."*

Aqueous film forming foam (AFFF) A fluorinated foaming agent which forms a film that floats on water or flammable liquids to smother the flames by excluding the oxygen.

Articulated boom A folding type hoist used to raise and lower the platform on certain elevating platform apparatus.

Atmospheric displacement The effect produced when water particles applied in a fire area are converted into steam, expanding some 1650 times in volume; this displaces heated atmosphere and carries heat outside of the burning building.

Atmospheric pressure Normally the weight of the air or atmosphere, which at sea level is approximately 14.7 psi; this is equivalent to 33.9-ft head of water. With fire department pumpers, pressures below atmospheric are measured by gage readings calibrated in inches of mercury, which closely approximate the suction lift in feet. Each 2.314 ft of lift requires 1 lb of pressure, plus any friction loss due to movement of water through the hose and strainers.

Attached A nozzle, hose, or other device carried or connected to hose or pump.

Attaching Connecting a nozzle to a hose line, or hose to pump or hydrant outlet.

Attack The actual physical fire fighting operation utilizing available fire crew and equipment. The implementation of tactical plans on the fire ground in an aggressive manner.

Attack line A line of hose, usually from a pump, used to directly fight or attack a fire; this is contrasted with supply or feeder lines connecting the water supply with the pumping apparatus.

Auxiliary Supplemental fire fighting equipment or personnel which may not be part of the regular first line fire fighting equipment or manpower; frequently this means makeshift or reconditioned equipment provided for abnormal emergencies or a secondary unit that supplements heavy duty equipment.

Auxiliary pump A pump used on fire apparatus, with a capacity rating less than the minimum standard rated capacity of 500 gpm at 150 psi. In the auxiliary classification there are two types, booster pumps and high-pressure pumps.

Axle locks Locks provided on tractor-trailer type aerial ladder trucks to immobilize the chassis and prevent spring action prior to raising an aerial ladder. The locks prevent movement of the axle when the ladder is elevated.

Back pressure A pressure (head) of 0.434 psi per foot of elevation, due to the weight of a column of water. In fire department practice, 5 psi is added to pump pressure for each floor (or 4½ psi for each 10 ft) of elevation of the nozzle above the pump to overcome back pressure caused by gravity.

Back flushing Flushing of the fire fighting equipment with plain water, from discharge outlets to suction inlets, to remove clogging debris, traces of salt water, or chemicals used in fire fighting.

Backup line Ordinarily a line of 2½-in. or larger hose laid in the event that the initial attack with small lines proves inadequate. Also, an additional hose line backing up the attack and protecting personnel using fog lines for a close attack on a flammable liquids fire.

Baffle A compartment provided in water tanks to prevent shifting of the water load when apparatus is in motion. Also, a partition dividing a hose body into compartments.

Basket The mobile platform of an aerial platform apparatus. A container, usually of heavy gauge wire or sheet steel, placed on fire apparatus for the purpose of carrying small hose or various tools. Also basket stretchers used for rescue purposes.

Beam The main structural members of a ladder supporting the rungs or rung blocks.

Bed ladder The lower section of an extension ladder into which the upper sections or fly ladders retract. The lower section of an aerial ladder which beds on to the truck frame.

Big line A line of 2½-in. or larger fire hose, especially when used as a handline.

Big stick The raised main ladder of an aerial ladder truck. The term originated in the time when aerial ladders were of wooden construction. Usually used in reference to the 85-ft aerial, which was the largest wooden aerial ladder.

Blaze A free-burning spectacular fire characterized by a generous amount of flame.

Bleed Drain water from hose or piping, to remove pressure preparatory to breaking connections.

Blitz attack A rapid and massive attack on a fire, usually using the booster tank to supply large hose lines, wagon battery, or a water tower to knock down a fire before it has a chance to accelerate to a major blaze.

Boat A fireboat. A boat with pumps, hose, monitor nozzles, and other fire fighting equipment and appliances and operated as a company in the fire department.

Body of fire An intense mass of flame accompanied by heavy smoke under pressure, indicating the heart of a fire; distinguished from light concentrations of smoke or small amounts of flame.

Boom One of the operating sections of an elevating platform or water tower.

Booster Small line equipment consisting of water tank and pump using ¾-in., 1-in., or 1½-in. hose and small nozzles.

Booster pump A pump of less than 500 gpm rating carried as an integral part of fire apparatus and used to supply streams through small hose lines.

Booster reel A reel for booster hose mounted on fire apparatus. Usually carries ¾-in. or 1-in. hose.

Booster tank The water tank on a triple combination pumper or other apparatus equipped with a pump and hose.

Brakes The long levers or bars by which a hand fire engine is pumped. Also, any braking devices used on vehicles.

Break a line Disconnect a line to insert a valve or attach a nozzle.

Break coupling Detach two pieces of hose by backing the swivel thread (female) off the nipple thread (male).

Bresnan nozzle A rotating type of distributor nozzle for use in cellar fires.

BTU (btu) British Thermal Unit. The amount of heat required to raise the temperature of one pound of water one degree Fahrenheit.

Butt One coupling of a fire hose; the end of a section of hose; the foot or lower end of a ladder.

Bypass valve Any of a number of valves provided to pass water or other extinguishing agents around part of the normal piping or equipment. For example, a tank fill valve may bypass water around a check valve. In particular, a pressure relief valve may bypass water to prevent building up pressure. A churn valve.

Cantilever position The position of an aerial ladder supported only at the base by the turntable and unsupported at the top. Recommended safe ladder loadings are less when the ladder is in a cantilever position than when it is supported at the top. A cantilever beam or truss is one that is rigidly connected to a fixed support and is free to move at the other end.

Capacity operation Operation of a pumper at or near its rated capacity and rated pressure. A multistage pressure-volume type of pump provides capacity during parallel, rather than during series, operation. Each pump impeller provides its rated volume at rated pressure when in the capacity (parallel or volume) position. *See also* **Parallel operation** and **Pressure operation.**

Cavitation A condition in which air cavities are formed in a pump, resulting in vibrations, loss of efficiency, and possible damage to the pump. It occurs chiefly in areas of high vacuum and may be referred to as *pump is running away from the water.* The impeller may move faster than the water can move at the existing input pressure, and water vapor may be released from the water due to low vapor pressure.

Cellar nozzle A special nozzle for attacking fires in basements, cellars, and other spaces below floors or decks. Not to be confused with a distributor nozzle used for the same service.

Centrifugal pump A pump that uses impellers to impact velocity to water by centrifugal force.

Changeover valve A transfer valve on a multistage pump or a pressure-volume type pump. The valve enables the pump to provide either rated capacity at specified test pressure (usually 150 psi net pump pressure) or smaller flows at higher pressures *See also* **Transfer valve.**

Characteristic curve A graph showing power or work output at various pressures or speeds. The characteristic curve of a fire pump could show the output of the pump at various discharge pressures.

Charged line A line of hose filled with water, ready for use and under pressure.

Chassis The vehicular frame of an apparatus on which the body, pump, aerial ladder, and other components are attached.

Check valve A valve designed to permit

flow in one direction only, but which closes to prevent flow in the opposite direction. For example, a check valve may be provided to permit water to flow on demand from a water tank on a pumper, and to close to prevent inadvertent overfilling of the tank from the fire pump. With fire fighting equipment, a *clapper valve* is used as a check valve.

Chemical foam A type of foam designed for smothering Class B (flammable liquid) fires. It is made by the reaction of an alkaline salt solution (usually bicarbonate of soda) and an acid salt solution (usually aluminum sulfate) to form a gas (carbon dioxide) in the presence of a foaming agent, which causes the gas to be trapped in tough, fire-resistant bubbles.

Chocks *See* **Wheel blocks.**

Churn valve A pressure relief valve, sometimes manually operated, which bypasses water from the discharge side to the suction side of the pump, thus reducing pressure.

Clapper A hinged valve that permits flow of water in only one direction. A check valve in a siamese connection that closes to prevent escape of water through an unused inlet.

Class A pumper The present performance standards for fire department pumpers require that Class A pumpers deliver rated capacity at draft at 150 psi net pump pressure at a lift not exceeding 10 ft; they must also deliver 70% of rated capacity at 200 psi and 50% of rated capacity at 250 psi.

Class B pumper An earlier performance standard for pumpers, adopted about 1912, required that Class B pumpers deliver rated capacity at draft at 120 psi net pump pressure, 50% of rated capacity at 200 psi, and 33⅓% of rated capacity at 250 psi. The pressure-volume requirements at 120 psi and 200 psi net pump pressure called for in this older standard are now applied to booster pumps of less than 500 gpm rated capacity.

Combination A piece of fire apparatus or equipment designed to perform two or more functions. Originally, a combination of hose wagon with a chemical engine. Later, the addition of an engine driven pump made a *triple combination.* The addition of long ladders made a quadruple combination or *quad,* the addition of an aerial ladder made a *quint.* Booster tank equipment replaced the chemical engine a long time ago.

Combination nozzle A nozzle designed to provide either a solid stream or a fixed-spray pattern. An all-purpose nozzle.

Combustible A material or structure that ignites and burns at temperatures ordinarily encountered at fires. Technically, a material which, when heated, gives off vapors that in the presence of oxygen (air) may be oxidized and consumed by fire.

Company A basic fire fighting organizational unit headed by an officer. A company is usually organized to operate certain types of apparatus: an engine company, a truck company, or a rescue company.

Compound gage A pressure gage on a pumper capable of recording pressures both above and below atmospheric. Pressures below atmospheric are usually measured in inches of mercury and occur when a pump is drafting; pressures above atmospheric are measured in pounds per square inch (psi).

Coupling For convenience in handling and replacement, fire hose is usually cut into lengths of 50 ft. These sections must be coupled together to produce a continuous hose line; each section of hose will have a male threaded shank on one end and a female threaded swivel on the other end, with a gasket to make a leakproof connection.

Crack a nozzle Open a nozzle slightly to allow a small amount of water to discharge or to clear air out of a line.

Crash truck Specialized fire fighting apparatus designed to handle fires and accidents involving aircraft. Such apparatus usually embodies special extinguishing capability for handling large fires involving flammable liquids.

Deadman control A pedal, lever, switch, or button that must be depressed before a device can be operated. Elevating platforms and water towers incorporate these devices so operation will cease if the operator is disturbed.

Dead-end main A water main that is supplied from one direction.

Deck gun Also referred to as a deck monitor, wagon battery, and turret. This term originated with the large monitors on the deck of a fireboat, but now it is also used to describe a heavy-stream appliance which is mounted midship on a pumper or hose wagon.

Deluge set A heavy-stream device.

Discharge gage A pressure gage recording psi discharge of water from a pump.

Discharge gate A gate valve provided to control the flow of water from an individual pump outlet to a hose line.

Distribution grid The interconnected network of water mains that most effectively furnish a water supply to an area.

Draft To draw water from static sources into a pump that is above the level of the water supply. This is done by removing the air from the pump and allowing atmospheric pressure (14.7 psi at sea level) to push water through a suction hose into the pump.

Drains Openings to the outside, usually closed by valves, through which water may be removed from pumps and hose lines.

Drill Practice of fire fighting procedures, such as laying hose, raising ladders, and operating pumps, in order to develop teamwork and proficiency.

Dry barrel A type of hydrant that is used in areas where freezing conditions are likely to occur. The valve that controls the water is buried below ground; opening the hydrant valve will fill the hydrant barrel. It is a good idea to place a valve on the unused outlets so they may later be used without shutting down the hydrant flow.

Dump valve A quick-opening valve that can be utilized between the booster tank and the pump or to rapidly dump the contents of a large tanker into a portable tank.

Eductor A device placed in a hose line to proportion liquid foam or wetting agents into the fire stream. They, like proportioners, operate on the venturi principle. Also, an ejector device that siphons water by creating a vacuum from the velocity of water passing through it.

Effective height The vertical distance that an aerial apparatus can effectively serve.

Effective length The distance that an aerial ladder can attain. This is measured from the top ladder rung to the ground when the aerial is elevated to its maximum angle.

Effective reach The maximum horizontal distance that an aerial apparatus can extend and effectively serve.

Elevated streams Large-capacity fire streams from a ladder pipe, water tower, or elevating platform. Occasionally, this term may be used to describe fire streams from handlines which are being directed on a blaze from an elevated position, such as a roof top.

Ejector A suction device that creates a vacuum from the velocity of a stream of gas or water passing through it. *See also* **Eductor, Suction booster,** and **Siphon.**

Elevating platform apparatus *See* **Aerial platform apparatus.**

Engine A fire department pumper. Also, the internal combustion engine used to propel vehicles and to provide power to operate pumps and other power consuming equipment. Sometimes termed a *motor:* in general, however, an engine generates power and a motor utilizes power.

Engine company A fire company equipped with a pumping engine. A pumper company.

Engineer A pumper operator or licensed fireboat engineer.

Engine house An earlier form of *fire house,* arising from the days when a hand engine or steam fire engine was the principle fire fighting apparatus.

Evolution An agreed operational sequence requiring teamwork and covering various basic fire fighting tasks, such as the placement of hose lines and ladders.

Excessive heat Temperatures, generally in excess of 300° F, which tend to vaporize exposed fuels and to enable fires to spread. The one abnormal part of the fire triangle.

Exposure Property that may be endangered by a fire in another structure or by an outside fire. In general, property within 40 ft may be considered to present exposure hazard, although in very large fires, danger may exist at much greater distances. Flying brands may present an exposure hazard for extensive distances if there are, for instance, wooden shingle roofs in the path of the brands. After the saving of life, the protection of exposures is the first duty of the fire service; the safety of the community comes before that of any individual property owner who should have provided private protection. However, in the vast majority of cases, the extinguishment of a fire is the most effective method of protecting exposures.

Extend A line may be extended beyond the nozzle by attaching hose to the nozzle tip thread or by removing the nozzle and attaching hose. A fire may extend to areas that were not burning when the alarm was given.

Extension ladder A ladder of two or more sections that may be extended to various heights.

Extension (of fire) Spread of fire, usually during the course of fire fighting operations, to areas not previously involved; an extension of fire through open partitions into an attic, or extension through openings into another room or building.

Extinguish To put out flames; essentially, to completely control the fire so that no abnormal heat or smoke remains. To quench.

Feeder line A line of hose, from water source, used to supply pumping apparatus stationed near a fire with water to be discharged through leader lines or wagon batteries. A supply line, as contrasted with a fire fighting or attack line.

Female coupling A swivel coupling made to receive a male hose nipple of approximately the same thread pitch and diameter.

Fire Rapid oxidation of combustible materials which results in light and heat. Burning.

Fire behavior The manner in which fuel ignites, flame develops, and fire spreads. Sometimes used to distinguish characteristics of a particular fire from typical fire characteristics.

Fireboat *See* **Boat.**

Fire flow The volume of water required in a specific section or area. The quantity should be stated both in gallons per minute (gpm) and in specific duration of time. Fire flow is the anticipated increased volume of water that will be required in addition to the normal average water consumption in the same area.

Fire ground The operational area at a fire at which the ranking fire officer is in charge of all operations and at which fire fighting is under way.

Fire hose *See* **Hose.**

Fire house A fire department station housing fire department apparatus and fire department members while on duty. A fire station.

Fire pump The main pump on a fire department pumper. A stationary water pump approved for private fire service.

Fire service The organization that supplies fire prevention and fire fighting service to the community; its members, individually and collectively.

Fire stream A stream of water of suitable shape and pressure, directed from a nozzle, and effective for the control of fires.

Fire truck Any piece of motorized fire fighting apparatus; a ladder truck.

First-in The first-arriving unit, company, or fire crew to an incident or reported fire.

First water Stream from the first unit to get water on a fire. Formerly considered a point of honor, but now frequently of less importance because of a greater emphasis on the proper placement of the fire stream.

Flame The light from burning gases and incandescent particles that accompany a fire.

Flow pressure The pressure indicated on the suction gage after the water has been placed in motion. *Flow pressure* and *residual pressure* are often used interchangeably.

Fly The upper section of an extension or aerial ladder. The top of the ladder.

Foam A fluffy foam formed by compounds introduced into a stream of water (by special nozzles or proportioning devices) to develop a stream of tenacious foam capable of smothering fires, especially those involving flammable liquids.

Foam applicator *See* **Applicator.**

Foam blanket The covering of foam applied over a burning surface for smothering effect.

Fog A jet or cloud of fine water spray discharged by fog or spray nozzles.

Fog cone The angle at which water spray leaves a fog nozzle; usually taken as two opposite sides of the jet with the orifice as the base of the angle.

Foot plates Also referred to as ground plates. Large metal plates placed beneath the outriggers and jacks to increase the ground contact area. The plates for aerial ladders and water towers should be not less than 320 sq. in.; the area of plates for elevating platforms should be not less than 575 sq. in. in size.

Forward lay Before the days of pumpers, the normal hose lay was from hydrant to fire so that advantage could be taken of the hydrant pressure. Now forward lay usually refers to the laying of hose in the manner envisioned when hose was loaded on the apparatus.

Four-way valve A hydrant connection valve permitting initial use of a direct hydrant stream and subsequent connection of the pumper without shutting down the hose stream.

Free burning Materials or structure lacking appreciable fire resistance and subject to quick hot fires. Fire resulting from the burning of such materials in the presence of adequate oxygen.

Friction The resistance to relative motion between two bodies in contact.

Friction loss A term commonly applied to loss of pressure in fire hose, pipe, or fittings. The actual friction is due to resistance to motion between the water and the inside of the hose or fitting.

Front As distinguished from the rear, it is usually the position in fire fighting assumed by the commanding officer and one of the first positions covered upon the arrival of fire fighting units. In general, the first-due company takes the front inside position if possible, taking care not to drive a fire towards the rear, but protecting vertical openings.

Front mount pump A fire apparatus pump located in front of the vehicle radiator; usually used in rural service.

Fuel Combustible material adding to the magnitude or intensity of a fire or combining with oxygen to contribute to the burning process. One of the three essentials in the fire triangle of oxygen, heat, and fuel.

Fully involved The entire area of a fire building so involved with heat, smoke, and flame that immediate access to the interior is not possible until some measure of control has been obtained with hose streams.

Gage (or Gauge) A device giving indication of pressure, speed, etc. *Gage* usually refers to an indicating instrument, while *gauge* describes thickness or the act of measuring.

Gallon A U.S. gallon consisting of 231 cu. in. One gallon of water weighs 8.336 lb. An Imperial gallon contains 1.201 U.S. gallons.

Gasket The rubber or synthetic sealing ring used in a female or swivel hose coupling to make a water-tight connection.

Gate A control valve for a hose, a pump outlet, or a large caliber nozzle.

Gate valve *See* **Gate.**

Gated inlets Suction connections to supply pumps, batteries, water towers, or other appliances. There is a valve to control the water; this will allow additional hose lines to be connected after water has been supplied by other inlets.

Governor A control device on an engine, used to prevent overspeeding. A device used to regulate pump discharge pressure by controlling the speed of the engine driving the pump. The governor is set to give a maximum desired pressure and to slow the engine down when nozzles are shut down; this keeps the pressure within safe limits on other hose lines and within the pump.

GPM (gpm) Gallons per minute, the measure of water flow in fire fighting; it is used to measure the output of fire department pumpers, hose streams, nozzles, hydrants, water mains, etc.

Gravity tank A water storage tank for fire protection. Sometimes used for community water service. A water level of 100 ft provides a static pressure head of 43.4 psi minus the friction losses in piping when water is flowing.

Gross vehicle weight (GVW) The actual vehicle weight, which is the sum of the weights of chassis, body, cab, equipment, water, fuel, crew, and all other load. *See also* **Rated gross vehicle weight.**

Ground jack A heavy jack attached to the frame or chassis of an aerial ladder truck to provide stability when the ladder is raised.

Ground monitor A portable monitor that can be manually carried and placed in the most strategic position to apply a large-capacity fire stream. During long-duration fires, or in hazardous locations, these appliances may be left untended.

Handline A line of small hose that is handled manually by hose operators, rather than through fixed heavy-stream devices.

Hard suction Noncollapsible suction hose for drafting water from static sources lower than the pump.

Head Pressure due to elevation of water, sometimes termed *back pressure* by fire fighters, and amounting to 0.434 psi for each foot of elevation.

Heat Temperatures above the normal atmospheric, as produced by the burning or oxidation process. While there is some heat at all temperatures above absolute zero (the point at which molecular action ceases), the fire fighter is concerned with the abnormal aspects of heat, which induct physical discomfort and which may cause exposed combustible materials to reach an ignition point.

Heavy stream A large-caliber fire stream too heavy for convenient or safe manual operation and therefore discharged through a monitor nozzle, deluge set, ladder pipe, portable monitor, wagon battery, or turret pipe. Usually a fire stream of 400 gpm or more. Also called *master stream.*

Heel The base of a ladder; the heel plates. To take a position at the base of the ladder for raising or lowering. To heel a ladder.

Higbee cut The removal of the first or outside thread of a hose coupling or nipple to prevent crossing or mutilation of threads.

High-pressure fog A small-capacity spray

jet produced at very high pressure and discharged through small hose with a gun-type nozzle.

High-pressure pumper A pumper with special high-pressure pumps and equipment used to supply small, high-pressure fire streams. Also, a pump purchased under specifications calling for a discharge of stated quantities at pressures higher than required for normal pump performance; used in areas where very tall buildings or other circumstances may make extremely high pressures desirable.

Hook and ladder Obsolete term for a ladder truck. In early days, large pulldown hooks were commonly used when fighting fires.

Hooking up Connecting a fire department pumper to a hydrant and connecting hose lines.

Hose A flexible pipe used to convey liquids. Fire hose is usually constructed of a rubber tube covered with one or two woven jackets to protect the inner liner. Double-jacketed hose is most commonly used in the fire service.

Hose coupling *See* **Coupling.**

Hose lay A method or sequence of laying hose from an apparatus.

Hose layout The arrangement of hose lines depended upon to supply the correct volume of water to the fire with the least amount of friction loss.

Hose line A line of hose. May be shortened to *line* or *hose.*

Hose reel A reel permanently mounted on fire apparatus for small fire hose; also called *booster reel.*

Hose roller A metal device with rollers used to protect hose while it is being pulled over cornices and window sills.

Hose thread *See* **Thread.**

Hose wagon A fire truck chiefly used to carry fire hose and usually equipped with a booster pump of less than 500 gpm capacity. Also, a pumping engine used as a hose wagon in a two-piece engine company.

Hydrant A valved outlet to a water supply system with one or more threaded outlets to supply water to fire department hose and pumpers.

Hydraulic hoist A mechanism for operating aerial ladders and platforms that supplies power by means of fluid under pressure.

Hydraulics The study of the use and movements of fluids, especially as pertaining to the extinguishment and control of fires.

Hydromechanical hoist An aerial ladder mechanism using hydraulic power to elevate the ladder but using other mechanical means to rotate the turntable and extend the fly ladders.

Impeller The rotating part of a centrifugal pump which imparts centrifugal force to the water.

Incipient A fire of minor consequence or in its initial stages.

Inclinometer A device on an aerial ladder to show the angle of elevation.

Increaser A coupling used on hose, pump, or fittings to permit connection of a larger size of hose.

Indirect application A technique of injecting water particles into the upper atmospheric level of a fire within a confined space. Its purpose is to generate steam and distribute unvaporized droplets of water that cool heated materials beyond the immediate reach of the fire stream.

Initial attack The first point of attack on a fire; the point where hose lines are used to prevent further extension of the fire and to safeguard life while additional lines are being laid and placed in position.

Inlet An intake water supply connection.

Inlet gage A gage showing pressure at the intake side of the pump. Usually a compound gage, since intake pressure may be above or below atmospheric.

In-line A device or fitting inserted in a hose line used for fire fighting.

Intercom A two-way communication system with a microphone and loudspeaker at each station.

Involved The building, area, room, or structure either actually enveloped in the flame and smoke of a fire, or in danger of such involvement.

Jackknife The spotting of a tractor-trailer type of ladder truck to provide maximum stability. An inadvertent skidding and doubling up of a tractor-trailer type of vehicle.

Jacks Ground jacks supporting the turntable of an aerial ladder. A lifting mechanism to raise heavy objects.

Jet A stream of water from a fire nozzle.

Kinetic energy Energy of an object in motion.

Kink Hose bent back upon itself so as to interfere with the flow of water.

Knock down Reduce flame and heat so as to prevent further extension of a fire. Bring a fire under control.

Ladder company A fire company operating a ladder truck and especially trained in ladder work, ventilation, rescue, forcible entry, and salvage work.

Ladder lock A control for engaging the pawls of an aerial ladder. A control for securing a ladder in its bed when not in use.

Ladder pipe A heavy-stream device attached to an aerial ladder and capable of directing a large stream of water from a height.

Ladder truck A fire department apparatus that carries ladders and other equipment.

Lateral strength The sturdiness of an aerial ladder that will enable it to withstand side thrust.

Lay A method or sequence of stretching hose from fire apparatus.

Lay in A command to stretch or lay out hose while approaching fire with apparatus.

Layout A distribution of hose on the fire ground to achieve certain tactical results. An order to make a hose lay. The act of laying hose from pumper.

Lead A line of hose starting toward a fire from a pump or hose bed.

Leader lines Lines from pump to nozzle, as distinguished from lines supplying the pump.

Length A 50-ft length of fire hose.

Lift Distance in feet of elevation between a static source of water and the suction chamber of a pumper.

Line Usually refers to a line of fire hose. May refer to a length of rope, such as a *life line*.

Loss of pressure Sudden loss or drop in pressure of hose streams used to fight fire.

Lugging A condition of engine overload resulting from failure to develop efficient engine speeds.

Male coupling The threaded hose nipple that fits in the thread of a swivel coupling of the same size thread. A coupling to which nozzles and other appliances are attached.

Master stream Any of a variety of heavy streams formed by utilizing large hose lines or siamesing two or more hose lines into a single heavy stream device. *See also* **Heavy stream.**

Mercury (Hg) A silver-white liquid metal used in thermometers and barometers. Mercury weighs 0.49 lb per cu. in.; thus, if a column of mercury stands at a height of 30 in. (sea level) it indicates a pressure of 14.7 psi (30 × 0.49) is present on the surface of the mercury in the barometer. Vacuum is indicated on gages in terms of inches of mercury with 1 in. of mercury (Hg) equal to approximately ½ psi of pressure (negative). *See also* **Atmospheric pressure.**

Midship pump A fire pump located under or behind the driver's seat and supported on the pumper frame between the front and rear wheels.

Monitor A heavy-stream device used to control large fire streams.

Mop-up A late stage of fire fighting in which remaining hot spots are quenched with small amounts of water.

National standard thread American (U.S.)

National Standard fire hose threads are as follows:

Nominal Hose Size, in.	Number of Threads Per Inch	Outside Diameter of Male Coupling, in.
4-1/2	4	5-3/4
3-1/2	6	4-1/4
3	6	3-5/8
2-1/2	7-1/2	3-1/16
1-1/2	8	2
1	9	1-3/8

Negative pressure A term for pressures below atmospheric; a suction or vacuum.

Net pump pressure The amount of pressure actually developed in the pump. This is determined by subtracting the intake pressure from the discharge pressure.

Nozzle A device attached to a hose or appliance to restrict the area of flow in such a manner as to increase the stream velocity and form a jet which will be effective in reaching and controlling the fire.

Nozzle operator A fire fighter assigned to operate a nozzle on a handline.

Nozzle pressure The amount of pressure measured in pounds per square inch (psi) that is created by the water being discharged from a nozzle.

Nozzle reaction As water leaves a nozzle under pressure, it causes a force in the opposite direction. The amount of this reaction depends on the size of the nozzle tip (gpm) and the pressure (psi), and is due to acceleration of water in the nozzle.

Nozzle tip The interchangeable tip of a nozzle, which will easily allow changing the nozzle discharge. *Tip* is often used interchangeably with the term *nozzle*.

Nurse tanker A water tank truck used to supply a pumper stationed at a fire.

Officer in charge The ranking officer in command at a fire.

Open up To ventilate a building filled with smoke and heat so that hose streams may be advanced to extinguish a fire. Also used in reference to forcible entry of a closed, burning building.

Operating suction lift The vertical distance in feet between the pump center line and the water level, plus the friction losses in suction hose, strainers, and fittings.

Outlets The fittings and valves from which the water is discharged from a pump or tank.

Outrigger jacks Jacks designed to extend outward from the sides of an aerial ladder or elevating platform to provide a wider base of support than provided by the width of the truck chassis.

Overhauling A late stage of the fire extinguishment process during which the area involved in the fire and the contents of the building are carefully scrutinized for any remaining trace of fire or embers; during this time an effort is made to protect the property against further damage from the elements.

Overload Gross vehicle weight in excess of the rated gross vehicle weight specified by the chassis manufacturer, or in excess of axle ratings or permissible tire and rim loadings.

Overload test A test required for new pumping engines to demonstrate ability to develop 10% excess power; this consists of discharging rated capacity at 165 psi net pump pressure for a short time.

Oxygen A gas present in the atmosphere in about 21% concentration and an essential part of the fire triangle (heat, fuel, and oxygen)

Oxygen deficiency Insufficient oxygen to support life or combustion. Where oxygen content of the air falls below 16%, flame production is reduced and asphyxiation of humans may occur.

Parallel Anything equal to or resembling another in all essential particulars; a counterpart.

Parallel operation Operation of a multistage pump with the impellers contributing volume rather than serving as pressure stages.

Pattern Adjustment of a fog or spray nozzle for a particular fog cone and reach.

Pay out To pull off or feed out hose or rope. To lay out.

Pawl The part of the mechanism of an extension ladder that secures the extensions to the rungs of the main ladder.

Pedestal The operator's stand located on the turntable of a fire department aerial ladder.

Performance test *See* **Acceptance test.**

Pick up To take up hose and other fire fighting equipment that has been used at a fire and place it back on the apparatus.

Piece, or **Pieces** One or more sections or lengths of fire hose. A single unit of fire apparatus.

Piezometer tube A tube inserted from a gage into a stream of water to measure the pressure of the stream. *See also* **Pitot tube.**

Piston pump A positive displacement pump using reciprocating pistons to force water from the pump chamber in conjunction with the appropriate action of inlet and discharge valves.

Pitot gage *See* **Pitot tube.**

Pitot tube A tube inserted from a pressure gage into a stream of water from a nozzle to measure pressure.

Platform The basket of an elevating platform, which is occupied and controlled by the operator.

Plug A term for a fire hydrant arising from the time when water for fires was obtained by removing a wooden plug from a water main.

Portable monitor A heavy-stream device that may be removed from an apparatus and placed in a position that cannot be reached by wagon batteries and ladder pipes.

Portable pump A small, gasoline-engine-driven pump designed to be carried on fire apparatus.

Position An area of operation assigned to a fire company or an area of command at a fire assigned to a chief officer.

Positioning Spotting or placing an apparatus, hose line, or nozzle in the most strategic spot where it can be utilized to its full advantage.

Positive displacement pump A pump of the piston, rotary vane or rotary gear type which moves a given quantity of water through a pump chamber with each stroke or cycle. A positive displacement pump is capable of pumping air and, therefore, is self-priming.

Power takeoff (PTO) A method of attaching a pump or other device to the transmission so it may be powered by the vehicle's engine. Commonly, the apparatus booster pumps and hydraulic pumps are driven by a PTO.

Preconnected Suction or discharge hose carried connected to the pump; saves time at the fire.

Prefire planning Surveys of target hazards and advance plans of possible fire fighting operations.

Pressure A force in the nature of a thrust. It is usually measured in pounds per square inch (psi).

Pressure control device A pressure-regulating valve, relief valve, or governor which is installed on pumping apparatus to stabilize discharge pressures.

Pressure loss Reduction in water pressure (between a pump, or hydrant, and a nozzle) due to friction loss and back pressure.

Pressure operation Operation of a multistage centrifugal pump with the impellers in series to provide increased pressure rather than volume.

Pressure surge Sudden opening and closing of valves and nozzles may cause a sudden rising and falling of pressure in the hose line which tends to travel rapidly and become serious in long hose lines.

Preventive maintenance Also referred to as

preventative maintenance. The routine inspection, lubrication, and repair of a vehicle and its components to keep it ready for duty and to anticipate and prevent breakdowns.

Priming Filling a pump with water to eliminate air locks or to allow drafting.

Priming pump A small, positive displacement pump used to remove the air from a centrifugal pump in order to bring water into the pump chamber and allow drafting.

Proportioner A device for proportioning foam stabilizer or wetting agent into fire streams in the correct ratios.

PSI (psi) Measurement of pressure in pounds per square inch.

Pumper A fire department pumping engine of at least 500-gpm capacity and carrying fire hose and other fire fighting equipment.

Pumper-ladder A combination apparatus that carries the standard equipment of both a pumper and a ladder truck.

Pumper test pit A large enclosed cistern used for testing pumps at draft and for the instruction of pump operators.

Pump operator A fire department member whose duty it is to operate a pumper.

Pump panel The pump operator's panel. The control position of a pumper where the control instruments and gages are located.

Pump pressure Water pressure maintained by the operation of a pump on the discharge gates.

Pump slippage A term used to describe leakage between the internal parts of a pump. The amount of slippage depends on the condition of the pump and the operating pressure.

Pump transmission In apparatus that use one engine to drive both the fire pump and the vehicle, the gear box, controlled by a lever, which changes the power drive from one to the other.

Quad A quadruple combination apparatus carrying (1) water tank, (2) hose, (3) pump, (4) ladders. A better term is *pumper-ladder* since this unit carries the standard equipment and appliances of both a pumper and a ladder truck.

Quadruple pumper *See* **Quad.**

Quarters The fire station to which a given fire company or individual is assigned.

Quint A pumper-ladder having an aerial ladder in addition to the four standard equipment components of a quad.

Quintuple pumper *See* **Quint.**

Raise Elevate and place in position any of the various fire ladders.

Range Refers to the distance that a fire stream from a nozzle, monitor, ladder pipe, or other appliance can reach and still be effective.

Rated capacity The quantity of water that will be discharged by a pump at a certain pressure; this rating is arrived at by tests and should be checked annually to be certain that the pumper has not deteriorated excessively.

Rated gross vehicle weight (Rated GVW) The maximum approved gross vehicle weight for a vehicle as specified by the manufacturer.

Reaction The counterforce or thrust directed backward against the hose operator or appliance, caused by acceleration of water in the nozzle.

Ready line A line of hose with nozzle attached, preconnected to a pump and ready for use.

Reducer A reducing coupling used on hose or fitting so that smaller couplings may be attached.

Reducing Attaching smaller line as a leader to large-diameter hose.

Reducing couplings Couplings used to extend a line of hose with hose of a smaller diameter.

Reel A hose reel, now chiefly used with booster hose.

Relay The use of two or more pumpers to move water at distances that would require excessive pressures to overcome friction loss if only one pump were used at the source.

Relief valve A pressure-controlling device

on a pump to prevent excessive pressures when a nozzle is shut down. Relief valves operate on the discharge side of a pump by automatically bypassing water when the pump pressure exceeds the desired amount. Relief valves are also used to relieve excessively high pressures in pressure vessels.

Rescue The saving of life and removal of endangered persons to a place of safety.

Reserve Apparatus not in the first-line duty, but ready for relief in case of large fires, accidents, or breakdown of regular apparatus.

Residual pressure Pressure remaining on the inlet side of a pumper, or in the water main, while water is being discharged.

Respond Answer an alarm in accordance with a prearranged assignment or upon instruction of a dispatcher. Proceed to the scene of a fire or alarm.

Response The act of responding to an alarm. The entire complement of fire crew and apparatus assigned to an alarm.

Reverse lay Laying of hose in a manner opposite to that normally envisioned when the hose was loaded on the apparatus, and thereby requiring use of double couplings to connect nozzle and pump.

Rig A piece of fire apparatus.

Riser A vertical water pipe used to carry water for fire protection to elevations above grade, such as a standpipe riser, sprinkler riser, etc.

Road transmission The gear box between the engine and the differential of a vehicle, which allows a choice of gear ratios for more efficient driving.

Rotary gear pump A positive displacement pump that employs closely fitting rotors or gears to force water through a pump chamber.

RPM (rpm) Revolutions per minute. Used in reference to the rotating speed of an engine crankshaft, pump impeller, etc.

Run Response to a fire or alarm. The term arises from the fact that in the days of the hand-drawn apparatus, fire fighters ran to the fire, pulling their apparatus.

Rung The round or step of a ladder.

Running away from the water Attempt to operate a pump so as to provide a greater volume and pressure than is obtainable from the water supply source being used. *See also* **Cavitation.**

Salvage Procedures to reduce incidental losses from smoke, water, fire, and weather during and following fires.

Salvage company An organized company or unit specializing in fire salvage or protective duties.

Seat of fire Area where the main body of fire is located, as determined by the outward movement of heat and gases.

Second-in The unit or company that will be the second to arrive at a reported fire or other incident.

Series operation Operation of a multistage centrifugal pump so that the water passes through each impeller consecutively to build up pressure.

Service aerial An aerial ladder mounted on a four-wheel truck type chassis; does not require a tiller operator.

Service test The test required annually and after major repairs to indicate whether a pumper is in good working condition and capable of providing its rated capacity.

Service truck Sometimes called a City Service Truck. A generally obsolete type of ladder truck that carries only ground ladders and other truck company equipment.

Shutoff nozzle A type of nozzle that permits the flow to be controlled by the hose operator. A controlling nozzle.

Siamese A hose fitting for combining the flow from two or more hose lines into a single stream.

Simple hose lay Hose consisting of consecutively coupled lengths of hose between pump and nozzle, without wyes, siameses, or other fittings.

Siphon A device that uses the suction developed by a jet of water to raise water.

Siren A warning device that sets up undulating sound waves.

Size-up The mental evaluation made by

the officer in charge to determine a course of action.

Sleeve A suction or supply hose.

Slip, Slippage *See* **Pump slippage.**

Smoke A combination of gases, carbon particles, and other products of incomplete combustion that hinders respiration, obscures visibility, and delays access to the seat of a fire.

Smooth bore tip A straight-stream nozzle tip.

Snorkel® A common term used to designate certain aerial platform fire apparatus.

®Trademark of Snorkel Fire Equipment Company.

Soft sleeve A short length of large-diameter fire hose used to supply pumpers with water from fire hydrants.

Soft suction *See* **Soft sleeve.**

Solid stream A stream of water used for fire fighting and discharged from an open, round orifice. Also called a straight stream.

Spaghetti An extensive snarl of charged hose lines from various fire companies in the street adjacent to the fire area.

Spanner A metal wrench device used in tightening and freeing hose couplings.

Spotting Placement of fire companies or fire fighting equipment for effective operation and attack on a fire.

Spotting a ladder Positioning a ladder with reference to the objective to be reached.

Spray Water applied through specially designed orifices in the form of finely divided particles so as to more readily absorb heat and smother fire by the generation of steam.

Spray nozzle A nozzle constructed to produce a water spray. Often called a fog nozzle.

Spurt test *See* **Overload test.**

Squad A fire company assigned to an apparatus carrying special equipment and assigned to assist pumper and ladder companies. Also a first aid squad to help fire fighters and the public.

Squirrel tail suction A long length of hard suction hose carried permanently connected to a pumper inlet and curved around the front of the apparatus; this makes it unnecessary to connect the suction hose to the pumper when the hydrant is reached.

Stabilizing The spotting of an aerial ladder truck, water tower, or elevating platform so it will be stable and secure. This will entail setting the brakes, chocking the wheels, and correctly placing the ground jacks and outriggers so the maximum stability is attained.

Standard No. 1901 Specifications for Automotive Fire Apparatus.

Standpipe A vertical water pipe riser used to supply fire hose outlets in buildings.

Static pressure Water pressure available at a specific location when no fire flow is being used and when, therefore, no pressure losses due to friction are being encountered.

Static water supply A supply of water at rest which does not provide a pressure head for fire fighting, but which may be employed as a suction source for fire pumps.

Steamer connection A siamese connection for pumping into an automatic sprinkler system or standpipe system to provide an added supply or suitable pressure. Originally provided for supply by steam fire engines.

Steamer outlet The large outlet on a hydrant provided for supplying pumpers, originally for steam fire engines.

Straight stream A hose stream provided by a solid stream orifice, or by adjusting a spray nozzle into a straight stream pattern.

Strainer A wire or metal guard used in pumps and on suction hose to keep foreign material from clogging or damaging the pump.

Strategy The overall method and plan for

controlling an outbreak of fire with the most advantageous use of available forces.

Stream A jet of water thrown from a nozzle and used to extinguish or to control a fire.

Stretch A command to lay out hose in a long continuous line.

Suction The practice of taking water from static sources located below the level of the pump by exhausting air from the pump chamber and using atmospheric pressure (14.7 psi) to push water through suction hose into the pump.

Suction booster A type of jet siphon device used to bring water from greater distances and to higher elevations than is possible with suction depending upon atmospheric pressure.

Suction eye The opening through which water enters the impeller of a centrifugal pump.

Suction hose A hose reinforced against collapse, which might occur when drafting water.

Suction lift The number of feet of vertical lift from the surface of the water to the center of the pump impeller when drafting.

Supplemental pumping Relaying water from a distant hydrant to a pumper that cannot obtain sufficient water from a hydrant adjacent to a fire. Therefore, the pumper supplying the hose lines is obtaining water from a hydrant as well as from another pumper.

Supply line The hose line supplying a pumper, water tower, or other apparatus with water.

Suppression The total work of extinguishing a fire from its discovery.

Tactics Methods of employing fire companies in an efficient, coordinated manner in the field so as to get satisfactory results with the forces employed and to deny the fire any potential avenues of extension.

Tailboard The backstep of a hose wagon or pumper on which fire fighters stand while riding on the apparatus.

Take a hydrant Order to connect to or stand by a hydrant with a pumper; to lay out from a designated hydrant.

Tandem With pumpers in tandem, a pumper at a hydrant supplies a second pumper by connecting suction inlets of the two pumps with a suction hose.

Tank apparatus Water tank fire apparatus that may be equipped with small hose and a pump. Normally, tank apparatus should not have a greater capacity than 1500 gal because the weight may cause overloading.

Tanker *See* **Tank apparatus.**

Tanker call A preplanned assignment of water tankers used to supply pumpers at rural fires or elsewhere where water supplies are limited.

Tanker shuttle A technique of using one or more tankers to transport water from the nearest source to the fire ground.

Telescopic boom A power-operated boom constructed with retracting sliding sections and used as a support for some aerial platforms.

Tender A hose-carrying vehicle assigned to serve in conjunction with a pumper or fire boat to increase the hose capacity of the unit and provide greater flexibility of operation.

Thread The specific dimensions of a screw thread employed to couple fire hose and equipment.

Throw Raise a ladder quickly. Throw water, as in a fire stream, from one or more nozzles. Spread salvage covers quickly by approved means.

Tied up A fire company engaged for a period of time and not available to cover its district. A pumper connected to a hydrant.

Tiller The rear steering wheel of a large tractor-trailer ladder truck or other vehicle requiring a separate steering wheel for the rear axle.

Tiller operator The fire crew member who steers the rear axle.

Tips Nozzle tips that change the size of the orifice of a hose stream.

Torque A method of defining or measuring usable engine power at the shaft.

Transfer valve A valve that places a multistage centrifugal pump in either volume or pressure operation.

Triple combination A combination pumper carrying hose, a water tank, and a pump.

Truck A fire department ladder apparatus.

Truck company A ladder truck company whose primary duties at a fire are rescue, ventilation, forcible entry, and salvage.

Truss A structural member of a ladder that joins and supports the ladder beams.

Turntable The rotating table or platform at the base of an aerial ladder upon which the ladder stands.

Turret A heavy-stream device mounted on a pumper or fire boat.

Two-piece engine company An engine company equipped with two apparatus; these may be two triple combination pumpers, a pumper and a hose wagon, a pumper and a water tower, or any two other types of fire fighting vehicles. One of the apparatus shall be a pumper.

Vacuum Pressure below atmospheric; shown in inches of mercury.

Vacuum gage An instrument calibrated to measure vacuum in inches of mercury.

Vacuum test A test using the priming mechanism to detect leaks and measure the efficiency of the priming system.

Volume operation Operation of the impellers in a multistage centrifugal pump so that the maximum capacity of the pump is discharged.

Volute A spiral casing, surrounding the impeller of a centrifugal pump, which collects water in a chamber of increasing area and leads to the pump discharge connections.

Wagon A hose truck, usually equipped with a water tank and hose. Sometimes the front pumper of a two-piece engine company is called the wagon and the rear pumper, the pump.

Wagon battery A heavy-stream device mounted on a hose wagon or pumper. Usually connected to the pump by internal piping.

Wagon pipe *See* **Wagon battery.**

Washer A round metal disk with a hole in it; often this term is incorrectly used in place of *gasket* to describe a sealing ring.

Water curtain A stream of water projected between a fire and exposed structures to protect against extension of the fire.

Water hammer A surge of impact energy due to the sudden shutting off of nozzles or valves.

Water head The water pressure caused by the height of a column of water.

Water tower A fire apparatus carrying an extendible mast having one or more heavy-stream nozzles. Used to apply master streams into upper floors of buildings.

Waterway The internal passage for water in hose or fire equipment.

Wet barrel A type of hydrant used in areas which are not subject to freezing temperatures. There is normally a separate valve to control the flow from each outlet.

Wetting agent A chemical additive used to modify water and aid greater penetration by reducing surface tension of water.

Wet water Water to which a wetting agent has been added.

Wheel blocks Blocks used to secure the wheels of vehicles to prevent movement. (Also called *chocks.*)

Wye A hose connection with two outlets. Permits two connections or hose lines to be taken from a single supply line.

Index

NOTE: Glossary references are italicized.